Chrysler
Sebring & 200
Dodge Avenger
Automotive Repair Manual

by Jeff Killingsworth and John H Haynes

Member of the Guild of Motoring Writers

Models covered:

Chrysler Sebring sedan - 2007 through 2010
Chrysler Sebring convertible - 2008 through 2010
Chrysler 200 - 2011 through 2017
Dodge Avenger - 2008 through 2014

(25041-10Z3)

J H Haynes & Co. Ltd.
Haynes North America, Inc.
www.haynes.com

Acknowledgements

Technical writers who contributed to this project include Jay Hayes, Tracy Martin, Demian Hurst, Dennis Gibb, Peter Sessler and Scott "Gonzo" Weaver.

A book in the Haynes Automotive Repair Manual Series

ISBN-13: 978-1-62092-330-6
ISBN-10: 1-62092-330-0

Library of Congress Control Number: 2019931573

Contents

Introductory pages

About this manual	0-5
Introduction	0-5
Vehicle identification numbers	0-6
Recall information	0-7
Buying parts	0-11
Maintenance techniques, tools and working facilities	0-11
Booster battery (jump) starting	0-18
Jacking and towing	0-19
Automotive chemicals and lubricants	0-20
Conversion factors	0-21
Fraction/decimal/millimeter equivalents	0-22
Safety first!	0-23
Troubleshooting	0-24

Chapter 1
Tune-up and routine maintenance — **1-1**

Chapter 2 Part A
Four-cylinder engine — **2A-1**

Chapter 2 Part B
2.7L V6 engine — **2B-1**

Chapter 2 Part C
3.5L V6 engine — **2C-1**

Chapter 2 Part D
3.6L V6 engine — **2D-1**

Chapter 2 Part E
General engine overhaul procedures — **2E-1**

Chapter 3
Cooling, heating and air conditioning systems — **3-1**

Chapter 4
Fuel and exhaust systems — **4-1**

Chapter 5
Engine electrical systems — **5-1**

Chapter 6
Emissions and engine control systems — **6-1**

Chapter 7 Part A
Automatic transaxle — **7A-1**

Chapter 7 Part B
Transfer case — **7B-1**

Chapter 8
Driveline — **8-1**

Chapter 9
Brakes — **9-1**

Chapter 10
Suspension and steering systems — **10-1**

Chapter 11
Body — **11-1**

Chapter 12
Chassis electrical system — **12-1**

Wiring diagrams — **12-19**

Index — **IND-1**

Haynes mechanic and photographer with a 2013 Dodge Avenger

About this manual

Its purpose

The purpose of this manual is to help you get the best value from your vehicle. It can do so in several ways. It can help you decide what work must be done, even if you choose to have it done by a dealer service department or a repair shop; it provides information and procedures for routine maintenance and servicing; and it offers diagnostic and repair procedures to follow when trouble occurs.

We hope you use the manual to tackle the work yourself. For many simpler jobs, doing it yourself may be quicker than arranging an appointment to get the vehicle into a shop and making the trips to leave it and pick it up. More importantly, a lot of money can be saved by avoiding the expense the shop must pass on to you to cover its labor and overhead costs. An added benefit is the sense of satisfaction and accomplishment that you feel after doing the job yourself.

Using the manual

The manual is divided into Chapters. Each Chapter is divided into numbered Sections, which are headed in bold type between horizontal lines. Each Section consists of consecutively numbered paragraphs.

At the beginning of each numbered Section you will be referred to any illustrations which apply to the procedures in that Section. The reference numbers used in illustration captions pinpoint the pertinent Section and the Step within that Section. That is, illustration 3.2 means the illustration refers to Section 3 and Step (or paragraph) 2 within that Section.

Procedures, once described in the text, are not normally repeated. When it's necessary to refer to another Chapter, the reference will be given as Chapter and Section number. Cross references given without use of the word "Chapter" apply to Sections and/or paragraphs in the same Chapter. For example, "see Section 8" means in the same Chapter.

References to the left or right side of the vehicle assume you are sitting in the driver's seat, facing forward.

Even though we have prepared this manual with extreme care, neither the publisher nor the author can accept responsibility for any errors in, or omissions from, the information given.

NOTE

A **Note** provides information necessary to properly complete a procedure or information which will make the procedure easier to understand.

CAUTION

A **Caution** provides a special procedure or special steps which must be taken while completing the procedure where the Caution is found. Not heeding a Caution can result in damage to the assembly being worked on.

WARNING

A **Warning** provides a special procedure or special steps which must be taken while completing the procedure where the Warning is found. Not heeding a Warning can result in personal injury.

Introduction to the Chrysler Sebring & 200, Dodge Avenger

These models feature transversely mounted engines equipped with electronic multi-port fuel injection. The engine drives the front wheels through a four-speed, six-speed or nine-speed automatic transaxle via independent driveaxles. Some models were available with all-wheel drive.

The fully-independent front suspension consists of coil spring/strut units, control arms and a stabilizer bar. The rear suspension uses a multi-link design that consists of a knuckle, stabilizer bar, upper control arms, trailing arms, toe links, lower control arms and coil-over shock absorber assemblies.

The power-assisted rack-and-pinion steering unit is mounted on the front suspension subframe.

Front brakes are disc-type. Rear brakes are either disc or drum-type. Power brake assist is standard, with an Antilock Brake System (ABS) optional.

Vehicle identification numbers

Modifications are a continuing and unpublicized process in vehicle manufacturing. Since spare parts manuals and lists are compiled on a numerical basis, the individual vehicle numbers are essential to correctly identify the component required.

Vehicle Identification Number (VIN)

This very important identification number is stamped on a plate attached to the left side of the dashboard just inside the windshield (see illustration). The VIN also appears on the Vehicle Certificate of Title and Registration. It contains information such as where and when the vehicle was manufactured, the model year and the body style.

VIN year and engine codes

Two particularly important pieces of information located in the VIN are the model year and engine codes. Counting from the left, the engine code is the eighth digit and the model year code is the 10th digit.

On the models covered by this manual the engine codes are:

K	2.4L (2007 and 2008)
J	2.4L PZEV (2008)
B	2.4L (2009 and later)
R	2.7L (2007 and 2008)
D	2.7L (2009 and 2010)
M	3.5L (2007 and 2008)
V	3.5L (2009 and 2010)
G	3.6L (2011 and later)

On the models covered by this manual the model year codes are:

8	2008
9	2009
A	2010
B	2011
C	2012
D	2013
E	2014
F	2015
G	2016
H	2017

Equipment identification plate

This plate is located on the inside of the hood. It contains valuable information concerning the production of the vehicle as well as information on all production or special equipment.

Safety Certification label

The Safety Certification label is affixed to the left front door (see illustration). The plate contains the name of the manufacturer, the month and year of production, the Gross Vehicle Weight Rating (GVWR) and the safety certification statement. This label also contains the paint code. It is especially useful for matching the color and type of paint during repair work.

Engine identification number

The engine identification number on V6 engines is stamped into the rear of the engine block, below the cylinder head. On four-cylinder engines, the ID number is stamed into a machined pad on the front left side of the engine block, behind the throttle body (see illustration).

Transaxle identification number

The ID number on the automatic transaxle is located on an adhesive label affixed to the transaxle bellhousing (see illustration). The number is also stamped into the left end of the transaxle, but the transfer gear cover (side cover) must be removed to view it.

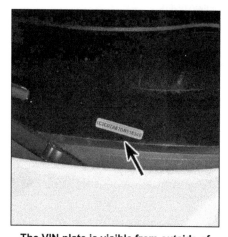

The VIN plate is visible from outside of the vehicle, through the driver's side of the windshield

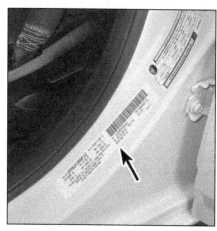

The Vehicle Safety Certification label is affixed to the end of the driver's door

Engine identification number location

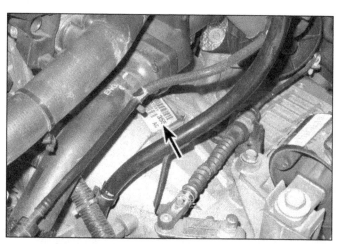

Automatic transaxle identification tag location

Recall information

Vehicle recalls are carried out by the manufacturer in the rare event of a possible safety-related defect. The vehicle's registered owner is contacted at the address on file at the Department of Motor Vehicles and given the details of the recall. Remedial work is carried out free of charge at a dealer service department.

If you are the new owner of a used vehicle which was subject to a recall and you want to be sure that the work has been carried out, it's best to contact a dealer service department and ask about your individual vehicle - you'll need to furnish them your Vehicle Identification Number (VIN).

The table below is based on informa-

tion provided by the National Highway Traffic Safety Administration (NHTSA), the body which oversees vehicle recalls in the United States. The recall database is updated constantly. For the latest information on vehicle recalls, check the NHTSA website at www.nhtsa.gov, www.safercar.gov or call the NHTSA hotline at 1-888-327-4236.

Recall date	Recall campaign number	Model(s) affected	Concern
DEC 29, 2006	06V494000	2007 Sebring	On some models equipped with automatic temperature control (ATC), the software programmed into the heating ventilation and air conditioning (HVAC) module may cause the windshield defrosting and defogging functions to become inoperative. This can decrease the driver's visibility under certain conditions and result in a crash.
MAR 15, 2007	07V104000	2008 Avenger	On some models, the front door latch cable may become partially unseated from the interior release handle housing causing the door latch to stick in the open position and not be secured to the striker. During certain conditions, the possibility of a door opening could cause an unbelted passenger to be ejected.
JUN 13, 2007	07V240000	2007, 2008 Sebring, Avenger	On some models, the front seat track position sensors utilized for the airbag system may not function properly. This could increase the risk of injury to the front seat occupants during certain crash conditions.
SEP 12, 2007	07V414000	2007, 2008 Sebring, Avenger	On some models, the front door latch cable may become partially unseated from the interior release handle housing possibly causing the latch to stick in the unlatched position or the lock function to become inoperative. Driving a vehicle with an unlatched door can result in an unbelted front seat occupant falling out of the vehicle or being ejected in a crash, causing increased risk of injury.

Recall date	Recall campaign number	Model(s) affected	Concern
SEP 13, 2007	07V426000	2008 Avenger	On some models with all wheel drive, the fuel tank straps were improperly manufactured and could separate. This can cause the fuel tank to loosen and leak fuel. Fuel leakage in the presence of an ignition source can result in a fire.
OCT 10, 2007	07V473000	2007, 2008 Sebring, Avenger	On some models equipped with 2.4L engines, the engine coolant may be drawn into the left radiator cooling fan motor connector. This can cause an electrical short circuit and result in an engine compartment fire.
APR 02, 2008	08V152000	2007, 2008 Sebring, Avenger	On some models equipped with a standard tire pressure monitoring system (TPMS). Unused electrical connectors for the TPMS may become corroded and could short circuit, which can cause a variety of conditions including engine no-start, dead battery, an inoperative cruise control or remote start system, and/or engine stalling. Engine stalling could cause a crash.
OCT 08, 2008	08V528000	2009 Sebring, 2008 Avenger	On some models, a new adhesive used in the power train control module (PCM) manufacturing process can cause the printed circuit board to break. This can cause the engine to stall and cause a crash.
FEB 04, 2009	09V047000	2009 Sebring, Avenger	On some models equipped with 2.4L engines, a compatibility issue between the instrument cluster software and the fuel pump module can result in the fuel gauge overstating the actual fuel tank level. As a result, the vehicle may run out of fuel when the gauge indicates that there is still fuel remaining in the tank. This could cause a crash.
JAN 11, 2010	10V009000	2010 Sebring, Avenger	Some models have been built with an improperly formed or missing brake booster input rod retaining clip. This could result in brake failure which could cause a crash.
OCT 07, 2010	10V475000	2010 Sebring, Avenger	Some models can experience a separation at the crimped end of the power steering pressure hose assembly. Leaked power steering fluid onto hot engine components could cause a fire.
JUN 08, 2011	11V315000	2011 Avenger, 200	Some models have been built with a missing or incorrectly installed steering column pivot rivet. A missing or incorrectly installed rivet could compromise the ability of the steering column to support the occupant loads in the event of a frontal crash, decreasing the effectiveness of the frontal impact safety system. As a result, the condition may increase the potential for injury in a frontal crash.

Recall date	Recall campaign number	Model(s) affected	Concern
SEP 29, 2011	11V487000	2012 200	Some models equipped with 3.6L engines may experience connecting rod bearing failure due to debris inside the engine block. Connecting rod failure may lead to engine seizure which may increase the risk of a crash.
FEB 06, 2013	13V043000	2013, Avenger, 200	Some models may have a gas tank that has a broken control valve in the fuel tank assembly. A broken control valve in the fuel tank assembly may lead to an engine stall or fuel leakage. An engine stall while driving may contribute to a vehicle crash. Fuel leakage, in the presence of an ignition source, may result in a fire.
JUL 02, 2013	13V282000	2011, 2012, 2013 Sebring, Avenger, 200	Some models may have electrical over-stress of a resistor in the occupant restraint control module may lead to the non-deployment of the active head restraints during a rear impact collision. In the event of a crash necessitating the deployment of the active head restraints, their non-deployment could increase the risk of injury.
NOV 06, 2013	13V552000	2013 Avenger, 200	Chrysler is recalling certain models equipped with 2.4L engines. Due to abrasive debris in the balance shaft bearings, these engines may have a loss of engine oil pressure, possibly resulting in an engine stall or engine failure. If the engine stalls while driving it may increase the risk of a crash.
JUL 2, 2014	14V392000	2015 200	The rear shock absorber may detach due to a faulty factory weld, which may result in injury in the event of a crash.
AUG 5, 2014	14V480000	2015 200	A wire leading into the driver's side door may be of the wrong wire gauge, causing excessive heat and potentially resulting in a fire.
FEB 13, 2015	15V090000	2015 200	The park pawl/park lock in the transmission may become contaminated and not engage when the shifter is in Park, resulting in the vehicle rolling away and causing injury or property damage.
JUL 23, 2015	15V461000	2015 200	The radio software may have vulnerabilities that could allow third party unauthorized access to the radio or certain vehicle control systems, resulting in theft or a crash.
JUL 28, 2015	15V470000	2015 200	The Power Distribution Center (PDC, fuse box assembly) may have a faulty connector that causes intermittent power signals to certain electrical components, and could increase the risk of a crash.

FEB 26, 2016	16V114000	2015 200	If certain airbag components were not replaced as an assembly, the passenger airbag may not deploy properly and result in injury in the event of a crash.
JUL 12, 2016	16V529000	2015 200	The transmission may unexpectedly shift into Neutral as a result of insufficient crimps in a transmission wiring harness, which could result in a crash.
SEP 15, 2016	16V668000	2011-2014 Chrysler 200, 2010 Sebring, 2010-2014 Avenger	The airbag control module may have a tendency to short circuit, causing the airbags (front and side) and/or seat belt pre-tensioners to be disabled, resulting in injury in the event of a crash.
OCT 10, 2017	17V640000	2012-2013 200 & Avenger	On models equipped with active headrests, the airbag control module may contain a faulty component that disables the headrest airbag from deploying in the event of a collision from the rear, resulting in potential injury.

Buying parts

Replacement parts are available from many sources, which generally fall into one of two categories - authorized dealer parts departments and independent retail auto parts stores. Our advice concerning these parts is as follows:

Retail auto parts stores: Good auto parts stores will stock frequently needed components which wear out relatively fast, such as clutch components, exhaust systems, brake parts, tune-up parts, etc. These stores often supply new or reconditioned parts on an exchange basis, which can save a considerable amount of money. Discount auto parts stores are often very good places to buy materials and parts needed for general vehicle maintenance such as oil, grease, filters, spark plugs, belts, touch-up paint, bulbs, etc. They also usually sell tools and general accessories, have convenient hours, charge lower prices and can often be found not far from home.

Authorized dealer parts department: This is the best source for parts which are unique to the vehicle and not generally available elsewhere (such as major engine parts, transmission parts, trim pieces, etc.).

Warranty information: If the vehicle is still covered under warranty, be sure that any replacement parts purchased - regardless of the source - do not invalidate the warranty!

To be sure of obtaining the correct parts, have engine and chassis numbers available and, if possible, take the old parts along for positive identification.

Maintenance techniques, tools and working facilities

Maintenance techniques

There are a number of techniques involved in maintenance and repair that will be referred to throughout this manual. Application of these techniques will enable the home mechanic to be more efficient, better organized and capable of performing the various tasks properly, which will ensure that the repair job is thorough and complete.

Fasteners

Fasteners are nuts, bolts, studs and screws used to hold two or more parts together. There are a few things to keep in mind when working with fasteners. Almost all of them use a locking device of some type, either a lockwasher, locknut, locking tab or thread adhesive. All threaded fasteners should be clean and straight, with undamaged threads and undamaged corners on the hex head where the wrench fits. Develop the habit of replacing all damaged nuts and bolts with new ones. Special locknuts with nylon or fiber inserts can only be used once. If they are removed, they lose their locking ability and must be replaced with new ones.

Rusted nuts and bolts should be treated with a penetrating fluid to ease removal and prevent breakage. Some mechanics use turpentine in a spout-type oil can, which works quite well. After applying the rust penetrant, let it work for a few minutes before trying to loosen the nut or bolt. Badly rusted fasteners may have to be chiseled or sawed off or removed with a special nut breaker, available at tool stores.

If a bolt or stud breaks off in an assembly, it can be drilled and removed with a special tool commonly available for this purpose. Most automotive machine shops can perform this task, as well as other repair procedures, such as the repair of threaded holes that have been stripped out.

Flat washers and lockwashers, when removed from an assembly, should always be replaced exactly as removed. Replace any damaged washers with new ones. Never use a lockwasher on any soft metal surface (such as aluminum), thin sheet metal or plastic.

Fastener sizes

For a number of reasons, automobile manufacturers are making wider and wider use of metric fasteners. Therefore, it is important to be able to tell the difference between standard (sometimes called U.S. or SAE) and metric hardware, since they cannot be interchanged.

All bolts, whether standard or metric, are sized according to diameter, thread pitch and length. For example, a standard 1/2 - 13 x 1 bolt is 1/2 inch in diameter, has 13 threads per inch and is 1 inch long. An M12 - 1.75 x 25 metric bolt is 12 mm in diameter, has a thread pitch of 1.75 mm (the distance between threads) and is 25 mm long. The two bolts are nearly identical, and easily confused, but they are not interchangeable.

In addition to the differences in diameter, thread pitch and length, metric and standard bolts can also be distinguished by examining the bolt heads. To begin with, the distance across the flats on a standard bolt head is measured in inches, while the same dimension on a metric bolt is sized in millimeters (the same is true for nuts). As a result, a standard wrench should not be used on a metric bolt and a metric wrench should not be used on a standard bolt. Also, most standard bolts have slashes radiating out from the center of the head to denote the grade or strength of the bolt, which is an indication of the amount of torque that can be applied to it. The greater the number of slashes, the greater the strength of the bolt. Grades 0 through 5 are commonly used on automobiles. Metric bolts have a property class (grade) number, rather than a slash, molded into their heads to indicate bolt strength. In this case, the higher the number, the stronger the bolt. Property class numbers 8.8, 9.8 and 10.9 are commonly used on automobiles.

Strength markings can also be used to distinguish standard hex nuts from metric hex nuts. Many standard nuts have dots stamped into one side, while metric nuts are marked with a number. The greater the number of

dots, or the higher the number, the greater the strength of the nut.

Metric studs are also marked on their ends according to property class (grade). Larger studs are numbered (the same as metric bolts), while smaller studs carry a geometric code to denote grade.

It should be noted that many fasteners, especially Grades 0 through 2, have no distinguishing marks on them. When such is the case, the only way to determine whether it is standard or metric is to measure the thread pitch or compare it to a known fastener of the same size.

Standard fasteners are often referred to as SAE, as opposed to metric. However, it should be noted that SAE technically refers to a non-metric fine thread fastener only. Coarse thread non-metric fasteners are referred to as USS sizes.

Since fasteners of the same size (both standard and metric) may have different

strength ratings, be sure to reinstall any bolts, studs or nuts removed from your vehicle in their original locations. Also, when replacing a fastener with a new one, make sure that the new one has a strength rating equal to or greater than the original.

Tightening sequences and procedures

Most threaded fasteners should be tightened to a specific torque value (torque is the twisting force applied to a threaded component such as a nut or bolt). Overtightening the fastener can weaken it and cause it to break, while undertightening can cause it to eventually come loose. Bolts, screws and studs, depending on the material they are made of and their thread diameters, have specific torque values, many of which are noted in the Specifications at the beginning of each Chapter. Be sure to follow the torque recommendations closely. For fasteners not assigned a

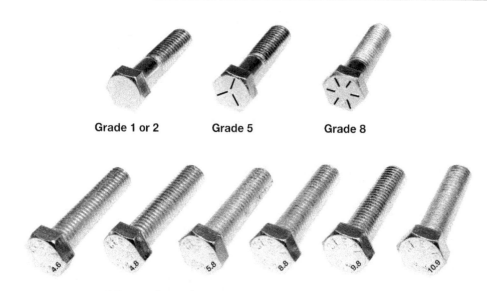

Grade 1 or 2   Grade 5   Grade 8

Bolt strength marking (standard/SAE/USS; bottom - metric)

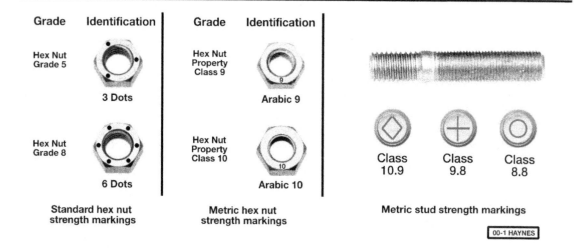

Grade	Identification	Grade	Identification
Hex Nut Grade 5	3 Dots	Hex Nut Property Class 9	Arabic 9
Hex Nut Grade 8	6 Dots	Hex Nut Property Class 10	Arabic 10

Standard hex nut strength markings

Metric hex nut strength markings

Class 10.9   Class 9.8   Class 8.8

Metric stud strength markings

00-1 HAYNES

specific torque, a general torque value chart is presented here as a guide. These torque values are for dry (unlubricated) fasteners threaded into steel or cast iron (not aluminum). As was previously mentioned, the size and grade of a fastener determine the amount of torque that can safely be applied to it. The figures listed here are approximate for Grade 2 and Grade 3 fasteners. Higher grades can tolerate higher torque values.

Fasteners laid out in a pattern, such as cylinder head bolts, oil pan bolts, differential cover bolts, etc., must be loosened or tightened in sequence to avoid warping the component. This sequence will normally be shown in the appropriate Chapter. If a specific pattern is not given, the following procedures can be used to prevent warping.

Initially, the bolts or nuts should be assembled finger-tight only. Next, they should be tightened one full turn each, in a criss-cross or diagonal pattern. After each one has been tightened one full turn, return to the first one and tighten them all one-half turn, following the same pattern. Finally, tighten each of them one-quarter turn at a time until each fastener has been tightened to the proper torque. To loosen and remove the fasteners, the procedure would be reversed.

Component disassembly

Component disassembly should be done with care and purpose to help ensure that

Metric thread sizes	Ft-lbs	Nm
M-6	6 to 9	9 to 12
M-8	14 to 21	19 to 28
M-10	28 to 40	38 to 54
M-12	50 to 71	68 to 96
M-14	80 to 140	109 to 154
Pipe thread sizes		
1/8	5 to 8	7 to 10
1/4	12 to 18	17 to 24
3/8	22 to 33	30 to 44
1/2	25 to 35	34 to 47
U.S. thread sizes		
1/4 - 20	6 to 9	9 to 12
5/16 - 18	12 to 18	17 to 24
5/16 - 24	14 to 20	19 to 27
3/8 - 16	22 to 32	30 to 43
3/8 - 24	27 to 38	37 to 51
7/16 - 14	40 to 55	55 to 74
7/16 - 20	40 to 60	55 to 81
1/2 - 13	55 to 80	75 to 108

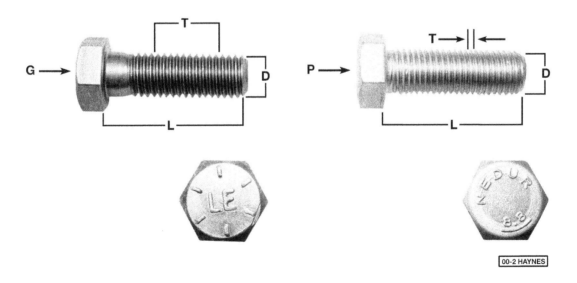

Standard (SAE and USS) bolt dimensions/grade marks

G   Grade marks (bolt strength)
L   Length (in inches)
T   Thread pitch (number of threads per inch)
D   Nominal diameter (in inches)

Metric bolt dimensions/grade marks

P   Property class (bolt strength)
L   Length (in millimeters)
T   Thread pitch (distance between threads in millimeters)
D   Diameter

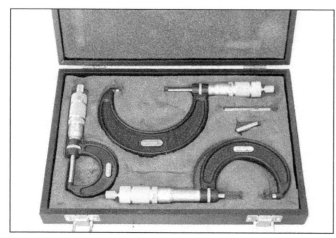

Micrometer set

Dial indicator set

the parts go back together properly. Always keep track of the sequence in which parts are removed. Make note of special characteristics or marks on parts that can be installed more than one way, such as a grooved thrust washer on a shaft. It is a good idea to lay the disassembled parts out on a clean surface in the order that they were removed. It may also be helpful to make sketches or take instant photos of components before removal.

When removing fasteners from a component, keep track of their locations. Sometimes threading a bolt back in a part, or putting the washers and nut back on a stud, can prevent mix-ups later. If nuts and bolts cannot be returned to their original locations, they should be kept in a compartmented box or a series of small boxes. A cupcake or muffin tin is ideal for this purpose, since each cavity can hold the bolts and nuts from a particular area (i.e. oil pan bolts, valve cover bolts, engine mount bolts, etc.). A pan of this type is especially helpful when working on assemblies with very small parts, such as the carburetor, alternator, valve train or interior dash and trim pieces. The cavities can be marked with paint or tape to identify the contents.

Whenever wiring looms, harnesses or connectors are separated, it is a good idea to identify the two halves with numbered pieces of masking tape so they can be easily reconnected.

Gasket sealing surfaces

Throughout any vehicle, gaskets are used to seal the mating surfaces between two parts and keep lubricants, fluids, vacuum or pressure contained in an assembly.

Many times these gaskets are coated with a liquid or paste-type gasket sealing compound before assembly. Age, heat and pressure can sometimes cause the two parts to stick together so tightly that they are very difficult to separate. Often, the assembly can be loosened by striking it with a soft-face hammer near the mating surfaces. A regular hammer can be used if a block of wood is placed between the hammer and the part. Do

not hammer on cast parts or parts that could be easily damaged. With any particularly stubborn part, always recheck to make sure that every fastener has been removed.

Avoid using a screwdriver or bar to pry apart an assembly, as they can easily mar the gasket sealing surfaces of the parts, which must remain smooth. If prying is absolutely necessary, use an old broom handle, but keep in mind that extra clean up will be necessary if the wood splinters.

After the parts are separated, the old gasket must be carefully scraped off and the gasket surfaces cleaned. Stubborn gasket material can be soaked with rust penetrant or treated with a special chemical to soften it so it can be easily scraped off. **Caution:** *Never use gasket removal solutions or caustic chemicals on plastic or other composite components.* A scraper can be fashioned from a piece of copper tubing by flattening and sharpening one end. Copper is recommended because it is usually softer than the surfaces to be scraped, which reduces the chance of gouging the part. Some gaskets can be removed with a wire brush, but regardless of the method used, the mating surfaces must be left clean and smooth. If for some reason the gasket surface is gouged, then a gasket sealer thick enough to fill scratches will have to be used during reassembly of the components. For most applications, a non-drying (or semi-drying) gasket sealer should be used.

Hose removal tips

Warning: *If the vehicle is equipped with air conditioning, do not disconnect any of the A/C hoses without first having the system depressurized by a dealer service department or a service station.*

Hose removal precautions closely parallel gasket removal precautions. Avoid scratching or gouging the surface that the hose mates against or the connection may leak. This is especially true for radiator hoses. Because of various chemical reactions, the rubber in hoses can bond itself to the metal spigot that the hose fits over. To remove

a hose, first loosen the hose clamps that secure it to the spigot. Then, with slip-joint pliers, grab the hose at the clamp and rotate it around the spigot. Work it back and forth until it is completely free, then pull it off. Silicone or other lubricants will ease removal if they can be applied between the hose and the outside of the spigot. Apply the same lubricant to the inside of the hose and the outside of the spigot to simplify installation.

As a last resort (and if the hose is to be replaced with a new one anyway), the rubber can be slit with a knife and the hose peeled from the spigot. If this must be done, be careful that the metal connection is not damaged.

If a hose clamp is broken or damaged, do not reuse it. Wire-type clamps usually weaken with age, so it is a good idea to replace them with screw-type clamps whenever a hose is removed.

Tools

A selection of good tools is a basic requirement for anyone who plans to maintain and repair his or her own vehicle. For the owner who has few tools, the initial investment might seem high, but when compared to the spiraling costs of professional auto maintenance and repair, it is a wise one.

To help the owner decide which tools are needed to perform the tasks detailed in this manual, the following tool lists are offered: *Maintenance and minor repair, Repair/overhaul* and *Special.*

The newcomer to practical mechanics should start off with the *maintenance and minor repair* tool kit, which is adequate for the simpler jobs performed on a vehicle. Then, as confidence and experience grow, the owner can tackle more difficult tasks, buying additional tools as they are needed. Eventually the basic kit will be expanded into the *repair and overhaul* tool set. Over a period of time, the experienced do-it-yourselfer will assemble a tool set complete enough for most repair and overhaul procedures and will add tools from the special category when it is felt that the expense is justified by the frequency of use.

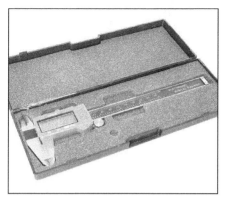

Dial caliper

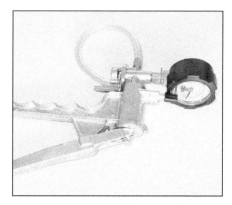

Hand-operated vacuum pump

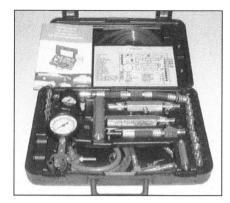

Fuel pressure gauge set

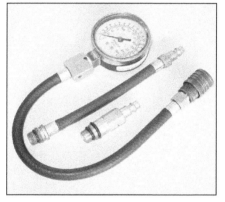

Compression gauge with spark plug hole adapter

Damper/steering wheel puller

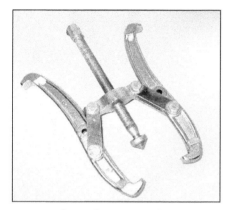

General purpose puller

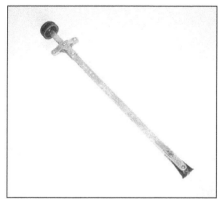

Hydraulic lifter removal tool

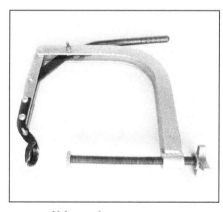

Valve spring compressor

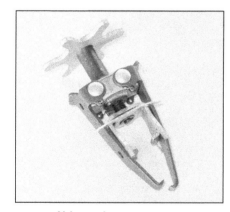

Valve spring compressor

Ridge reamer

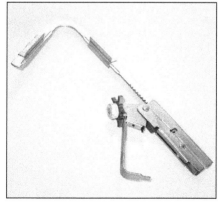

Piston ring groove cleaning tool

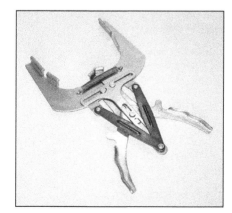

Ring removal/installation tool

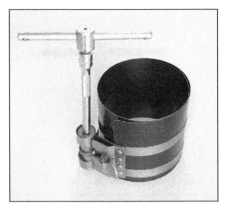

Ring compressor

Cylinder hone

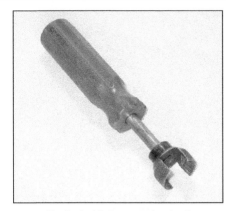

Brake hold-down spring tool

Torque angle gauge

Clutch plate alignment tool

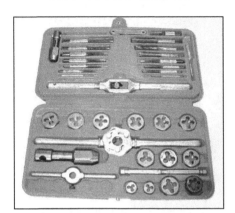

Tap and die set

Maintenance and minor repair tool kit

The tools in this list should be considered the minimum required for performance of routine maintenance, servicing and minor repair work. We recommend the purchase of combination wrenches (box-end and open-end combined in one wrench). While more expensive than open end wrenches, they offer the advantages of both types of wrench.

Combination wrench set (1/4-inch to
1 inch or 6 mm to 19 mm)
Adjustable wrench, 8 inch
Spark plug wrench with rubber insert
Spark plug gap adjusting tool
Feeler gauge set
Brake bleeder wrench
Standard screwdriver (5/16-inch x
6 inch)
Phillips screwdriver (No. 2 x 6 inch)
Combination pliers - 6 inch
Hacksaw and assortment of blades
Tire pressure gauge
Grease gun
Oil can
Fine emery cloth
Wire brush
Battery post and cable cleaning tool
Oil filter wrench
Funnel (medium size)
Safety goggles
Jackstands (2)
Drain pan

Note: *If basic tune-ups are going to be part of routine maintenance, it will be necessary to purchase a good quality stroboscopic timing light and combination tachometer/dwell meter. Although they are included in the list of special tools, it is mentioned here because they are absolutely necessary for tuning most vehicles properly.*

Repair and overhaul tool set

These tools are essential for anyone who plans to perform major repairs and are in addition to those in the maintenance and minor repair tool kit. Included is a comprehensive set of sockets which, though expensive, are invaluable because of their versatility, especially when various extensions and drives are available. We recommend the 1/2-inch drive over the 3/8-inch drive. Although the larger drive is bulky and more expensive, it has the capacity of accepting a very wide range of large sockets. Ideally, however, the mechanic should have a 3/8-inch drive set and a 1/2-inch drive set.

Socket set(s)
Reversible ratchet
Extension - 10 inch
Universal joint
Torque wrench (same size drive as
sockets)
Ball peen hammer - 8 ounce
Soft-face hammer (plastic/rubber)
Standard screwdriver (1/4-inch x 6 inch)

Standard screwdriver (stubby -
5/16-inch)
Phillips screwdriver (No. 3 x 8 inch)
Phillips screwdriver (stubby - No. 2)
Pliers - vise grip
Pliers - lineman's
Pliers - needle nose
Pliers - snap-ring (internal and external)
Cold chisel - 1/2-inch
Scribe
Scraper (made from flattened copper
tubing)
Centerpunch
Pin punches (1/16, 1/8, 3/16-inch)
Steel rule/straightedge - 12 inch
Allen wrench set (1/8 to 3/8-inch or
4 mm to 10 mm)
A selection of files
Wire brush (large)
Jackstands (second set)
Jack (scissor or hydraulic type)

Note: *Another tool which is often useful is an electric drill with a chuck capacity of 3/8-inch and a set of good quality drill bits.*

Special tools

The tools in this list include those which are not used regularly, are expensive to buy, or which need to be used in accordance with their manufacturer's instructions. Unless these tools will be used frequently, it is not very economical to purchase many of them. A consideration would be to split the cost and use between yourself and a friend or friends. In addition, most of these tools can be obtained from a tool rental shop on a temporary basis.

This list primarily contains only those tools and instruments widely available to the public, and not those special tools produced by the vehicle manufacturer for distribution to dealer service departments. Occasionally, references to the manufacturer's special tools are included in the text of this manual. Generally, an alternative method of doing the job without the special tool is offered. However, sometimes there is no alternative to their use. Where this is the case, and the tool cannot be purchased or borrowed, the work should be turned over to the dealer service department or an automotive repair shop.

> *Valve spring compressor*
> *Piston ring groove cleaning tool*
> *Piston ring compressor*
> *Piston ring installation tool*
> *Cylinder compression gauge*
> *Cylinder ridge reamer*
> *Cylinder surfacing hone*
> *Cylinder bore gauge*
> *Micrometers and/or dial calipers*
> *Hydraulic lifter removal tool*
> *Balljoint separator*
> *Universal-type puller*
> *Impact screwdriver*
> *Dial indicator set*
> *Stroboscopic timing light (inductive*
> *   pick-up)*
> *Hand operated vacuum/pressure pump*
> *Tachometer/dwell meter*
> *Universal electrical multimeter*
> *Cable hoist*
> *Brake spring removal and installation*
> *   tools*
> *Floor jack*

Buying tools

For the do-it-yourselfer who is just starting to get involved in vehicle maintenance and repair, there are a number of options available when purchasing tools. If maintenance and minor repair is the extent of the work to be done, the purchase of individual tools is satisfactory. If, on the other hand, extensive work is planned, it would be a good idea to purchase a modest tool set from one of the large retail chain stores. A set can usually be bought at a substantial savings over the individual tool prices, and they often come with a tool box. As additional tools are needed, add-on sets, individual tools and a larger tool box can be purchased to expand the tool selection. Building a tool set gradually allows the cost of the tools to be spread over a longer period of time and gives the mechanic the freedom to choose only those tools that will actually be used.

Tool stores will often be the only source of some of the special tools that are needed, but regardless of where tools are bought, try to avoid cheap ones, especially when buying screwdrivers and sockets, because they won't last very long. The expense involved in replacing cheap tools will eventually be greater than the initial cost of quality tools.

Care and maintenance of tools

Good tools are expensive, so it makes sense to treat them with respect. Keep them clean and in usable condition and store them properly when not in use. Always wipe off any dirt, grease or metal chips before putting them away. Never leave tools lying around in the work area. Upon completion of a job, always check closely under the hood for tools that may have been left there so they won't get lost during a test drive.

Some tools, such as screwdrivers, pliers, wrenches and sockets, can be hung on a panel mounted on the garage or workshop wall, while others should be kept in a tool box or tray. Measuring instruments, gauges, meters, etc. must be carefully stored where they cannot be damaged by weather or impact from other tools.

When tools are used with care and stored properly, they will last a very long time. Even with the best of care, though, tools will wear out if used frequently. When a tool is damaged or worn out, replace it. Subsequent jobs will be safer and more enjoyable if you do.

How to repair damaged threads

Sometimes, the internal threads of a nut or bolt hole can become stripped, usually from overtightening. Stripping threads is an all-too-common occurrence, especially when working with aluminum parts, because aluminum is so soft that it easily strips out.

Usually, external or internal threads are only partially stripped. After they've been cleaned up with a tap or die, they'll still work. Sometimes, however, threads are badly damaged. When this happens, you've got three choices:

1) *Drill and tap the hole to the next suitable oversize and install a larger diameter bolt, screw or stud.*
2) *Drill and tap the hole to accept a threaded plug, then drill and tap the plug to the original screw size. You can also buy a plug already threaded to the original size. Then you simply drill a hole to the specified size, then run the threaded plug into the hole with a bolt and jam nut. Once the plug is fully seated, remove the jam nut and bolt.*
3) *The third method uses a patented thread repair kit like Heli-Coil or Slimsert. These*

easy-to-use kits are designed to repair damaged threads in straight-through holes and blind holes. Both are available as kits which can handle a variety of sizes and thread patterns. Drill the hole, then tap it with the special included tap. Install the Heli-Coil and the hole is back to its original diameter and thread pitch.

Regardless of which method you use, be sure to proceed calmly and carefully. A little impatience or carelessness during one of these relatively simple procedures can ruin your whole day's work and cost you a bundle if you wreck an expensive part.

Working facilities

Not to be overlooked when discussing tools is the workshop. If anything more than routine maintenance is to be carried out, some sort of suitable work area is essential.

It is understood, and appreciated, that many home mechanics do not have a good workshop or garage available, and end up removing an engine or doing major repairs outside. It is recommended, however, that the overhaul or repair be completed under the cover of a roof.

A clean, flat workbench or table of comfortable working height is an absolute necessity. The workbench should be equipped with a vise that has a jaw opening of at least four inches.

As mentioned previously, some clean, dry storage space is also required for tools, as well as the lubricants, fluids, cleaning solvents, etc. which soon become necessary.

Sometimes waste oil and fluids, drained from the engine or cooling system during normal maintenance or repairs, present a disposal problem. To avoid pouring them on the ground or into a sewage system, pour the used fluids into large containers, seal them with caps and take them to an authorized disposal site or recycling center. Plastic jugs, such as old antifreeze containers, are ideal for this purpose.

Always keep a supply of old newspapers and clean rags available. Old towels are excellent for mopping up spills. Many mechanics use rolls of paper towels for most work because they are readily available and disposable. To help keep the area under the vehicle clean, a large cardboard box can be cut open and flattened to protect the garage or shop floor.

Whenever working over a painted surface, such as when leaning over a fender to service something under the hood, always cover it with an old blanket or bedspread to protect the finish. Vinyl covered pads, made especially for this purpose, are available at auto parts stores.

Booster battery (jump) starting

Observe the following precautions when using a booster battery to start a vehicle:

a) *Before connecting the booster battery, make sure the ignition switch is in the Off position.*

b) *Turn off the lights, heater and other electrical loads.*

c) *Your eyes should be shielded. Safety goggles are a good idea.*

d) *Make sure the booster battery is the same voltage as the dead one in the vehicle.*

e) *The two vehicles MUST NOT TOUCH each other!*

f) *Make sure the transmission is in Neutral (manual) or Park (automatic).*

g) *If the booster battery is not a maintenance-free type, remove the vent caps and lay a cloth over the vent holes.*

Connect one jumper lead between the positive (+) terminals of the two batteries **(see illustration)**.

Note: *2014 and earlier models are equipped with a remote positive terminal located on the left strut tower* **(see illustration)**.

Connect the other jumper lead first to the negative (-) terminal of the booster battery, then to a good engine ground on the vehicle to be started.

Note: *2014 and earlier models are also equipped with a remote grounding point, located on the left strut tower (near the remote positive terminal). Attach the lead at least 18 inches from the battery, if possible. Make sure the that the jumper leads will not contact the fan, drivebelt or other moving parts of the engine.*

Start the engine using the booster battery, then, with the engine running at idle speed, disconnect the jumper cables in the reverse order of connection.

The remote battery terminals are located on the left strut tower (2014 and earlier models)

1   Positive terminal   2   Negative terminal

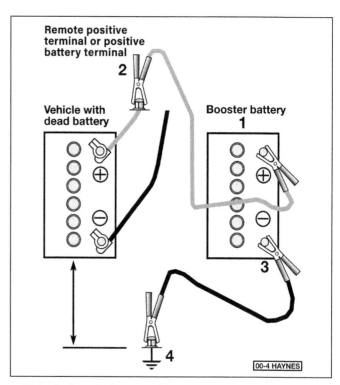

Make the booster battery cable connections in the numerical order shown (note that the negative cable of the booster battery is NOT attached to the negative terminal of the dead battery)

Jacking and towing

Jacking

The jack supplied with the vehicle should only be used for raising the vehicle for changing a tire or placing jackstands under the frame.

Warning: *Never crawl under the vehicle or start the engine when the jack is being used as the only means of support.*

All vehicles are supplied with a scissors-type jack. When jacking the vehicle, it should be engaged with the rocker panel flange, in the relief area approximately 10 inches from the wheel opening **(see illustration).**

The vehicle should be on level ground with the wheels blocked and the transmission in Park. Pry off the hub cap (if equipped) using the tapered end of the lug wrench. Loosen the lug nuts one-half turn and leave them in place until the wheel is raised off the ground.

Place the jack under the side of the vehicle in the indicated position. Use the supplied wrench to turn the jackscrew clockwise until the wheel is raised off the ground. Remove the lug nuts, pull off the wheel and install the spare.

With the beveled side in, install the lug nuts and tighten them until snug. Lower the vehicle by turning the jackscrew counterclockwise. Remove the jack and tighten the nuts in a diagonal pattern to the torque listed in the Chapter 1 Specifications. If a torque wrench is not available, have the torque checked by a service station as soon as possible. Install the hubcap by placing it in position and using the heel of your hand or a rubber mallet to seat it.

Towing

Front-wheel drive models

As a general rule, the vehicle should be towed with the front (drive) wheels off the ground or, preferably, on a flat bed car carrier. If the front wheels can't be raised or a carrier isn't available, place them on a dolly. The ignition key must be in the ACC position, since the steering lock mechanism isn't strong enough to hold the front wheels straight while towing.

In emergency situations the vehicle can be towed from the front with all four wheels on the ground, provided that speeds don't exceed 40 mph and the distance is not over 100 miles. Before towing, check the transaxle fluid level (see Chapter 1). If the level is below the HOT mark on the dipstick, add fluid.

Towing equipment specifically designed for this purpose should be used and should be attached to the main structural members of the vehicle, not the bumper or brackets.

Safety is a major consideration when towing and all applicable state and local laws must be obeyed. A safety chain system must be used for all towing.

While towing, the parking brake must be released and the transmission must be in Neutral. The steering must be unlocked (ignition switch in the Off position). Remember that power steering and power brakes will not work with the engine off.

All-wheel drive models

A flatbed car carrier must be used to transport these vehicles.

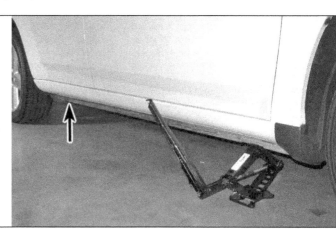

The jack fits over the rocker panel flange (there are two jacking points on each side of the vehicle)

Automotive chemicals and lubricants

A number of automotive chemicals and lubricants are available for use during vehicle maintenance and repair. They include a wide variety of products ranging from cleaning solvents and degreasers to lubricants and protective sprays for rubber, plastic and vinyl.

Cleaners

Carburetor cleaner and choke cleaner is a strong solvent for gum, varnish and carbon. Most carburetor cleaners leave a dry-type lubricant film which will not harden or gum up. Because of this film it is not recommended for use on electrical components.

Brake system cleaner is used to remove brake dust, grease and brake fluid from the brake system, where clean surfaces are absolutely necessary. It leaves no residue and often eliminates brake squeal caused by contaminants.

Electrical cleaner removes oxidation, corrosion and carbon deposits from electrical contacts, restoring full current flow. It can also be used to clean spark plugs, carburetor jets, voltage regulators and other parts where an oil-free surface is desired.

Demoisturants remove water and moisture from electrical components such as alternators, voltage regulators, electrical connectors and fuse blocks. They are non-conductive and non-corrosive.

Degreasers are heavy-duty solvents used to remove grease from the outside of the engine and from chassis components. They can be sprayed or brushed on and, depending on the type, are rinsed off either with water or solvent.

Lubricants

Motor oil is the lubricant formulated for use in engines. It normally contains a wide variety of additives to prevent corrosion and reduce foaming and wear. Motor oil comes in various weights (viscosity ratings) from 0 to 50. The recommended weight of the oil depends on the season, temperature and the demands on the engine. Light oil is used in cold climates and under light load conditions. Heavy oil is used in hot climates and where high loads are encountered. Multi-viscosity oils are designed to have characteristics of both light and heavy oils and are available in a number of weights from 0W-20 to 20W-50.

Gear oil is designed to be used in differentials, manual transmissions and other areas where high-temperature lubrication is required.

Chassis and wheel bearing grease is a heavy grease used where increased loads and friction are encountered, such as for wheel bearings, balljoints, tie-rod ends and universal joints.

High-temperature wheel bearing grease is designed to withstand the extreme temperatures encountered by wheel bearings in disc brake equipped vehicles. It usually contains molybdenum disulfide (moly), which is a dry-type lubricant.

White grease is a heavy grease for metal-to-metal applications where water is a problem. White grease stays soft under both low and high temperatures (usually from -100 to +190-degrees F), and will not wash off or dilute in the presence of water.

Assembly lube is a special extreme pressure lubricant, usually containing moly, used to lubricate high-load parts (such as main and rod bearings and cam lobes) for initial start-up of a new engine. The assembly lube lubricates the parts without being squeezed out or washed away until the engine oiling system begins to function.

Silicone lubricants are used to protect rubber, plastic, vinyl and nylon parts.

Graphite lubricants are used where oils cannot be used due to contamination problems, such as in locks. The dry graphite will lubricate metal parts while remaining uncontaminated by dirt, water, oil or acids. It is electrically conductive and will not foul electrical contacts in locks such as the ignition switch.

Moly penetrants loosen and lubricate frozen, rusted and corroded fasteners and prevent future rusting or freezing.

Heat-sink grease is a special electrically non-conductive grease that is used for mounting electronic ignition modules where it is essential that heat is transferred away from the module.

Sealants

RTV sealant is one of the most widely used gasket compounds. Made from silicone, RTV is air curing, it seals, bonds, waterproofs, fills surface irregularities, remains flexible, doesn't shrink, is relatively easy to remove, and is used as a supplementary sealer with almost all low and medium temperature gaskets.

Anaerobic sealant is much like RTV in that it can be used either to seal gaskets or to form gaskets by itself. It remains flexible, is solvent resistant and fills surface imperfections. The difference between an anaerobic sealant and an RTV-type sealant is in the curing. RTV cures when exposed to air, while an anaerobic sealant cures only in the absence of air. This means that an anaerobic sealant cures only after the assembly of parts, sealing them together.

Thread and pipe sealant is used for sealing hydraulic and pneumatic fittings and vacuum lines. It is usually made from a Teflon compound, and comes in a spray, a paint-on liquid and as a wrap-around tape.

Chemicals

Anti-seize compound prevents seizing, galling, cold welding, rust and corrosion in fasteners. High-temperature anti-seize, usually made with copper and graphite lubricants, is used for exhaust system and exhaust manifold bolts.

Anaerobic locking compounds are used to keep fasteners from vibrating or working loose and cure only after installation, in the absence of air. Medium strength locking compound is used for small nuts, bolts and screws that may be removed later. High-strength locking compound is for large nuts, bolts and studs which aren't removed on a regular basis.

Oil additives range from viscosity index improvers to chemical treatments that claim to reduce internal engine friction. It should be noted that most oil manufacturers caution against using additives with their oils.

Gas additives perform several functions, depending on their chemical makeup. They usually contain solvents that help dissolve gum and varnish that build up on carburetor, fuel injection and intake parts. They also serve to break down carbon deposits that form on the inside surfaces of the combustion chambers. Some additives contain upper cylinder lubricants for valves and piston rings, and others contain chemicals to remove condensation from the gas tank.

Miscellaneous

Brake fluid is specially formulated hydraulic fluid that can withstand the heat and pressure encountered in brake systems. Care must be taken so this fluid does not come in contact with painted surfaces or plastics. An opened container should always be resealed to prevent contamination by water or dirt.

Weatherstrip adhesive is used to bond weatherstripping around doors, windows and trunk lids. It is sometimes used to attach trim pieces.

Undercoating is a petroleum-based, tar-like substance that is designed to protect metal surfaces on the underside of the vehicle from corrosion. It also acts as a sound-deadening agent by insulating the bottom of the vehicle.

Waxes and polishes are used to help protect painted and plated surfaces from the weather. Different types of paint may require the use of different types of wax and polish. Some polishes utilize a chemical or abrasive cleaner to help remove the top layer of oxidized (dull) paint on older vehicles. In recent years many non-wax polishes that contain a wide variety of chemicals such as polymers and silicones have been introduced. These non-wax polishes are usually easier to apply and last longer than conventional waxes and polishes.

Conversion factors

Length (distance)

Inches (in)	X	25.4	= Millimeters (mm)	X 0.0394	= Inches (in)
Feet (ft)	X	0.305	= Meters (m)	X 3.281	= Feet (ft)
Miles	X	1.609	= Kilometers (km)	X 0.621	= Miles

Volume (capacity)

Cubic inches (cu in; in^3)	X	16.387	= Cubic centimeters (cc; cm^3)	X 0.061	= Cubic inches (cu in; in^3)
Imperial pints (Imp pt)	X	0.568	= Liters (l)	X 1.76	= Imperial pints (Imp pt)
Imperial quarts (Imp qt)	X	1.137	= Liters (l)	X 0.88	= Imperial quarts (Imp qt)
Imperial quarts (Imp qt)	X	1.201	= US quarts (US qt)	X 0.833	= Imperial quarts (Imp qt)
US quarts (US qt)	X	0.946	= Liters (l)	X 1.057	= US quarts (US qt)
Imperial gallons (Imp gal)	X	4.546	= Liters (l)	X 0.22	= Imperial gallons (Imp gal)
Imperial gallons (Imp gal)	X	1.201	= US gallons (US gal)	X 0.833	= Imperial gallons (Imp gal)
US gallons (US gal)	X	3.785	= Liters (l)	X 0.264	= US gallons (US gal)

Mass (weight)

Ounces (oz)	X	28.35	= Grams (g)	X 0.035	= Ounces (oz)
Pounds (lb)	X	0.454	= Kilograms (kg)	X 2.205	= Pounds (lb)

Force

Ounces-force (ozf; oz)	X	0.278	= Newtons (N)	X 3.6	= Ounces-force (ozf; oz)
Pounds-force (lbf; lb)	X	4.448	= Newtons (N)	X 0.225	= Pounds-force (lbf; lb)
Newtons (N)	X	0.1	= Kilograms-force (kgf; kg)	X 9.81	= Newtons (N)

Pressure

Pounds-force per square inch (psi; lbf/in^2; lb/in^2)	X	0.070	= Kilograms-force per square centimeter (kgf/cm^2; kg/cm^2)	X 14.223	= Pounds-force per square inch (psi; lbf/in^2; lb/in^2)
Pounds-force per square inch (psi; lbf/in^2; lb/in^2)	X	0.068	= Atmospheres (atm)	X 14.696	= Pounds-force per square inch (psi; lbf/in^2; lb/in^2)
Pounds-force per square inch (psi; lbf/in^2; lb/in^2)	X	0.069	= Bars	X 14.5	= Pounds-force per square inch (psi; lbf/in^2; lb/in^2)
Pounds-force per square inch (psi; lbf/in^2; lb/in^2)	X	6.895	= Kilopascals (kPa)	X 0.145	= Pounds-force per square inch (psi; lbf/in^2; lb/in^2)
Kilopascals (kPa)	X	0.01	= Kilograms-force per square centimeter (kgf/cm^2; kg/cm^2)	X 98.1	= Kilopascals (kPa)

Torque (moment of force)

Pounds-force inches (lbf in; lb in)	X	1.152	= Kilograms-force centimeter (kgf cm; kg cm)	X 0.868	= Pounds-force inches (lbf in; lb in)
Pounds-force inches (lbf in; lb in)	X	0.113	= Newton meters (Nm)	X 8.85	= Pounds-force inches (lbf in; lb in)
Pounds-force inches (lbf in; lb in)	X	0.083	= Pounds-force feet (lbf ft; lb ft)	X 12	= Pounds-force inches (lbf in; lb in)
Pounds-force feet (lbf ft; lb ft)	X	0.138	= Kilograms-force meters (kgf m; kg m)	X 7.233	= Pounds-force feet (lbf ft; lb ft)
Pounds-force feet (lbf ft; lb ft)	X	1.356	= Newton meters (Nm)	X 0.738	= Pounds-force feet (lbf ft; lb ft)
Newton meters (Nm)	X	0.102	= Kilograms-force meters (kgf m; kg m)	X 9.804	= Newton meters (Nm)

Vacuum

Inches mercury (in. Hg)	X	3.377	= Kilopascals (kPa)	X 0.2961	= Inches mercury
Inches mercury (in. Hg)	X	25.4	= Millimeters mercury (mm Hg)	X 0.0394	= Inches mercury

Power

Horsepower (hp)	X	745.7	= Watts (W)	X 0.0013	= Horsepower (hp)

Velocity (speed)

Miles per hour (miles/hr; mph)	X	1.609	= Kilometers per hour (km/hr; kph)	X 0.621	= Miles per hour (miles/hr; mph)

Fuel consumption*

Miles per gallon, Imperial (mpg)	X	0.354	= Kilometers per liter (km/l)	X 2.825	= Miles per gallon, Imperial (mpg)
Miles per gallon, US (mpg)	X	0.425	= Kilometers per liter (km/l)	X 2.352	= Miles per gallon, US (mpg)

Temperature

Degrees Fahrenheit = (°C x 1.8) + 32   Degrees Celsius (Degrees Centigrade; °C) = (°F - 32) x 0.56

*It is common practice to convert from miles per gallon (mpg) to liters/100 kilometers (l/100km), where mpg (Imperial) x l/100 km = 282 and mpg (US) x l/100 km = 235

DECIMALS to MILLIMETERS

Decimal	mm	Decimal	mm
0.001	0.0254	0.500	12.7000
0.002	0.0508	0.510	12.9540
0.003	0.0762	0.520	13.2080
0.004	0.1016	0.530	13.4620
0.005	0.1270	0.540	13.7160
0.006	0.1524	0.550	13.9700
0.007	0.1778	0.560	14.2240
0.008	0.2032	0.570	14.4780
0.009	0.2286	0.580	14.7320
		0.590	14.9860
0.010	0.2540		
0.020	0.5080		
0.030	0.7620		
0.040	1.0160	0.600	15.2400
0.050	1.2700	0.610	15.4940
0.060	1.5240	0.620	15.7480
0.070	1.7780	0.630	16.0020
0.080	2.0320	0.640	16.2560
0.090	2.2860	0.650	16.5100
		0.660	16.7640
0.100	2.5400	0.670	17.0180
0.110	2.7940	0.680	17.2720
0.120	3.0480	0.690	17.5260
0.130	3.3020		
0.140	3.5560		
0.150	3.8100	0.700	17.7800
0.160	4.0640	0.710	18.0340
0.170	4.3180	0.720	18.2880
0.180	4.5720	0.730	18.5420
0.190	4.8260	0.740	18.7960
0.200	5.0800	0.750	19.0500
0.210	5.3340	0.760	19.3040
0.220	5.5880	0.770	19.5580
0.230	5.8420	0.780	19.8120
0.240	6.0960	0.790	20.0660
0.250	6.3500		
0.260	6.6040		
0.270	6.8580	0.800	20.3200
0.280	7.1120	0.810	20.5740
0.290	7.3660	0.820	21.8280
		0.830	21.0820
0.300	7.6200	0.840	21.3360
0.310	7.8740	0.850	21.5900
0.320	8.1280	0.860	21.8440
0.330	8.3820	0.870	22.0980
0.340	8.6360	0.880	22.3520
0.350	8.8900	0.890	22.6060
0.360	9.1440		
0.370	9.3980		
0.380	9.6520		
0.390	9.9060	0.900	22.8600
0.400	10.1600	0.910	23.1140
0.410	10.4140	0.920	23.3680
0.420	10.6680	0.930	23.6220
0.430	10.9220	0.940	23.8760
0.440	11.1760	0.950	24.1300
0.450	11.4300	0.960	24.3840
0.460	11.6840	0.970	24.6380
0.470	11.9380	0.980	24.8920
0.480	12.1920	0.990	25.1460
0.490	12.4460	1.000	25.4000

FRACTIONS to DECIMALS to MILLIMETERS

Fraction	Decimal	mm	Fraction	Decimal	mm
1/64	0.0156	0.3969	33/64	0.5156	13.0969
1/32	0.0312	0.7938	17/32	0.5312	13.4938
3/64	0.0469	1.1906	35/64	0.5469	13.8906
1/16	0.0625	1.5875	9/16	0.5625	14.2875
5/64	0.0781	1.9844	37/64	0.5781	14.6844
3/32	0.0938	2.3812	19/32	0.5938	15.0812
7/64	0.1094	2.7781	39/64	0.6094	15.4781
1/8	0.1250	3.1750	5/8	0.6250	15.8750
9/64	0.1406	3.5719	41/64	0.6406	16.2719
5/32	0.1562	3.9688	21/32	0.6562	16.6688
11/64	0.1719	4.3656	43/64	0.6719	17.0656
3/16	0.1875	4.7625	11/16	0.6875	17.4625
13/64	0.2031	5.1594	45/64	0.7031	17.8594
7/32	0.2188	5.5562	23/32	0.7188	18.2562
15/64	0.2344	5.9531	47/64	0.7344	18.6531
1/4	0.2500	6.3500	3/4	0.7500	19.0500
17/64	0.2656	6.7469	49/64	0.7656	19.4469
9/32	0.2812	7.1438	25/32	0.7812	19.8438
19/64	0.2969	7.5406	51/64	0.7969	20.2406
5/16	0.3125	7.9375	13/16	0.8125	20.6375
21/64	0.3281	8.3344	53/64	0.8281	21.0344
11/32	0.3438	8.7312	27/32	0.8438	21.4312
23/64	0.3594	9.1281	55/64	0.8594	21.8281
3/8	0.3750	9.5250	7/8	0.8750	22.2250
25/64	0.3906	9.9219	57/64	0.8906	22.6219
13/32	0.4062	10.3188	29/32	0.9062	23.0188
27/64	0.4219	10.7156	59/64	0.9219	23.4156
7/16	0.4375	11.1125	15/16	0.9375	23.8125
29/64	0.4531	11.5094	61/64	0.9531	24.2094
15/32	0.4688	11.9062	31/32	0.9688	24.6062
31/64	0.4844	12.3031	63/64	0.9844	25.0031
1/2	0.5000	12.7000	1	1.0000	25.4000

Safety first!

Regardless of how enthusiastic you may be about getting on with the job at hand, take the time to ensure that your safety is not jeopardized. A moment's lack of attention can result in an accident, as can failure to observe certain simple safety precautions. The possibility of an accident will always exist, and the following points should not be considered a comprehensive list of all dangers. Rather, they are intended to make you aware of the risks and to encourage a safety conscious approach to all work you carry out on your vehicle.

Essential DOs and DON'Ts

DON'T rely on a jack when working under the vehicle. Always use approved jackstands to support the weight of the vehicle and place them under the recommended lift or support points.

DON'T attempt to loosen extremely tight fasteners (i.e. wheel lug nuts) while the vehicle is on a jack - it may fall.

DON'T start the engine without first making sure that the transmission is in Neutral (or Park where applicable) and the parking brake is set.

DON'T remove the radiator cap from a hot cooling system - let it cool or cover it with a cloth and release the pressure gradually.

DON'T attempt to drain the engine oil until you are sure it has cooled to the point that it will not burn you.

DON'T touch any part of the engine or exhaust system until it has cooled sufficiently to avoid burns.

DON'T siphon toxic liquids such as gasoline, antifreeze and brake fluid by mouth, or allow them to remain on your skin.

DON'T inhale brake lining dust - it is potentially hazardous (see *Asbestos* below).

DON'T allow spilled oil or grease to remain on the floor - wipe it up before someone slips on it.

DON'T use loose fitting wrenches or other tools which may slip and cause injury.

DON'T push on wrenches when loosening or tightening nuts or bolts. Always try to pull the wrench toward you. If the situation calls for pushing the wrench away, push with an open hand to avoid scraped knuckles if the wrench should slip.

DON'T attempt to lift a heavy component alone - get someone to help you.

DON'T *rush or take unsafe shortcuts to finish a job.*

DON'T allow children or animals in or around the vehicle while you are working on it.

DO wear eye protection when using power tools such as a drill, sander, bench grinder, etc. and when working under a vehicle.

DO keep loose clothing and long hair well out of the way of moving parts.

DO make sure that any hoist used has a safe working load rating adequate for the job.

DO get someone to check on you periodically when working alone on a vehicle.

DO carry out work in a logical sequence and make sure that everything is correctly assembled and tightened.

DO keep chemicals and fluids tightly capped and out of the reach of children and pets.

DO remember that your vehicle's safety affects that of yourself and others. If in doubt on any point, get professional advice.

Steering, suspension and brakes

These systems are essential to driving safety, so make sure you have a qualified shop or individual check your work. Also, compressed suspension springs can cause injury if released suddenly - be sure to use a spring compressor.

Airbags

Airbags are explosive devices that can **CAUSE** injury if they deploy while you're working on the vehicle. Follow the manufacturer's instructions to disable the airbag whenever you're working in the vicinity of airbag components.

Asbestos

Certain friction, insulating, sealing, and other products - such as brake linings, brake bands, clutch linings, torque converters, gaskets, etc. - may contain asbestos or other hazardous friction material. Extreme care must be taken to avoid inhalation of dust from such products, since it is hazardous to health. If in doubt, assume that they do contain asbestos.

Fire

Remember at all times that gasoline is highly flammable. Never smoke or have any kind of open flame around when working on a vehicle. But the risk does not end there. A spark caused by an electrical short circuit, by two metal surfaces contacting each other, or even by static electricity built up in your body under certain conditions, can ignite gasoline vapors, which in a confined space are highly explosive. Do not, under any circumstances, use gasoline for cleaning parts. Use an approved safety solvent.

Always disconnect the battery ground (-) cable at the battery before working on any part of the fuel system or electrical system. Never risk spilling fuel on a hot engine or exhaust component. It is strongly recommended that a fire extinguisher suitable for use on fuel and electrical fires be kept handy in the garage or workshop at all times. Never try to extinguish a fuel or electrical fire with water.

Fumes

Certain fumes are highly toxic and can quickly cause unconsciousness and even death if inhaled to any extent. Gasoline vapor falls into this category, as do the vapors from some cleaning solvents. Any draining or pouring of such volatile fluids should be done in a well ventilated area.

When using cleaning fluids and solvents, read the instructions on the container carefully. Never use materials from unmarked containers.

Never run the engine in an enclosed space, such as a garage. Exhaust fumes contain carbon monoxide, which is extremely poisonous. If you need to run the engine, always do so in the open air, or at least have the rear of the vehicle outside the work area.

The battery

Never create a spark or allow a bare light bulb near a battery. They normally give off a certain amount of hydrogen gas, which is highly explosive.

Always disconnect the battery ground (-) cable at the battery before working on the fuel or electrical systems.

If possible, loosen the filler caps or cover when charging the battery from an external source (this does not apply to sealed or maintenance-free batteries). Do not charge at an excessive rate or the battery may burst.

Take care when adding water to a non maintenance-free battery and when carrying a battery. The electrolyte, even when diluted, is very corrosive and should not be allowed to contact clothing or skin.

Always wear eye protection when cleaning the battery to prevent the caustic deposits from entering your eyes.

Household current

When using an electric power tool, inspection light, etc., which operates on household current, always make sure that the tool is correctly connected to its plug and that, where necessary, it is properly grounded. Do not use such items in damp conditions and, again, do not create a spark or apply excessive heat in the vicinity of fuel or fuel vapor.

Secondary ignition system voltage

A severe electric shock can result from touching certain parts of the ignition system (such as the spark plug wires) when the engine is running or being cranked, particularly if components are damp or the insulation is defective. In the case of an electronic ignition system, the secondary system voltage is much higher and could prove fatal.

Hydrofluoric acid

This extremely corrosive acid is formed when certain types of synthetic rubber, found in some O-rings, oil seals, fuel hoses, etc. are exposed to temperatures above 750-degrees F (400-degrees C). The rubber changes into a charred or sticky substance containing the acid. *Once formed, the acid remains dangerous for years. If it gets onto the skin, it may be necessary to amputate the limb concerned.*

When dealing with a vehicle which has suffered a fire, or with components salvaged from such a vehicle, wear protective gloves and discard them after use.

Troubleshooting

Contents

Symptom	Section

Engine and performance
Engine will not rotate when attempting to start 1
Engine rotates but will not start 2
Engine hard to start when cold 3
Engine hard to start when hot 4
Starter motor noisy or excessively rough in engagement 5
Engine starts but stops immediately 6
Oil puddle under engine 7
Engine lopes while idling or idles erratically 8
Engine misses at idle speed 9
Engine misses throughout driving speed range 10
Engine stumbles on acceleration 11
Engine surges while holding accelerator steady 12
Engine stalls 13
Engine lacks power 14
Engine backfires 15
Pinging or knocking engine sounds during
   acceleration or uphill 16
Engine runs with oil pressure light on 17
Engine continues to run after switching off 18

Engine electrical system
Battery will not hold a charge 19
Voltage warning light fails to go out 20
Voltage warning light fails to come on when key
   is turned on 21

Fuel system
Excessive fuel consumption 22
Fuel leakage and/or fuel odor 23

Cooling system
Overheating 24
Overcooling 25
External coolant leakage 26
Internal coolant leakage 27
Coolant loss 28
Poor coolant circulation 29

Automatic transaxle
Fluid leakage 30
Transaxle fluid brown or has a burned smell 31

Symptom	Section

General shift mechanism problems 32
Engine will start in gears other than Park or Neutral 33
Transaxle slips, shifts roughly, is noisy or has no drive
   in forward or reverse gears 34

Driveaxles
Clicking noise in turns 35
Knock or clunk when accelerating after coasting 36
Shudder or vibration during acceleration 37

Brakes
Vehicle pulls to one side during braking 38
Noise (grinding or high-pitched squeal) when the brakes
   are applied) 39
Brake roughness or chatter (pedal pulsates) 40
Excessive pedal effort required to stop vehicle 41
Excessive brake pedal travel 42
Dragging brakes 43
Grabbing or uneven braking action 44
Brake pedal feels spongy when depressed 45
Brake pedal travels to the floor with little resistance 46
Parking brake does not hold 47

Suspension and steering systems
Vehicle pulls to one side 48
Abnormal or excessive tire wear 49
Wheel makes a thumping noise 50
Shimmy, shake or vibration 51
Hard steering 52
Steering wheel does not return to center position correctly 53
Abnormal noise at the front end 54
Wander or poor steering stability 55
Erratic steering when braking 56
Excessive pitching and/or rolling around corners or
   during braking 57
Suspension bottoms 58
Cupped tires 59
Excessive tire wear on outside edge 60
Excessive tire wear on inside edge 61
Tire tread worn in one place 62
Excessive play or looseness in steering system 63
Rattling or clicking noise in steering gear 64

This section provides an easy reference guide to the more common problems which may occur during the operation of your vehicle. Various symptoms and their possible causes are grouped under headings denoting components or systems, such as Engine, Cooling system, etc. They also refer to the Chapter and/or Section that deals with the problem.

Remember that successful troubleshooting isn't a mysterious art practiced only by professional mechanics. It's simply the result of knowledge combined with an intelligent, systematic approach to a problem. Always use a process of elimination, starting with the simplest solution and working through to the most complex - and never overlook the obvious. Anyone can run the gas tank dry or leave the lights on overnight, so don't assume that you're exempt from such oversights.

Finally, always establish a clear idea why a problem has occurred and take steps to ensure that it doesn't happen again. If the electrical system fails because of a poor connection, check all other connections in the system to make sure they don't fail as well. If a particular fuse continues to blow, find out why - don't just go on replacing fuses. Remember, failure of a small component can often be indicative of potential failure or incorrect functioning of a more important component or system.

Engine and performance

1 Engine will not rotate when attempting to start

1   Battery terminal connections loose or corroded (Chapter 1).
2   Battery discharged or faulty (Chapter 1).
3   Automatic transaxle not completely engaged in Park (Chapter 7A).
4   Broken, loose or disconnected wiring in the starting circuit (Chapters 5 and 12).
5   Starter motor pinion jammed in flywheel ring gear (Chapter 5).
6   Starter solenoid faulty (Chapter 5).
7   Starter motor faulty (Chapter 5).
8   Ignition switch faulty (Chapter 12).
9   Transmission Range (TR) sensor faulty (Chapter 6).
10   Starter pinion or driveplate teeth worn or broken (Chapter 5).

2 Engine rotates but will not start

1   Fuel tank empty.
2   Battery discharged (engine rotates slowly) (Chapter 5).
3   Battery terminal connections loose or corroded (Chapter 1).
4   Leaking fuel injector(s), fuel pump, pressure regulator, etc. (Chapter 4).
5   Fuel not reaching fuel injection system (Chapter 4).
6   Broken timing belt or chain (Chapter 2A, 2B, 2C or 2D).
7   Ignition system problem (Chapter 5).
8   Defective crankshaft sensor or camshaft sensor (Chapter 6).

3 Engine hard to start when cold

1   Battery discharged or low (Chapter 1).
2   Fuel system malfunctioning (Chapter 4).
3   Emissions or engine control system malfunctioning (Chapter 6).

4 Engine hard to start when hot

1   Air filter clogged (Chapter 1).
2   Fuel not reaching the fuel injection system (Chapter 4).
3   Corroded battery connections, especially ground (Chapter 1).
4   Emissions or engine control system malfunctioning (Chapter 6).

5 Starter motor noisy or excessively rough in engagement

1   Pinion or driveplate gear teeth worn or broken (Chapter 5).

2   Starter motor mounting bolts loose or missing (Chapter 5).

6 Engine starts but stops immediately

1   Insufficient fuel reaching the fuel injectors (Chapter 4).
2   Vacuum leak at the gasket between the intake manifold/plenum and throttle body (Chapters 1 and 4).
3   Restricted exhaust system (most likely the catalytic converter) (Chapters 4 and 6).

7 Oil puddle under engine

1   Oil pan gasket and/or oil pan drain bolt seal leaking (Chapters 1 and 2).
2   Oil pressure sending unit leaking (Chapter 2E).
3   Valve cover gaskets leaking (Chapter 2).
4   Engine oil seals leaking (Chapter 2).

8 Engine lopes while idling or idles erratically

1   Vacuum leakage (Chapter 4).
2   Leaking EGR valve or plugged PCV valve (Chapter 6).
3   Air filter clogged (Chapter 1).
4   Fuel pump not delivering sufficient fuel to the fuel injection system (Chapter 4).
5   Leaking head gasket (Chapter 2).
6   Camshaft lobes worn (Chapter 2).

9 Engine misses at idle speed

1   Spark plugs worn or not gapped properly (Chapter 1).
2   Faulty coil(s) or spark plug wires (Chapter 5).
3   Vacuum leaks (Chapters 1 and 4).
4   Uneven or low compression (Chapter 2E).

10 Engine misses throughout driving speed range

1   Fuel filter clogged and/or impurities in the fuel system (Chapter 4).
2   Low fuel pressure (Chapter 4).
3   Faulty or incorrectly gapped spark plugs (Chapter 1).
4   Leaking spark plug wires (Chapter 1).
5   Faulty emission system components (Chapter 6).
6   Low or uneven cylinder compression pressures (Chapter 2E).
7   Weak or faulty ignition system (Chapter 5).
8   Vacuum leak (Chapters 2 and 4).

11 Engine stumbles on acceleration

1   Spark plugs fouled (Chapter 1).
2   Fuel injection system problem (Chapter 4).
3   Fuel filter clogged (Chapter 4).
4   Intake manifold air leak (Chapter 4).
5   Problem with emissions/engine control system (Chapter 6).

12 Engine surges while holding accelerator steady

1   Intake air leak (Chapter 4).
2   Fuel pump faulty (Chapter 4).
3   Problem with the fuel injection system (Chapter 4).
4   Problem with the emission or engine control system (Chapter 6).

13 Engine stalls

1   Fuel filter clogged and/or water and impurities in the fuel system (Chapter 4).
2   Faulty emissions or engine control system components (Chapter 6).
3   Faulty or incorrectly gapped spark plugs (Chapter 1).
4   Faulty spark plug wires (Chapter 1).
5   Vacuum leak (Chapters 2 and 4).

14 Engine lacks power

1   Faulty or incorrectly gapped spark plugs (Chapter 1).
2   Restricted exhaust system (most likely the catalytic converter) (Chapters 4 and 6).
3   Fuel injection system malfunctioning (Chapter 4).
4   Faulty coil(s) (Chapter 5).
5   Brakes binding (Chapter 9).
6   Automatic transaxle fluid level incorrect (Chapter 1).
7   Fuel filter clogged and/or impurities in the fuel system (Chapter 1).
8   Emission control system not functioning properly (Chapter 6).
9   Low or uneven cylinder compression pressures (Chapter 2E).

15 Engine backfires

1   Emissions system not functioning properly (Chapter 6).
2   Fuel injection system malfunctioning (Chapter 4).
3   Vacuum leak at fuel injectors, intake manifold or vacuum hoses (Chapter 4).
4   Valves sticking (Chapter 2).
5   Timing belt or chain worn (Chapter 2).

16 Pinging or knocking engine sounds during acceleration or uphill

1   Incorrect grade of fuel.
2   Fuel injection system malfunctioning (Chapter 4).
3   Improper or damaged spark plugs (Chapter 1).
4   Worn or damaged ignition components (Chapter 5).
5   Faulty emissions or engine control system (Chapter 6).
6   Vacuum leak (Chapters 2 and 4).

17 Engine runs with oil pressure light on

1   Low oil level (Chapter 1).
2   Short in wiring circuit (Chapter 12).
3   Faulty oil pressure sender (Chapter 2E).
4   Oil viscosity too low or oil diluted.
5   Worn engine bearings and/or oil pump (Chapter 2).

18 Engine continues to run after switching off

1   Excessive engine operating temperature (Chapter 3).
2   Excessive carbon deposits on valves and pistons.
3   Leaking fuel injector(s).

Engine electrical system

19 Battery will not hold a charge

1   Drivebelt or tensioner defective (Chapter 1).
2   Battery terminals loose or corroded (Chapter 1).
3   Alternator not charging properly (Chapter 5).
4   Loose, broken or faulty wiring in the charging circuit (Chapter 5).
5   Internally defective battery (Chapters 1 and 5).

20 Voltage warning light fails to go out

1   Faulty alternator or charging circuit (Chapter 5).
2   Drivebelt or tensioner defective (Chapter 1).
3   Alternator voltage regulator inoperative (Chapter 5).

21 Voltage warning light fails to come on when key is turned on

1   Warning light bulb defective (Chapter 12).
2   Fault in the printed circuit, dash wiring or bulb holder (Chapter 12).

Fuel system

22 Excessive fuel consumption

1   Dirty or clogged air filter element (Chapter 1).
2   Emissions or engine control system not functioning properly (Chapter 6).
3   Fuel injection system malfunctioning (Chapter 4).
4   Low tire pressure or incorrect tire size (Chapter 1).

23 Fuel leakage and/or fuel odor

1   Leak in a fuel feed or vent line (Chapter 4).
2   Tank overfilled.
3   Evaporative emissions control canister defective (Chapter 6).
4   Fuel injector seals faulty (Chapter 4).

Cooling system

24 Overheating

1   Insufficient coolant in system (Chapter 1).
2   Drivebelt or tensioner defective (Chapter 1).
3   Radiator core blocked or grille restricted (Chapter 3).
4   Thermostat faulty (Chapter 3).
5   Electric cooling fan blades broken or cracked (Chapter 3).
6   Radiator cap not maintaining proper pressure (Chapter 3).
7   Faulty water pump (Chapter 3).

25 Overcooling

Incorrect (opening temperature too low) or faulty thermostat (Chapter 3).

26 External coolant leakage

1   Deteriorated/damaged hoses or loose clamps (Chapters 1 and 3).
2   Water pump seal defective (Chapter 3).
3   Leakage from radiator core (Chapter 3).
4   Engine drain or water jacket core plugs leaking (Chapter 1).

27 Internal coolant leakage

1   Leaking cylinder head gasket (Chapter 2).
2   Cracked cylinder bore or cylinder head (Chapter 2).

28 Coolant loss

1   Too much coolant in system (Chapter 1).
2   Coolant boiling away because of overheating (Chapter 3).
3   Internal or external leakage (Chapter 3).
4   Faulty radiator cap (Chapter 3).

29 Poor coolant circulation

1   Inoperative water pump (Chapter 3).
2   Restriction in cooling system (Chapters 1 and 3).
3   Drivebelt or tensioner defective or out of adjustment (Chapter 1).
4   Thermostat sticking (Chapter 3).

Automatic transaxle

30 Fluid leakage

1   Automatic transmission fluid is a deep red color. Fluid leaks should not be confused with engine oil, which can easily be blown by airflow to the transaxle.
2   To pinpoint a leak, first remove all built-up dirt and grime from the transaxle housing with degreasing agents and/or steam cleaning. Drive the vehicle at low speeds so air flow will not blow the leak far from its source. Raise the vehicle and determine where the leak is coming from. Common areas of leakage are:
a)   *Fluid pan*
b)   *Fluid cooler lines (Chapter 7A)*
c)   *Vehicle Speed Sensor (Chapter 6)*

31 Transaxle fluid brown or has a burned smell

Transaxle overheated. Change fluid (Chapter 1).

32 General shift mechanism problems

1   Chapter 7A deals with checking and adjusting the shift linkage on automatic transaxles. Common problems which may be attributed to a poorly adjusted linkage are:
a)   *Engine starting in gears other than Park or Neutral.*

b) Indicator on shifter pointing to a gear other than the one actually being used.

c) Vehicle moves when in Park.

2   Refer to Chapter 7A for the shift linkage adjustment procedure.

33   Engine will start in gears other than Park or Neutral

Transmission Range (TR) sensor malfunctioning (Chapter 6).

34   Transaxle slips, shifts roughly, is noisy or has no drive in forward or reverse gears

1   There are many probable causes for the above problems, but the home mechanic should be concerned with only one possibility - fluid level. Before taking the vehicle to a repair shop, check the level and condition of the fluid as described in Chapter 1.

2   Correct the fluid level as necessary or change the fluid and filter if needed. If the problem persists, have a professional diagnose the probable cause.

Driveaxles

35   Clicking noise in turns

Worn or damaged outer CV joint. Check for cut or damaged boots (Chapter 1). Repair as necessary (Chapter 8).

36   Knock or clunk when accelerating after coasting

Worn or damaged CV joint. Check for cut or damaged boots (Chapter 1). Repair as necessary (Chapter 8).

37   Shudder or vibration during acceleration

1   Worn or damaged CV joints. Repair or replace as necessary (Chapter 8).

2   Sticking inner joint assembly. Correct or replace as necessary (Chapter 8).

Brakes

38   Vehicle pulls to one side during braking

1   Incorrect tire pressures (Chapter 1).

2   Front end out of alignment (have the front end aligned).

3   Unmatched tires on same axle.

4   Restricted brake lines or hoses (Chapter 9).

5   Sticking caliper or wheel cylinder piston (Chapter 9).

6   Loose suspension parts (Chapter 10).

7   Contaminated brake pad material (Chapter 9).

39   Noise (grinding or high-pitched squeal) when the brakes are applied

Disc brake pads worn out. Replace pads with new ones immediately (Chapter 9).

40   Brake roughness or chatter (pedal pulsates)

1   Excessive brake disc lateral runout.

2   Parallelism of disc not within specifications (Chapter 9).

3   Defective brake disc (Chapter 9).

41   Excessive pedal effort required to stop vehicle

1   Malfunctioning power brake booster (Chapter 9).

2   Partial system failure (Chapter 9).

3   Excessively worn pads (Chapter 9).

4   One or more caliper or wheel cylinder pistons seized or sticking (Chapter 9).

5   Brake pads contaminated with oil or grease (Chapter 9).

6   New pads or shoes installed and not yet seated. It will take a while for the new material to seat.

42   Excessive brake pedal travel

1   Partial brake system failure (Chapter 9).

2   Insufficient fluid in master cylinder (Chapters 1 and 9).

3   Air trapped in system (Chapter 9).

4   Faulty master cylinder (Chapter 9).

43   Dragging brakes

1   Master cylinder pistons not returning correctly (Chapter 9).

2   Restricted brake lines or hoses (Chapters 1 and 9).

3   Incorrect parking brake adjustment (Chapter 9).

4   Defective brake calipers (Chapter 9).

44   Grabbing or uneven braking action

Contaminated brake pads (Chapter 9).

45   Brake pedal feels spongy when depressed

1   Air in hydraulic lines (Chapter 9).

2   Master cylinder mounting bolts loose (Chapter 9).

3   Master cylinder defective (Chapter 9).

46   Brake pedal travels to the floor with little resistance

Little or no fluid in the master cylinder reservoir caused by leaking caliper, or loose, damaged or disconnected brake lines (Chapter 9).

47   Parking brake does not hold

Parking brake cables improperly adjusted (Chapter 9).

Suspension and steering systems

48   Vehicle pulls to one side

1   Mismatched or uneven tires (Chapter 10).

2   Broken or sagging coil springs (Chapter 10).

3   Wheel alignment incorrect.

4   Front brakes dragging (Chapter 9).

49   Abnormal or excessive tire wear

1   Front wheel alignment incorrect.

2   Sagging or broken springs (Chapter 10).

3   Tire out-of-balance (Chapter 10).

4   Worn strut or shock absorber (Chapter 10).

5   Overloaded vehicle.

6   Tires not rotated regularly.

50   Wheel makes a thumping noise

1   Blister or bump on tire (Chapter 1).

2   Improper strut or shock absorber action (Chapter 10).

51   Shimmy, shake or vibration

1   Tire or wheel out-of-balance or out-of-round (Chapter 10).

2   Worn wheel bearings (Chapter 10).

3   Worn tie-rod ends (Chapter 10).

4   Worn balljoints (Chapter 10).

5   Excessive wheel runout (Chapter 10).

6   Blister or bump on tire (Chapter 1).

52   Hard steering

1   Balljoints, tie-rod ends or steering gear worn (Chapter 10).
2   Front wheel alignment incorrect.
3   Low tire pressure (Chapter 1).

53   Steering wheel does not return to center position correctly

1   Balljoints or tie-rod ends worn (Chapters 1 and 10).
2   Defective rack-and-pinion assembly (Chapter 10).
3   Front wheel alignment problem.

54   Abnormal noise at the front end

1   Balljoints or tie-rod ends worn (Chapter 1).
2   Loose upper strut mount (Chapter 10).
3   Worn tie-rod ends (Chapter 10).
4   Loose stabilizer bar (Chapter 10).
5   Loose wheel lug nuts (Chapter 1).
6   Loose suspension bolts (Chapter 10).

55   Wander or poor steering stability

1   Mismatched or uneven tires (Chapter 10).
2   Balljoints or tie-rod ends worn (Chapters 1 and 10).
3   Worn struts or shock absorbers (Chapter 10).
4   Broken or sagging springs (Chapter 10).
5   Front wheel alignment incorrect.
6   Worn steering gear clamp bushing (Chapter 10).

56   Erratic steering when braking

1   Wheel bearings worn (Chapter 10).
2   Broken or sagging springs (Chapter 10).
3   Leaking caliper (Chapter 9).
4   Warped brake discs (Chapter 9).
5   Worn steering gear clamp bushing (Chapter 10).
6   Wheel alignment incorrect.

57   Excessive pitching and/or rolling around corners or during braking

1   Loose stabilizer bar (Chapter 10).
2   Worn struts/shock absorbers or mounts (Chapter 10).
3   Broken or sagging coil springs (Chapter 10).
4   Overloaded vehicle.

58   Suspension bottoms

1   Overloaded vehicle.
2   Worn struts or shock absorbers (Chapter 10).
3   Incorrect, broken or sagging springs (Chapter 10).

59   Cupped tires

1   Front wheel alignment incorrect.
2   Worn struts or shock absorbers (Chapter 10).
3   Wheel bearings worn (Chapter 10).
4   Excessive tire or wheel runout (Chapter 10).
5   Worn balljoints (Chapter 10).

60   Excessive tire wear on outside edge

1   Inflation pressures incorrect (Chapter 1).
2   Excessive speed in turns.
3   Wheel alignment incorrect (excessive toe-in or positive camber). Have professionally aligned.
4   Suspension arm bent (Chapter 10).

61   Excessive tire wear on inside edge

1   Inflation pressures incorrect (Chapter 1).
2   Wheel alignment incorrect (toe-out or excessive negative camber). Have professionally aligned.
3   Loose or damaged steering components (Chapter 10).

62   Tire tread worn in one place

1   Tires out-of-balance.
2   Damaged wheel.
3   Defective tire (Chapter 1).

63   Excessive play or looseness in steering system

1   Wheel bearings worn (Chapter 10).
2   Tie-rod end loose or worn (Chapter 10).
3   Steering gear loose (Chapter 10).

64   Rattling or clicking noise in steering gear

1   Steering gear mounting bolts loose (Chapter 10).
2   Steering gear defective (Chapter 10).

Chapter 1
Tune-up and routine maintenance

Contents

	Section		Section
Air filter check and replacement	13	Introduction	2
Automatic transaxle fluid and filter change	24	Maintenance schedule	1
Battery check, maintenance and charging	8	Positive Crankcase Ventilation (PCV) valve check	
Brake fluid change	20	and replacement	22
Brake system check	14	Seat belt check	10
Cabin air filter replacement	18	Spark plug check and replacement	21
Cooling system check	12	Steering, suspension and driveaxle boot check	15
Cooling system servicing (draining, flushing and refilling)	23	Tire and tire pressure checks	5
Drivebelt check and replacement	19	Tire rotation	9
Engine oil and filter change	6	Tune-up general information	3
Exhaust system check	16	Underhood hose check and replacement	11
Fluid level checks	4	Windshield wiper blade inspection and replacement	7
Fuel system check	17		

Specifications

Recommended lubricants and fluids

Engine oil	
Type	API "Certified for gasoline engines"
Viscosity	
2.4L (2014 and earlier) and 2.7L engines	SAE 5W-20
2.4L engine (2015 and later)	SAE 0W-20
3.5L engine	SAE 10W-30
3.6L engine	SAE 5W-30
Automatic transaxle fluid	
2014 and earlier models	Mopar® ATF +4 automatic transmission fluid or equivalent
2015 and later models	Mopar® ZF 8&9 Speed ATF
Power steering fluid	Mopar® ATF +4 automatic transmission fluid or equivalent
Brake fluid	DOT type 3 brake fluid
Engine coolant*	
2012 and earlier models	50/50 mixture of Mopar® 5 year/100,000 Mile Formula (MS-9769) antifreeze/coolant with HOAT (Hybrid Organic Additive Technology) and water
2013 and later models	50/50 mixture of Mopar® 10 year/150,000 Mile Formula (MS-12106) antifreeze/coolant with OAT (Organic Additive Technology) and water
Door and liftgate latches	Multi-purpose grease
Fuel filler door remote control latch mechanism	Multi-purpose grease
Hood, door and liftgate hinge lubricant	Engine oil
Key lock cylinder lubricant	Graphite spray
Parking brake mechanism grease	Mopar Spray White Lube or equivalent
Rear differential lubricant**	AAM fluid 70W-80 meeting Chrysler specification MS-2395
Transfer case lubricant	70W-90 gear lubricant

These vehicles are filled with a 50/50 mixture of Mopar 5 year/100,000 mile or Mopar 10 year/150,000 mile coolant that shouldn't be mixed with each other or other coolants. Refer to the coolant reservoir label under the hood to determine what type coolant you have. Always refill with the correct coolant.Capacities

**The differential lubricant contains a special additive specific to these models - it should not be mixed with any other type of lubricant or fluid. Use of any other fluid will cause damage to the differential.

Capacities*

2.4L engine (2014 and earlier)	4.5 quarts	
2.4L (2015 and later), 2.7L and 3.5L engines	5.5 quarts	
3.6L engines	6.0 quarts	

Automatic transaxle**
  Drain and fill

40TE and 41TE models	4.0 quarts	
62TE models	5.5 quarts	
48TE models	2.0 quarts	

  Dry fill

40TE and 41TE models	9.2 quarts	
62TE models	9.0 quarts	
48TE models	4.8 quarts	

Cooling system***

2.4L engines (2014 and earlier)	7.7 quarts	
2.4L engines (2015 and later)	7.2 quarts	
2.7L engines	9.8 quarts	
3.5L engines	11.6 quarts	
3.6L engines	11.4 quarts	
Rear differential	0.94 quarts	
Transfer case	0.74 quarts	

*All capacities approximate. Add as necessary to bring to appropriate level.

**The best way to determine the amount of fluid to add during a routine fluid change is to measure the amount drained. It's important to not overfill the transaxle.

***Includes heater and coolant reservoir.

Brakes

Disc brake pad wear limit (minimum)	1/16 inch	1.5 mm
Drum brake shoe wear limit	1/8 inch	3 mm

Ignition system

Spark plug type and gap
  Type

2.4L without PZEV engines (2014 and earlier)	NGK (Nickel) - ZFR5F-11
2.4L PZEV engines (2014 and earlier)	NGK (single platinum) - ZFR5AP
2.4L (2015 and later)	NGK (Iridium) - 4469
2.7L engines	Champion (single platinum) - TE10PMC5
3.5L engines	NGK (single platinum) - ZFR5LP13G
3.6L engines	Champion - RER8ZWYCB4

  Gap

2.4L without PZEV engines	0.043 inch
2.4L PZEV engines	0.030 inch
2.4L engines (2015 and later)	0.047 inch
2.7L engines	0.050 inch
3.5L engines	0.051 inch
3.6L engines	0.040 inch

Firing order

Four cylinder engines	1-3-4-2
V6 engines	1-2-3-4-5-6

Cylinder location diagram -
four-cylinder engine

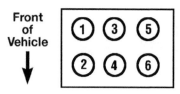

Cylinder location diagram - V6 engines

Torque specifications

	Ft-lbs (unless otherwise indicated)	**Nm**

Note: *One foot-pound (ft-lb) of torque is equivalent to 12 inch-pounds (in-lbs) of torque. Torque values below approximately 15 ft-lbs are expressed in inch-pounds, since most foot-pound torque wrenches are not accurate at these smaller values.*

	Ft-lbs	**Nm**
Automatic transaxle oil pan mounting bolts		
40TE and 41TE models	19	26
62TE models	105 in-lbs	12
Automatic transaxle oil filter mounting nuts - 62TE transaxle	40 in-lbs	4.5
Automatic transaxle fluid check/fill plug (948TE transaxle)	17	23
Drivebelt tensioner mounting bolt		
2.4L engines		
2014 and earlier	18	24
2015 and later (all bolts)	19	26
2.7L engines	40	54
3.5L engines	45	61
3.6L engines		
2014 and earlier	30	41
2015 and later	41	55
Engine oil drain plug		
2.4L engines (2014 and earlier)	30	40
V6 engines and 2015 and later 2.4L engines	20	27
Engine oil filter cap (3.6L engine)	18	24
Spark plugs		
2.4L and 3.5L engines	20	27
2.7L engines	15	20
3.6L engines	156 in-lbs	17.5
Wheel lug nuts	100	136

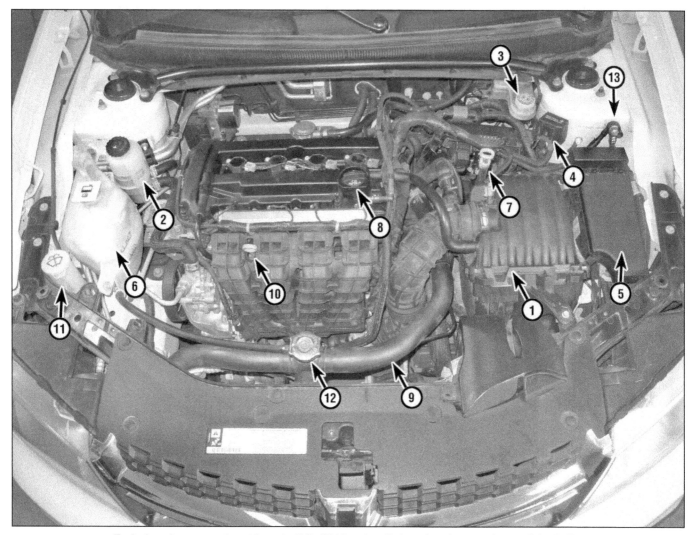

Typical engine compartment layout – 2.4L (2014 and earlier) engine shown, other models similar

1	Air filter housing	6	Coolant reservoir	11	Windshield washer fluid reservoir
2	Power steering fluid reservoir	7	Transaxle fluid dipstick tube	12	Cooling system pressure cap
3	Brake fluid reservoir	8	Engine oil filler cap	13	Remote negative battery terminal
4	Remote positive battery terminal	9	Upper radiator hose		
5	Underhood fuse/relay block	10	Engine oil dipstick		

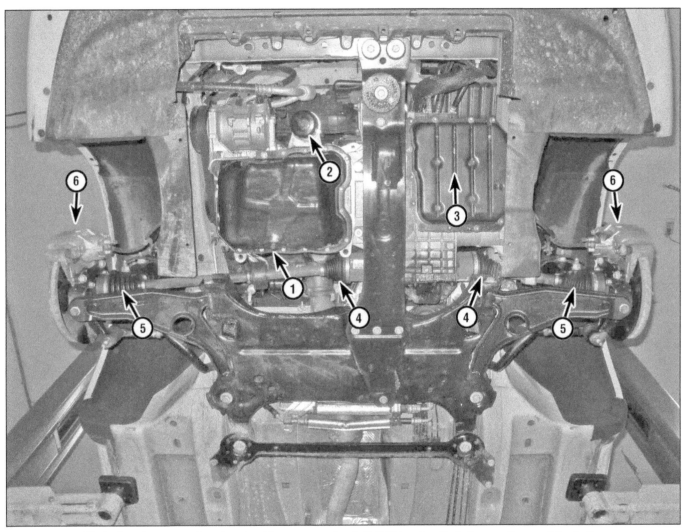

Typical engine compartment underside components - 2.4L (2014 and earlier) shown, other models similar

| 1 | Engine oil drain plug | 3 | Automatic transaxle fluid pan | 5 | Outer driveaxle boots |
| 2 | Oil filter | 4 | Inner driveaxle boots | 6 | Brake calipers |

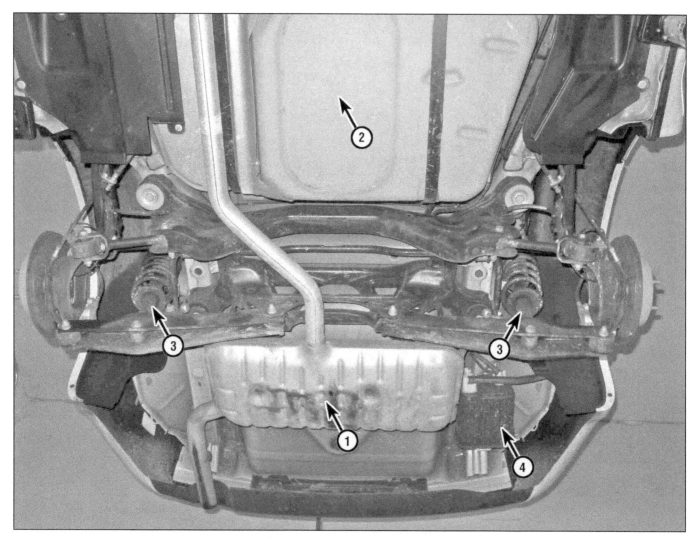

Typical rear underside components

1	Muffler	3	Rear shock absorbers
2	Fuel tank	4	EVAP canister

1 Maintenance schedule

The following maintenance intervals are based on the assumption that the vehicle owner will be doing the maintenance or service work, as opposed to having a dealer service department do the work. Although the time/mileage intervals are loosely based on factory recommendations, most have been shortened to ensure, for example, that such items as lubricants and fluids are checked/changed at intervals that promote maximum engine/driveline service life. Also, subject to the preference of the individual owner interested in keeping his or her vehicle in peak condition at all times, and with the vehicle's ultimate resale in mind, many of the maintenance procedures may be performed more often than recommended in the following schedule. We encourage such owner initiative.

When the vehicle is new it should be serviced initially by a factory authorized dealer service department to protect the factory warranty. In many cases the initial maintenance check is done at no cost to the owner (check with your dealer service department for more information).

Every 250 miles or weekly, whichever comes first

Check the engine oil level (Section 4)
Check the engine coolant level (Section 4)
Check the windshield washer fluid level (Section 4)
Check the brake fluid level (Section 4)
Check the power steering fluid level (Section 4)
Check the tires and tire pressures (Section 5)
Check the operation of all lights
Check the horn operation

Every 3,000 miles or 3 months, whichever comes first

All items listed above, plus:
Change the engine oil and filter (Section 6)

Every 6,000 miles or 6 months, whichever comes first

All items listed above, plus:
Check the wiper blade condition (Section 7)
Check and clean the battery and terminals (Section 8)
Rotate the tires (Section 9)
Check the seatbelts (Section 10)

Check the condition of all underhood hoses and connections (Section 11)
Check the cooling system hoses and connections for leaks and damage (Section 12)
Check and replace, if necessary, the air filter element (Section 13)

Every 12,000 miles or 12 months, whichever comes first

All items listed above, plus:
Check the automatic transaxle fluid level (Section 4)
Check the brake system (Section 14)
Check the suspension components and driveaxle boots (Section 15)
Check the exhaust pipes and hangers (Section 16)
Check the fuel system hoses and connections for leaks and damage (Section 17)
Replace the cabin air filter (Section 18)

Every 30,000 miles or 30 months, whichever comes first

All items listed above, plus:
Check the drivebelts (Section 19)
Replace the air filter element (Section 13)*
Replace the brake fluid (Section 20)

Every 60,000 miles or 60 months, whichever comes first

All items listed above, plus:
Check and replace, if necessary, the PCV valve (Section 22)*

Every 60 months (regardless of mileage)

Service the cooling system (drain, flush and refill) (Section 23)

Every 100,000 miles

Replace the spark plugs (Section 21)
Change the automatic transaxle fluid and filter (Section 24)*

Every 120,000 miles or 120 months, whichever comes first

Replace the timing belt (3.5L engine only) (Chapter 2C)

* This item is affected by severe operating conditions as described below. If the vehicle in question is operated under severe conditions, perform all maintenance procedures marked with an asterisk (*) at the intervals specified by the mileage headings below. Consider the conditions severe if most driving is done. . .

a) *In dusty areas*
b) *Towing a trailer*
c) *Idling for extended periods and/or low-speed operation*
   When outside temperatures remain below freezing and most trips are less than four miles
d) *In heavy city traffic where outside temperatures regularly reach 90-degrees F or higher*

2   Introduction

1   This Chapter is designed to help the home mechanic maintain the Chrysler Sebring, Dodge Avenger and Chrysler 200 with the goals of maximum performance, economy, safety and reliability in mind.
2   Included is a master maintenance schedule, followed by procedures dealing specifically with each item on the schedule. Visual checks, adjustments, component replacement and other helpful items are included. Refer to the accompanying illustrations of the engine compartment and the underside of the vehicle for the locations of various components.
3   Adhering to the mileage/time maintenance schedule and following the step-by-step procedures, which is simply a preventive maintenance program, will result in maximum reliability and vehicle service life. Keep in mind that it's not possible for this comprehensive program to produce the same results if you maintain some items at the specified intervals but not others.
4   As you service the vehicle, you'll discover that many of the procedures can - and should - be grouped together because of the nature of the particular procedure you're performing or because of the close proximity of two otherwise unrelated components to one another.
5   For example, if the vehicle is raised, you should inspect the exhaust, suspension, steering and fuel systems while you're under the vehicle. When you're rotating the tires, it makes good sense to check the brakes, since the wheels are already removed. Finally, let's suppose you have to borrow or rent a torque wrench. Even if you only need it to tighten the spark plugs, you might as well check the torque of as many critical fasteners as time allows.

6   The first step in this maintenance program is to prepare before the actual work begins. Read through all the procedures you're planning, then gather together all the parts and tools needed. If it looks like you might run into problems during a particular job, seek advice from a mechanic or an experienced do-it-yourselfer.

Owner's Manual and VECI label information

7   Your vehicle owner's manual was written for your year and model and contains very specific information on component locations, specifications, fuse ratings, part numbers, etc. The Owner's Manual is an important resource for the do-it-yourselfer to have; if one was not supplied with your vehicle, it can generally be ordered from a dealer parts department.
8   Among other important information, the Vehicle Emissions Control Information (VECI) label contains specifications and procedures for applicable tune-up adjustments and, in some instances, spark plugs (see Chapter 6 for more information on the VECI label). The information on this label is the exact maintenance data recommended by the manufacturer. This data often varies by intended operating altitude, local emissions regulations, month of manufacture, etc.
9   This Chapter contains procedural details, safety information and more ambitious maintenance intervals than you might find in manufacturer's literature. However, you may also find procedures or specifications in your Owner's Manual or VECI label that differ with what's printed here. In these cases, the Owner's Manual or VECI label can be considered correct, since it is specific to your particular vehicle.

3   Tune-up general information

1   The term tune-up is used in this manual to represent a combination of individual operations rather than one specific procedure.
2   The engine will be kept in relatively good running condition and the need for additional work will be minimized if the routine maintenance schedule is followed closely and frequent checks are made of fluid levels and high wear items, as suggested throughout this manual from the time the vehicle is new.
3   More likely than not, however, there will be times when the engine is running poorly due to lack of regular maintenance. This is even more likely if a used vehicle, which hasn't received regular and frequent maintenance checks, is purchased. In such cases, an engine tune-up will be needed outside of the regular routine maintenance intervals.
4   The first step in any tune-up or diagnostic procedure to help correct a poor running engine is a cylinder compression check. A compression check (see Chapter 2E) will help determine the condition of internal engine components and should be used as a guide for tune-up and repair procedures. For instance, if a compression check indicates serious internal engine wear, a conventional tune-up will not improve the performance of the engine and would be a waste of time and money. Because of its importance, someone with the right equipment and the knowledge to use it properly should do the compression check.
5   The following procedures are those most often needed to bring a generally poor running engine back into a proper state of tune:

4.2 The engine oil dipstick is located at the front of the engine and is clearly marked

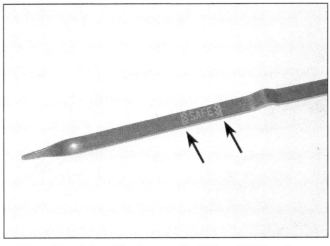

4.4 The oil level should be between the MIN and MAX marks, near the top of the cross-hatched area on the dipstick - if it isn't, add enough oil to bring the level up to or near the upper mark (do not overfill)

Minor tune-up

Check all engine related fluids (see Section 4)
Clean and inspect the battery (see Section 8)
Check all underhood hoses (see Section 11)
Check and adjust the drivebelts (see Section 19)
Check the air filter (see Section 13)
Service the cooling system (see Section 23)
Replace the spark plugs (see Section 21)
Check the PCV valve (see Section 22)

Major tune-up

All items listed under Minor tune-up plus. . .
Check the fuel system (see Section 17)
Replace the air filter (see Section 13)
Check the charging system (see Chapter 5)

4.5 Turn the oil filler cap counterclockwise to remove it

4   Fluid level checks (see Maintenance Schedule for service intervals)

Note: *The following are fluid level checks to be done on a 250 mile or weekly basis. Additional fluid level checks can be found in specific maintenance procedures that follow. Regardless of the intervals, develop the habit of checking under the vehicle periodically for evidence of fluid leaks.*

1   Fluids are an essential part of the lubrication, cooling, brake and window washer systems. Because the fluids gradually become depleted and/or contaminated during normal operation of the vehicle, they must be replenished periodically. See *Recommended lubricants and fluids* in this Chapter's Specifications before adding fluid to any of the following components.

Note: *The vehicle must be on level ground when fluid levels are checked.*

Engine oil

2   Engine oil level is checked with a dipstick that is located on the side of the engine facing the front of the vehicle **(see illustration)**. The dipstick extends through a tube and into the oil pan at the bottom of the engine.

3   The oil level should be checked before the vehicle has been driven, or about 5 minutes after the engine has been shut off. If the oil is checked immediately after driving the vehicle, some of the oil will remain in the upper engine components, resulting in an inaccurate reading on the dipstick.

4   Pull the dipstick out of the tube and wipe all the oil off the end with a clean rag or paper towel. Insert the clean dipstick all the way back into the tube, then pull it out again. Note the oil level at the end of the dipstick. Add oil as necessary to bring the oil level to the top of the cross-hatched area, or MAX mark **(see illustration)**.

5   Oil is added to the engine after removing a cap located on the valve cover **(see illus-** tration). Use a funnel to prevent spills as the oil is added.

6   Don't allow the level to drop below the MIN mark on the dipstick or engine damage may occur. On the other hand, don't overfill the engine by adding too much oil - it may result in oil aeration and loss of oil pressure and also could result in oil fouled spark plugs, oil leaks or seal failures.

7   Checking the oil level is an important preventive maintenance step. A consistently low oil level indicates oil leakage through damaged seals, defective gaskets or past worn rings or valve guides. If the oil looks milky in color or has water droplets in it, the block or head may be cracked and leaking coolant is entering the crankcase. The engine should be checked immediately. The condition of the oil should also be checked. Each time you check the oil level, slide your thumb and index finger up the dipstick before wiping off the oil. If you see small dirt or metal particles clinging to the dipstick, the oil should be changed (see Section 6).

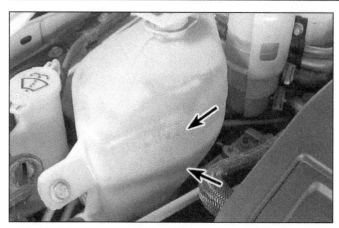

4.9 Maintain the coolant level between the ADD and FULL HOT marks on the reservoir (non-pressurized system)

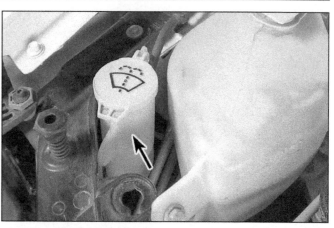

4.17 The windshield washer fluid reservoir is located in the right (passenger's) side of the engine compartment

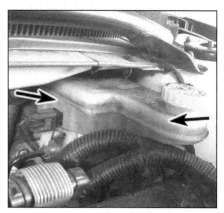

4.20 Brake fluid level, indicated on the translucent white plastic brake fluid reservoir, should be kept between the lower ADD and upper (FULL) mark

Engine coolant

Warning: *Do not allow antifreeze to come in contact with your skin or painted surfaces of the vehicle. Flush contaminated areas immediately with plenty of water. Don't store new coolant or leave old coolant lying around where it's accessible to children or pets – they're attracted by its sweet smell. Ingestion of even a small amount of coolant can be fatal! Wipe up garage floor and drip pan spills immediately. Keep antifreeze containers covered and repair cooling system leaks as soon as they're noticed.*

Non-pressurized system

8   A white plastic coolant reservoir is located at the right side of the engine compartment and is connected by a hose to the cooling system filler neck. If the coolant heats up sufficiently during operation, in excess of the pressure cap rating, it can escape past the cap and into the reservoir. As the engine cools, the coolant is drawn back into the cooling system to maintain the correct level.

Warning: *Do not remove the radiator cap to check the coolant level when the engine is warm!*

9   The coolant level in the reservoir should be checked regularly. The level in the reservoir varies with the temperature of the engine. When the engine is cold, the coolant level should be mid-way between the ADD and FULL HOT marks on the reservoir. Once the engine has warmed up, the level should be at or near the FULL HOT mark. If it isn't, allow the engine to cool, then remove the cap from the tank and add a 50/50 mixture of ethylene glycol based antifreeze and water **(see illustration)**.

10   If you have to remove the radiator cap, wait until the engine has cooled completely, then wrap a thick cloth around the cap and turn it to the first stop. If coolant or steam escapes, or if you hear a hissing noise, let the engine cool down longer, then remove the cap.

Pressurized system

Warning: *Do not remove the expansion tank cap to check the coolant level when the engine is warm!*

11   Some models are equipped with a pressurized coolant recovery system. On these models, a plastic expansion tank replaces the traditional coolant overflow reservoir, and is connected by a pair of hoses to the cooling system. As the engine heats up during operation, the expanding coolant fills the tank to a higher level.

12   The coolant level in the tank should be checked regularly, and the idea for checking the fluid level is similar to that of a traditional coolant reservoir setup **(see illustration 4.9)**, except the cap should only be removed with the engine completely cool.

13   The level in the tank varies with the temperature of the engine. When the engine is cold, the coolant level should be at or slightly above the ADD mark on the reservoir. If it isn't, wrap a thick cloth around the expansion tank cap, then slowly unscrew and remove it. If coolant or steam escapes, let the engine cool down longer, then remove the cap. Add a 50/50 mixture of the specified type of antifreeze and water. Once the engine has reached normal operating temperature, the level will rise to the FULL mark.

All models

14   Drive the vehicle and recheck the coolant level. If only a small amount of coolant is required to bring the system up to the proper level, water can be used. However, repeated additions of water will dilute the antifreeze and water solution. In order to maintain the proper ratio of antifreeze and water, always top up the coolant level with the correct mixture. Don't use rust inhibitors or additives. An empty plastic milk jug or bleach bottle makes an excellent container for mixing coolant.

15   If the coolant level drops consistently, there may be a leak in the system. Inspect the radiator, hoses, filler cap, drain plugs and water pump (see Section 12). If no leaks are noted, have the pressure cap pressure tested by a service station.

16   Check the condition of the coolant as well. It should be relatively clear. If it's brown or rust colored, the system should be drained, flushed and refilled. Even if the coolant appears to be normal, the corrosion inhibitors wear out, so it must be replaced at the specified intervals.

Windshield and rear window washer fluid

17   The fluid for the windshield and rear window washer system is stored in a plastic reservoir located at the right front corner of the engine compartment **(see illustration)**. The reservoir level should be maintained about one inch (25 mm) below the filler cap.

18   In milder climates, plain water can be used in the reservoir, but it should be kept no more than two-thirds full to allow for expansion if the water freezes. In colder climates, use windshield washer system antifreeze, available at any auto parts store, to lower the freezing point of the fluid. Mix the antifreeze with water in accordance with the manufacturer's directions on the container.

Caution: *DO NOT use cooling system antifreeze - it will damage the vehicle's paint. To help prevent icing in cold weather, warm the windshield with the defroster before using the washer.*

4.27 Location of the power steering fluid reservoir

4.29 At normal operating temperature, the power steering fluid level should be between the FILL RANGE marks

Brake fluid

19   The brake fluid reservoir is located on top of the brake master cylinder on the driver's side of the engine compartment near the firewall.

20   The fluid level should be maintained between the lower ADD mark and the upper (FULL or MAX) mark on reservoir **(see illustration)**.

21   If additional fluid is necessary to bring the level up, use a rag to clean all dirt off the top of the reservoir to prevent contamination of the system. Also, make sure all painted surfaces around the reservoir are covered, since brake fluid will ruin paint. Carefully pour new, clean brake fluid obtained from a sealed container into the reservoir. Be sure the specified fluid is used; mixing different types of brake fluid can cause damage to the system. See *Recommended lubricants and fluids* in this Chapter's Specifications or your owner's manual.

22   At this time the fluid and the master cylinder should be inspected for contamination. Normally the brake hydraulic system won't need periodic draining and refilling, but if rust deposits, dirt particles or water droplets are observed in the fluid, the system should be dismantled, cleaned and refilled with fresh fluid. Over time brake fluid will absorb moisture from the air. Moisture in the fluid lowers the fluid boiling point; if the fluid boils, the brakes will become ineffective. Normal brake fluid is clear in color. If the brake fluid is dark brown in color, it's a good idea to replace it (see Chapter 9).

23   Reinstall the fluid reservoir cap.

24   The brake fluid in the master cylinder will drop slightly as the brake lining material at each wheel wears down during normal operation. If the master cylinder requires repeated replenishing to maintain the correct level, there is a leak in the brake system that should be corrected immediately. Check all brake lines and connections, along with the calipers and power brake booster (see Section 14 and Chapter 9 for more information).

25   If you discover that the reservoir is empty or nearly empty, the system should be thoroughly inspected, refilled and then bled (see Chapter 9 for brake system bleeding).

4.34 The automatic transaxle dipstick is located at the left end of the engine – 40TE transaxle shown 41TE similar

Power steering fluid (2014 and earlier models)

26   Check the power steering fluid level periodically to avoid steering system problems, such as damage to the pump.

Caution: *DO NOT hold the steering wheel against either stop (extreme left or right turn) for more than five seconds. If you do, the power steering pump could be damaged.*

27   The power steering reservoir is located in the right side of the engine compartment **(see illustration)**.

28   For the check, the front wheels should be pointed straight ahead and the engine should be off.

29   The reservoir has ADD and FILL RANGE fluid level marks on the side. The fluid level can be seen without removing the reservoir cap **(see illustration)**.

30   If additional fluid is required, pour the specified type directly into the reservoir, using a funnel to prevent spills.

31   If the reservoir requires frequent fluid additions, all power steering hoses, hose connections, steering gear and the power steering pump should be carefully checked for leaks.

Automatic transaxle

Note: *It isn't necessary to check the transaxle fluid weekly; every 12,000 miles*

or 12 months will be adequate (unless a fluid leak is noticed).

40TE and 41TE (4-speed) models

32   The automatic transaxle fluid level should be carefully maintained. Low fluid level can lead to slipping or loss of drive, while overfilling can cause foaming and loss of fluid.

33   With the parking brake set, start the engine, then move the shift lever through all the gear ranges, ending in Neutral. The fluid level must be checked with the vehicle level and the engine running at idle.

Note: *Incorrect fluid level readings will result if the vehicle has just been driven at high speeds for an extended period, in hot weather in city traffic, or if it has been pulling a trailer. If any of these conditions apply, wait until the fluid has cooled (about 30 minutes).*

34   With the transaxle at normal operating temperature, remove the dipstick from the filler tube. The dipstick is located on the left side of the engine compartment **(see illustration)**.

Note: *Normal operating temperature is reached after a few miles of driving.*

35   Wipe the fluid from the dipstick with a clean rag and push it back into the filler tube until the cap seats.

36   Pull the dipstick out again and note the fluid level.

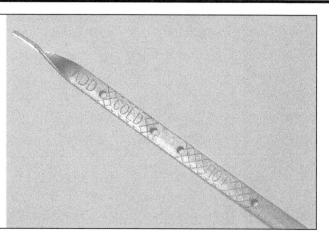

4.37 Check the fluid with the transaxle at normal operating temperature - the level should be kept in the HOT range (40TE and 41TE transaxles)

37 At normal operating temperature, the fluid level should be between the two upper reference holes (HOT) **(see illustration)**. If additional fluid is required, add it directly into the tube using a funnel. Add the fluid a little at a time and keep checking the level until it's correct.
Note: *Wait at least two minutes before rechecking the fluid level to allow the fluid to fully drain into the transaxle.*
38 The condition of the fluid should also be checked along with the level. If the fluid at the end of the dipstick is a dark reddish-brown color, or if it smells burned, it should be changed. If you are in doubt about the condition of the fluid, purchase some new fluid and compare the two for color and smell.

62TE (6-speed) models
Note: *The 62TE transaxles require the use of a scan tool to check transaxle fluid temperature and special tool #9336A (or a homemade equivalent, see Step 38) to measure the fluid level. If you do not have both tools to make the proper temperature-to-fluid level comparisons, we do not recommend attempting this procedure.*

39 Make sure the vehicle is parked on a level area.
Warning: *Make sure to set the parking brake and block the front wheels to prevent the vehicle from moving when the engine is running.*
40 Apply the parking brake, start the engine and allow it to idle for a minute, then move the shift lever through each gear position, ending in Park or Neutral.
41 Special oil dipstick tool no. 9336A will be required to check the fluid level. An alternative to this tool can be fabricated from a straightened-out coat hanger **(see illustrations)**.
42 Allow the transaxle to warm up, waiting at least two minutes, then remove the dipstick tube cap **(see illustration)**.
43 Insert the tool into the transaxle fill tube until the tip of the tool contacts the stop in the oil pan (or up to the tape if you're using a homemade dipstick) then pull it out **(see illustration)**. It may be necessary to repeat this several times to get an accurate reading.
Caution: *Do not use too much force when inserting the tool; there is a stop in the oil pan that the tool can be pushed beyond giving incorrect measurements. The approximate length that the tool should be inserted into the dipstick/fill tube is 16-11/16 inches (424 mm).*
Note: *The dipstick tool should stick out from the fill tube when it is installed.*

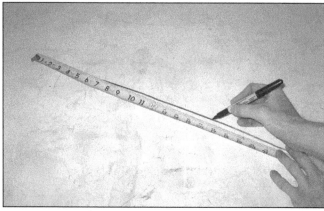

4.41a Using a standard non-painted coat hanger, straighten out the hanger then measure 16-11/16 inches from the tip. . .

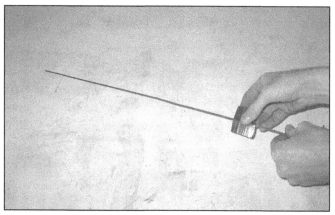

4.41b. . . and attach a piece of tape to mark the point to which the dipstick will be inserted

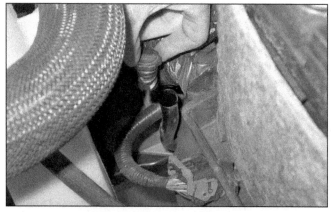

4.42 Locate the dipstick tube at the front of the transaxle (near the battery) and remove the cap

4.43 Insert the dipstick tool into the tube, then pull it out and measure how far the fluid goes up the dipstick

44   With the engine warmed up and running, check transaxle oil temperature with a scan tool.

45   Compare the reading on the dipstick tool with the transaxle fluid temperature reading on the scan tool.

46   Match the two readings with the transaxle fluid level chart **(see illustration)** to make sure the transaxle oil level is correct

47   Add or remove transaxle fluid as necessary, then recheck the fluid level and install the dipstick tube cap.

948TE/9HP48 (9-speed) models

Note: *There is no recommended service interval for the 948TE/9HP48 9-speed automatic transaxle when it is operated under normal conditions. These transaxles come pre-filled from the factory and do not require fluid checking unless there has been a fluid change, repair or a leak.*

Note: *The 9-speed transaxle requires the use of a scan tool to check transaxle fluid temperature and special tool #10323A (or a home-made equivalent) to measure the fluid level. If you do not have both tools to make the proper temperature-to-fluid level comparisons, we do not recommend attempting this procedure.*

Note: *The transaxle fluid must be between 122 and 194-degrees F for this check.*

48   If you do not have access to the special tool, fabricate a dipstick from a coat hanger (see illustrations 4.41a and 4.41b), but apply the correct measurement from the tip of the coat hanger to the applied tape, which is 5-1/2 inches (140mm).

49   Loosen the driver's side wheel lug nuts. Raise the vehicle and support it securely on jackstands, then remove the wheel. The vehicle must be level from the front of the vehicle to the rear for an accurate check.

50   Remove the transaxle check/fill plug,

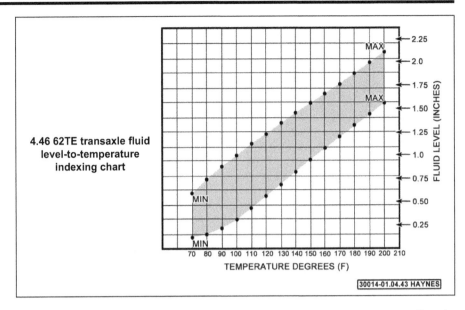

4.46 62TE transaxle fluid level-to-temperature indexing chart

which is located at the rear of the differential portion of the transaxle and is accessed by working through the driver's side fenderwell. It might also be easier to remove the plug with the steering wheel turned to the full-left position.

51   Start the engine and allow it to idle for a minute, then move the shift lever through each gear position, ending in Park. Pause for two seconds in each position.

52   Connect a scan tool to the OBD II diagnostic connector (see Chapter 6) and locate the temperature display on the scan tool.

53   Insert the tool into the fluid level check hole and allow the cap of the tool to rest on the transaxle case (or, if you made a tool, up to the tape), then pull it out and measure the level of fluid (from the bottom of the dipstick). It may be necessary to repeat this several times to get an accurate reading.

Note: *Be careful not to burn yourself on the hot components.*

54   Compare the reading on the dipstick tool with the transaxle fluid temperature reading on the scan tool.

55   Match the two readings with the transaxle fluid level chart **(see illustration)** to make sure the transaxle fluid level is correct.

56   Add or remove transaxle fluid as necessary, through the check plug hole, then recheck the fluid level and install the check plug, tightening it to the torque listed in this Chapter's Specifications. Turn off the engine. If too much fluid is added, loosen the drain plug and allow some fluid to drain until the level is correct, or remove it through the check/fill plug hole with a suction gun. Tighten the wheel lug nuts to the torque listed in this Chapter's Specifications.

Fluid temperature (degrees Fahrenheit)	Minimum level (mm)	Desired level (mm)	Maximum level (mm)
122	13	16	19
131	13.5	17.5	20
140	16	20	23
149	18	21.5	24
158	19	22.5	25
167	20	24	27
176	21	25	28
185	22	26	29
194	24	27.5	31

4.55 Transaxle fluid level-chart

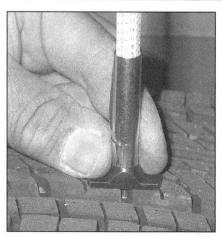

5.2 A tire tread depth indicator should be used to monitor tire wear - they are available at auto parts stores and service stations and cost very little

UNDERINFLATION   CUPPING   OVERINFLATION

Cupping may be caused by:
• Underinflation and/or mechanical irregularities such as out-of-balance condition of wheel and/or tire, and bent or damaged wheel.
• Loose or worn steering tie-rod or steering idler arm.
• Loose, damaged or worn front suspension parts.

INCORRECT TOE-IN OR EXTREME CAMBER   FEATHERING DUE TO MISALIGNMENT

5.3 This chart will help you determine the condition of your tires, the probable cause(s) of abnormal wear and the corrective action necessary

5   Tire and tire pressure checks (every 250 miles or weekly)

1   Periodic inspection of the tires may spare you the inconvenience of being stranded with a flat tire. It can also provide you with vital information regarding possible problems in the steering and suspension systems before major damage occurs.

2   The original tires on this vehicle are equipped with 1/2-inch wide bands that will appear when tread depth reaches 1/16-inch, at which point they can be considered worn out. Tread wear can be monitored with a simple, inexpensive device known as a tread depth indicator **(see illustration)**.

3   Note any abnormal tread wear **(see illustration)**. Tread pattern irregularities such as cupping, flat spots and more wear on one side than the other are indications of front end alignment and/or balance problems. If any of these conditions are noted, take the vehicle to a tire shop or service station to correct the problem.

4   Look closely for cuts, punctures and embedded nails or tacks. Sometimes a tire will hold air pressure for a short time or leak down very slowly after a nail has embedded itself in the tread. If a slow leak persists, check the valve stem core to make sure it is tight **(see illustration)**. Examine the tread for an object that may have embedded itself in the tire or for a plug that may have begun to leak (radial tire punctures are repaired with a plug that is installed in a puncture). If a puncture is suspected, it can be easily verified by spraying a solution of soapy water onto the puncture area **(see illustration)**. The soapy solution will bubble if there is a leak. Unless the puncture is unusually large, a tire shop or service station can usually repair the tire.

5   Carefully inspect the inner sidewall of each tire for evidence of brake fluid leakage. If you see any, inspect the brakes immediately.

6   Correct air pressure adds miles to the life span of the tires, improves mileage and enhances overall ride quality. Tire pressure cannot be accurately estimated by looking at

a tire, especially if it's a radial. A tire pressure gauge is essential. Keep an accurate gauge in the glove compartment. The pressure gauges attached to the nozzles of air hoses at gas stations are often inaccurate.

7   Always check tire pressure when the tires are cold. Cold, in this case, means the vehicle has not been driven over a mile in the three hours preceding a tire pressure check. A pressure rise of four to eight pounds is not uncommon once the tires are warm.

8   Unscrew the valve cap protruding from the wheel or hubcap and push the gauge firmly onto the valve stem **(see illustration)**. Note the reading on the gauge and compare the figure to the recommended tire pressure shown on the tire placard on the driver's side door. Be sure to reinstall the valve cap to keep dirt and moisture out of the valve stem mechanism. Check all four tires and, if necessary, add enough air to bring them up to the recommended pressure.

9   Don't forget to keep the spare tire inflated to the specified pressure (refer to the pressure molded into the tire sidewall).

5.4a If a tire loses air on a steady basis, check the valve core first to make sure it's snug (special inexpensive wrenches are commonly available at auto parts stores)

5.4b If the valve core is tight, raise the corner of the vehicle with the low tire and spray a soapy water solution onto the tread as the tire is turned slowly - slow leaks will cause small bubbles to appear

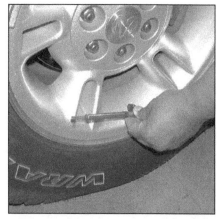

5.8 To extend the life of your tires, check the air pressure at least once a week with an accurate gauge (don't forget the spare!)

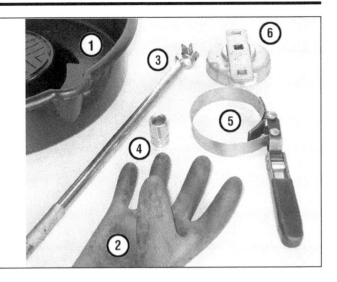

6.2 These tools are required when changing the engine oil and filter

1   **Drain pan** - It should be fairly shallow in depth, but wide to prevent spills
2   **Rubber gloves** - When removing the drain plug and filter, you will get oil on your hands (the gloves will prevent burns)
3   **Breaker bar** - Sometimes the oil drain plug is tight, and a long breaker bar is needed to loosen it
4   **Socket** – To be used with the breaker bar or a ratchet (must be the correct size to fit the drain plug - six-point preferred)
5   **Filter wrench** - This is a metal band-type wrench, which requires clearance around the filter to be effective
6   **Filter wrench** - This type fits on the bottom of the filter and can be turned with a ratchet or breaker bar (different-size wrenches are available for different types of filters)

6 Engine oil and filter change (every 3000 miles or 3 months)

1   Frequent oil changes are the best preventive maintenance the home mechanic can give the engine, because aging oil becomes diluted and contaminated, which leads to premature engine wear.

2   Make sure you have all the necessary tools before you begin this procedure **(see illustration)**. You should also have plenty of rags or newspapers handy for mopping up any spills.

3   Access to the underside of the vehicle is greatly improved if the vehicle can be lifted on a hoist, driven onto ramps or supported by jackstands. **Warning:** *Do not work under a vehicle which is supported only by a bumper, hydraulic or scissors-type jack.*

4   If this is your first oil change, get under the vehicle and familiarize yourself with the locations of the oil drain plug and the oil filter. The engine and exhaust components will be warm during the actual work, so try to anticipate any potential problems before the engine and accessories are hot.

5   Park the vehicle on a level spot. Start the engine and allow it to reach its normal operating temperature. Warm oil and sludge will flow out more easily. Turn off the engine when it's warmed up. Remove the filler cap from the valve cover.

6   Raise the vehicle and support it securely on jackstands. **Warning:** *Never get beneath the vehicle when it is supported only by a jack. The jack provided with your vehicle is designed solely for raising the vehicle to remove and replace the wheels. Always use jackstands to support the vehicle when it becomes necessary to place your body underneath the vehicle.*

7   Being careful not to touch the hot exhaust components, place the drain pan under the drain plug in the bottom of the pan and remove the plug **(see illustration)**. You may want to wear gloves while unscrewing the plug the final few turns if the engine is hot. **Note:** *2015 and later 2.4L engines have removable panels in the splash shield to allow*

for the removal of the drain plug and oil filter. They are removed by inserting a flat-head screwdriver into the fastener slot, and turning 90-degrees.

8   Allow the old oil to drain into the pan. It may be necessary to move the pan farther under the engine as the oil flow slows to a trickle. Inspect the old oil for the presence of metal shavings and chips. **Note:** *On 3.6L engines, the oil filler cap must be removed to allow all the oil to properly drain out.*

9   After all the oil has drained, wipe off the drain plug with a clean rag. Even minute metal particles clinging to the plug would immediately contaminate the new oil.

10   Clean the area around the drain plug opening, reinstall the plug and tighten it to the torque listed in this Chapter's Specifications.

11   Move the drain pan into position under the oil filter.

2.4L, 2.7L and 3.5L engines

12   Loosen the oil filter by turning it counterclockwise with an oil filter wrench **(see illustration)**. Once the filter is loose, use your

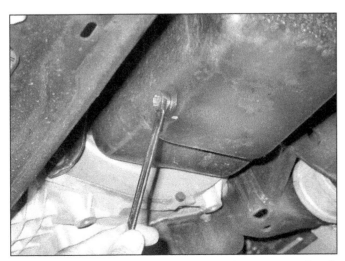

6.7 Use a proper size box-end wrench or socket to remove the oil drain plug and avoid rounding it off

6.12 Use an oil filter wrench to remove the filter (2.4L, 2.7L and 3.5L engines)

6.14 Lubricate the oil filter gasket with clean engine oil before installing the filter on the engine

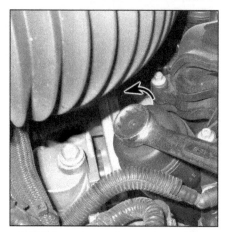

6.17 Using a box end wrench or socket, turn the filter cap counterclockwise to remove it (3.6L engine)

hands to unscrew it from the block. Keep the open end pointing up to prevent the oil inside the filter from spilling out.

Warning: *The exhaust system may still be hot, so be careful.*

13   With a clean rag, wipe off the mounting surface on the block. If a residue of old oil is allowed to remain, it will smoke when the block is heated up. Also make sure that none of the old gasket remains stuck to the mounting surface. It can be removed with a scraper if necessary.

14   Compare the old filter with the new one to make sure they are the same type. Smear some clean engine oil on the rubber gasket of the new filter **(see illustration)**.

15   Attach the new filter to the engine, following the tightening directions printed on the filter canister or packing box. Most filter manufacturers recommend against using a filter wrench due to the possibility of overtightening and damaging the seal.

3.6L engines

16   Lift up on the outer edges of the engine cover to disengage the rubber mounts from

the ballstuds, and remove the engine cover.

17   Place a rag at the base of the filter housing, then loosen the filter cap by turning it counterclockwise **(see illustration)**.

18   Remove the cap and filter from the engine, then pull the filter out of the cap.

19   Remove and discard the O-ring from the filter cap.

20   Insert a new filter into the cap, making sure the filter clips lock into the cap **(see illustration)**.

21   Install a new O-ring onto the cap and apply a light amount of engine oil to the O-ring.

22   Place the filter assembly into the filter housing and carefully thread the cap into the housing. Tighten the cap to the torque listed in this Chapter's Specifications.

All models

23   Remove all tools, rags, etc. from under the vehicle, being careful not to spill the oil in the drain pan, then lower the vehicle.

24   Add new oil to the engine through the oil filler cap in the valve cover. Use a funnel, if necessary, to prevent oil from spilling onto the

top of the engine. Pour four quarts of fresh oil into the engine. Wait a few minutes to allow the oil to drain into the pan, then check the level on the oil dipstick (see Section 4). If the oil level is at or near the FULL mark on the dipstick, install the filler cap, start the engine and allow the new oil to circulate.

25   Allow the engine to run for about a minute. While the engine is running, look under the vehicle and check for leaks at the oil pan drain plug and around the oil filter. If either is leaking, stop the engine and tighten the plug or filter.

26   Wait a few minutes to allow the oil to trickle down into the pan, recheck the level on the dipstick and, if necessary, add enough oil to bring the level to the FULL mark.

27   During the first few trips after an oil change, make it a point to check frequently for leaks and proper oil level.

28   The old oil drained from the engine cannot be reused in its present state and should be disposed of. Check with your local auto parts store, disposal facility or environmental agency to see if they will accept the oil for recycling. After the oil has cooled it can be drained into a container (capped plastic jugs, topped bottles, milk cartons, etc.) for transport to one of these disposal sites. Don't dispose of the oil by pouring it on the ground or down a drain!

Oil change indicator resetting

Note: *It is possible that, driving under the best possible conditions, the oil life monitoring system may not indicate the oil needs to be changed. The manufacturer states that the oil and filter must be changed at least once every year and the oil life monitor reset.*

Note: *If the "Oil Change Required" message comes on when the vehicle is immediately restarted, the oil life monitor was not reset and the reset procedure must be done again.*

Note: *If the message is not reset, it will continue to show up each time you turn the ignition switch On or start the vehicle. It is possible to temporarily turn off the message by pressing and releasing the "Menu" button.*

29   Turn the ignition key to the "ON" position but DO NOT start the engine.

30   Slowly depress the accelerator pedal all the way to the floor, three times within 10 seconds.

31   Turn the ignition key to the "OFF" or "LOCK" position, then start the vehicle.

---

7   Windshield wiper blade inspection and replacement (every 6,000 miles or 6 months)

---

1   The windshield wiper and blade assembly should be inspected periodically for damage, loose components and cracked or worn blade elements.

2   Road film can build up on the wiper blades and affect their efficiency, so they should be washed regularly with a mild detergent solution.

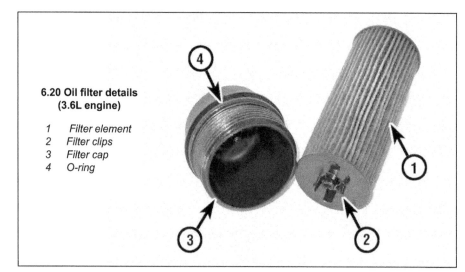

6.20 Oil filter details (3.6L engine)

1   *Filter element*
2   *Filter clips*
3   *Filter cap*
4   *O-ring*

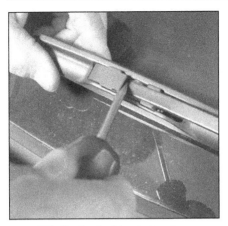

7.5a Pry the locking lever up. . .

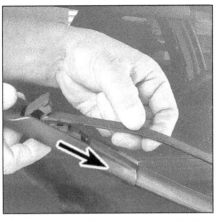

7.5b. . . then slide the wiper blade in the direction of the arrow to separate it from the arm

3   The action of the wiping mechanism can loosen bolts, nuts and fasteners, so they should be checked and tightened, as necessary, at the same time the wiper blades are checked.

4   If the wiper blade elements are cracked, worn or warped, or no longer clean adequately, they should be replaced with new ones.

2010 and earlier models

5   Lift the arm assembly away from the glass for clearance. Pry the locking lever up, then slide the wiper blade assembly out of the hook at the end of the arm **(see illustrations)**.

Caution: *Do not allow the wiper arm to so snap back against the windshield without the wiper blade installed or the arm could damaged the window.*

6   Attach the new wiper to the arm. Connection can be confirmed by an audible click.

2011 and later models

Note: *Some later models are equipped with the earlier style wiper blade (see Steps 5 and 6).*

7   Lift the arm from the windshield and rotate the blade until it is centered on the arm.

8   Depress the release tab, then slide the blade away from the tip of the arm enough to disengage the blade assembly.

Caution: *Do not allow the wiper arm to snap back against the windshield without the wiper blade installed or the arm could damaged the window.*

9   Attach the new wiper to the arm. Connection can be confirmed by an audible click of the release tab into the pivot block.

---

8   Battery check, maintenance and charging (every 6,000 miles or 6 months)

Warning: *Certain precautions must be followed when checking and servicing the battery. Hydrogen gas, which is highly flammable, is always present in the battery cells, so keep lighted tobacco and all other open flames and sparks away from the battery. The electro-*

lyte inside the battery is actually diluted sulfuric acid, which will cause injury if splashed on your skin or in your eyes. It will also ruin clothes and painted surfaces. When removing the battery cables, always detach the negative cable first and hook it up last!

Note: *The battery on these vehicles is located at the right front corner of the vehicle, ahead of the wheel and behind the inner fender splash shield. Refer to Chapter 5 for the access procedure.*

1   A routine preventive maintenance program for the battery in your vehicle is the only way to ensure quick and reliable starts. But before performing any battery maintenance, make sure that you have the proper equipment necessary to work safely around the battery **(see illustration)**.

2   There are also several precautions that should be taken whenever battery maintenance is performed. Before servicing the battery, always turn the engine and all accessories off and disconnect the cable from the negative terminal of the battery (see Chapter 5).

3   The battery produces hydrogen gas, which is both flammable and explosive. Never create a spark, smoke or light a match around the battery. Always charge the battery in a ventilated area.

4   Electrolyte contains poisonous and corrosive sulfuric acid. Do not allow it to get in your eyes, on your skin on your clothes. Never ingest it. Wear protective safety glasses when working near the battery. Keep children away from the battery.

5   Note the external condition of the battery. If the positive terminal and cable clamp on your vehicle's battery is equipped with a rubber protector, make sure that it's not torn or damaged. It should completely cover the terminal. Look for any corroded or loose connections, cracks in the case or cover or loose hold-down clamps. Also check the entire length of each cable for cracks and frayed conductors.

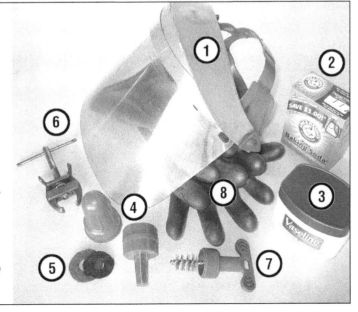

8.1 Tools and materials required for battery maintenance

1   **Face shield/safety goggles** - *When removing corrosion with a brush, the acidic particles can easily fly up into your eyes*

2   **Baking soda** - *A solution of baking soda and water can be used to neutralize corrosion*

3   **Petroleum jelly** - *A layer of this on the battery posts will help prevent corrosion*

4   **Battery post/cable cleaner** - *This wire brush cleaning tool will remove all traces of corrosion from the battery posts and cable clamps*

5   **Treated felt washers** - *Placing one of these on each post, directly under the cable clamps, will help prevent corrosion*

6   **Puller** - *Sometimes the cable clamps are very difficult to pull off the posts, even after the nut/bolt has been completely loosened. This tool pulls the clamp straight up and off the post without damage*

7   **Battery post/cable cleaner** - *Here is another cleaning tool which is a slightly different version of Number 4 above, but it does the same thing*

8   **Rubber gloves** - *Another safety item to consider when servicing the battery; remember that's acid inside the battery!*

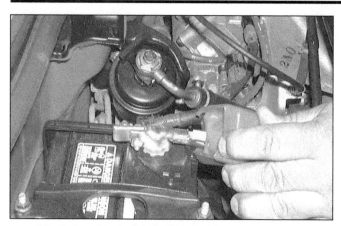

8.6a Battery terminal corrosion usually appears as light, fluffy powder

8.6b Removing a cable from the battery post with a wrench - sometimes a pair of special battery pliers are required for this procedure if corrosion has caused deterioration of the nut hex (always remove the ground (-) cable first and hook it up last!)

6   If corrosion, which looks like white, fluffy deposits **(see illustration)** is evident, particularly around the terminals, the battery should be removed for cleaning. Loosen the cable clamp bolts with a wrench, being careful to remove the ground cable first, and slide them off the terminals **(see illustration)**. Then disconnect the hold-down clamp bolt and nut, remove the clamp and lift the battery from the engine compartment.

7   Clean the cable clamps thoroughly with a battery brush or a terminal cleaner and a solution of warm water and baking soda **(see illustration)**. Wash the terminals and the top of the battery case with the same solution but make sure that the solution doesn't get into the battery. When cleaning the cables, terminals and battery top, wear safety goggles and rubber gloves to prevent any solution from coming in contact with your eyes or hands. Wear old clothes too - even diluted, sulfuric acid splashed onto clothes will burn holes in them. If the terminals have been extensively corroded, clean them up with a terminal cleaner **(see illustration)**. Thoroughly wash all cleaned areas with plain water.

8   Make sure that the battery tray is in good condition and the hold-down clamp fasteners are tight. If the battery is removed from the tray, make sure no parts remain in the bottom of the tray when the battery is reinstalled. When reinstalling the hold-down clamp bolts, do not overtighten them.

9   Information on removing and installing the battery can be found in Chapter 5. If you disconnected the cable(s) from the negative and/or positive battery terminals, see Chapter 5. Information on jump starting can be found at the front of this manual. For more detailed battery checking procedures, refer to the *Haynes Automotive Electrical Manual*.

Cleaning

10   Corrosion on the hold-down components, battery case and surrounding areas can be removed with a solution of water and baking soda. Thoroughly rinse all cleaned areas with plain water.

11   Any metal parts of the vehicle damaged by corrosion should be covered with a zinc-based primer, then painted.

Charging

Warning: *When batteries are being charged, hydrogen gas, which is very explosive and flammable, is produced. Do not smoke or allow open flames near a charging or a recently charged battery. Wear eye protection when near the battery during charging. Also, make sure the charger is unplugged before connecting or disconnecting the battery from the charger.*

12   Slow-rate charging is the best way to restore a battery that's discharged to the point where it will not start the engine. It's also a good way to maintain the battery charge in a vehicle that's only driven a few miles between starts. Maintaining the battery charge is particularly important in the winter when the battery must work harder to start the engine and electrical accessories that drain the battery are in greater use.

13   It's best to use a one or two-amp battery charger (sometimes called a trickle charger). They are the safest and put the least strain on the battery. They are also the least expensive. For a faster charge, you can use a higher amperage charger, but don't use one

8.7a When cleaning the cable clamps, all corrosion must be removed

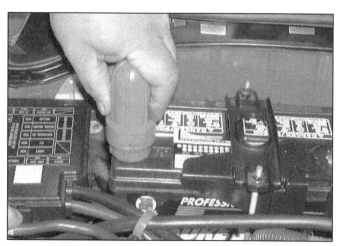

8.7b Regardless of the type of tool used to clean the battery posts, a clean, shiny surface should be the result

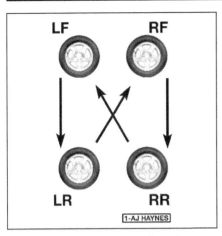

9.2 The recommended four-tire rotation pattern

rated more than 1/10th the amp/hour rating of the battery. Rapid boost charges that claim to restore the power of the battery in one to two hours are hardest on the battery and can damage batteries not in good condition. This type of charging should only be used in emergency situations.

14   The average time necessary to charge a battery should be listed in the instructions that come with the charger. As a general rule, a trickle charger will charge a battery in 12 to 16 hours.

9 Tire rotation (every 6,000 miles or 6 months)

1   The tires should be rotated at the specified intervals and whenever uneven wear is noticed.

2   Refer to the **accompanying illustration** for the preferred tire rotation pattern.

3   Refer to the information in *Jacking and towing* for the proper procedures to follow when ra ising the vehicle and changing a tire. If the brakes are to be checked, don't apply the parking brake as stated. Make sure the tires are blocked to prevent the vehicle from rolling as it's raised.

4   Preferably, the entire vehicle should be raised at the same time. This can be done on a hoist or by jacking up each corner and then lowering the vehicle onto jackstands placed under the frame rails. Always use four jackstands and make sure the vehicle is safely supported.

5   After rotation, check and adjust the tire pressures as necessary. Tighten the lug nuts to the torque listed in this Chapter's Specifications.

10 Seat belt check (every 6,000 miles or 6 months)

1   Check the seat belts, buckles, latch plates and guide loops for obvious damage and signs of wear.

2   Where the seat belt receptacle bolts to the floor of the vehicle, check that the bolts are secure.

3   See if the seat belt reminder light comes on when the key is turned to the Run or Start position.

11 Underhood hose check and replacement (every 6,000 miles or 6 months)

General

Caution: *Never remove air conditioning components or hoses until the system has been depressurized by a licensed air conditioning technician.*

1   High temperatures in the engine compartment can cause the deterioration of the rubber and plastic hoses used for engine, accessory and emission systems operation. Periodic inspection should be made for cracks, loose clamps, material hardening and leaks. Information specific to the cooling system hoses can be found in Section 12.

2   Some, but not all, hoses are secured to their fittings with clamps. Where clamps are used, check to be sure they haven't lost their tension, allowing the hose to leak. If clamps aren't used, make sure the hose has not expanded and/or hardened where it slips over the fitting, allowing it to leak.

Vacuum hoses

3   It's quite common for vacuum hoses, especially those in the emissions system, to be color-coded or identified by colored stripes molded into them. Various systems require hoses with different wall thickness, collapse resistance and temperature resistance. When replacing hoses, be sure the new ones are made of the same material.

4   Often the only effective way to check a hose is to remove it completely from the vehicle. If more than one hose is removed, be sure to label the hoses and fittings to ensure correct installation.

5   When checking vacuum hoses, be sure to include any plastic T-fittings in the check. Inspect the fittings for cracks and the hose where it fits over the fitting for distortion, which could cause leakage.

6   A small piece of vacuum hose (1/4-inch inside diameter) can be used as a stethoscope to detect vacuum leaks. Hold one end of the hose to your ear and probe around vacuum hoses and fittings, listening for the "hissing" sound characteristic of a vacuum leak. **Warning:** *When probing with the vacuum hose stethoscope, be very careful not to come into contact with moving engine components such as the drivebelt, cooling fan, etc.*

Fuel hose

Warning: *There are certain precautions that must be taken when inspecting or servicing fuel system components. Work in a well-*

ventilated area and do not allow open flames (cigarettes, appliances, etc.) or bare light bulbs near the work area. Mop up any spills immediately and do not store fuel soaked rags where they could ignite. The fuel system is under high pressure, so if any fuel lines are to be disconnected, the pressure in the system must be relieved first (see Chapter 4 for more information).

7   Check all rubber fuel lines for deterioration and chafing. Check especially for cracks in areas where the hose bends and just before fittings, such as where a hose attaches to the fuel filter.

8   High quality fuel line, made specifically for high-pressure fuel injection systems, must be used for fuel line replacement. Never, under any circumstances, use unreinforced vacuum line, clear plastic tubing or water hose for fuel lines.

9   Spring-type clamps are commonly used on fuel lines. These clamps often lose their tension over a period of time, and can be "sprung" during removal. Replace all spring-type clamps with screw clamps whenever a hose is replaced.

Metal lines

10   Sections of metal line are routed along the frame, between the fuel tank and the engine. Check carefully to be sure the line has not been bent or crimped and that cracks have not started in the line.

11   If a section of metal fuel line must be replaced, only seamless steel tubing should be used, since copper and aluminum tubing don't have the strength necessary to withstand normal engine vibration.

12   Check the metal brake lines where they enter the master cylinder and brake proportioning unit for cracks in the lines or loose fittings. Any sign of brake fluid leakage calls for an immediate and thorough inspection of the brake system.

12 Cooling system check (every 6,000 miles or 6 months)

1   Many major engine failures can be attributed to a faulty cooling system. Since the vehicle is equipped with an automatic transaxle, the cooling system also cools the transaxle fluid and thus plays an important role in prolonging transaxle life.

2   The cooling system should be checked with the engine cold. Do this before the vehicle is driven for the day or after it has been shut off for at least three hours.

3   Remove the cooling system pressure cap and thoroughly clean the cap, inside and out, with clean water. Also clean the filler neck. All traces of corrosion should be removed. The coolant in the system should be relatively transparent. If it is rust-colored, the system should be drained, flushed and refilled (see Section 23). If the coolant level is not up to the top, add additional antifreeze/coolant mixture (see Section 4).

Check for a chafed area that could fail prematurely.

Check for a soft area indicating the hose has deteriorated inside.

Overtightening the clamp on a hardened hose will damage the hose and cause a leak.

Check each hose for swelling and oil-soaked ends. Cracks and breaks can be located by squeezing the hose.

12.4 Hoses, like drivebelts, have a habit of failing at the worst possible time - to prevent the inconvenience of a blown radiator or heater hose, inspect them carefully as shown here

4   Carefully check the large upper and lower radiator hoses along with the smaller diameter heater hoses that run from the engine to the firewall. Inspect each hose along its entire length, replacing any hose that is cracked, swollen or shows signs of deterioration. Cracks may become more apparent if the hose is squeezed **(see illustration)**. Regardless of condition, it's a good idea to replace hoses with new ones every two years.

5   Make sure all hose connections are tight. A leak in the cooling system will usually show up as white or rust-colored deposits on the areas adjoining the leak. If wire-type clamps are used at the ends of the hoses, it may be a good idea to replace them with more secure screw-type clamps.

6   Use compressed air or a soft brush to remove bugs, leaves, etc. from the front of the radiator or air conditioning condenser. Be

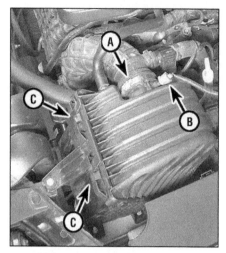

13.3a Loosen the clamp (A) and disconnect the IAT sensor connector (B) (only if you are removing the cover completely). Detach the intake duct, carefully pull the locking clips back off of the tabs (C). . .

careful not to damage the delicate cooling fins or cut yourself on them.

7   Every other inspection, or at the first indication of cooling system problems, have the cap and system pressure tested. If you don't have a pressure tester, most repair shops will do this for a minimal charge.

13   Air filter check and replacement (every 6,000 miles or 6 months)

Note: *On 2015 and later models, the air filter housing must be removed first (see Chapter 4) in order to access the filter cover mounting screws.*

1   The air filter is located inside a housing at the right (passenger's) side of the engine compartment.

2   If you're removing the top cover, disconnect the Intake Air Temperature (IAT) sensor electrical connector.

3   To remove the air filter on 2014 and earlier models, pull the locking clips off of the housing tabs that secure the two halves of the air filter housing together, then separate the cover halves and remove the air filter element **(see illustrations)**. On 2015 and later models, remove the air filter housing (see Chapter 4) then remove the cover screws and separate the cover from the housing to access the filter element.

Note: *On 3.6L and 2015 and later engines, the air filter cover is held in place by four mounting screws.*

4   Inspect the outer surface of the filter element. If it is dirty, replace it. If it is only moderately dusty, it can be reused by blowing it clean from the back to the front surface with compressed air. Because it is a pleated paper type filter, it cannot be washed or oiled. If it

13.3b. . . then lift the air filter housing cover and remove the air filter element – 2014 and earlier 2.4L engine shown, other models similar

cannot be cleaned satisfactorily with compressed air, discard and replace it. While the cover is off, be careful not to drop anything down into the housing.

5   Wipe out the inside of the air filter housing.

6   Place the new filter into the air filter housing, making sure it seats properly.

7   On 2015 and later models, join the cover halves and install/tighten the screws securely, then install the air filter housing.

8   On 2014 and earlier models reinstall the cover, making sure the locking clips are seated against the locking tabs (or tighten the mounting screws).

14   Brake system check (every 12,000 miles or 12 months)

Warning: *The dust created by the brake system is harmful to your health. Never blow it out with compressed air and don't inhale any of it. An approved filtering mask should be worn when working on the brakes. Do not, under any circumstances, use petroleum-based solvents to clean brake parts. Use brake system cleaner only!*

Note: *For detailed photographs of the brake system, refer to Chapter 9.*

1   In addition to the specified intervals, the brakes should be inspected every time the wheels are removed or whenever a defect is suspected.

2   Any of the following symptoms could indicate a potential brake system defect: The vehicle pulls to one side when the brake pedal is depressed; the brakes make squealing or dragging noises when applied; brake pedal travel is excessive; the pedal pulsates; or brake fluid leaks, usually onto the inside of the tire or wheel.

3   Disc brakes can be visually checked without removing any parts except the wheels. To check drum brakes, the rear

14.6 With the wheel off, check the thickness of the pads through the inspection hole

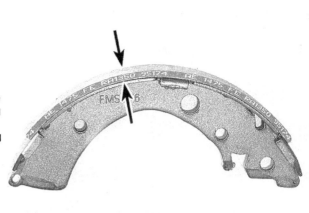

14.13 If the lining is bonded to the brake shoe, measure the lining thickness from the outer surface to the metal shoe; if the lining is riveted to the shoe, measure from the lining outer surface to the rivet head

wheels and brake drums will have to be removed. Remove the hub caps (if applicable) and loosen the wheel lug nuts a quarter turn each.

4   Loosen the wheel lug nuts, then raise the vehicle and place it securely on jackstands. **Warning:** *Never work under a vehicle that is supported only by a jack!*

Disc brakes

5   Remove the wheels. Now visible is the disc brake caliper which contains the pads. There is an outer brake pad and an inner pad. Both must be checked for wear.

6   Measure the thickness of the pads through the inspection hole in the caliper body **(see illustration)**. Compare the measurement with the limit given in this Chapter's Specifications; if any brake pad thickness is less than specified, then all brake pads must be replaced (see Chapter 9).

7   If you're in doubt as to the exact pad thickness or quality, remove them for measurement and further inspection (see Chapter 9).

8   Check the disc for score marks, wear and burned spots. If any of these conditions exist, the disc should be removed for servicing or replacement (see Chapter 9).

9   Before installing the wheels, check all the brake lines and hoses for damage, wear, deformation, cracks, corrosion, leakage, bends and twists, particularly in the vicinity of the rubber hoses and calipers.

10   Install the wheels, lower the vehicle and tighten the wheel lug nuts to the torque given in this Chapter's Specifications.

Drum brakes

11   Remove the brake drum (see Chapter 9).

12   With the drum removed, carefully clean off any accumulations of dirt and dust using brake system cleaner.

13   Measure the thickness of the lining

material on both leading and trailing brake shoes **(see illustration)**. Compare the measurement with the limit given in this Chapter's Specifications. if any brake shoe thickness is less than specified, then all brake shoes must be replaced (see Chapter 9).

14   Inspect the brake shoes for uneven wear patterns, cracks, glazing and delamination and replace if necessary. If the shoes have been saturated with brake fluid, oil or grease, this also necessitates replacement (see Chapter 9).

15   Make sure all the brake assembly springs are connected and in good condition.

16   Check the brake wheel cylinder for signs of fluid leakage. Carefully pry back the rubber dust boots on the wheel cylinder **(see illustration)**. Any leakage here is an indication that the wheel cylinders must be overhauled immediately (see Chapter 9). Also, check all hoses and connections for signs of leakage.

17   Clean the inside of the drum with brake system cleaner. Again, be careful not to breathe the dust.

18   Inspect the inside of the drum for cracks, score marks, deep scratches and hard spots which will appear as small discolored areas. If imperfections cannot be removed with fine emery cloth, the drum must be taken to an automotive machine shop for resurfacing.

19   Repeat the procedure for the remaining wheel.

20   Install the wheels, lower the vehicle and tighten the wheel lug nuts to the torque given in this Chapter's Specifications.

Brake booster check

21   Sit in the driver's seat and perform the following sequence of tests.

22   With the brake fully depressed, start the engine - the pedal should move down a little when the engine starts.

23   With the engine running, depress the brake pedal several times - the travel distance should not change.

24   Depress the brake, stop the engine and hold the pedal in for about 30 seconds - the

14.16 Pull the boot away from the cylinder and check for fluid leakage

pedal should neither sink nor rise.

25   Restart the engine, run it for about a minute and turn it off. Then firmly depress the brake several times - the pedal travel should decrease with each application.

26   If your brakes do not operate as described, the brake booster has failed. Refer to Chapter 9 for the replacement procedure.

15   Steering, suspension and driveaxle boot check (every 12,000 miles or 12 months)

Note: *For detailed illustrations of the steering and suspension components, refer to Chapter 10.*

With the wheels on the ground

1   With the vehicle stopped and the front wheels pointed straight ahead, rock the steering wheel gently back and forth. If freeplay is excessive, a front wheel bearing, steering shaft universal joint, lower arm balljoint or steering gear is worn. Refer to Chapter 10 for the appropriate repair procedure.

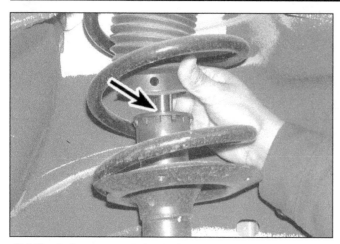

15.4 Check the struts and shocks for leakage at the indicated area

15.10 To check the balljoint for wear, try to pry the control arm up and down to make sure there is no play in the balljoint (if there is, replace it)

2   Other symptoms, such as excessive vehicle body movement over rough roads, swaying (leaning) around corners and binding as the steering wheel is turned, may indicate faulty steering and/or suspension components.

3   Check the shock absorbers by pushing down and releasing the vehicle several times at each corner. If the vehicle does not come back to a level position within one or two bounces, the shocks/struts are worn and must be replaced. When bouncing the vehicle up and down, listen for squeaks and noises from the suspension components.

4   Check the struts and shock absorbers for evidence of fluid leakage **(see illustration)**. A light film of fluid is no cause for concern. Make sure that any fluid noted is from the struts/shocks and not from some other source. If leakage is noted, replace the struts/shocks as a set.

5   Check the struts and shocks to be sure they are securely mounted and undamaged. Check the upper mounts for damage and wear. If damage or wear is noted, replace the shocks as a set (front and rear).

6   If the struts or shocks must be replaced, refer to Chapter 10 for the procedure.

Under the vehicle

7   Raise the vehicle and support it securely on jackstands.

8   Check the tires for irregular wear patterns and proper inflation. See Section 5 in this Chapter for information regarding tire wear and Chapter 10 for information on wheel bearing replacement.

9   Inspect the universal joint between the steering shaft and the steering gear housing. Check the steering gear housing for lubricant leakage. Make sure that the dust seals and boots are not damaged and that the boot clamps are not loose. Check the steering linkage for looseness or damage. Check the tie-rod ends for excessive play. Look for loose bolts, broken or disconnected parts and deteriorated rubber bushings on all suspension and steering components. While an assistant turns the steering wheel from side to side, check the steering components for free movement, chafing and binding. If the steering components do not seem to be reacting with the movement of the steering wheel, try to determine where the slack is located.

10   Check the balljoints for wear by trying to move each control arm up and down with a prybar **(see illustration)** to ensure that its balljoint has no play. If any balljoint does have play, replace it. Try to turn the balljoint grease fitting with your fingers. If you can move the grease fitting, the balljoint is worn out. See Chapter 10 for the balljoint replacement procedure.

11   Inspect the balljoint boots for damage and leaking grease **(see illustration)**. Replace the balljoints with new ones if they are damaged (see Chapter 10).

12   At the rear of the vehicle, inspect the suspension arm bushings for deterioration. Additional information on suspension components can be found in Chapter 10.

Driveaxle boot check

Note: *For detailed illustrations of the driveaxles, refer to Chapter 8.*

13   The driveaxle boots are very important because they prevent dirt, water and foreign material from entering and damaging the constant velocity (CV) joints. Oil and grease can cause the boot material to deteriorate prema-

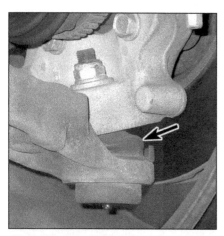

15.11 Check the balljoint boot for damage

15.14 Flex the driveaxle boots by hand to check for cracks and leaking grease

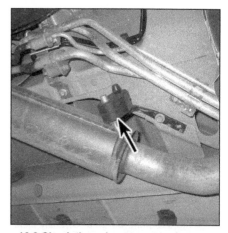

16.2 Check the exhaust system for rust, damage, or broken rubber hangers

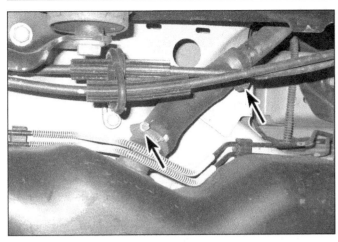

17.6 Check the fuel system hoses and clamps for damage and deterioration

18.3 Press the tabs towards each other to disengage the clips

turely, so it's a good idea to wash the boots with soap and water. Because it constantly pivots back and forth following the steering action of the front hub, the outer CV boot wears out sooner and should be inspected regularly.

14   Inspect the boots for tears and cracks as well as loose clamps **(see illustration)**. If there is any evidence of cracks or leaking lubricant, they must be replaced as described in Chapter 8.

16   Exhaust system check (every 12,000 miles or 12 months)

1   With the engine cold (at least three hours after the vehicle has been driven), check the complete exhaust system from the engine to the end of the tailpipe. Ideally, the inspection should be done with the vehicle on a hoist to permit unrestricted access. If a hoist isn't available, raise the vehicle and support it securely on jackstands.

2   Check the exhaust pipes and connections for evidence of leaks, severe corrosion and damage. Make sure that all brackets and hangers are in good condition and tight **(see illustration)**.

3   At the same time, inspect the underside of the body for holes, corrosion, open seams, etc. which may allow exhaust gases to enter the passenger compartment. Seal all body openings with silicone or body putty.

4   Rattles and other noises can often be traced to the exhaust system, especially the mounts and hangers. Try to move the pipes, muffler and catalytic converter. If the components can come in contact with the body or suspension parts, secure the exhaust system with new mounts.

5   Check the running condition of the engine by inspecting inside the end of the tailpipe. The exhaust deposits here are an indication of engine state-of-tune. If the pipe is black and sooty or coated with white deposits, the engine may need a tune-up, including a thorough fuel system inspection and adjustment.

17   Fuel system check (every 12,000 miles or 12 months)

Warning: *Gasoline is flammable, so take extra precautions when you work on any part of the fuel system. Don't smoke or allow open flames or bare light bulbs near the work area, and don't work in a garage where a gas-type appliance (such as a water heater or clothes dryer) is present. Since fuel is carcinogenic, wear fuel-resistant gloves when there's a possibility of being exposed to fuel, and, if you spill any fuel on your skin, rinse it off immediately with soap and water. Mop up any spills immediately and do not store fuel-soaked rags where they could ignite. When you perform any kind of work on the fuel system, wear safety glasses and have a Class B type fire extinguisher on hand. The fuel system is under constant pressure, so, before any lines are disconnected, the fuel system pressure must be relieved (see Chapter 4).*

1   If you smell gasoline while driving or after the vehicle has been sitting in the sun, inspect the fuel system immediately.

2   Remove the fuel filler cap and inspect it for damage and corrosion. The gasket should have an unbroken sealing imprint. If the gasket is damaged or corroded, install a new cap.

3   Inspect the fuel feed line for cracks. Make sure that the connections between the fuel lines and the fuel injection system are tight.

Warning: *Your vehicle is fuel injected, so you must relieve the fuel system pressure before servicing fuel system components. The fuel system pressure relief procedure is outlined in Chapter 4.*

4   Since some components of the fuel system - the fuel tank and the fuel lines, for example - are underneath the vehicle, they can be inspected more easily with the vehicle raised on a hoist. If that's not possible, raise the vehicle and support it on jackstands.

5   With the vehicle raised and safely supported, inspect the gas tank and filler neck for punctures, cracks and other damage. The connection between the filler neck and the tank is particularly critical. Sometimes a rubber filler neck will leak because of loose clamps or deteriorated rubber. Inspect all fuel tank mounting brackets and straps to be sure that the tank is securely attached to the vehicle.

Warning: *Do not, under any circumstances, try to repair a fuel tank (except rubber components). A welding torch or any open flame can easily cause fuel vapors inside the tank to explode.*

6   Carefully check all hoses and lines leading away from the fuel tank. Check for loose connections, deteriorated hoses, crimped lines and other damage **(see illustration)**. Repair or replace damaged sections as necessary (see Chapter 4).

18   Cabin air filter replacement (every 12,000 miles or 12 months)

Warning: *The models covered by this manual are equipped with a Supplemental Restraint System (SRS), more commonly known as airbags. Always disable the airbag system before working in the vicinity of any airbag system component to avoid the possibility of accidental deployment of the airbag, which could cause personal injury (see Chapter 12).*

1   Disconnect the cable from the remote ground terminal or battery (see Chapter 5).

2014 and earlier models

2   Disengage the clips on each side of the glove box. Pull the sides inwards until the stops are clear and lower glove box (see Chapter 11).

3   Disengage the two clips at the ends of the cover **(see illustration)** and remove the cover.

18.4 Remove the filter from the housing, noting the position of the filter as it is removed

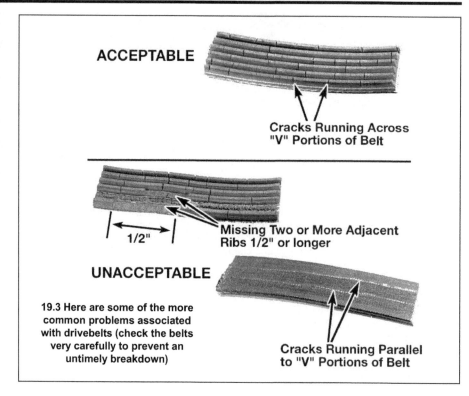

19.3 Here are some of the more common problems associated with drivebelts (check the belts very carefully to prevent an untimely breakdown)

4   Remove the filter from the housing (**see illustration**).
5   Install the filter, making sure the arrow on the filter is pointing towards the floor.
Note: *The cabin air filter is labeled with an arrow and the word "Airflow" on it. The filter should be installed with the arrow pointing towards the floor.*
6   Installation is the reverse of removal.

2015 and later models

Note: *On 2015 and later models, the cabin air filter is mounted vertically instead of horizontally, unlike the previous generation's setup. Note the direction in which the old filter was installed.*

7   Pry out the clips, disconnect the light electrical connector and remove the panel from beneath the glove box.
8   Remove the center console side panel from the passenger footwell.
9   Taking care not to damage the tabs, remove the upper and lower filter door retaining tabs and pull the door straight off the HVAC unit.
Note: *If the tab(s) were damaged during removal, there are pre-made holes in the filter door that will accept a 16mm long, M4 self-tapping screw as an alternative means to secure it.*
10   Noting the direction of airflow, remove the cabin filter element.
11   Install the cabin air filter into the housing in the proper direction.
12   The remainder of installation is the reverse of removal.

---

19   Drivebelt check and replacement (every 30,000 miles or 30 months)

Warning: *The electric cooling fan(s) on these models can activate at any time when the ignition switch is in the ON position. Make sure the ignition is OFF when working in the vicinity of the fan(s).*
1   The drivebelt is located at the front of

the engine and plays an important role in the operation of the vehicle and its components.
2   Due to its function and material makeup, the belt is prone to failure after a period of time, and should be inspected periodically to prevent major damage.

Check

3   With the engine off, open the hood and use your fingers (and a flashlight, if necessary) to move along the belt checking for cracks and separation of the belt plies. Also check for fraying and glazing, which gives the belt a shiny appearance. Also check the ribs on the underside of the belt. They should all be the same depth, with none of the surface uneven (**see illustration**).
4   The serpentine belt tension is adjusted by an automatic tensioner.

Replacement (except power steering pump drivebelt on 2007 through 2010 2.7L V6 models)

5   Apply the parking brake, loosen the right front wheel lug nuts, raise the front of the vehicle and support it securely on jackstands. Remove the wheel, then remove the drivebelt splash shield (**see illustration**).
6   On 2015 and later 3.6L models, drain the cooling system (see Section 23), then disconnect the radiator/heater hoses from the thermostat housing (see Chapter 3). On 2015 and later 2.4L models, remove the air filter housing (see Chapter 4).
7   The automatic tensioner must be released to allow drivebelt replacement.

Place a wrench on the tensioner pulley center bolt and rotate it until the belt can be removed (**see illustration**).
8   On 2.7L and 3.6L engines, insert a 3/8-inch drive ratchet into the tensioner (**see illustration**) and rotate the tensioner away (counterclockwise) from the belt until the belt can be removed, then slowly release the tensioner.
Warning: *Damage to the tensioner or possible injury can occur if the tensioner snaps or springs back without the belt in place.*
9   Installation is the reverse of removal. When installing the belt, make sure the belt is centered on the pulleys.
10   Install the drivebelt splash shield, wheel and lug nuts. Lower the vehicle and tighten the lug nuts to the torque listed in this Chapter's Specifications.

Automatic tensioner replacement

11   Remove the wheel, then remove the drivebelt splash shield (**see illustration 19.5**)
12   Remove the drivebelt (see previous Steps).

2015 and later 2.4L models

13   Remove the tensioner lower bolt.
14   Remove the tensioner upper bolt, accessing it with a socket and 12-inch extension placed between the catalytic converter and engine block. Remove the tensioner from the engine.

All other models

15   Unscrew the mounting bolt from the center of the tensioner and remove the tensioner (**see illustration**).

19.5 Remove the drivebelt splash shield fasteners and shield

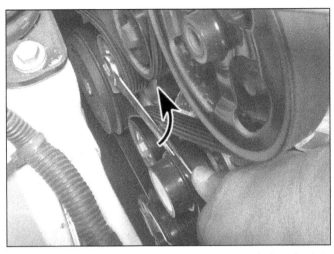

19.7 Place a wrench on the tensioner pulley center bolt and rotate it to relieve belt tension - 2.4L engine shown

All models

16   Installation is the reverse of removal. Tighten the mounting bolt(s) to the torque listed in this Chapter's Specifications. Lower the vehicle and tighten the lug nuts to the torque listed in this Chapter's Specifications.

Power steering pump drivebelt (2007 through 2010 2.7L V6 models)

Note: *The power steering pump belt doesn't use a tensioner and must be cut to be removed. Make sure you have a new belt before cutting the old belt off.*

17   Remove the drivebelt (see Steps 5 and 8).
18   Using a pair of diagonal cutters, cut the power steering belt to remove it.
19   Place the new belt on the crankshaft pulley, making sure the belt is fully seated in the pulley grooves.
20   Start the belt on the power steering pump pulley and secure it to the pulley with a plastic wire tie.

21   Rotate the crankshaft pulley clockwise - the belt should pull up over the power steering pump pulley until it is seated. Continue rotating the pulley until the wire tie breaks.
22   The remainder of installation is the reverse of removal.

20   Brake fluid change (every 30,000 miles or 30 months)

Warning: *Brake fluid can harm your eyes and damage painted surfaces, so use extreme caution when handling or pouring it. Do not use brake fluid that has been standing open or is more than one year old. Brake fluid absorbs moisture from the air. Excess moisture can cause a dangerous loss of braking effectiveness.*

1   At the specified intervals, the brake fluid should be drained and replaced. Since the brake fluid may drip or splash when pouring it, place plenty of rags around the master cylinder to protect any surrounding painted surfaces.

2   Before beginning work, purchase the specified brake fluid (see *Recommended lubricants and fluids* in this Chapter's Specifications).
3   Remove the cap from the master cylinder reservoir.
4   Using a hand suction pump or similar device, withdraw the fluid from the master cylinder reservoir.
5   Add new fluid to the master cylinder until it rises to the base of the filler neck.
6   Bleed the brake system at all four brakes until new and uncontaminated fluid is expelled from the bleeder screw (see Chapter 9). Be sure to maintain the fluid level in the master cylinder as you perform the bleeding process. If you allow the master cylinder to run dry, air will enter the system.
7   Refill the master cylinder with fluid and check the operation of the brakes. The pedal should feel solid when depressed, with no sponginess.
Warning: *Do not operate the vehicle if you are in doubt about the effectiveness of the brake system.*

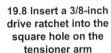

19.8 Insert a 3/8-inch drive ratchet into the square hole on the tensioner arm

19.15 The drivebelt tensioner is secured by a single bolt in the center of the tensioner – 3.6L engine shown, other models similar (except 2015 and later 2.4L)

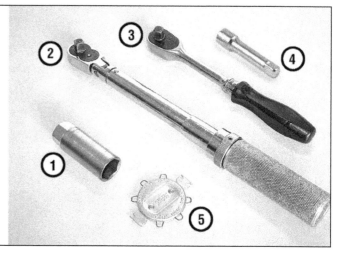

21.2 Tools required for changing spark plugs

1   **Spark plug socket** - This will have special padding inside to protect the spark plug's porcelain insulator
2   **Torque wrench** - Although not mandatory, using this tool is the best way to ensure the plugs are tightened properly
3   **Ratchet** - Standard hand tool to fit the spark plug socket
4   **Extension** - Depending on model and accessories, you may need special extensions and universal joints to reach one or more of the plugs
5   **Spark plug gap gauge** - This gauge for checking the gap comes in a variety of styles. Make sure the gap for your engine is included

21.5 Spark plug manufacturers recommend using a wire-type gauge when checking the gap - the wire should slide between the electrodes with a slight drag

21   Spark plug check and replacement (every 100,000 miles)

1   The spark plugs are located in the center of the valve cover(s).

2   In most cases the tools necessary for spark plug replacement include a spark plug socket which fits onto a ratchet (this special socket is padded inside to protect the porcelain insulators on the new plugs and hold them in place), various extensions and a feeler gauge to check and adjust the spark plug gap **(see illustration)**. Since these engines are equipped with aluminum cylinder heads, a torque wrench should be used when tightening the spark plugs.

3   The best approach when replacing the spark plugs is to purchase the new spark plugs beforehand, adjust them to the proper gap and then replace each plug one at a time. When buying the new spark plugs, be sure to obtain the correct plug for your specific engine. This information can be found in this Chapter's Specifications or in your owner's manual.

4   Allow the engine to cool completely before attempting to remove any of the plugs. During this cooling off time, each of the new spark plugs can be inspected for defects and the gaps can be checked.

5   The gap is checked by inserting the proper thickness gauge between the electrodes at the tip of the plug **(see illustration)**. The gap between the electrodes should be

as listed in this Chapter's Specifications or in your owner's manual. The wire should touch each of the electrodes. Also, at this time check for cracks in the spark plug body (if any are found, the plug must not be used).

Caution: *The manufacturer recommends against checking the gap on platinum-tipped spark plugs; the platinum coating could be scraped off.*

6   Cover the fender to prevent damage to the paint. Fender covers are available from auto parts stores but an old blanket will work just fine.

7   On 2.4L engines, pull the engine cover upwards to disengage the cover from the ball-studs and remove the cover to gain access to the ignition coils **(see illustration)**.

8   On 2.7L, 3.5L and 3.6L engines, remove the upper intake manifold (see Chapter 2B, 2C or 2D) to gain access to all the ignition coils **(see illustration)**.

9   If compressed air is available, use it to blow any dirt or foreign material away from the spark plug area.

Warning: *Wear eye protection!*

10   Place the spark plug socket over the plug and remove it from the engine by turning

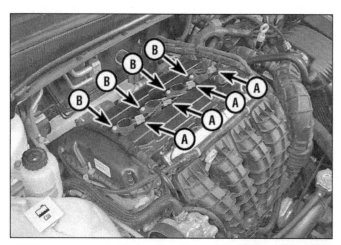

21.7 To remove an ignition coil(s), depress the tab (A), disconnect the electrical connector, remove the coil retaining bolt (B), then pull the coil straight up (2.4L engine)

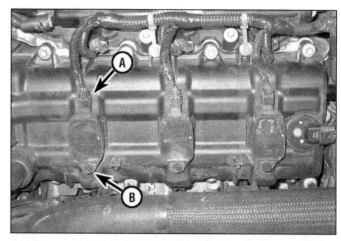

21.8 To remove an ignition coil(s), depress the tab (A), disconnect the electrical connector, remove the coil retaining bolt (B), then pull the coil straight up (V6 engines)

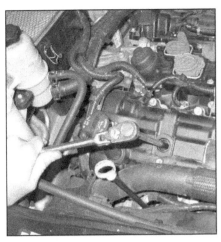

21.10 Use a ratchet, socket and long extension to remove the spark plugs (3.6L engine shown, all other engines similar)

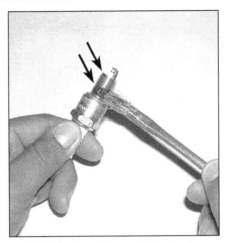

21.12 Apply a thin coat of anti-seize compound to the spark plug threads, but be careful not to get any of it near the electrodes

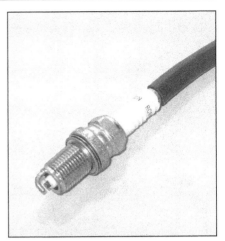

21.13 A length of snug-fitting rubber hose will save time and prevent damaged threads when installing the spark plugs

it in a counterclockwise direction (see illustration).

11   Compare each old spark plug with those shown on the inside back cover of this manual to determine the overall running condition of the engine.

12   It's a good idea to lightly coat the threads of the spark plugs with an anti-seize compound (see illustration) to insure that the spark plugs do not seize in the aluminum cylinder head.

13   It's often difficult to insert spark plugs into their holes without cross-threading them. To avoid this possibility, fit a piece of rubber hose over the end of the spark plug (see illustration). The flexible hose acts as a universal joint to help align the plug with the plug hole. Should the plug begin to cross-thread, the hose will slip on the spark plug, preventing thread damage. Install the spark plug and tighten it to the torque listed in this Chapter's Specifications.

14   Using a slight twisting motion, install the ignition coils.

15   Follow the above procedure for the remaining spark plugs, replacing them one at a time to prevent mixing up the ignition coils.

22   Positive Crankcase Ventilation (PCV) valve check and replacement (every 60,000 miles or 60 months)

Warning: *Do not attempt to clean the PCV valve.*

Note: *If replacing the valve, compare the old and new valves to make sure they're identical.*

2.4L engines

1   The PCV valve is located on the top of the valve cover.

2   Detach the PCV hose and remove the valve (see illustrations).

3   Install the PCV valve and tighten the valve securely.

2.7L and 3.5L engines

4   The PCV valve is located on the right end of the rear valve cover under the upper intake manifold.

5   Disconnect the PCV hose, then unscrew the valve from the valve cover.

6   Install the PCV valve and tighten the valve securely.

3.6L engines

7   The PCV valve is located on the end of the rear valve cover.

8   Lift up on the outer edges of the engine cover to disengage the rubber mounts from the ball studs, and remove the engine cover.

9   Disconnect the camshaft sensor electrical connector.

10   Use a screwdriver to pry the sensor electrical harness retainer up and off of the valve cover stud (see illustration).

22.2a Disconnect the PCV hose. . .

22.2b. . . then use a wrench to unscrew the PCV valve from the valve cover

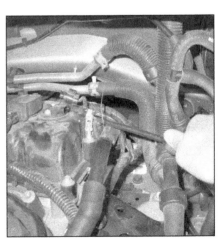

22.10 Carefully pry the harness retainer up to release it (intake manifold removed for clarity)

22.11a Disconnect the hose from the valve. . .

22.11b. . . then remove the screws and pull the valve out of the cover

11   Detach the PCV hose, then remove the valve mounting screws and the valve **(see illustrations)**.

All engines

12   Shake the valve. The valve should rattle freely - if it doesn't, replace it.

13   Replace the PCV valve with the correct one for your specific vehicle and engine size and tighten the mounting screws securely.

14   Installation is the reverse of removal.

---

23   Cooling system servicing (draining, flushing and refilling) (every 60 months)

Warning: *Do not allow antifreeze to come in contact with your skin or painted surfaces of the vehicle. Flush contaminated areas immediately with plenty of water. Do not store new coolant or leave old coolant lying around where it's accessible to children or pets - they're attracted by its sweet smell. Ingestion of even a small amount of coolant*

can be fatal! Wipe up garage floor and drip pan spills immediately. Keep antifreeze containers covered and repair cooling system leaks as soon as they're noticed. Check with local authorities about the disposal of used antifreeze. Many communities have collection centers which will see that antifreeze is disposed of properly.

Warning: *The electric cooling fan(s) on these models can activate at any time when the ignition switch is in the ON position. Make sure the ignition is OFF when working in the vicinity of the fan(s). As an added precaution, disconnect the negative battery cable from the remote ground terminal (see Chapter 5).*

Note: *These vehicles are originally filled with Mopar 5 year/100,000 mile coolant that shouldn't be mixed with other coolants. Always refill with the correct coolant.*

Note: *Be sure to avoid getting coolant on the drivebelt or pulleys by shielding the area with shop rags - if coolant accidentally contacts the belt/pulleys, hose the area off with clean water when finished.*

Note: *If the coolant is clean and free of debris, it can be saved and reused.*

1   Periodically, the cooling system should be drained, flushed and refilled to replenish the antifreeze mixture and prevent formation of rust and corrosion, which can impair the performance of the cooling system and cause engine damage.

2   At the same time the cooling system is serviced, all hoses and the radiator (pressure) cap should be inspected, tested and replaced if faulty (see Section 12).

Draining

Warning: *Wait until the engine is completely cool before beginning this procedure.*

3   With the engine cold, remove the radiator or expansion tank cap and set the heater control to maximum heat.

4   Move a large container, capable of holding at least 12 quarts, under the radiator drain fitting to catch the coolant mixture as it's drained.

5   Open the drain fitting located at the bottom of the radiator **(see illustration)**. Allow the coolant to completely drain out.

6   After the coolant stops flowing out of the radiator, move the container under the engine

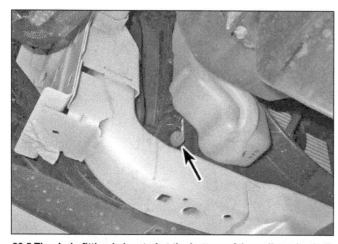

23.5 The drain fitting is located at the bottom of the radiator (typical)

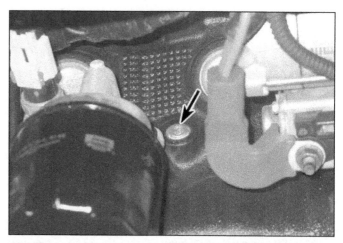

23.6 The block drain plugs are generally located about one to two inches above the oil pan - there is one on each side of the engine block

block drain plugs and allow the coolant in the block to drain **(see illustration)**.

7 While the coolant is draining, check the condition of the radiator hoses, heater hoses and clamps. Replace any damaged clamps or hoses (see Section 12).

Flushing

8 Close the radiator and engine block drain plugs. Fill the cooling system with clean water, following the *Refilling* procedure (see Step 15).

9 Start the engine and allow it to reach normal operating temperature, then rev up the engine a few times.

10 Turn the engine off and allow it to cool completely, then drain the system as described earlier.

11 Repeat Steps 9 and 10 until the water being drained is free of contaminants.

12 Severe cases of radiator contamination or clogging will require removing the radiator (see Chapter 3) and reverse flushing it. This involves inserting a hose in the bottom radiator outlet to allow the clean water to run against the normal flow, draining out through the top. A radiator repair shop should be consulted if further cleaning or repair is necessary.

13 When the coolant is regularly drained and the system refilled with the correct coolant mixture, there should be no need to employ chemical cleaners or descalers.

14 Disconnect the coolant reservoir hose, remove the reservoir from the vehicle and flush it with clean water (see Chapter 3). Inspect it for damage and replace if necessary.

Refilling

2014 and earlier models

15 Install the coolant reservoir, reconnect the hose and close the radiator drain fitting.

16 Remove the radiator cap. Add the correct mixture of the proper type of antifreeze/coolant and water, in the ratio specified on the antifreeze container or in this Chapter's Specifications, through the filler neck until it reaches the radiator cap seat.

17 Add the same coolant mixture to the reservoir until the level is between the FULL HOT and ADD marks. Install the radiator cap.

18 Run the engine until normal operating temperature is reached (the fans will cycle on, then off), then allow the engine to cool. With the engine cold, add coolant as necessary to bring it up to the correct level.

2015 and later models

2.4L engines

Note: *The refilling part of this procedure is more complex than usual on these models. To ensure correct refilling and to allow the system to bleed properly, the special funnel tool kit - Mopar #8195A or equivalent - should be used. A length of clear hose and a small hose clamp will also be needed.*

19 With the expansion tank cap removed, install the special funnel tool 8195A onto the expansion tank, choosing the correct adaption system to create a reliable seal.

20 Cut a length of clear hose of the correct diameter, to be fitted onto the upper radiator vent fitting and fed into the funnel.

21 Cut the crimp-type clamp, then disconnect the coolant return hose from the expansion tank. Set this hose aside - it will act as a bleeder hose when refilling. Plug the fitting at the expansion tank.

22 With the clear hose running from the vent fitting into the funnel, open the radiator vent knob a half-turn.

Note: *The radiator vent knob is located at the upper corner of the radiator.*

23 Add coolant to the top of the funnel, making sure to keep topping it off as the cooling system fills. The funnel should not become empty during refilling.

Note: *Make sure there is coolant flowing through the clear hose when adding coolant.*

24 Continue to add coolant until the level in the funnel settles (comes to a stop) and is free of bubbles.

25 Disconnect the heater core return hose quick-connect fitting, which is located behind the battery and between the fuse box and engine block. Wait until coolant becomes present and spills out of the fitting. Once this happens, reconnect the quick-connect fitting.

26 Loosen the cylinder head drain plug about five full rotations, using a 6mm Allen wrench. The plug is located in front of the thermostat housing. Once fluid emerges from the plug that is free of air bubbles, tighten the plug securely.

27 Close and tighten the radiator vent valve. Don't remove the clear hose just yet.

28 With fluid still being in the funnel, install the stopping handle into the funnel to prevent spillage as the funnel is removed.

29 Fill the expansion tank to the FULL mark, then remove the funnel and adapter assembly.

30 Unplug the fitting at the expansion tank and connect the return hose, securing it with a new clamp. When connecting this hose, the level in the tank might drain slightly - so be sure to top it off to the FULL mark.

31 Install the expansion tank cap.

32 Start the engine and allow it to idle for a minute.

33 Raise the engine speed to 3,000 RPM and have an assistant check the coolant return hose port at the expansion tank to confirm that there is a flow of coolant.

Caution: *If there is no coolant flow at the port, the procedure must be performed again from the beginning or engine damage may occur.*

34 Once the flow of fluid has been confirmed, feel the return hose until it starts getting warm, then return the engine back to idle.

35 Run the engine until normal operating temperature is reached - the upper radiator hose will become hot.

36 Wearing heat resistant gloves, pinch the upper radiator hose as much as possible by hand, while at the same time, an assistant runs the engine at 3,000 RPM for one minute. This will force any trapped air out of the oil cooler lines.

37 Allow the engine to idle for another minute, then shut it off. Let the engine cool down.

38 Remove the expansion tank cap. Open the radiator vent valve once more until there is a solid (bubble-free) stream of coolant in the clear hose. Once accomplished, close the valve and remove the clear hose.

39 Top up the expansion tank to the FULL line, if necessary, then install the cap.

3.6L engines

40 With the expansion tank cap removed, insert a funnel into the tank.

41 Loosen the bleeder screw on the thermostat housing, then slowly add coolant into the expansion tank until there is a solid stream present at the bleeder screw (free of bubbles). Once coolant is present, close and tighten the screw.

42 Disconnect the upper heater core hose on the firewall. Continue filling until there is coolant present at the heater core hose fitting, then reconnect the hose.

43 Continue filling until the coolant level is at the FULL mark on the tank, then install the expansion tank cap.

44 Start the engine and place the heater temperature control at the hottest setting, with the blower motor switch on low. Allow the engine to idle for 10 minutes, then turn off the engine.

45 Using a rag and being extra careful to avoid burns, slowly loosen and remove the expansion tank cap, then top off the level as necessary. Install the cap.

46 Start and run the engine at 4,000 RPM until the engine reaches normal operating temperature - the thermostat will open and the cooling fan will operate.

47 Let the engine idle for 25 minutes, then turn off the engine and allow it to cool.

48 Slowly remove the expansion tank cap, top off the coolant to the correct level, then install the cap.

49 Once again, start the engine and run it at 4,000 RPM until the thermostat opens, then let the engine idle for a few minutes. Make sure the fan operates. Turn off the engine and let it cool down.

50 Top off the coolant in the tank to the correct level (FULL mark).

All models

51 Keep a close watch on the coolant level and the various cooling system hoses during the first few miles of driving and check for any coolant leaks. Tighten the hose clamps and add more coolant mixture as necessary.

24 Automatic transaxle fluid and filter change (every 100,000 miles)

1 The automatic transaxle fluid and filter should be changed and the magnet cleaned at the recommended intervals.

2 Raise the front of the vehicle, support it securely on jackstands, and apply the parking brake.

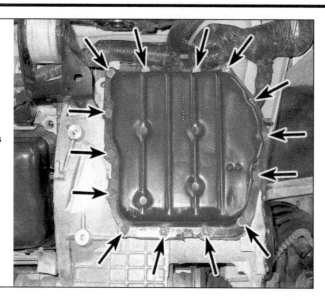

**24.3 Transaxle pan bolts
– 41TE transaxle**

40TE and 41TE transaxles

3   Place a container under the transaxle pan and loosen the pan bolts **(see illustration)**. Completely remove all of the the bolts except the ones along the rear of the pan. Tap the corner of the pan to break the seal and allow the fluid to drain into the container (the remaining bolts will prevent the pan from separating from the transaxle). Remove the remaining bolts.
4   Remove the old sealant from the pan and transaxle body (don't nick or gouge the sealing surfaces) and clean the pan magnet with a clean, lint-free cloth.
5   Pry the filter from the valve body **(see illustration)**. If the O-ring does not come out with the filter, be sure to retrieve it from the bore in the valve body **(see illustration)**.
6   Install a new filter and O-ring and reinstall the magnet(s).
7   Apply a 1/8-inch bead of RTV sealant to the pan sealing surface and position it on the transaxle (see illustration). Install the bolts and tighten them to the torque in this Chap-

ter's Specifications using a criss-cross pattern. Work up to the final torque in three or four steps.
8   Lower the vehicle and add four quarts of the specified transaxle fluid (see Recommended lubricants and fluids in this Chapter's Specifications). Start the engine and allow it to idle for a minute, then move the shift lever through each gear position, ending in Park. Check the fluid level (see Section 4). Also check for fluid leaks around the pan.
9   If necessary, add more fluid (a little at a time) until the level is between the Add and Full marks (be careful not to overfill it).
10   Make sure the dipstick is completely seated or dirt could get into the transaxle.

62TE transaxle

Note: *The 62TE transaxles require the use of a scan tool to check transaxle fluid temperature and special tool #9336A (or a homemade equivalent) to measure the fluid level. If you do not have both tools to make*

the proper temperature-to-fluid level comparisons, we do not recommend attempting this repair procedure.
11   Place a container under the transaxle pan and loosen the pan bolts (see illustration). Completely remove all of the the bolts except the ones along the rear of the pan. Tap the corner of the pan to break the seal and allow the fluid to drain into the container (the remaining bolts will prevent the pan from separating from the transaxle). Remove the remaining bolts.
12   Remove the filter retaining nuts and filter.
13   Install a new filter and tighten the fasteners to the torque listed in this Chapter's Specifications.
14   Remove the old sealant from the pan and transaxle body (don't nick or gouge the sealing surfaces) and clean the pan magnet with a clean, lint-free cloth.
15   Apply a 1/8-inch bead of RTV to the pan sealing surface and position it on the transaxle **(see illustration 24.7)**. Install the bolts and tighten them to the torque in this Chapter's Specifications using a criss-cross pattern. Work up to the final torque in three or four steps.
16   Lower the vehicle and add four quarts of the specified transaxle fluid (see Recommended lubricants and fluids in this Chapter's Specifications). Start the engine and allow it to idle for a minute, then move the shift lever through each gear position, ending in Park.
17   Check the fluid level using the fluid level check procedure (see Section 4).

948TE/9HP48 (9-speed) transaxles

Note: *The manufacturer states that this transaxle is a "sealed for life" design and does not require periodic maintenance unless there are signs of a fluid leak.*
Note: *The 9-speed transaxle requires the use of a scan tool to check transaxle fluid temper-*

24.5a Pry the filter from the valve body. . .

24.5b. . . and retrieve the O-ring

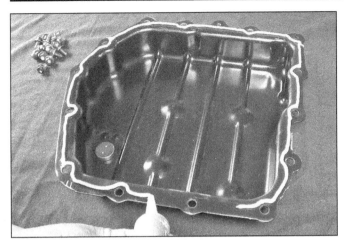

24.7 Apply a 1/8-inch bead of RTV sealant to the pan flange, inboard of the bolt holes

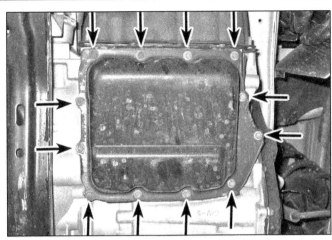

24.11 Transaxle pan bolts – 62TE transaxle

ature and special tool #10323A (or a home-made equivalent) to measure the fluid level. If you do not have both tools to make the proper temperature-to-fluid level comparisons, we do not recommend attempting this procedure.
Note: *The fluid filter in the 948TE/9HP48 9-speed automatic transaxle is not periodically serviced. To replace the fluid filter the transaxle*

must be removed and disassembled.

18   Place a container under the transaxle and remove the drain plug. Allow the fluid to drain into the container.

19   Once the fluid is done draining, install the drain plug and tighten it securely.

20   Measure the amount of fluid drained to allow for an accurate refill.

21   Remove the transaxle check/fill plug (see Section 4) and add the same quantity of new fluid as measured in the previous Step. Be sure to use only the type of fluid mentioned in this Chapter's Specifications.

22   Check the fluid level as described in Section 4, adding or removing fluid as necessary.

Notes

Chapter 2 Part A
Four-cylinder engine

Contents

	Section
Camshaft(s) – removal, inspection and installation	10
Crankshaft front oil seal - replacement	8
Crankshaft pulley - removal and installation	7
Cylinder head - removal and installation	11
Driveplate - removal and installation	15
Engine mounts - check and replacement	17
Exhaust manifold - removal, inspection and installation	6
General information	1
Intake manifold - removal and installation	5
Oil pan - removal and installation	13

	Section
Oil pump/balance shaft module and chain - removal, inspection and installation	14
Rear main oil seal - replacement	16
Repair operations possible with the engine in the vehicle	2
Timing chain cover, chain and sprockets - removal, inspection and installation	9
Top Dead Center (TDC) for number one piston - locating	3
Valve clearance - check and adjustment	12
Valve cover - removal and installation	4
Variable Valve Actuation Assembly (VVAA) - removal and installation	18

Specifications

General

Bore	3.465 inches
Stroke	
2014 and earlier	3.819 inches
2015 and later	3.228 inches
Compression ratio	
2014 and earlier	10.5:1
2015 and later	10.0:1
Displacement	146.5 cubic inches (2.4 liters)
Firing order	1-3-4-2

Camshaft

Bearing bore	
2014 and later	
Front intake	1.1810 to 1.1819 inches
Front exhaust	1.5747 to 1.5756 inches
Remaining bores	0.9448 to 0.9457 inch
2015 and later	
Number 1 bore	1.4181 to 1.4189 inches
Remaining bores (2 through 6)	1.0039 to 1.0047 inches
Camshaft journal diameter	
2014 and earlier	
Front intake	1.1797 to 1.1803 inches
Front exhaust	1.4166 to 1.4173 inches
Remaining journals	0.943 to 0.944 inches
2015 and later	
Number 1 bore	1.4162 to 1.4169 inches
Remaining bores (2 through 6)	1.0021 to 1.0028 inches

Camshaft (continued)
Bearing clearance
  2014 and earlier
    Front intake.. 0.0008 to 0.0022 inch
    Front exhaust... 0.0007 to 0.0020 inch
    Remaining journals... 0.0008 to 0.0026 inch
  2015 and later
    Bearing journal clearance (all).. 0.0008 to 0.0026 inch
Endplay
  2014 and earlier.. 0.004 to 0.009 inch
  2015 and later... 0.006 to 0.010 inch
Lobe lift
  Intake @ 0.007 inch lash
    2014 and earlier.............................. .... 0.362 inch
    2015 and later.. 0.2032 inch
  Exhaust @ 0.011 inch lash
    2014 and earlier.. 0.331 inch
    2015 and later.. 0.2105 inch
Valve clearance (2014 and earlier)
  Intake .... 0.006 to 0.009 inch
  Exhaust .... 0.010 to 0.012 inch

Cylinder head
Head gasket surface warpage limit .... 0.004 inch maximum
Exhaust manifold mounting surface warpage limit .... 0.006 inch maximum

Intake and exhaust manifolds
Warpage limit... 0.006 inch maximum

Torque specifications   **Ft-lbs** (unless otherwise indicated)
Note: *One foot-pound (ft-lb) of torque is equivalent to 12 inch-pounds (in-lbs) of torque. Torque values below approximately 15 ft-lbs are expressed in inch-pounds, because most foot-pound torque wrenches are not accurate at these smaller values.*

Drivebelt idler pulley bolt
  2014 and earlier................................ 37
  2015 and later... 21
Camshaft bearing cap bolts (in sequence)
  2014 and earlier (DOHC engine - **see illustration 10.41**)
    M6 bolts... 106 in-lbs
    M8 bolts... 22
  2015 and later (SOHC engine - **see illustration 10.53**)
    Step 1.. 44 in-lbs
    Step 2.. 89 in-lbs
Phaser-to-camshaft bolt (2014 and earlier models)........ 44
Camshaft sprocket bolts (2015 and later models)
  Step 1... 89 in-lbs
  Step 2... ..... Tighten an additional 92-degrees
Crankshaft pulley/damper bolt... 155
Cylinder head bolts (in sequence - **see illustration 11.28**)
  2014 and earlier models
    Short-head (5/16-inch) bolts
      Step 1.. 25
      Step 2.. 45
      Step 3.. 45
      Step 4.. Tighten an additional 1/4-turn (90-degrees)
    Long-head (1/2-inch) bolts
      Step 1.. 25
      Step 2.. 54
      Step 3.. 54
      Step 4.. Tighten an additional 1/4-turn (90-degrees)
  2015 and later models
    Step 1.. 84 in-lbs
    Step 2.. 144 in-lbs
    Step 3.. 26
    Step 4.. Tighten an additional 160-degrees
Driveplate-to-crankshaft bolts*
  Step 1 22
  Step 2 .... Tighten an additional 51-degrees
Exhaust manifold-to-cylinder head bolts.............................. 25
Exhaust manifold-to-exhaust pipe bolts.............................. 21

On 2015 and later models, new fasteners must be used.

Torque specifications (continued)

Ft-lbs (unless otherwise indicated)

Note: *One foot-pound (ft-lb) of torque is equivalent to 12 inch-pounds (in-lbs) of torque. Torque values below approximately 15 ft-lbs are expressed in inch-pounds, because most foot-pound torque wrenches are not accurate at these smaller values.*

Intake manifold bolts
    2014 and earlier models... 18
    2015 and later models.. 108 in-lbs
Oil pan bolts
    M6 bolts.. 80 in-lbs
    M8 bolts.. 18
Balance shaft/oil pump module
    New bolts - 180 mm
        Step 1 .... 132 in-lbs
        Step 2 24
        Step 3 .... Tighten an additional 1/4-turn (90-degrees)
    New bolts - 185 mm (these are the only size bolts used on 2015 and later models)
        Step 1 .... 132 in-lbs
        Step 2 22
        Step 3 .... Tighten an additional 1/4-turn (90-degrees)
Timing chain cover bolts
    M6 bolts.. 80 in-lbs
    M8 bolts.. 19
    M10 bolts (2015 and later)... 48
    Single stud bolt (2015 and later)... 19
Timing chain tensioner bolts
    2014 and earlier models.. 89 in-lbs
    2015 and later models... 80 in-lbs
Timing chain guide bolts
    2014 and earlier models.. 106 in-lbs
    2015 and later models... 21
Valve cover bolts
    2014 and earlier models
        Step 1 .... 44 in-lbs
        Step 2 .... 90 in-lbs
    2015 and later models... 89 in-lbs
Variable Valve Actuator Asssembly (VVAA) (2015 and later models) (in sequence - **see illustration 18.20**)
    Step 1.. 89 in-lbs
    Step 2.. 16
    Step 3.. 16

1 General information

1   This Part of Chapter 2 is devoted to in-vehicle engine repair procedures. Information concerning engine removal and installation can be found in Chapter 2E.

2   These engines utilize an aluminum in-line four cylinder block. The aluminum cylinder head is heat treated and is equipped with replaceable valve guides, seats, mechanical lifters and four valves per cylinder.

3   2014 and earlier 2.4L models are equipped with a dual overhead camshaft (DOHC) system; there is a dedicated camshaft to control the intake valves and exhaust valves separately. 2015 and later 2.4L models utilize a single overhead camshaft (SOHC). This system has a Variable Valve Actuator Assembly (VVAA) that replaces the traditional intake camshaft. The intake valves are actuated by hydraulic pumping elements via solenoid-operated hydraulic ports, controlled by the Powertrain Control Module (PCM).

4   All models incorporate a balance shaft module. The oil pump is integrated into the module and is serviced as a unit with the balance shaft that is installed below the crankshaft. For service information on the balance shaft, refer to Chapter 2E. This engine is not a free wheeling engine; if the timing chain fails, the pistons will contact the valves.

5   The following repair procedures are based on the assumption that the engine is installed in the vehicle. If the engine has been removed from the vehicle and mounted on a stand, many of the steps outlined in this Part of Chapter 2 will not apply.

6   The Specifications included in this Part of Chapter 2 apply only to the procedures contained in this Part.

2 Repair operations possible with the engine in the vehicle

1   Many major repair operations can be accomplished without removing the engine from the vehicle.

2   Clean the engine compartment and the exterior of the engine with some type of degreaser before any work is done. It will make the job easier and help keep dirt out of the internal areas of the engine.

3   Depending on the components involved, it may be helpful to remove the hood to improve access to the engine as repairs are performed (refer to Chapter 11 if necessary). Cover the fenders to prevent damage to the paint. Special pads are available, but an old bedspread or blanket will also work.

4   If vacuum, exhaust, oil or coolant leaks develop, indicating a need for gasket or seal replacement, the repairs can generally be made with the engine in the vehicle. The intake and exhaust manifold gaskets, oil pan gasket, camshaft and crankshaft oil seals and cylinder head gasket are all accessible with the engine in place.

5   Exterior engine components, such as the intake and exhaust manifolds, the oil pan, the oil pump, the water pump, the starter motor, the alternator, the distributor and the fuel system components can be removed for repair with the engine in place.

6   Since the camshaft(s) and cylinder head can be removed without pulling the engine, valve component servicing can also be accomplished with the engine in the vehicle. Replacement of the timing chain and sprockets is also possible with the engine in the vehicle.

7   In extreme cases caused by a lack of necessary equipment, repair or replacement of piston rings, pistons, connecting rods and rod bearings is possible with the engine in the vehicle. However, this practice is not recommended because of the cleaning and preparation work that must be done to the components involved.

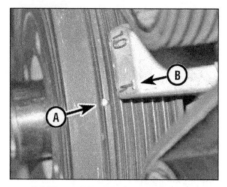

3.1 The notched timing mark on the crankshaft pulley (A) aligns with the "T" mark of the indicator on the timing cover (B) when TDC for number one piston is reached (2014 and earlier models. 2015 and later models have only a single alignment notch on the timing cover, instead of a "T" mark)

3 Top Dead Center (TDC) for number one piston - locating

1   Top Dead Center (TDC) is the highest point in the cylinder that each piston reaches as it travels up-and-down when the crankshaft turns. Each piston reaches TDC on the compression stroke and again on the exhaust stroke, but TDC generally refers to piston position on the compression stroke. When the notched timing mark on the crankshaft pulley is aligned with the proper mark of the indicator on the timing cover, number one piston is at TDC **(see illustration)**.
2   Positioning a specific piston at TDC is an essential part of many procedures such as camshaft(s) removal, timing chain and sprocket replacement.
3   Remove all of the spark plugs, as this will make it easier to rotate the engine by hand.
4   Insert a compression gauge (screw-in type with a hose) in the number 1 spark plug hole. Place the gauge dial where you can see it while turning the crankshaft pulley bolt.
Note: *The number one cylinder is located at the front (timing chain end) of the engine.*
5   Turn the crankshaft clockwise with a socket and large breaker bar until you see compression building up on the gauge, indi-

4.8 Install new inner and outer gaskets on the valve cover

cating that you are on the compression stroke for that cylinder. If you did not see compression build up, continue with one more complete revolution to achieve TDC for the number one cylinder.
6   Continue turning the crankshaft until the notch in the crankshaft pulley is aligned with the mark on the timing chain cover indicator **(see illustration 3.1)**. This is TDC compression for cylinder number one.
7   After the number one piston has been positioned at TDC on the compression stroke, TDC for any of the remaining cylinders can be located by turning the crankshaft 180-degrees (1/2-turn) at a time and following the firing order (see this Chapter's Specifications).

4 Valve cover - removal and installation

Removal

1   Disconnect the negative battery cable from the remote ground terminal or battery (see Chapter 5).
2   Remove the engine cover by pulling the cover up and off of the ballstuds.
3   Remove the ignition coils (see Chapter 5).
4   Clearly label then disconnect any electrical wiring harnesses which connect to, or cross over, the valve cover.
5   Disconnect the PCV valve hose and breather hose from the valve cover.
6   On 2014 and earlier models, remove the ballstuds for the engine cover from the valve cover stud bolts **(see illustration)**. Remove the valve cover bolts in the reverse order of the tightening sequence **(see illustrations 4.10a and 4.10b)**, clean any debris from the cover, then lift off the cover. If the cover sticks to the cylinder head, tap on it with a soft-face hammer or place a wood block against the cover and tap on the wood with a hammer.
Caution: *If you have to pry between the valve cover and the cylinder head, be extremely careful not to gouge or nick the gasket surfaces of either part. A leak could develop after reassembly.*
7   Remove the valve cover perimeter rubber seal and spark plug tube seal. Thoroughly

clean the valve cover and remove all traces of old gasket or sealant material. Gasket removal solvents are available from auto parts stores and may prove helpful. After cleaning the surfaces, degrease them with a rag saturated in brake cleaner.

Installation

8   Install new gaskets into the channels on the cover **(see illustration)**.
9   Before installing the valve cover, apply a small amount of RTV sealant to the seams where the cylinder head meets the timing chain cover. On 2015 and later models, sealant must be applied in a few more spots: on the seams where the Variable Valve Actuator Assembly (VVAA) meets the cylinder head, and at the bases of the rear camshaft bearing cap. Once the sealant has been applied at the indicated locations, carefully place the valve cover onto the cylinder head and install the bolts within 10 minutes. Be careful not to smudge the sealant when positioning the cover.
10   Tighten the bolts, in the proper sequence **(see illustration)**, to the torque listed in this Chapter's Specifications.
11   The remainder of installation is the reverse of removal. Run the engine and check for oil leaks.

5 Intake manifold - removal and installation

Note: *On 2014 and earlier models, the intake manifold is mounted to the front side of the cylinder head. On 2015 and later models it is mounted at the rear.*

Removal

1   Relieve the fuel system pressure (see Chapter 4), then disconnect the negative battery cable from the battery or remote ground terminal (see Chapter 5).

2014 and earlier models

2   Remove the engine cover by pulling the cover up and off of the ballstuds.
3   Remove the air inlet resonator (see Chapter 4).
4   Remove the ignition coil pack from the valve cover (see Chapter 5).

4.6 Remove the ballstuds from the valve cover stud bolts (2014 and earlier models) - the stud bolts/ballstuds on 2015 and later models are one piece

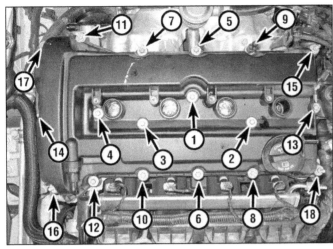

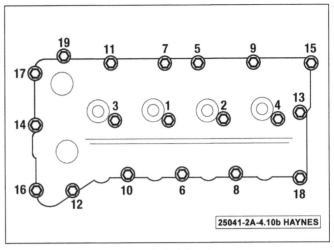

4.10a Valve cover bolt tightening sequence (2014 and earlier models)

4.10b Valve cover bolt tightening sequence (2015 and later)

5   Remove the fuel rail, fuel pressure regulator and fuel injectors as a single assembly (see Chapter 4).

6   Disconnect the oil temperature sensor (see Chapter 6).

7   Unplug the electrical connectors from the Variable Valve Timing (VVT) solenoid, the Manifold Absolute Pressure (MAP) sensor and the intake Camshaft Position (CMP) sensor (see Chapter 6), and move the electrical harness out of the way.

8   Disconnect the electrical connector to the throttle body. Remove the fasteners from the transaxle-to-throttle body support bracket at the throttle body end of the bracket, and loosen - but don't remove - the fastener at the transaxle end of the bracket.

9   Remove the wiring harness retainer from the intake manifold and move the harness out of the way.

10   Label then disconnect the vacuum lines to the manifold.

11   If you're planning to replace or service the intake manifold, remove the throttle body (see Chapter 4). If you're simply removing the intake manifold to remove or service the cylinder head, it's not necessary to remove the throttle body from the intake manifold.

12   Remove the intake manifold fasteners and washers **(see illustration 5.29)**, then remove the intake manifold and the manifold gasket **(see illustration)**.

2015 and later models

13   Raise the front of the vehicle and support it securely on jackstands.

14   Remove the front suspension subframe (see Chapter 10).

15   Remove the right driveaxle (see Chapter 8).

16   Unscrew the bolt/nuts and remove the intake manifold lower bracket.

17   Remove the intake tube from the top of the engine (see Chapter 4).

18   Remove the intake manifold upper support bracket bolts and only loosen the nuts to allow for a slight amount of movement. The bracket doesn't have to be completely removed from the engine.

19   Remove the throttle body bracket and throttle body (see Chapter 4).

20   Using a razor blade, cut through the fuel injector foam silencing pad at the marked score line and remove the silencing pad.

21   Disconnect the VVAA solenoid electrical connectors (see Chapter 6).

22   Remove the four intake manifold silencer cover screws, then remove the cover.

23   Disconnect the fuel injector electrical connectors (see Chapter 4).

24   Disconnect the various intake manifold vacuum hoses, labeling them as necessary for correct installation.

25   Disconnect the Manifold Air Pressure

(MAP) sensor electrical connector (see Chapter 6). Release the wiring harness clips from the intake manifold and position the wiring harness aside.

26   Remove the intake manifold bolts and nuts, pull the manifold off of the studs, then lower and remove it from beneath the engine. Replace the manifold gaskets with new ones.

Inspection

27   Using a straightedge and feeler gauge, check the intake manifold mating surface for warpage. Check the intake manifold surface on the cylinder head also. If the warpage on either surface exceeds the limit listed in this Chapter's Specifications, the intake manifold and/or the cylinder head must be resurfaced at an automotive machine shop or, if the warpage is too excessive for resurfacing, replaced.

Installation

Note: *Clean all gasket surfaces before installing the intake manifold. On 2015 and later models, position the foam pad evenly over the coolant transfer tube before the manifold is installed.*

28   Using a new manifold gasket(s), install the intake manifold onto the manifold studs **(see illustration)**.

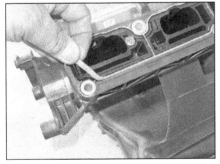

5.12 Remove the intake manifold gasket from the manifold

5.28 Slide the manifold onto the studs (2014 and earlier models shown, 2015 and later models similar)

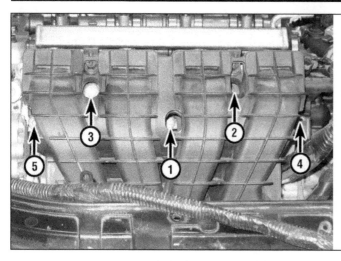

5.29 Intake manifold fastener tightening sequence (2014 and earlier models shown, 2015 and later models similar)

29 Tighten the intake manifold fasteners gradually and evenly, in the indicated sequence **(see illustrations)**, to the torque listed in this Chapter's Specifications.
30 Installation is otherwise the reverse of removal.

---

6   Exhaust manifold - removal, inspection and installation

Warning: *Allow the engine to cool completely before beginning this procedure.*
Note: *On 2014 and earlier models, the exhaust manifold is mounted to the rear side of the cylinder head. On 2015 and later models it is mounted at the front.*

Removal

2014 and earlier models

1 Raise the vehicle and place it securely on jackstands.
2 Disconnect the exhaust pipe from the exhaust manifold (see Chapter 6).

Note: *On some models, it might be necessary to remove the exhaust system to create sufficient clearance to remove the manifold.*
3 Remove the exhaust manifold heat shield mounting bolts and shield. Also remove the manifold brace.
4 Unplug the electrical connector for the oxygen sensor, then remove the oxygen sensor from the exhaust manifold (see Chapter 6).
5 Unscrew the mounting bolts **(see illustrations)** and remove the exhaust manifold and the old manifold gasket.

2015 and later models

6 Remove the engine cover and disconnect the cable from the negative battery terminal (see Chapter 5).
7 Disconnect the electrical connectors from both oxygen sensors (see Chapter 6).
8 Raise the vehicle and support it securely on jackstands, then remove the engine under cover.
9 Remove the catalytic converter-to-exhaust manifold nuts. When separated, discard the gasket and replace it with a new one.

10 Lower the vehicle.
11 Remove the exhaust manifold heat shield, which is secured by four bolts.
12 Remove the exhaust manifold mounting bolts **(see illustrations 6.5a and 6.5b)**. Guide the manifold upward and out of the engine compartment.

Inspection

13 Inspect the exhaust manifold for cracks and any other obvious damage. If the manifold is cracked or damaged in any way, replace it.
14 Using a wire brush, clean up the threads of the exhaust manifold bolts and inspect the threads for damage. Replace any bolts that have thread damage.
15 Using a scraper, remove all traces of gasket material from the mating surfaces and inspect them for wear and cracks.
Caution: *When removing gasket material from any surface, especially aluminum, be very careful not to scratch or gouge the gasket surface. Any damage to the surface may result in a leak after reassembly. Gasket removal solvents are available from auto parts stores and may prove helpful.*
16 Using a straightedge and feeler gauge, inspect the exhaust manifold mating surface for warpage. Check the exhaust manifold surface on the cylinder head also. If the warpage on any surface exceeds the limits listed in this Chapter's Specifications, the exhaust manifold and/or cylinder head must be replaced or resurfaced at an automotive machine shop.

Installation

17 Using a new exhaust manifold gasket (and NO sealant), install the exhaust manifold and tighten the mounting bolts, a little at a time and working from the center outwards, to the torque listed in this Chapter's Specifications.
18 The remainder of installation is the reverse of removal.

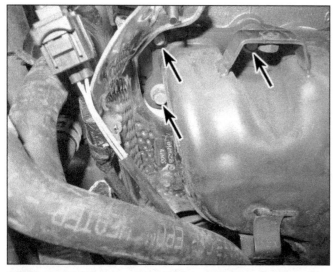

6.5a To detach the exhaust manifold, remove these fasteners from the left end. . .

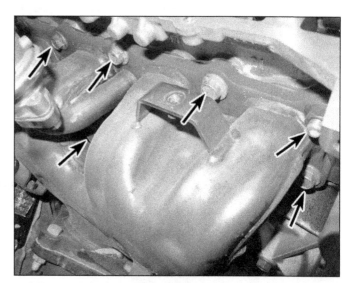

6.5b. . . and these fasteners from the center and the right-end

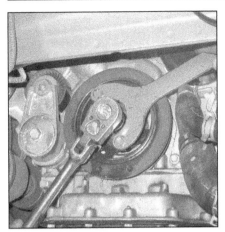

7.5 Using a special holding tool to prevent the crankshaft from turning, loosen, then remove the bolt

7.6 Slide the pulley from the end of the crankshaft; a puller shouldn't be required

8.2 Use a hook tool and pry the seal from the timing cover

7 Crankshaft pulley - removal and installation

Removal

1 Disconnect the negative battery cable from the remote ground terminal or battery (see Chapter 5).
2 Loosen the lug nuts on the right front wheel, raise the vehicle, and support it securely on jackstands.
3 Remove the right front wheel.
4 Remove the drivebelt (see Chapter 1).
5 The crankshaft pulley bolt is incredibly tight; using a breaker bar, socket and special tool #9707 or equivalent **(see illustration)**, hold the balancer from turning while loosening the bolt.
6 Pull the crankshaft pulley off the crankshaft **(see illustration)**.

Installation

7 Apply clean engine oil or multi-purpose grease to the seal contact surface of the pulley hub (if it isn't lubricated, the seal lip could

be damaged and oil leakage would result).
8 Install the crankshaft pulley, aligning the keyway on the crankshaft with the slot in the pulley hub. Install the bolt and tighten it by hand.
9 Prevent the engine from rotating (see Step 5) then tighten the bolt to the torque listed in this Chapter's Specifications.
10 The remainder of installation is the reverse of removal.

8 Crankshaft front oil seal - replacement

1 Remove the crankshaft balancer (see Section 7).
2 Use a screwdriver or hook tool to carefully pry out the seal **(see illustration)**.
Note: *Be careful not to damage the oil pump cover bore where the seal is seated or the nose and sealing surface of the crankshaft.*
3 Another procedure for removing the seal is to drill a small hole on each side of the seal and place a self-tapping screw in each hole **(see illustration)**. Use these screws as a

means of pulling the seal out without having to pry on it.
4 If the seal is being replaced when the timing chain cover is removed, support the cover on top of two blocks of wood and drive the seal out from the backside with a hammer and punch.
Caution: *Be careful not to scratch, gouge or distort the area that the seal fits into or a leak will develop.*
5 Apply clean engine oil or multi-purpose grease to the outer edge of the new seal, then install it in the cover with the lip (spring side) facing IN. Drive the seal into place with a large socket and a hammer **(see illustration)**. Make sure the seal enters the bore squarely and stop when the front face is at the proper depth.
Note: *If a large socket isn't available, a piece of pipe will also work.*
6 Check the surface on the balancer hub that the oil seal rides on. If the surface has been grooved from long-time contact with the seal, the balancer will need to be replaced.
7 Lubricate the balancer hub with clean engine oil and install the crankshaft balancer (see Section 7).
8 The remainder of installation is the reverse of the removal.

9 Timing chain cover, chain and sprockets - removal, inspection and installation

Warning: *Wait until the engine is completely cool before beginning this procedure.*
Caution: *The timing system is complex, and severe engine damage will occur if you make any mistakes. Do not attempt this procedure unless you are highly experienced with this type of repair. If you are at all unsure of your abilities, be sure to consult an expert. Double-check all your work and be sure everything is correct before you attempt to start the engine.*
Caution: *Do not rotate the crankshaft or cam-shafts separately during this procedure (with*

8.3 Another way of removing an old oil seal is to screw a self-tapping screw partially into the seal, then use pliers as a lever to pull it from the engine

8.5 Drive the seal squarely into the cover using a socket and hammer

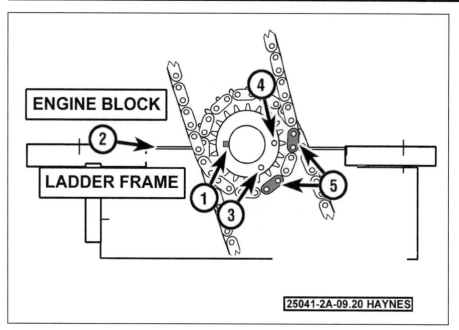

9.20a Align the keyway of the crankshaft to the 9 o'clock position; at this position, the keyway will be pointing to the line where the engine block and ladder frame meet (2004 and earlier models)

1   *Crankshaft keyway at the 9 o'clock position*
2   *Line made between the engine block and ladder frame*
3   *Early production timing mark*
4   *Late production timing mark*
5   *Timing chain plated link - there will only be one plated link on the chain - illustration shows both approximate link locations*

9.20b Align the keyway of the crankshaft to the 9 o'clock position; at this position, the keyway will be pointing to the line where the engine block and ladder frame meet (2015 and later)

1   *Crankshaft keyway at the 9 o'clock position*
2   *Crankshaft sprocket timing mark*
3   *Timing chain plated link*

the timing chains removed), as damage to the valves may occur.
Note: *Several special tools are required to complete these procedures, so read through the entire Section and obtain the special tools before beginning work.*

Removal

Timing chain cover

1   Relieve the fuel system pressure (see Chapter 4), then disconnect the negative battery cable from the battery terminal or remote ground terminal (see Chapter 5).
2   Loosen the lug nuts on the right front wheel, raise the front of the vehicle and support it securely on jackstands.
3   Drain the engine coolant (see Chapter 1).
4   Drain the engine oil (see Chapter 1).
5   Remove the right-front wheel and drivebelt splash shield (see Chapter 1).
6   Remove the engine cover by pulling the cover up and off of the ballstuds. On 2015 and later models, also remove the intake tube and air filter housing (see Chapter 4), then remove and position aside the coolant expansion tank (see Chapter 3). On 2014 and earlier models, remove the coolant reservoir (see Chapter 3).
7   Remove the drivebelt, drivebelt tensioner and idler pulleys (see Chapter 1).

8   Position the number one piston at Top Dead Center (see Section 3).
9   On 2014 and earlier models, remove the power steering fluid reservoir (see Chapter 10) without disconnecting the lines and secure it out of the way.
10   On 2014 and earlier models, remove the power steering hose bracket bolt from the front mount. Remove the power steering pump mounting bolts and secure the pump out of the way without disconnecting the hoses (see Chapter 10).
11   Remove the crankshaft pulley (see Section 7).
12   Remove the ignition coils (see Chapter 5).
13   Remove the valve cover (see Section 4).
Caution: *Once the valve covers are removed, the magnetic timing wheels are exposed. The magnetic timing wheels on the camshafts must not come in contact with any type of magnet or magnetic field. If contact is made, the timing wheels will need to be replaced.*
14   Remove the water pump pulley, and on 2014 and earlier models, remove the air conditioning compressor and bracket (see Chapter 3).
Warning: *Only reposition the compressor out of the way of the timing chain cover, supporting it with a length of wire - don't disconnect the refrigerant lines.*
15   On 2014 and earlier models, remove

the oil pan-to-timing chain cover lower bolts, but don't remove the other oil pan bolts. On 2015 and later models, remove all the oil pan bolts except for the bolts at the opposite end of the timing chain cover, which should only be loosened a few turns. Once the bolts are removed/loosened, the pan must be briefly tilted from the rear and separated off the timing chain cover to fully break the oil pan-to-timing chain cover seal. Once the seal is broken, reinstall the pan bolts except the pan-to-cover bolts.
16   Support the engine with a floor jack and block of wood, slightly raise the engine, then remove the right-side engine mount (see Section 17).
17   Remove the right-side engine mount bracket bolts and bracket from the timing chain cover.
18   Remove the timing chain cover mounting bolts, noting the locations of the mounting bolts; there are several different types of bolts used that must be installed in their original locations. There are three indented prying points, one upper and one lower on the right side, and one lower left pry point; carefully pry the cover free of the engine block and cylinder head and remove the wcover from the bottom of the vehicle. If it still sticks, slip a putty knife between the engine block and cover to break the bond (but be careful not to scratch the surfaces).

**9.21a Camshaft phaser timing mark locations
(2014 and earlier models)**

1   *Timing marks on the camshaft sprockets facing each other and parallel with the cylinder head mating surface*
2   *Intake camshaft phaser timing mark*
3   *Exhaust camshaft phaser timing mark*
4   *Plated timing chain links*

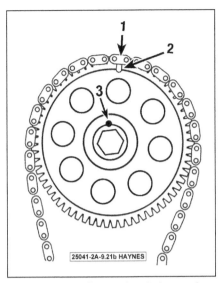

**9.21b Camshaft sprocket timing marks
(2015 and later)**

1   *Timing chain plated link*
2   *Mark on camshaft sprocket*
3   *Camshaft dowel/notch*

19   Once the cover is removed, RTV gasket material must be thoroughly cleaned off from the cylinder head, engine block and back side of the timing chain cover.

Timing chain

Warning: *When the timing chains are removed, do not rotate the camshafts or crankshaft; the valves and pistons can be damaged if contact is made.*

20   If the engine has moved from TDC, temporarily install the crankshaft pulley bolt. Turn the crankshaft with the bolt to TDC number 1 to align the timing marks on the crankshaft and camshaft sprockets. Rotate the engine clockwise only, until the crankshaft keyway aligns with the line where the engine block and ladder frame meets **(see illustrations)**.

Note: *On 2014 and earlier engines, there are early and late production runs that have different timing marks on the crankshaft sprocket. On the early production run, the crankshaft sprocket timing mark is between the 5 and 6 o'clock positions. On the later run, the sprocket timing mark is closer to the 3 o'clock position. On both production runs, the crankshaft keyway will always be aligned at the 9 o'clock position, pointing to the line made where the engine block and ladder frame meet.*

Note: *If the timing chain plated links are faded or can no longer be seen, mark the links to the corresponding timing marks before removing the chain (if you plan to reuse the old chain).*

21   2014 and earlier models: The timing chain plated links should be aligned with the camshaft phaser timing marks. On the intake (front) side camshaft phaser, the plated link should be aligned with the machined dot, which should be facing away from the opposite camshaft phaser dot - about the 1 o'clock position. The machined scribe line

should be pointing towards the other camshaft phasers machined line, in a parallel line with the gasket surface of the cylinder head - about the 9 o'clock position. On the exhaust (rear) side camshaft phaser, the timing chain plated link should be aligned with the machined dot, which should be facing away from each other phaser mark - about the 11 o'clock position. The machined scribe line should be pointing towards the phaser scribe line, in a parallel line with the gasket surface of the cylinder head - at the 3 o'clock position **(see illustration)**.

Note: *On 2014 and earlier models, each of the camshaft phasers have two timing marks - a machined dot and machined scribe line. When they are properly timed to No.1 cylinder (TDC) the dots are facing away from each other and the lines are pointing towards each other in a parallel line. On 2015 and later models, the camshaft dowel at the hub of the sprocket should be facing 90-degrees upwards from the horizontal cylinder head surface.*

22   Verify the phaser or sprocket marks are aligned with the plated links **(see illustrations 9.21a or 9.21b)**; if the plated links cannot be distinguished but the camshaft mark(s) are in the proper position(s), mark the plated links to the corresponding camshaft timing marks before removing the chain or the chain will need to be replaced with a new one that has the identifying plated links.

Note: *Use paint or a permanent marker to mark the direction of rotation on all chains before removing them so they can be installed in the same direction.*

23   On 2014, and earlier engines, remove the timing chain tensioner mounting bolts and remove the tensioner and guide from the left side of the timing chain; the tensioner will not come apart when it is removed.

24   On 2014 and earlier engines, remove the timing chain from the camshaft phasers and crankshaft sprocket.

25   On 2015 and later engines, remove the timing chain tensioner.

26   On 2015 and later engines, prevent the camshaft from rotating with a wrench placed on the wrench engagement flats built into the camshaft. While holding the camshaft in place, remove the camshaft sprocket bolt, then slide off the camshaft sprocket and remove the timing chain.

27   If necessary, remove the oil pump/balance shaft chain (see Section 14), then remove the crankshaft gear (see Section 8).

28   If necessary, remove the chain guide fasteners and guide.

Inspection

29   Inspect the timing chain dampener (guide) for cracks and wear and replace it, if necessary.

30   Clean the timing chain and sprockets with solvent and dry them with compressed air (if available).

Warning: *Wear eye protection when using compressed air.*

31   Inspect the components for wear and damage. Look for teeth that are deformed, chipped, pitted, and cracked.

32   The timing chain and sprockets should be replaced with new ones if the engine has high mileage, the chain has visible damage, or total freeplay midway between the sprockets exceeds one inch. Failure to replace a worn timing chain and sprockets may result in erratic engine performance, loss of power, and decreased fuel mileage. Loose chains can jump timing. In the worst case, chain jumping or breakage will result in severe engine damage.

Installation

Caution: *Before starting the engine, carefully rotate the crankshaft by hand through at least two full revolutions (use a socket and breaker bar on the crankshaft pulley center bolt). If you feel any resistance, STOP! There is something wrong - most likely, valves are contacting the pistons. You must find the problem before proceeding. Check your work and see if any updated repair information is available.*

33   Use a plastic gasket scraper to remove all traces of old gasket material and sealant from the cover, engine block and cylinder heads. The components are all made of aluminum, so be careful not to nick or gouge them. Only clean the gasket sealing surfaces with rubbing alcohol (isopropyl) - do not use any oil based fluids.

34   If removed, install the crankshaft sprocket and oil pump/balance shaft module chain (see Section 14).

35   If removed, install the right side chain guide.

36   Make sure the keyway is installed on the crankshaft and is at TDC, with the keyway pointing towards the 9 o'clock position in line with the line made where the engine block and ladder frame meet **(see illustrations 9.20a or 9.20.b)**.

37   Verify the camshaft phasers or sprockets are at TDC **(see illustrations 9.21a or 9.21b)**.

38   Using clean engine oil, coat the sprockets and chain.

39   On 2014 and earlier models, place the chain onto the camshaft sprockets, aligning the plated links with the machined dots on the phasers. Once aligned, loop the chain down and around the crankshaft sprocket, aligning the plated link with the timing mark on the crankshaft sprocket **(see illustration 9.20a)**. Double-check the alignment of the plate link marks on the camshaft sprockets.

40   On 2015 and later models, pre-align the chain with the corresponding mark on the camshaft sprocket, then lower the chain and sprocket and loop the chain around the crankshaft sprocket. Be sure the crankshaft sprocket marks and plated link are aligned properly **(see illustration 9.20b)**. Install the camshaft sprocket (with the chain in place) onto the camshaft. Make sure the dowel, sprocket and plate link marks are aligned. While holding the camshaft with a wrench on the flats, tighten the camshaft sprocket bolt to the torque listed in this Chapter's Specifications.

41   While holding the timing chain tensioner with light pressure against the plunger, use a pick to lift up on the plunger ratchet through the front hole until the plunger can be pressed in and special tool #8514 or a 3 mm Allen wrench can be inserted through the rear hole of the tensioner body, holding the plunger in the compressed position.

42   Install the chain guide and tensioner, then tighten the fasteners to the torque listed in this Chapter's Specifications. Remove the special tool from the tensioner plunger.

43   Temporarily reinstall the crankshaft pulley and rotate the engine two complete turns, with the line made where the engine block and ladder frame meet as the reference point. Verify all the marks line up **(see illustrations 9.20 and 9.21)**; if the marks are off, rotate the engine two more complete turns and check again.

44   Once the timing marks are correct, apply a continuous bead of RTV sealant along the left- and right-side timing chain cover sealing surfaces (engine side). Also apply sealant around the bolt contact surfaces that are located inside of the cover.

45   On 2014 and earlier engines, apply a 1/8-inch wide by 1/16-inch high bead of RTV sealant to the sealing surface of the oil pan, making sure to make a complete circle around each bolt hole on the pan, including the corners where the timing cover meets the oil pan and cylinder block. On 2015 and later models, remove the oil pan, then reinstall it using RTV sealant (see Section 13) after the timing chain cover has been installed.

46   Working from under the vehicle, install the cover from the bottom towards the top.

47   On 2014 and earlier engines, install the timing cover bolts and tighten them in a criss-cross pattern, in three steps, to the torque listed in this Chapter's Specifications. On 2015 and later models, install the timing chain cover bolts in their proper locations **(see illustration)** and tighten them to the torque values listed in this Chapter's Specifications.

48   The remainder of installation is the reverse of removal.

49   Add oil and coolant (see Chapter 1), start the engine and check for leaks.

---

10   Camshaft(s) – removal, inspection and installation

Warning: *Wait until the engine is completely cool before beginning this procedure.*

Caution: *The timing system is complex, and severe engine damage will occur if you make any mistakes. Do not attempt this procedure unless you are highly experienced with this type of repair. If you are at all unsure of your abilities, be sure to consult an expert. Double-check all your work and be sure everything is correct before you attempt to start the engine.*

Note: *The timing chain can only be removed from the camshafts, using the tools outlined in this Section. If the tools are not available, the timing chain cover will need to be removed before the camshafts can be removed (see Section 9).*

Removal

1   Remove the engine cover by pulling the cover up and off of the ballstuds.

2   Disconnect the negative battery cable from the remote ground terminal or battery (see Chapter 5).

3   Loosen the lug nuts on the right front wheel, raise the front of the vehicle and support it securely on jackstands. Remove the

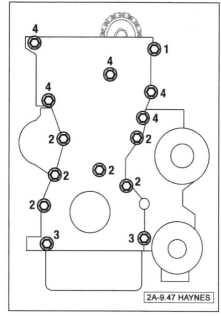

9.47 Timing chain cover bolts (2015 and later models)

1   *Single stud bolt*
2   *M6 bolts*
3   *M8 bolts*
4   *M10 bolts*

right front wheel and the drivebelt splash shield.

4   Drain the engine oil and coolant, then remove the drivebelt (see Chapter 1).

5   Remove the air filter housing (see Chapter 4) and resonator.

6   Disconnect the ignition coils and remove the spark plugs (see Chapter 1). Label the ignition coils to simplify reinstallation.

7   Remove the valve cover (see Section 4).

8   Rotate the crankshaft clockwise and place the no. 1 piston at TDC (see Section 3).

2014 and earlier models

9   Verify the timing marks are in alignment (see Section 9), then use a permanent marker or paint to mark the camshaft phasers to the timing chains for reinstallation.

10   Locate, then remove the timing chain tensioner plug from the timing chain cover (when looking straight at the crankshaft pulley, the plug is located at approximately the 10 o'clock position).

11   With the plug removed you should be able to see the tensioner. Working through the small hole in the side of the tensioner, use a pick to lift up on the plunger ratchet through the front hole then insert special tool #8514 or a 3 mm Allen wrench through the rear hole of the tensioner body, holding the plunger in this position for the reminder of the repair.

12   Insert the special locking tool #9701 between the two camshaft phasers lightly tapping the tool down until it is locked between the chain and the cover.

Caution: *The chain holding tool must remain in*

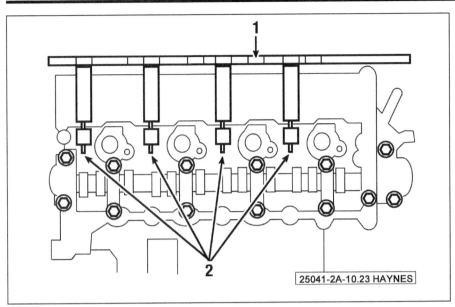

10.23 Install the special tool (1) onto the cylinder head, then tighten the compressor assemblies (2) until the valve springs are fully compressed (2015 and later models)

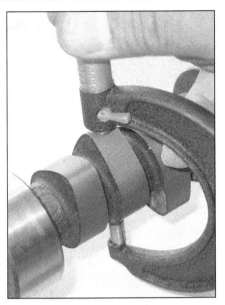

10.28 Use a micrometer to measure cam lobe height

place while the phasers are removed or the timing chain will fall off into the timing chain cover.

Note: *The camshaft bearing caps are marked with a number and letter code; " 1I " is for the number one intake camshaft bearing cap. The notch on the caps should always be installed towards the front.*

13   Loosen the camshaft bearing cap bolts in the reverse order of the tightening sequence **(see illustration 10.41)**.

14   Remove the front (or joined) camshaft bearing cap and upper half of the bearing insert from the exhaust side of the cap first, then carefully remove the remaining camshaft bearing caps.

Note: *Only the front (or joined) bearing cap has bearing inserts on the exhaust side of the cap.*

15   Carefully remove the intake camshaft first by lifting the rear of the intake camshaft upwards from the journals while slightly rotating the camshaft clockwise then lift the chain off of the phaser sprocket.

16   Once the intake camshaft is removed, lift the chain up and remove the exhaust camshaft, then secure the chain from falling into the timing chain cover.

17   Remove the lower half of the bearing from the exhaust camshaft journal.

18   Using a large wrench on the camshaft flats and a socket and ratchet on the phaser sprocket bolt, loosen, then remove the bolt and phaser from the end of the camshaft.

Caution: *Do not remove the phaser lock or try to disassemble the phasers.*

19   Mark the lifters so they can be installed in the same locations, then remove them from the cylinder head.

2015 and later models

Caution: *This procedure requires the use of Mopar special tool #10259B - attempting to perform this procedure without the use of this*

special tool will cause damage to the valves, cylinder head or the VVAA system.

20   Remove the brake vacuum pump (see Chapter 9).

21   Remove the timing chain cover and camshaft sprocket/timing chain (see Section 9).

22   Prepare the special valve spring compressor tool 10259B - pre-position the compressor assemblies and stops at the "2.4L" markings on the bracket.

Note: *Remove the first four outer VVAA bolts (counting from the timing chain end of the engine) before installing the special tool (when the tool is installed you won't be able to access these bolts).*

23   Install the special tool onto the cylinder head, adjusting the compressor assemblies to align over the valve springs. Tighten the compressor assemblies until all the valve springs are fully compressed (see illustration).

24   Place match marks on the camshaft bearing caps to identify them properly for location and direction when installing.

25   Remove the camshaft bearing caps, loosening the bolts one turn at a time, in the reverse order of the tightening sequence **(see illustration 10.53)**.

26   Carefully remove the camshaft from the engine.

Inspection

27   Check the camshaft bearing surfaces for pitting, score marks, galling, and abnormal wear. If the bearing surfaces are damaged, the cylinder head will have to be replaced.

28   Compare the camshaft lobe height by measuring each lobe with a micrometer **(see illustration)**. Measure each of the intake lobes and record the measurements and relative positions. Then measure each of the exhaust lobes and record the measurements and relative positions also. This will

let you compare all of the intake lobes to one another and all of the exhaust lobes to one another. If the difference between the lobes exceeds 0.005 inch, the camshaft should be replaced. Do not compare intake lobe heights to exhaust lobe heights as lobe lift may be different. Only compare intake lobes to intake lobes and exhaust lobes to exhaust lobes for this comparison.

29   Check the lifters for abnormal wear, pits, galling, score marks, and rough spots. Replace defective parts.

Installation

Caution: *Before starting the engine, carefully rotate the crankshaft by hand through at least two full revolutions (use a socket and breaker bar on the crankshaft pulley center bolt). If you feel any resistance, STOP! There is something wrong - most likely, valves are contacting the pistons. You must find the problem before proceeding. Check your work and see if any updated repair information is available.*

2014 and earlier models

30   Carefully slide both the intake or exhaust phaser onto the camshafts and verify the marks are aligned. Install the bolt, then tighten the bolt to the torque listed in this Chapter's Specifications.

31   Make sure to prevent the camshafts from turning by holding the camshaft with a large wrench on the camshaft flats.

32   Dip the lifters in clean engine oil and install them into their original locations.

33   Lubricate the camshaft bearing journals and lobes with moly-base grease or engine assembly lube.

34   The front (or joined) cap has a number stamped into the cap. The stamped numbers 1, 2 or 3 correspond to the exhaust bearing insert needed to be used. If replacing the

10.39 Location of the exhaust bearing insert number location (A) and installation direction (B) of the front cap

10.40a Install the exhaust camshaft bearing caps in the correct order (A) and direction (B) . . .

10.40b . . . then install the intake camshaft bearing caps in the correct order (A) and direction (B)

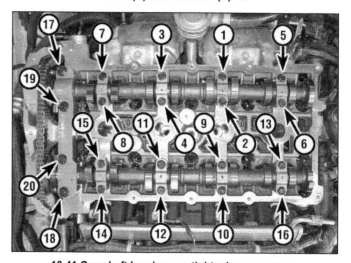

10.41 Camshaft bearing cap tightening sequence - bolts 1 through 16 first, then bolts 17 through 20 (2014 and earlier models)

bearing select the proper bearing and insert the lower bearing insert into the cylinder head.

35   Starting with the exhaust camshaft, set the timing chain over the exhaust phaser sprocket making sure the plate link or mark is aligned with the dot on the phaser sprocket.

36   Set the exhaust camshaft onto the cylinder head making sure to the machined scribe line on the phaser is in a parallel line with the gasket surface of the cylinder head at the 3 o'clock position **(see illustration 9.21)**.

37   Install the intake camshaft at an angle with the rear of the camshaft upward then roll the sprocket into the timing chain making sure the plate link or mark is aligned with the dot on the phaser sprocket.

38   Set the intake camshaft onto the cylinder head making sure to the machined scribe line on the phaser is in a parallel line with the gasket surface of the cylinder head at the 9 o'clock position **(see illustration 9.21)**.

39   The front (or joined) cap has a number stamped into the cap **(see illustration)**. The stamped numbers 1, 2 or 3 correspond to the exhaust bearing insert required. If replacing

the bearing, select the proper bearing and install the lower bearing insert into the cylinder head before the camshafts are installed.

40   Install the camshaft bearing caps **(see illustrations)** and front cap, then install the mounting bolts and finger tighten them.

41   Tighten bearing cap bolts 1 through 16 in sequence first, then bolts 17 through 20 in sequence **(see illustration)** to the torque listed in this Chapter's Specifications.

Caution: *Prevent the camshafts from turning by holding the camshaft with a large wrench on the camshaft flats.*

42   Verify all the timing marks are aligned **(see illustrations 9.20 and 9.21)**.

43   Remove the Allen wrench from the timing chain tensioner.

44   Remove the locking wedge tool from the timing chain cover.

45   Slowly rotate the engine two complete turns (720-degrees) and verify the alignment marks are correct **(see illustrations 9.20 and 9.21)**.

46   Check the valve adjustment (see Chapter 1).

47   Coat the timing chain tensioner plug with thread sealant then install the plug and tighten it securely.

48   The remainder of installation is the reverse of removal.

2015 and later models

49   With special tool 10259A still in position, coat the camshaft bearing surfaces, camshaft caps, and camshaft with clean engine oil.

50   Carefully install the camshaft onto the bearing surface - the dowel pin for the sprocket should be pointing straight up.

Note: If you cannot lay the camshaft easily onto the bearing surfaces because the lifters are contacting it, you must adjust/compress the lifters further inward with the special tool.

51   With the rearmost camshaft cap mating surfaces clean and dry, apply RTV sealant to the cap end where it mates with the cylinder head surface.

52   Lubricate the camshaft bearing journals with clean engine oil, then install them in their correct locations on the camshaft - noting their match marks made prior to removal.

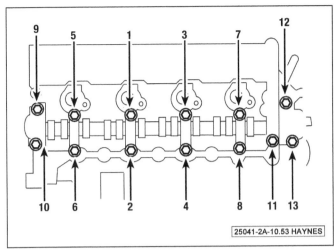

10.53 Camshaft bearing cap tightening sequence
(2015 and later models)

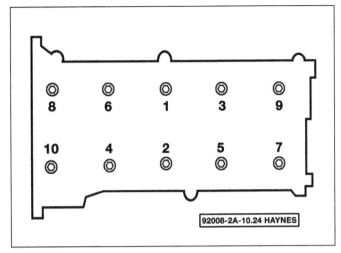

11.29 Cylinder head bolt TIGHTENING sequence

53 Tighten the camshaft bearing cap bolts, a little at a time, in sequence **(see illustration)**, to the torque settings listed in this Chapter's Specifications.
54 Loosen the special tool's valve spring compressor assemblies completely, then remove the special tool from the cylinder head.
55 Install the timing chain and sprocket (see Section 9).
56 The remainder of installation is the reverse of removal. Fill the engine with oil (see Chapter 1).

11 Cylinder head - removal and installation

Warning: *Allow the engine to cool completely before beginning this procedure.*

Removal

Note: *On 2015 and later models, the intake manifold bracket must be removed (along with hoses/connectors) and the throttle body connector must be disconnected.*
1 Loosen the lug nuts on the right front wheel, raise the front of the vehicle and support it securely on jackstands.
2 Remove the right front wheel, then remove the splash shields and inner fender liner (see Chapter 11).
3 Position the number one piston at Top Dead Center (see Section 3).
4 Disconnect the negative battery cable from the remote ground terminal (see Chapter 5).
5 Remove the engine cover by pulling the cover up and off of the ball studs.
6 Drain the cooling system and remove the spark plugs (see Chapter 1).
7 Remove the air filter housing (see Chapter 4).
8 Remove the coolant reservoir (see Chapter 3).
9 Remove the power steering reservoir without disconnecting the lines and secure it out of the way.

10 Remove the drivebelts (see Chapter 1).
11 Remove the intake manifold (see Section 5). Cover the intake ports with duct tape to keep out debris.
12 Remove the power steering hose bracket bolt from the front mount. Remove the power steering pump mounting bolts and secure the pump out of the way without disconnecting the hoses (see Chapter 10).
13 Remove the exhaust manifold (see Section 6).
14 Remove the ignition coils (see Chapter 5) and spark plugs (see Chapter 1).
15 Remove the valve cover (see Section 4).
16 Remove the water pump pulley, the air conditioning compressor and bracket (see Chapter 3).
17 Remove the timing chain cover and timing chain (see Section 9).
18 Remove the camshaft bearing caps, camshafts and lifters from the bores in the cylinder head (see Section 10). Store the lifters so they can be reinstalled in their original locations.
19 On 2015 and later models, remove the VVAA assembly (see Section 18).
Caution: *On 2014 and earlier engines, the lifters must be reinstalled in their original locations or the valve adjustment will be incorrect.*
20 Loosen the cylinder head bolts, 1/4-turn at a time, in the reverse of the tightening sequence **(see illustration 11.29)** until they can be removed by hand from the cylinder head.
Note: *Take notes on the bolt locations so they can be reinstalled in the correct locations.*
Note: *The cylinder head bolt washers are "captured washers," meaning they will stay on the bolts after the bolts have been removed, except for the first lower two cylinder head bolts on the timing chain side. The first two bolts have removable washers; one side is flat and the other side is beveled.*
21 Carefully lift the cylinder head straight up and place the head on wood blocks to prevent damage to the sealing surfaces. If the head sticks to the engine block, dislodge it by placing a wood block against the head casting

and tapping the wood with a hammer, or by prying the head with a prybar placed carefully on a casting protrusion.
Caution: *The cylinder head is aluminum, so you must be very careful not to gouge the sealing surfaces.*
Note: *It's a good idea to have the head checked for warpage, even if you're just replacing the gasket.*
22 Once the cylinder head is removed, use a pair of needle nose pliers to remove the VVT filter from the engine block.
23 Remove all traces of old gasket material from the block and head. Special gasket removal solvents that soften gaskets and make removal much easier are available at auto parts stores. Do not allow anything to fall into the engine. Clean and inspect all threaded fasteners and be sure the threaded holes in the block are clean and dry.

Installation

24 Install a new or cleaned VVT filter into the hole in the right front corner of the cylinder block.
25 Apply two small drops of RTV sealant at the timing chain side ends of the cylinder block face.
26 Place a new gasket and the cylinder head in position on the engine block.
27 Place a straightedge or ruler against the threads of each head bolt; if there is a gap or space between the edges of the threads and the ruler, the head bolt(s) must be replaced. Apply clean engine oil to the cylinder head bolt threads prior to installation.
28 Place the washers, beveled side up, on the first two head bolts.
29 Install the head bolts and tighten them in several stages, in the recommended sequence **(see illustration)**, to the torque listed in this Chapter's Specifications.
Caution: *Two different types of cylinder head bolts have been used on these engines - you must determine which ones your engine has in order to torque them properly. One style has a*

short head (5/16-inch from the washer to the top of the bolt head). The other style has a long bolt head (1/2-inch from the washer to the top of the bolt head).

Note: *The final step in the tightening procedure requires you to tighten the bolts a specific number of degrees. An angle-torque gauge, that fits on your torque wrench, is available at most auto parts stores and is highly recommended for this procedure. If the tool is not available, paint marks on the bolt heads and tighten them in sequence until the mark is the specified number of degrees from the starting point.*

30   Reinstall the timing chain and cover (see Section 9).

31   The remainder of installation is the reverse of removal.

32   Refill the cooling system and change the engine oil and filter (see Chapter 1). Rotate the crankshaft clockwise slowly by hand through six complete revolutions. Recheck the camshaft timing marks (see Section 9).

33   Start the engine and run it until normal operating temperature is reached. Check for leaks and proper operation.

12   Valve clearance check and adjustment

Note: *2015 and later models do not require valve adjustment.*

1   Position the number 1 piston at TDC on the compression stroke (see Section 3).

2   Remove the engine cover by pulling the cover up and off of the ball studs.

3   Disconnect the negative battery cable from the remote ground terminal (see Chapter 5).

4   Blow out the recessed area around the spark plug openings with compressed air, if available, to remove any debris that might fall into the cylinders, then remove the spark plugs (see Chapter 1).

5   Remove the valve cover (see Section 4).

Check

6   The lobes must be measured one set at a time. Rotate the camshaft(s) until two lobes on any one cylinder are pointing straight up and measure the clearances of those valves with feeler gauges **(see illustration)**. Record any measurements that don't fall within this Chapter's Specifications. These measurements will be used later to determine the required replacement lifters.

7   Rotate the camshaft(s) until all the lobes have been checked.

Adjustment

8   Remove the camshaft(s) for the valve(s) that you intend to adjust (see Section 10).

9   Remove and measure each lifter (whose clearance is not correct) with a micrometer **(see illustration)**. Put each lifter back into its bore in the cylinder head before moving on to the next lifter. Record the measurement for each lifter.

12.6 Check the clearance of each valve with a feeler gauge of the specified thickness - if the clearance is correct, you should feel a slight drag on the gauge as you pull it out

10   To calculate the correct thickness of a replacement lifter that will put the valve clearance within the specified range, use the following formula:

S - C = change, where:
S = specified valve clearance (see this Chapter's Specifications)
C = measured valve clearance

11   Remove the lifter and read the size from the bottom of the lifter then reduce the lifter by subtracting the specified valve clearance from the measured valve clearance or as close as possible to the calculated size.

12   Install the camshaft(s) (see Section 10).

13   Check the valve clearances again to verify that they're now within the range listed in this Chapter's Specifications.

14   The remainder of installation is the reverse of removal.

13   Oil pan - removal and installation

Removal

1   Disconnect the negative battery cable from the remote ground terminal or battery (see Chapter 5).

2   Raise the vehicle and support it securely on jackstands.

3   Drain the engine oil (see Chapter 1).

4   Remove the drivebelt splash shield.

5   Remove the air conditioning compressor (see Chapter 3) and set it out of the way, then remove the compressor bracket.

6   Remove the oil pan bolts, then carefully separate the oil pan from the block. Use a putty knife or gasket scraper to loosen the seal around the pan, but don't pry between the block and the pan or damage to the sealing surfaces could occur and oil leaks may develop.

7   Thoroughly clean the oil pan and sealing surfaces on the block and pan. Use a scraper to remove all traces of old gasket material.

12.9 Measure the thickness of each lifter head with a micrometer

Gasket removal solvents are available at auto parts stores and may prove helpful. Check the oil pan sealing surface for distortion. Straighten or replace as necessary, then wipe the gasket surfaces of the pan and block with a rag soaked in brake system cleaner.

Installation

8   Apply a 1/8-inch bead of RTV sealant at the cylinder block-to-front cover joint at the oil pan flange.

9   Apply a 1/8-inch wide by 1/16-inch high bead of RTV sealant to the sealing surface of the pan. Install the pan and the bolts, then tighten the bolts finger-tight.

10   Working side-to-side from the center out, tighten the oil pan bolts to the torque listed in this Chapter's Specifications.

11   The remainder of installation is the reverse of removal.

12   Refill the crankcase with the correct quantity and grade of oil, run the engine and check for leaks.

13   Road test the vehicle and check for leaks again.

14   Oil pump/balance shaft module and chain - removal, inspection and installation

Note: *The oil pump is an integral component of the balance shaft module and can't be removed or disassembled from the module. If there is a problem with the oil pump, the balance shaft module must be replaced.*

Removal

1   Relieve the fuel system pressure (see Chapter 4), then disconnect the negative battery cable from the remote ground terminal or battery (see Chapter 5).

2   Rotate the engine to Top Dead Center (TDC) for #1 cylinder on the compression stroke (see Section 3).

3   Loosen the lug nuts on the right front wheel, raise the vehicle, and support it securely on jackstands.

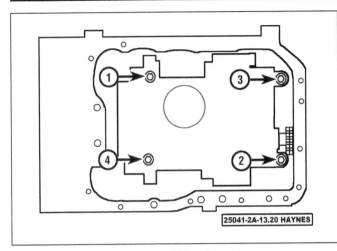

14.20 Balance Shaft Module (BSM) bolt TIGHTENING sequence

15.2 Mark the relative position of the driveplate to the crankshaft and, using an appropriate tool to hold the driveplate, remove the bolts

4   Remove the right front wheel, then remove the splash shield and inner fender shield (see Chapter 11).
5   Remove the valve cover (see Section 4).
6   Remove the oil pan (see Section 13).
7   Remove the timing chain cover, timing chain and crankshaft sprocket (see Section 9).
8   Align the drive chain plated links with the timing marks on the crankshaft sprocket and the Balance Shaft Module (BCM) drive gear.
Note: *If the marks or plated links can't be found, mark the oil pump/Balance Shaft Module (BSM) drive chain to the BSM drive sprocket and to the crankshaft gear, so the chain can be installed in the same position.*
9   Press the oil pump/BSM tensioner piston back into the tensioner body. While holding the piston in, insert special tool #9703 or a 3 mm drill bit into the hole in the side of the tensioner to retain the piston in the locked position.
Caution: *Do not remove the oil pump/BSM drive sprocket.*
10   Remove the BSM mounting bolts, in the reverse of the tightening sequence **(see illustration 14.20)**, in several steps. There are two different length bolts that can be used.
11   Lower the rear of the balance shaft module, remove the timing chain from the sprocket, then remove the BSM.
12   Remove the chain from the crankshaft sprocket.

Inspection

13   If necessary, replace the crankshaft front seal within the oil pump body (see Section 8).
14   Two different lengths of mounting bolts can be used on the BSM: 180 mm and 185 mm length bolts. The 180 mm bolts must be discarded and replaced with new bolts. Measure the BSM mounting bolts and replace the bolts as needed.
15   Place a straightedge or ruler against the threads of each 185 mm mounting bolt; if there is a gap or space between the edges of

the threads and the ruler, the mounting bolt(s) must be replaced. Apply clean engine oil to the mounting bolt threads prior to installation.

Installation

16   Clean the bolt holes for the BSM mounting bolts.
17   Place the timing chain over the crankshaft sprocket and align the plated link with the timing mark on the gear, or marks made prior to removal (see Step 8).
18   Lift up the BSM and place the drive sprocket into the timing chain, aligning the plated link with the timing mark on the drive gear, or marks made prior to removal (see Step 8). Pivot the BSM up into place against the ladder frame on the engine block.
19   While holding the BSM in place, insert the mounting bolts and tighten by hand in several even stages.
20   Tighten the bolts in several stages, in the recommended sequence **(see illustration),** to the torque listed in this Chapter's Specifications.
Note: *The final step in the tightening procedure requires you to tighten the bolts a specific number of degrees. An angle-torque gauge, that fits on your torque wrench, is available at most auto parts stores and is highly recommended for this procedure. If the tool is not available, paint marks on the bolt heads and tighten them in sequence until the mark is the specified number of degrees from the starting point.*
21   Remove the pin from the tensioner and release the piston then verify that the timing mark are aligned.
22   Install the timing chain and timing cover (see Section 9).
23   Install the oil pan (see Section 13).
24   Install the valve cover (see Section 4).
25   The remainder of installation is the reverse of removal.
26   Install a new oil filter and engine oil (see Chapter 1).
27   Start the engine and check for oil pressure and leaks.
28   Recheck the engine oil level.

15   Driveplate - removal and installation

Removal

1   Raise the vehicle and support it securely on jackstands, then remove the transaxle assembly (see Chapter 7A).
2   To ensure correct alignment during reinstallation, mark the position of the driveplate to the crankshaft before removal **(see illustration)**.
3   Remove the bolts that secure the driveplate to the crankshaft **(see illustration)**. A tool is available a most auto parts stores to hold the driveplate while loosening the bolts. If the tool is not available, wedge a screwdriver in the ring gear teeth to jam the driveplate.
4   Remove the driveplate from the crankshaft.
5   Clean and inspect the mating surfaces of the driveplate and the crankshaft. If the crankshaft rear main seal is leaking, replace it before reinstalling the driveplate (see Section 16).

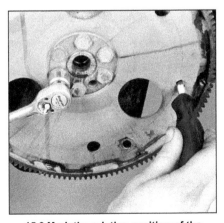

15.3 Mark the relative position of the driveplate to the crankshaft and, using an appropriate tool to hold the driveplate, remove the bolts

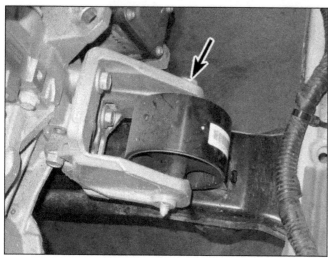

17.10 Front mount through-bolt

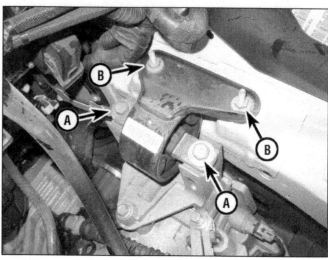

17.18 Remove the mount-to-transaxle bolts (A) then the mount-to-body nuts/bolts (B)

Installation

6   Position the driveplate against the crankshaft. Align the previously applied match marks. Before installing the bolts, apply thread locking compound to the threads.

7   Hold the driveplate with the holding tool, or wedge a screwdriver in the ring gear teeth to keep the driveplate from turning as you tighten the bolts to the torque listed in this Chapter's Specifications.

8   The remainder of installation is the reverse of removal.

---

16   Rear main oil seal - replacement

1   Remove the transaxle (see Chapter 7A).

2   Remove the driveplate (see Section 15).

3   Use a screwdriver wrapped with tape to pry out the seal, being careful not to gouge or nick the housing.

4   Lubricate the crankshaft seal journal and the lip of the new seal with multi-purpose grease.

5   Install the seal with the seal lip toward the engine and the dust seal toward the transaxle.

6   Tap the seal into place using a seal driver to make sure that it doesn't become tilted.

7   Install the seal so that its rear edge is flush with the face of the engine block, or up to 0.020 inch recessed.

8   The remainder of installation is the reverse of removal.

---

17   Engine mounts - check and replacement

1   There are four powertrain mounts: the front mount attaches the transaxle to the crossmember; the left mount attaches the driver's side of the engine to the subframe; the right engine mount attaches the engine

block to the passenger's side of the subframe; and a rear mount attaches the engine block to the subframe.

Check

2   During the check, the engine must be raised slightly to remove the weight from the mounts.

3   Raise the vehicle and support it securely on jackstands. Remove the front wheels and tires. Position two jacks, one under the crankshaft pulley and the other under the transaxle bellhousing. Place a block of wood between a floor jack head and the crankshaft pulley or bellhousing, then carefully raise the engine/transaxle just enough to take the weight off the mounts.

Warning: *DO NOT place any part of your body under the engine when it's supported only by a jack!*

4   Check the mounts to see if the rubber is cracked, hardened or separated from the metal plates. Sometimes the rubber will split right down the center.

5   Check for relative movement between the mount plates and the engine or subframe (use a large screwdriver or prybar to attempt to move the mounts). If movement is noted, lower the engine and tighten the mount fasteners.

6   Rubber preservative should be applied to the mounts to slow deterioration.

Replacement

Front mount

7   Raise the vehicle and support it securely on jackstands.

8   Remove the splash shield fasteners and remove the shield.

9   Place a block of wood between the floor jack head and the oil pan and support the engine.

10   Remove the front mount through-bolt **(see illustration)**.

11   Remove the crossmember-to-mount bolts from under the crossmember.

12   Remove the mount from the crossmember. If the mount bracket needs to be replaced, remove the bracket-to-block bolts and the bracket.

13   Installation is the reverse of removal.

Left mount

14   Disconnect the negative battery cable from the remote ground terminal (see Chapter 5), then raise the vehicle and support it securely on jackstands.

15   Remove the throttle body air inlet tube and the air filter housing (see Chapter 4).

16   Support the transaxle with a floor jack and wooden block placed between the jack and the transaxle.

17   Raise the transaxle enough to take the weight off of the mount.

18   Working in the engine compartment, remove the mount-to-transaxle bolts, then the mount-to-body bolts and the mount **(see illustration)**. Replace the left mount with a new one and tighten the bolts to the torque listed in this Chapter's Specifications.

19   The remainder of installation is the reverse of removal.

Right mount

20   Remove the coolant reservoir (see Chapter 3) and set it out of the way.

21   Remove the power steering line bracket fastener from the engine mount bracket. Also detach the ground strap.

22   Raise the vehicle and support it securely on jackstands. Remove the under-vehicle splash shield.

23   Support the engine with a floor jack and wooden block placed between the jack and the engine.

24   Remove the three engine mount bracket bolts **(see illustration)**.

25   Remove the two engine mount bolts and remove the mount.

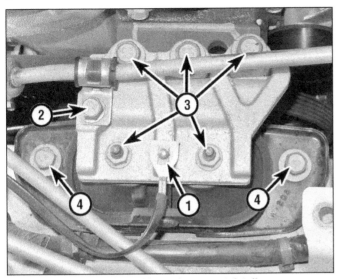

17.24 Right-side engine mount details

**17.34 Remove the rear mount bracket-to-subframe bolts (A)
then the through bolt (B)**

1 Ground strap	3 Mount bracket fasteners
2 Power steering line bracket bolt	4 Mount-to-body bolts

26 Remove the two engine mount bracket-to-mount nuts and separate the bracket from the mount.

27 Replace the engine mount with the new one. Tighten the mount-to-body bolts to the torque listed in this Chapter's Specifications.

28 Lower the engine and tighten the mount-to-bracket bolts to the torque listed in this Chapter's Specifications.

29 The remainder of installation is the reverse of removal.

Rear mount

30 Raise the vehicle and support it securely on jackstands.

31 Remove the engine splash shield.

32 Remove the downstream oxygen sensor (see Chapter 6).

33 Support the engine with a floor jack and wooden block placed between the jack and the engine.

34 Remove the rear mount bracket-to-subframe bolts **(see illustration)**.

35 Remove the mount through-bolt **(see illustration 17.34)** and separate the mount from the bracket.

36 Remove the mount towards the rear of the vehicle, slightly twisting it as you remove it.

37 Install the new mount into the bracket, then install the through-bolt and tighten it by hand.

38 Align the mount to the subframe, then install and tighten the bolts to the torque listed in this Chapter's Specifications.

39 Tighten the through bolt to the torque listed in this Chapter's Specifications.

40 The remainder of installation is the reverse of removal.

18 Variable Valve Actuation Assembly (VVAA) (2015 and later models) - removal and installation

Caution: *This procedure requires the use of Mopar special tool #10259B along with alignment pins #2025300090 - attempting to perform this procedure without the use of this special tool will cause damage to the valves, cylinder head or VVAA system.*

Note: *The VVAA is made up of a series of complex, non-serviceable parts. It also houses the hydraulic lifters, rocker arms, and oil temperature sensor, which are serviceable. If any further components of the VVAA assembly are malfunctioning, it must be replaced with an entire new assembly.*

1 Remove the valve cover (see Section 4).

2 Disconnect the oil temperature sensor harness connector (near the end of the fuel rail).

3 Disconnect the VVAA electrical connectors (they are adjacent to the fuel injector connectors and have yellow locking tabs).

4 Remove the first four outer VVAA bolts (counting from the timing chain end of the engine) before installing the special tool (when the tool is installed you won't be able to access these bolts).

5 Prepare and install the special tool 10259B onto the VVAA assembly **(see illustration 10.23).**

6 Tighten the special tool to compress the valve springs completely off of the camshaft (see Section 10).

7 Once compressed, remove the remaining four bolts from the camshaft side of the VVAA assembly.

8 Install the special dowel pins #2025300090 in place of the bolts that were removed at each end, diagonally from each other.

9 Carefully lift the VVAA assembly up to break the seal, but be careful to avoid gouging the mating surfaces. Remove the VVAA assembly together with the special tool installed. Discard the old gasket.

Caution: *While the VVAA assembly is removed, keep it level and don't tilt it - oil displacement inside the assembly must be kept at a minimal level.*

10 Loosen the compressor assemblies and remove the special tool from the VVAA assembly.

11 Set the VVAA across two blocks of wood, while removed from the engine.

12 If desired, the rocker arms can now be replaced.

13 Clean the mating surfaces of the cylinder head and VVAA assembly of all old gasket material.

14 Install the tool onto the VVAA, then compress the valve springs fully. Ensure that the locating pins are still installed in the VVAA.

15 With three of the special alignment pins #2025300090 in place on the cylinder head, using a new gasket, install the VVAA assembly onto the pins. The pins MUST be installed to ensure centering of the VVAA over the valve stems. The VVAA should also be seated evenly over the cylinder head with only a minimal amount of clearance between the two.

16 By visually inspecting through the openings on the camshaft side, make sure that the VVAA plungers are contacting the valve stems evenly before installing the VVAA mounting bolts.

17   Install the four camshaft-side VVAA mounting bolts hand tight.
18   Remove the special tool, then remove the special alignment pins.
19   Install the remaining VVAA mounting bolts hand tight.
20   Tighten the VVAA mounting bolts, in sequence **(see illustration)**, to the torque settings listed in this Chapter's Specifications.
21   Connect the electrical connectors for the VVAA assembly and oil temperature sensor.
Note: *Be careful not to mix-up the electrical connectors for the oil temperature sensor and any one of the VVAA connectors.*
22   Install the valve cover (see Section 4).

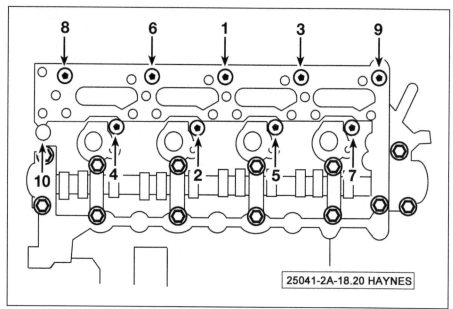

18.20 VVAA assembly bolt tightening sequence (2015 and later engines)

Chapter 2 Part B
2.7L V6 engine

Contents

	Section			Section
Camshafts - removal, inspection and installation	7		Oil pump - removal, inspection and installation	14
Crankshaft front oil seal - replacement	12		Rear main oil seal - replacement	16
Crankshaft pulley - removal and installation	11		Repair operations possible with the engine in the vehicle	2
Cylinder head(s) - removal and installation	10		Rocker arms and hydraulic lash adjusters - removal,	
Driveplate - removal and installation	15		inspection and installation	5
Engine mounts - check and replacement	17		Timing chain and sprockets - removal, inspection	
Exhaust manifold - removal and installation	9		and installation	6
General information	1		Top Dead Center (TDC) for number one piston - locating	3
Intake manifolds - removal and installation	8		Valve cover - removal and installation	4
Oil pan - removal and installation	13			

Specifications

General
Firing order	1-2-3-4-5-6	
Bore and stroke	3.386 x 3.091 inches	86.004 x 78.511 mm
Displacement	167 cubic inches	
Cylinder numbers (front to rear)		
Left bank	2-4-6	
Right bank	1-3-5	
Compression pressure	See Chapter 2E	

Cylinder head
Warpage limit	0.008 inch	0.203 mm

Camshaft
Endplay	0.0051 to 0.0110 inch	0.1295 to 0.2794 mm
Lobe wear limit		
Standard	0.001 inch	0.0254 mm
Service limit	0.010 inch	0.254 mm
Camshaft bearing oil clearance		
Standard	0.0020 to 0.0035 inch	0.0508 to 0.0889 mm
Service limit	0.0040 inch	0.1 mm
Camshaft journal diameter	0.9441 to 0.9449 inch	23.98 to 24.00 mm
Camshaft bore diameter	0.9469 to 0.9476 inch	24.05 to 24.07 mm

Front of Vehicle

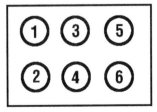

Cylinder identification diagram

30013-1-specs HAYNES

Oil pump

Cover warpage limit...	0.001 inch	0.0254 mm
Inner and outer rotor thickness..	0.3731 to 0.3741 inch	9.477 to 9.502 mm
Outer rotor diameter (minimum)...	3.5109 inches	89.177 mm
Outer rotor-to-housing clearance (maximum)........................	0.015 inch	0.381 mm
Inner rotor-to-outer rotor lobe clearance.............................	0.008 inch	0.203 mm
Oil pump housing-to-rotor side clearance............................	0.003 inch	0.076 mm

Torque specifications

	Ft-lbs (unless otherwise indicated)	**Nm**

Note: One foot-pound (ft-lb) of torque is equivalent to 12 inch-pounds (in-lbs) of torque. Torque values below approximately 15 ft-lbs are expressed in inch-pounds, because most foot-pound torque wrenches are not accurate at these smaller values.

	Ft-lbs	Nm
Camshaft sprocket bolts..	21	28
Camshaft bearing cap bolts (in sequence - **see illustration 7.19)**	105 in-lbs	12
Camshaft timing chain tensioner bolts (secondary)...............................	105 in-lbs	12
Crankshaft pulley bolt..	125	169
Cylinder head bolts (in sequence - **see illustration 10.19**)		
Step 1 (tighten bolts 1 through 8)...................................	35	47
Step 2 (tighten bolts 1 through 8)...................................	55	75
Step 3 (tighten bolts 1 through 8)...................................	55	75
Step 4 (tighten bolts 1 through 8)...................................	Tighten an additional 90 degrees	
Step 5 (tighten bolts 9 through 11).................................	21	28
Drivebelt tensioner...	40	54
Driveplate bolts..	70	95
Exhaust manifold bolts	16.5	22
Exhaust manifold heat shield bolts......................................	105 in-lbs	12
Engine mount heat shield bolts	97 in-lbs	11
Engine mount-to-bracket bolts...	55	75
Engine mount retaining nuts..	55	75
Engine rear mount-to-transaxle bolts	35	47
Idler pulley bolt	21	28
Intake manifold bolts		
Upper intake manifold bolts (in sequence - **see illustration 8.8)**	105 in-lbs	12
Lower intake manifold bolts (in sequence - **see illustration 8.16)**	105 in-lbs	12
Oil pan bolts		
6 mm bolts..	105 in-lbs	12
6 mm nuts	105 in-lbs	12
8 mm bolts..	21	28
Oil pan drain plug	20	27
Oil pan support brace **(see illustration 13.6)**		
Step 1		
Support brace-to-oil pan bolts....................................	10 in-lbs	1
Support brace-to-transaxle bolts................................	40	54
Step 2		
Support brace-to-oil pan bolts....................................	40	54
Oil pick-up tube mounting bolts	21	28
Oil pump mounting bolts...	21	28
Oil pump cover screws..	105 in-lbs	12
Rear main oil seal retainer bolts	105 in-lbs	12
Timing chain cover bolts		
6 mm	105 in-lbs	12
10 mm	40	54
Timing chain guide bolts...	21	28
Timing chain guide access plugs...	15	20
Timing chain tensioner arm pivot bolt..................................	21	28
Timing chain tensioner (primary)...	105 in-lbs	12
Timing chain tensioner (secondary)	105 in-lbs	12
Transaxle crossmember bolts	50	68
Valve cover bolts	105 in-lbs	12
Water outlet housing mounting bolts	105 in-lbs	12
Water outlet tube-to-housing bolts.......................................	30 in-lbs	3

1 General information

1   This Part of Chapter 2 is devoted to in-vehicle repair procedures for the 2.7L V6 engine. These engines utilize an aluminum block with six cylinders arranged in a "V" shape at a 60-degree angle between the two banks. The overhead camshaft aluminum cylinder heads are equipped with replaceable valve guides and seats. Stamped steel rocker arms with an integral roller bearing actuate the valves.

2   Information concerning engine removal and installation and engine overhaul can be found in Chapter 2E.

3   The following repair procedures are based on the assumption that the engine is installed in the vehicle. If the engine has been removed from the vehicle and mounted on a stand, many of the steps outlined in this Part of Chapter 2 will not apply.

2 Repair operations possible with the engine in the vehicle

1   Many major repair operations can be accomplished without removing the engine from the vehicle.

2   Clean the engine compartment and the exterior of the engine with some type of degreaser before any work is done. It will make the job easier and help keep dirt out of the internal areas of the engine.

3   Depending on the components involved, it may be helpful to remove the hood to improve access to the engine as repairs are performed (refer to Chapter 11, if necessary). Cover the fenders to prevent damage to the paint. Special pads are available, but an old bedspread or blanket will also work.

4   If vacuum, exhaust, oil or coolant leaks develop, indicating a need for gasket or seal replacement, the repairs can generally be made with the engine in the vehicle. The intake and exhaust manifold gaskets, oil pan gasket, crankshaft oil seals and cylinder head gaskets are all accessible with the engine in place.

5   Exterior engine components, such as the intake and exhaust manifolds, the oil pan, the oil pump, the water pump (see Chapter 3), the starter motor, the alternator and the fuel system components (see Chapter 4) can be removed for repair with the engine in place.

6   Since the cylinder heads can be removed without pulling the engine, valve component servicing can also be accomplished with the engine in the vehicle. Replacement of the camshafts, timing chains and sprockets are also possible with the engine in the vehicle.

7   In extreme cases caused by a lack of necessary equipment, repair or replacement of piston rings, pistons, connecting rods and rod bearings is possible with the engine in the vehicle. However, this practice is not recommended because of the cleaning and preparation work that must be done to the components involved.

3.6 A compression gauge can be used in the number one plug hole to assist in finding TDC

3 Top Dead Center (TDC) for number one piston - locating

1   Top Dead Center (TDC) is the highest point in the cylinder that each piston reaches as it travels up the cylinder bore. Each piston reaches TDC on the compression stroke and again on the exhaust stroke, but TDC generally refers to piston position on the compression stroke.

2   Positioning the piston(s) at TDC is an essential part of many procedures such as valve timing, camshaft and timing chain/belt sprocket removal.

3   Before beginning this procedure, be sure to place the shift lever in Park and apply the parking brake or block the rear wheels. Disconnect the negative battery cable from the remote ground terminal, then remove the ignition coils (see Chapter 5) and the spark plugs (see Chapter 1).

4   In order to bring any piston to TDC, the crankshaft must be turned. When looking at the front of the engine, normal crankshaft rotation is clockwise.

5   Turn the crankshaft with a socket and ratchet attached to the bolt threaded into the front of the crankshaft. Turn the bolt in a clockwise direction.

6   Install a compression pressure gauge in the number one spark plug hole (see Chapter 2E). It should be a gauge with a screw-in fitting and a hose at least six inches long **(see illustration)**.

7   Rotate the crankshaft while observing for pressure on the compression gauge. The moment the gauge shows pressure indicates that the number one cylinder is on the compression stroke.

8   Once the compression stroke has begun, TDC for the compression stroke of the number one cylinder is reached by bringing the piston to the top of the cylinder.

9   Insert a long dowel into the number one spark plug hole until it rests on top of the piston crown. Continue rotating the crankshaft slowly until the dowel levels off (piston

reaches top of travel). This will be approximate TDC for number 1 piston.

10   These engines are not equipped with components (vibration damper, flywheel, timing hole, etc.) that are marked to identify the position of number 1 TDC. Therefore the only method to double-check the exact location of TDC number 1 is to remove the timing chain cover to access timing chain sprockets (see Section 6), or with the use of a degree wheel on the crankshaft vibration damper and a positive stop threaded into the spark plug hole, as you would use in the process of degreeing a camshaft (this procedure is described in detail in the *Haynes Chrysler Engine Overhaul Manual*).

11   After the No. 1 piston has been positioned at TDC on the compression stroke, TDC for any of the remaining pistons can be located by turning the crankshaft and following the firing order. Divide the crankshaft pulley into three equal sections with chalk marks at each point, each indicating 120-degrees of crankshaft rotation. Rotating the engine past TDC no. 1 to the next mark will place the engine at TDC for cylinder no. 2.

4 Valve cover - removal and installation

Removal

1   Disconnect the negative battery cable from the remote ground terminal (see Chapter 5).

2   If you're removing the rear valve cover, remove the upper intake manifold (see Section 8) and cover the lower intake manifold with rags to keep out dirt.

3   Remove the ignition coils (see Chapter 1) and ignition coil capacitor mounting fastener.

4   Remove the PCV hose from the grommet on the right valve cover (see Chapter 6). Remove the intake hose from the left valve cover.

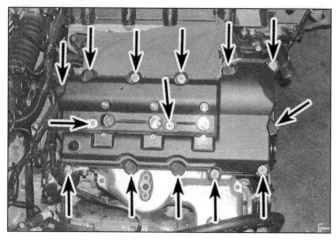

4.5 Valve cover mounting bolts

5.5 Using a shaft-type valve spring compressor, depress the valve spring just enough to remove the rocker arm

5.7 Pull the lash adjuster up and out of its bore to remove it from the cylinder head

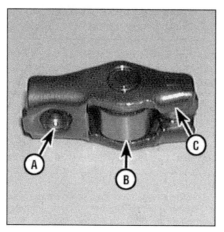

5.8 Inspect the rocker arms at the following locations

A   Lash adjuster pocket
B   Roller
C   Valve stem seat

5   Pull the wire harness up from the valve cover studs and remove the valve cover studs/nuts and bolts **(see illustration)**.
6   Detach the valve cover.
Note: *If the cover sticks to the cylinder head, use a block of wood and a hammer to dislodge it. If the cover still won't come loose, pry on it carefully, but don't distort the sealing flange.*

Installation

7   The mating surfaces of each cylinder head and valve cover must be perfectly clean when the covers are installed. Inspect the rubber gaskets; if they're in good condition and there were no oil leaks, they can be re-used.
8   Inspect the spark plug tube seals. Replace them if they're cracked or flattened, or if the rubber has hardened. Make sure the spark plug tube seals are in position before installing the valve cover.
9   Clean the mounting bolt threads with a die if necessary to remove any corrosion and restore damaged threads. Use a tap to clean the threaded holes in the heads.
10   Place the valve cover in position, then

install the bolts. Tighten the bolts evenly, in several steps, to the torque listed in this Chapter's Specifications.
11   The remainder of installation is the reverse of removal. Start the engine and check carefully for oil leaks.

---

5   Rocker arms and hydraulic lash adjusters - removal, inspection and installation

---

Note: *A universal shaft-type valve spring compressor (available from most aftermarket specialty tool manufacturers) will be required for this procedure. The only other alternative to accomplishing this task without the use of this special tool is to remove the timing chains and the camshafts which requires major disassembly of the engine and surrounding components.*

1   Before beginning this procedure, place the shift lever in Park and apply the parking brake or block the rear wheels. Disable the ignition system by disconnecting the primary electrical connectors at the ignition coils and removing the spark plugs (see Chapter 1).
2   Remove the upper intake manifold (see Section 8) and the valve cover(s) (see Section 4).
3   Rotate the engine with a socket and ratchet attached to the crankshaft pulley bolt until the cam lobe for the rocker arm to be removed is located on its base circle. Turn the crankshaft in a clockwise direction.
4   Before the rocker arms and lash adjusters are removed, arrange to label and store them, so they can be kept separate and reinstalled in their original locations.
5   Mount the valve spring compressor on the cylinder head. Depress the valve spring just enough to release tension on the rocker arm to be removed. Once tension on the rocker arm is relieved, the rocker arm can be removed by simply pulling it out **(see illustration)**.
6   If you're replacing or removing all of the rocker arms or lash adjusters, begin with cylinder number one and work on the rocker arms for one cylinder at a time. Move from cylinder-to-cylinder following the firing order sequence (see this Chapter's Specifications). Remember to keep the rocker arm and lash adjuster for each valve together so they can be reinstalled in the same locations
7   Once the rocker arms are removed, the lash adjusters can be pulled out of the cylinder head and stored with the corresponding rocker arm **(see illustration)**.
8   Inspect each rocker arm for wear and other damage. Make sure the rollers turn freely and show no signs of wear; also check the pivot area for wear, cracks and galling **(see illustration)**.
9   Inspect the lash adjuster contact surfaces wear or damage. Make sure the lash adjusters move up and down freely in their bores on the cylinder head without excessive side to side play.

10   Installation is the reverse of removal with the following exception: Always install the lash adjuster first and make sure they're at least partially full of oil before installation. This is indicated by little or no lash adjuster plunger travel.

6   Timing chain and sprockets - removal, inspection and installation

Note: *Special tools are necessary to complete this procedure. Read through the entire procedure and obtain the special tools before beginning work.*
Note: *Because of work necessary to get at the timing chain and replace it, and because the water pump is in this area, it is recommended that the water pump be thoroughly inspected and replaced if necessary during this procedure (see Chapter 3).*
Note: *The 2.7L engine utilizes three timing chains to produce proper valve timing. The primary timing chain runs around the crankshaft sprocket, the water pump and two intake camshaft sprockets. This chain synchronizes the valve timing with the crankshaft and pistons, while two secondary timing chains run around separate intake and exhaust camshaft sprockets to synchronize the intake and exhaust camshaft events. Refer to Step 22 for primary timing chain inspection prior to timing chain removal.*

Timing chain

Removal
1   Disconnect the negative battery cable from the remote ground terminal (see Chapter 5).
2   Relieve the fuel system pressure (see Chapter 4) and drain the cooling system (see Chapter 1).

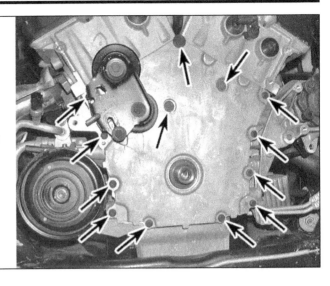
6.13 Timing chain cover retaining bolts

3   Loosen the lug nuts on the right front wheel, raise the front of the vehicle and support it securely on jackstands.
4   Remove the right front wheel and inner fender liner (see Chapter 11).
5   Disconnect the heater hose from the heater supply pipe at the bottom of the subframe. Remove the heater supply pipe fasteners and move the pipe out of the way.
Note: *It's a good idea to plug or cap the heater supply pipe to prevent coolant from dripping while the hose is disconnected.*
6   Remove the accessory drivebelts and tensioner (see Chapter 1), then remove the crankshaft pulley (see Section 11).
7   Remove the power steering pump mounting bolts and position the power steering pump off to the side without disconnecting the power steering system fluid lines (see Chapter 10).
8   Remove the upper intake manifold (see Section 8) and the valve covers (see Section 4).
9   Remove the spark plugs (see Chapter 1)

and position the number one piston at TDC on the compression stroke (see Section 3).
10   Remove the water outlet housing from the engine block (see Chapter 3).
11   Place a block of wood between a floor jack head and the oil pan and support the engine.
12   Remove the right engine mount (see Section 17).
13   Remove the timing chain cover **(see illustration)**. Note that various types and sizes of bolts are used. They must be reinstalled in their original locations. Mark each bolt or make a sketch to help remember where they go.
14   Rotate the engine until the crankshaft sprocket timing mark is aligned with the mark on the oil pump housing, and the colored links on the chain are aligned with the timing marks on the camshaft and crankshaft sprockets **(see illustrations)**. This position is approximately 60-degrees after TDC.
15   Remove the primary timing chain tensioner from the right cylinder head.

6.14a Make sure the mark on the crankshaft sprocket (A) aligns with the colored link on the primary timing chain and the mark on the oil pump housing (B)

6.14b Also verify that the mark on the front bank camshaft sprocket is flanked by the two colored links...

6.14c... and the marks on the rear bank camshaft sprocket align with the colored link on the primary chain

6.16a A 3/8-inch drive extension and ratchet inserted into the end of the camshaft hub is used as leverage while the sprocket bolts are loosened - make note of the vibration damper-to-camshaft sprocket alignment hole

6.16b Front cylinder bank camshaft sprocket retaining bolts

16   Remove the retaining bolts from the primary camshaft sprockets **(see illustrations)**. **Warning:** *Pressure from the valve springs will make the camshafts rotate clockwise as the bolts are removed. Do not rotate the crankshaft or camshaft separately after the primary timing chain is loosened or removed as piston or valve damage may occur. The only exception to this rule is when the camshafts must be rotated counterclockwise slightly, to realign the primary camshaft sprockets with the camshafts during installation.*

17   Pull the primary sprockets off the camshaft hubs one at a time, lower the sprocket(s) into the cylinder head opening until the chain can be displaced from around the sprocket, then remove the primary camshaft sprockets from the engine. Note that the left (primary) camshaft sprocket is identified by the camshaft position sensor ring which is mounted on the front of the sprocket by two nuts. It is not necessary to remove the sensor ring from the sprocket during this procedure unless damage to the sensor ring or sprocket has occurred. Also note that the right (primary)

camshaft sprocket is identified by a vibration damper which is mounted in front of the sprocket. Make note of the alignment holes on the damper and the right (primary) sprocket as these two components will separate as the sprocket is removed from the camshaft. Always install the damper in the same position from which it was removed.

18   Remove the access plugs from the front of the cylinder heads and detach the primary timing chain guides and tensioner arm **(see illustrations)**.

19   Remove the primary timing chain.

20   The secondary timing chains are removed as an assembly with the camshafts and are therefore covered in camshaft removal (see Section 7).

21   If the crankshaft sprocket needs to be replaced after thorough inspection or the sprocket simply needs to be removed for other procedures such as oil pump removal, proceed to Step 39.

Inspection

22   Inspect the camshaft, water pump and crankshaft sprockets for wear on the teeth

and keyways. Inspect the chains for cracks or excessive wear of the rollers. Inspect the facing of the chain guides and tensioner arm for excessive wear. If any of the components show signs of excessive wear, they must be replaced.

Installation

23   If removed, install the crankshaft sprocket as outlined in Steps 41 and 42.

24   If you purchased a new timing chain, verify that you have the correct timing chain for your vehicle by counting the number of links the chain has and comparing the new chain with the old chain. Also compare the position of the colored links in the new chain with the position of the colored links in the old chain.

25   Verify the crankshaft sprocket is still aligned with the mark on the oil pump housing **(see illustration 6.14a)** and install the upper chain guides back into position on the engine **(see illustration 6.18b)**.

26   Prepare to install the primary timing chain by aligning the left (primary) camshaft sprocket mark between the light colored links on the chain **(see illustration 6.16b)**. Then

6.18a Primary timing chain guide access plugs

6.18b Primary timing chain guide mounting details

1   *Upper rear timing chain guide mounting bolts*
2   *Upper front timing chain guide mounting bolts*
3   *Tensioner arm pivot bolt*
4   *Lower timing chain guide mounting bolts*

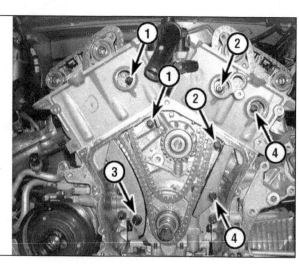

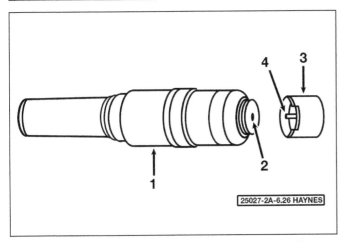

6.29 Primary timing chain tensioner - oil purging details

1	*Tensioner body*	3	*Special tool*
2	*Check ball*	4	*Pin*

6.35 With all slack removed from the left side of the chain, and the colored links on the chain aligned with their respective marks on the sprockets, engage the tensioner by pushing the tensioner arm inward slightly, then release the tensioner arm to extend the tensioner

lower the sprocket and chain assembly down through the opening in the left cylinder head and reposition the left camshaft sprocket over the camshaft hub.

27   Install the right (primary) camshaft sprocket, vibration damper and timing chain onto the engine by looping the chain around the crankshaft sprocket and aligning the light colored chain link with the mark on the crankshaft sprocket. Place the chain around the water pump sprocket and install the right (primary) camshaft sprocket over the right camshaft hub, making sure the colored links align with their respective marks on the sprockets **(see illustrations 6.14a, 6.14b and 6.14c)**.

28   Install the lower chain guide and tensioner arm **(see illustration 6.18b)**.

29   Use the special tool to purge the oil from the tensioner. Place the check ball end of the tensioner into the shallow end of the tool over the pin, apply hand pressure and slowly depress the tensioner until the oil has been purged from the tensioner **(see illustration)**.

30   After the oil has been purged, the tensioner must be reset before installation. Place the tensioner plunger into the deep end of the tool and depress the plunger until it locks into place. Once the tensioner is reset it will be about 1-1/2 inches shorter.

31   Install the reset tensioner and tighten the tensioner retaining plate bolts to the torque listed in this Chapter's Specifications. Always use a new O-ring or gasket on the tensioner housing or tensioner retaining plate.

Caution: *Align the tensioner retaining plate dowel pin with the hole in the cylinder head.*

32   Install the camshaft sprocket retaining bolts by inserting a 3/8-inch drive extension and ratchet into the primary camshaft hub and slightly rotating the camshafts counterclockwise until the camshaft sprocket bolt holes align with the camshaft hub bolt holes. Perform the procedure on the right (primary) camshaft sprocket first then proceed to the left (primary) camshaft sprocket.

33   Leaving the 3/8-inch extension and ratchet inserted into the camshaft hub as leverage, tighten the camshaft sprocket bolts to the torque listed in this Chapter's Specifications.

34   Rotate the engine clockwise just enough remove any slack from the left side of the timing chain.

35   Reconfirm that the timing marks on the camshaft and crankshaft sprockets are aligned with the colored links on the chain and release the primary timing chain tensioner from its reset position **(see illustration)**.

36   Remove all traces of old sealant from the timing chain cover and the cover bolts.

37   Apply a bead of RTV sealant to the timing chain cover gasket and sealing surfaces. Place the timing cover in position on the engine, install the bolts in their original locations, and tighten the bolts to the torque listed in this Chapter's Specifications.

38   The remainder of the installation is the reverse of removal. Refill the cooling system (see Chapter 1).

Crankshaft sprocket

39   If the crankshaft sprocket is to be removed and installed, it will require the use of several special tools, a three jaw puller, a propane torch or crankshaft sprocket installation tool and a machinist ruler or dial caliper.

40   After the primary timing chain has been removed, the crankshaft sprocket can be removed with a three jaw puller **(see illustration)**.

41   To install the crankshaft sprocket, it will be necessary to purchase a sprocket installation tool or a common household propane torch. If a sprocket installation tool is purchased, follow the tool manufacturer's installation instructions, and install the sprocket to the proper depth (see Step 42). If a propane torch is used, place the sprocket in a vise and heat the sprocket hub until the sprocket has expanded enough to slide over the crankshaft.

6.40 The crankshaft sprocket can be removed with a conventional three jaw puller

Caution: *After heating the sprocket always handle the sprocket with a pair of pliers or other insulated tool to avoid serious injury and never heat an object when gasoline or other volatile chemicals are present.*

42   Install the sprocket until it bottoms on the shoulder on the crankshaft. Using a machinist ruler or a caliper, make sure it has been installed completely by measuring from the end of the crankshaft to the face of the crankshaft sprocket; it should measure 1.537 inches +/- 0.020 inch (39.05 +/- 0.5 mm).

---

7   Camshafts - removal, inspection and installation

Removal

1   Disconnect the negative battery cable from the remote ground terminal (see Chapter 5).

7.4 Secondary timing chain tensioner retaining bolts (right cylinder head shown, left cylinder head similar)

7.5 Verify that the camshaft bearing caps are marked to ensure correct reinstallation

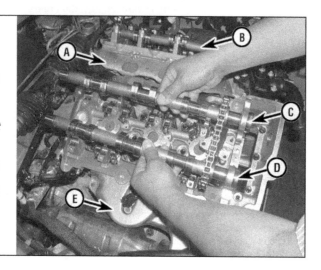

7.6 Camshaft installation details

A   *Intake manifold*
B   *Intake camshaft (left cylinder head)*
C   *Intake camshaft (right cylinder head)*
D   *Exhaust camshaft (right cylinder head)*
E   *Exhaust manifold*

2   Remove the valve covers (see Section 4).
3   Remove the primary timing chain and the primary camshaft sprockets (see Section 6).
4   Remove the retaining bolts from the secondary timing chain tensioner **(see illustration)**.
5   Verify the markings on the camshaft bearing caps. The caps should be marked from 1 to 5, and with an "I" or an "E," to indicate intake or exhaust. Also verify that there are arrow marks on the caps indicating the front of the engine **(see illustration)**. Loosen the camshaft bearing caps in two or three steps, in the reverse order of the tightening sequence **(see illustration 7.19)**.
Caution: *Keep the caps in order. They must go back in the same location they were removed from.*
6   Detach the bearing caps, then remove the camshafts, secondary timing chain and the secondary tensioner as an assembly from the cylinder head. Make a note that the intake camshafts have a flanged hub at the front and the exhaust camshafts do not. Also note that the intake camshafts are installed in the cylinder head (towards the center of the engine)

next to the intake manifold and the exhaust camshafts are installed (towards the sides of the engine) next to the exhaust manifolds **(see illustration)**.
7   Remove the secondary tensioner and the timing chain from the camshafts.

8   Inspect the camshaft secondary sprockets for wear on the teeth. Inspect the chains for cracks or excessive wear of the rollers. Inspect the facing of the secondary chain tensioners for excessive wear. If any of the components show signs of excessive wear they must be replaced.

Inspection
9   After the camshaft has been removed from the engine, cleaned with solvent and dried, inspect the bearing journals for uneven wear, pitting and evidence of seizure. If the camshaft journals are damaged, inspect the cylinder head and the camshaft bearing caps.
10   Measure the bearing journals with a micrometer to determine if they are excessively worn or out-of-round **(see illustration)**. Compare the measurements with this Chapter's Specifications.
11   Measure the lobe height of each cam lobe on the intake camshaft and record your measurements **(see illustration)**. Compare the measurements for excessive variations. If the lobe heights vary more than 0.010 inch

7.10 Check the diameter of each camshaft bearing journal to pinpoint excessive wear and out-of-round conditions

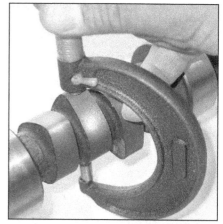

7.11 Measure the camshaft lobe height (greatest dimension) with a micrometer

(0.254 mm), replace the camshaft. Compare the lobe height measurements on the exhaust camshaft and follow the same procedure. Do not compare intake camshaft lobe heights with exhaust camshaft lobe heights, as they are different. Only compare intake lobes with intake lobes and exhaust lobes with other exhaust lobes.

12   Check the camshaft lobes for heat discoloration, score marks, chipped areas, pitting and uneven wear **(see illustration)**. If the lobes are in good condition and if the lobe lift variation measurements recorded earlier are within the limits, the camshaft can be reused.

Installation

13   Install the secondary timing chain(s) over the camshaft sprockets while aligning the colored links on the chain with the marks on the secondary camshaft sprockets in a 12 o'clock position.

Note: *The colored links on the chain must face outward toward the front of the engine.*

14   Using a paperclip, fabricate a U-shaped tool to use as a tensioner locking pin. Then place the secondary tensioner in a vise and compress the tensioner until the U-shaped tool can be inserted into the locking holes on the tensioner. This places the tensioner in the locked position so it can be reinstalled **(see illustration)**.

15   Once the tensioner is locked into place it can be installed back into the space between the camshafts and the secondary timing chain.

16   Apply moly-based engine assembly lubricant to the camshaft lobes and journals. Make sure the rocker arms are properly seated on their respective lash adjuster and the valve stem tip.

17   Install the camshafts, secondary tensioner and the timing chain as an assembly in their original position, with the marks

7.12 Check the cam lobes for pitting, excessive wear and scoring. If scoring is excessive, as shown here, replace the camshaft

on the sprockets and the colored links on the chain facing up (90-degrees from the valve cover mating surface) and inline with the cylinder bank **(see illustration)**.

Note: *When installed correctly, there should be 12 timing chain pins between the intake and exhaust camshaft marks.*

18   Install the bearing caps and bolts and tighten them hand tight.

19   Tighten the bearing cap bolts in several steps, to the torque listed in this Chapter's Specifications, using the proper tightening sequence **(see illustration)**.

20   Tighten the tensioner mounting bolts to the torque listed in this Chapter's Specifications and remove the tensioner locking pin.

21   Install the primary timing chain and camshaft sprockets (see Section 6).

22   The remainder of installation is the reverse of removal.

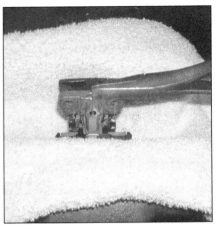

7.14 Compress the secondary tensioner in a vise until a paperclip can be inserted into the secondary tensioner locking holes

8   Intake manifolds - removal and installation

Upper intake manifold

1   Relieve the fuel system pressure (see Chapter 4).

2   Disconnect the negative battery cable from the remote ground terminal (see Chapter 5).

3   Remove the air intake duct and the air filter housing (see Chapter 4).

4   Label and disconnect the vacuum hoses and electrical connectors attached to the upper intake manifold and throttle body. Disconnect the Electronic Throttle Control (ETC) system from the throttle body (see Chapter 4).

5   Remove the EGR pipe from the upper intake manifold (see Chapter 6).

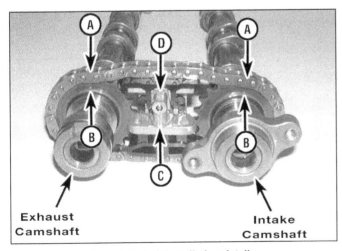

7.17 Camshaft installation details

A	Secondary timing chain colored links
B	Secondary camshaft sprocket marks
C	Tensioner locking pin
D	Secondary timing chain tensioner

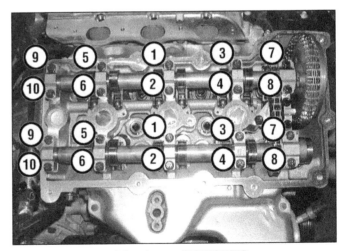

7.19 Camshaft bearing cap TIGHTENING sequence

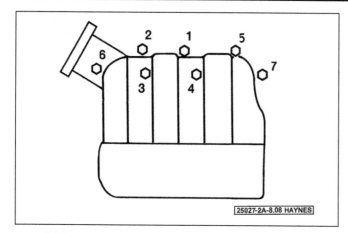

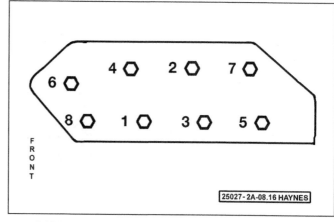

8.8 Upper intake manifold bolt TIGHTENING sequence   **8.16 Lower intake manifold bolt TIGHTENING sequence**

6   Remove the upper intake manifold support brackets. Loosen the upper intake manifold bolts in the reverse order of the tightening sequence **(see illustration 8.8)**. Remove the upper intake manifold and the foam insulator pad.

7   Clean and inspect the upper intake manifold-to-lower intake manifold sealing surfaces. Inspect the gaskets for tears or cracks, replacing them if necessary. The gaskets can be reused if not damaged.

8   Install the upper intake manifold onto the lower intake manifold and tighten the bolts, in the recommended tightening sequence **(see illustration)**, to the torque listed in this Chapter's Specifications. The remainder of installation is the reverse of removal.

Lower intake manifold

9   Remove the upper intake manifold (see Steps 1 through 6).

10   Label and detach any remaining hoses which would interfere with the removal of the lower intake manifold.

11   Remove the fuel rail and injectors from the lower intake manifold (see Chapter 4).

12   Loosen the lower intake manifold mounting bolts/nuts in 1/4-turn increments, in the reverse order of the tightening sequence **(see illustration 8.16)**, until they can be removed by hand,

13   The manifold will probably be stuck to the cylinder heads and force may be required to break the gasket seal.

Caution: *Don't pry between the manifold and the heads, or damage to the gasket sealing surfaces may occur, leading to vacuum leaks.*

14   Clean and inspect the lower intake manifold-to-cylinder head sealing surfaces. Inspect the gaskets for tears or cracks, replacing them if necessary. The gaskets can be reused if they aren't damaged.

15   Position the lower intake manifold onto the engine, making sure the gaskets and manifold are aligned correctly over the cylinder heads. Install the fuel rail and injectors onto the cylinder heads and the lower manifold. Loosely install the fuel rail retaining bolts into the manifold to ensure correct gasket/

manifold alignment, then install the remaining lower intake manifold bolts.

16   Following the recommended tightening sequence **(see illustration)**, tighten the bolts, in several steps, to the torque listed in this Chapter's Specifications.

17   The remainder of installation is the reverse of removal. Run the engine and check for fuel, vacuum and coolant leaks.

9   Exhaust manifold - removal and installation

Warning: *The engine must be completely cool before beginning this procedure.*

Removal

1   Disconnect the negative battery cable from the remote ground terminal (see Chapter 5).

2   Block the rear wheels, set the parking brake, raise the front of the vehicle and support it securely on jackstands. Disconnect the downstream oxygen sensor connectors (see Chapter 6).

3   Remove the under-vehicle splash shield, if equipped.

4   Unbolt the exhaust pipes at the exhaust manifolds.

5   Lower the vehicle. Working in the engine compartment, disconnect the oxygen sensor's electrical connector. If removing the rear exhaust manifold, detach the EGR pipe from the top of the manifold. If removing the front manifold, remove the air filter housing (see Chapter 4) and the engine oil dipstick tube.

6   Remove the exhaust manifold upper heat shield mounting bolts and remove the heat shields.

7   Remove the mounting bolts and detach the manifold from the cylinder head. Be sure to spray penetrating lubricant onto the bolts and threads before attempting to remove them.

Installation

8   Clean the mating surfaces to remove all traces of old gasket material, then inspect the

manifold for distortion and cracks. Warpage can be checked with a precision straightedge held against the mating flange. If a feeler gauge thicker than 0.030-inch can be inserted between the straightedge and flange surface, take the manifold to an automotive machine shop for resurfacing.

9   Place the exhaust manifold in position with a new gasket and the lower heat shields and install the mounting bolts finger tight.

Note: *Be sure to identify the exhaust manifold gaskets by the correct cylinder designation and the position of the exhaust ports on the gasket.*

10   Starting in the middle and working out toward the ends, tighten the mounting bolts in several increments, to the torque listed in this Chapter's Specifications.

11   The remainder of installation is the reverse of removal.

12   Start the engine and check for exhaust leaks between the manifold and cylinder head and between the manifold and exhaust pipe.

10   Cylinder head(s) - removal and installation

Caution: *The engine must be completely cool before beginning this procedure.*

Removal

1   Remove the primary timing chain and sprockets (see Section 6).

Caution: *Be careful not to disturb the crankshaft from its alignment marks during the remainder of this procedure.*

2   Remove the camshafts from the cylinder head (see Section 7).

3   Remove the rocker arms and hydraulic lash adjusters from the cylinder head (see Section 5). Before the rocker arms and lash adjusters are removed, arrange to label and store them, so they can be kept separate and reinstalled in their original locations.

4   Remove the lower intake manifold (see Section 8) and the exhaust manifold(s) (see Section 9).

5   Label and remove any remaining items attached to the cylinder head, such as coolant fittings, ground straps, cables, hoses, wires or brackets **(see illustration)**.

6   Using a breaker bar and the appropriate sized socket, loosen the cylinder head bolts in 1/4-turn increments until they can be removed by hand. Loosen the bolts in the reverse order of the tightening sequence **(see illustration 10.19)** to avoid warping or cracking the head.

7   Lift the cylinder head off the engine block with the exhaust manifold attached. If it's stuck, very carefully pry up at the transaxle end, beyond the gasket surface, at a casting protrusion.

8   Remove all external components from the cylinder head to allow for thorough cleaning and inspection before having the cylinder head serviced at a qualified automotive machine shop.

Installation

9   The mating surfaces of the cylinder head and block must be perfectly clean when the head is installed.

10   Use a gasket scraper to remove all traces of carbon and old gasket material from the cylinder head and engine block being careful not to gouge the aluminum, then clean the mating surfaces with lacquer thinner or acetone. If there's oil on the mating surfaces when the head is installed, the gasket may not seal correctly and leaks could develop. When working on the block, stuff the cylinders with clean shop rags to keep out debris. Use a vacuum cleaner to remove material that falls into the cylinders.

11   Check the block and head mating surfaces for nicks, deep scratches and other damage. If damage is slight, it can be removed with a file; if it's excessive, machining may be the only alternative.

12   Use a tap of the correct size to chase the threads in the head bolt holes, then clean the holes with compressed air - make sure that nothing remains in the holes.

Warning: *Wear eye protection when using compressed air!*

13   Check each cylinder head bolt for stretching by laying a metal ruler or a straightedge against the threads of the bolts. If the diameter of the bolt threads has necked down anywhere in the threaded area, the bolts have exceeded the maximum amount of stretch

10.5 Remove the water outlet housing by detaching the heater supply tube retaining bolts (A), then slide the heater supply tube out of the water outlet housing and remove the water outlet housing retaining bolts (B)

and will need to be replaced (all of the threads must touch the straightedge).

Note: *It's a good idea to replace the head bolts with new ones regardless of their condition.*

14   Using a precision straightedge and feeler gauges, check the cylinder head for warpage. If warpage exceeds the amount listed in this Chapter's Specifications, have the head machined at an automotive machine shop.

15   Install the components that were removed from the head.

16   Position the new cylinder head gasket over the dowel pins on the block, noting which direction on the gasket faces up.

17   Carefully set the head over the dowels on the block without disturbing the gasket.

18   Before installing the head bolts, apply a small amount of clean engine oil to the threads and hardened washers (if equipped). The chamfered side of the washers must face the bolt heads.

19   Install the bolts in their original locations and tighten them finger tight. Then tighten all the bolts in several steps, following the proper sequence **(see illustration)**, to the torque listed in this Chapter's Specifications.

20   Install the lash adjusters and the rocker arms in the cylinder head (see Section 5), then install the camshafts (see Section 7),

21   Install the primary timing chain and

sprockets (see Section 6). The remainder of installation is the reverse of removal.

22   Refill the cooling system and change the engine oil and filter (see Chapter 1).

23   Start the engine and check for oil and coolant leaks.

11   Crankshaft pulley - removal and installation

1   Disconnect the negative battery cable from the remote ground terminal (see Chapter 5).

2   Loosen the lug nuts on the right front wheel, raise the front of the vehicle and support it securely on jackstands.

3   Remove the right front wheel and inner fender liner (see Chapter 11).

4   Remove the drivebelts (see Chapter 1).

5   Use a strap wrench around the crankshaft pulley to hold it while using a breaker bar and socket to remove the crankshaft pulley center bolt **(see illustration)**.

11.5 Use a strap wrench to hold the crankshaft pulley while removing the center bolt (a chain-type wrench may be used if you wrap a section of old drivebelt or rag around the crankshaft pulley first)

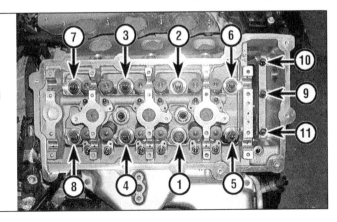

10.19 Cylinder head bolt TIGHTENING sequence

11.6 The use of a three jaw puller will be necessary to remove the crankshaft pulley - always place the puller jaws around the pulley hub, not the outer ring

11.7 If the sealing surface of the pulley hub has a wear groove from contact with the seal, repair sleeves are available at most auto parts stores

12.2 Pry the seal out very carefully with a seal removal tool or screwdriver, being careful not to nick or gouge the seal bore or the crankshaft

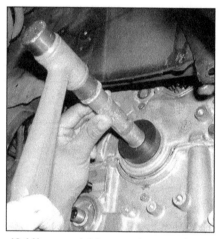

12.4 Use a seal driver or a large socket to drive the new seal into the cover

6   Insert special tool #8194 into the center of the crankshaft where the bolt was (or use a smaller bolt that can be inserted into the bolt hole), then pull the damper off the crankshaft with a puller **(see illustration)**, making sure the puller is pressing against the special tool or bolt.

Caution: *The jaws of the puller must only contact the hub of the pulley - not the outer ring.*

Note: *The crankshaft pulley covers the end of the crankshaft, so a special pin #8194 should be inserted into the end to the crankshaft or a long Allen-head bolt should be inserted into the crankshaft nose for the puller's tapered tip to push against to prevent damage to the crankshaft pulley.*

7   Check the surface on the pulley hub that the oil seal rides on. if the surface has been grooved from long-time contact with the seal, a press-on sleeve may be available to renew the sealing surface **(see illustration)**. This sleeve is pressed into place with a hammer

and a block of wood and is commonly available at auto parts stores for various applications.

8   Lubricate the pulley hub with clean engine oil and reinstall the crankshaft pulley. Use a vibration damper installation tool to press the pulley onto the crankshaft.

9   Install the crankshaft pulley retaining bolt and tighten it to the torque listed in this Chapter's Specifications.

10   The remainder of installation is the reverse of removal.

12   Crankshaft front oil seal - replacement

1   Remove the crankshaft pulley from the engine (see Section 11).

2   Carefully pry the seal out of the cover with a seal removal tool or a large screwdriver **(see illustration)**.

Caution: *Be careful not to scratch, gouge or distort the area that the seal fits into, or an oil leak will develop.*

3   Clean the bore to remove any old seal material and corrosion. Position the new seal in the bore with the seal lip (usually the side with the spring) facing IN (toward the engine). A small amount of oil applied to the outer edge of the new seal will make installation easier - but don't overdo it!

4   Drive the seal into the bore with a seal driver or a large socket and hammer until it's completely seated **(see illustration)**. Select a socket that's the same outside diameter as the seal and make sure the new seal is pressed into place until it bottoms against the cover flange.

5   Lubricate the seal lips with engine oil and reinstall the crankshaft pulley.

6   The remainder of installation is the reverse of removal. Run the engine and check for oil leaks.

13   Oil pan - removal and installation

Removal

1   Disconnect the negative battery cable from the remote ground terminal (see Chapter 5).

2   Remove the engine oil dipstick then remove the oil dipstick tube mounting bolt and pull the tube out of the oil pan.

3   Raise the front of the vehicle and support it securely on jackstands. Apply the parking brake and block the rear wheels to keep the vehicle from rolling off the stands.

4   Remove the under-vehicle splash shield, if equipped, and the right lower splash shield (see Chapter 1, Section 19). Drain the engine oil and remove the oil filter (see Chapter 1).

5   Remove the through-bolt from the front engine mount, then remove the longitudinal crossmember **(see illustration)**.

6   Remove the transaxle-to-oil pan support brace at the rear of the pan **(see illustration)**.

7   Detach any clips securing the transaxle oil cooler lines that would interfere with removal of the oil pan (see Chapter 7A). Also remove the lower bolt securing the air conditioning compressor to the oil pan and the bolt securing the alternator bracket to the oil pan.

8   Remove the bolts and nuts, then carefully separate the oil pan from the block. Don't pry between the block and the pan, or damage to the sealing surfaces could occur and oil leaks may develop. Instead, pry at the casting protrusion at the front of the pan **(see illustration)**.

Installation

9   Clean the pan with solvent and remove all old sealant and gasket material from the block and pan mating surfaces. Clean the mating surfaces with lacquer thinner or ace-

13.5 Longitudinal crossmember bolts

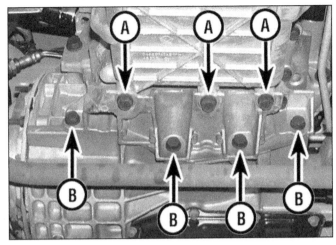

13.6 Transaxle-to-oil pan support brace

A   *Vertical bolts*   B   *Horizontal bolts*

tone and make sure the bolt holes in the block are clear. Check the oil pan flange for distortion, particularly around the bolt holes.

10   Apply a bead of RTV sealant to the oil pan rail parting lines at the front cover and at the rear main oil seal retainer. Install the gasket on the block.

11   Place the oil pan in position on the block and install the nuts/bolts.

12   After the fasteners are installed, tighten them to the torque listed in this Chapter's Specifications. Starting at the center, follow a criss-cross pattern and work up to the final torque in three steps.

13   Place the transmission-to-oil pan support brace in position and install the support brace-to-oil pan bolts. Tighten the bolts to the torque listed in this Chapter's Specifications, in two stages, starting with the center bolts, followed by the outer bolts.

14   The remainder of installation is the reverse of removal.

15   Refill the engine with oil (see Chapter 1), replace the filter, run it until normal operating temperature is reached and check for leaks.

14   Oil pump - removal, inspection and installation

Note: *The replacement oil pumps on early models may experience oil pressure problems because of a collapsing oil pressure relief valve and spring. These oil pumps are equipped with an expansion plug instead of the usual threaded plug to seal the oil pressure relief valve and spring inside the oil pump. Consult with a dealer parts department or other qualified automotive repair facility for the correct oil pump.*

Removal

1   Remove the primary timing chain and the crankshaft sprocket (see Section 6).
Note: *Before removing the crankshaft sprocket, verify that the crankshaft sprocket marks align with the marks on the oil pump housing and do not rotate the crankshaft from this position at any time during this procedure. This position is approximately 60 degrees after TDC.*

2   Remove the oil pan (see Section 13).

13.8 Insert a flat-head screwdriver or small prybar on the casting protrusion at the front of the oil pan to break it loose

3   Remove the oil pump pick-up tube **(see illustration)**.

4   Remove the oil pump-to-engine block bolts from the front of the engine **(see illustration)**.

14.3 Oil pump pick-up tube mounting bolts

14.4 Oil pump housing retaining bolts

14.6 Remove the screws and lift the cover off

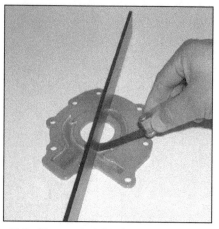

14.9a Place a straightedge across the oil pump cover and check it for warpage with a feeler gauge

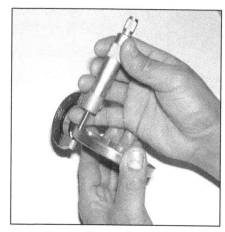

14.9b Use a micrometer or dial caliper to check the thickness and the diameter of the outer rotor

5   Gently pry the oil pump housing outward enough to clear the dowel pins on the engine block and remove the housing from the engine.

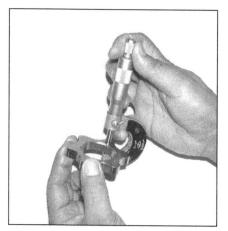

14.9c Use a micrometer or dial caliper to check the thickness of the inner rotor

Inspection

6   Remove the screws holding the front cover to the oil pump housing, and remove the cover **(see illustration)**.

7   Clean all components with solvent, then inspect them for wear and damage.

8   Remove the oil pressure regulator cap, washer, spring and valve. Check the oil pressure relief valve sliding surface and valve spring. If either the spring or the valve is damaged, they must be replaced as a set. Small burrs can be removed with 400-grit wet sandpaper and oil.

9   Check the clearance of the following oil pump components with a feeler gauge and a micrometer or dial caliper **(see illustrations)** and compare the measurements to the clearance listed in this Chapter's Specifications :

> Cover flatness
> Outer rotor diameter and thickness
> Inner rotor thickness
> Outer rotor-to-body clearance
> Inner rotor-to-outer rotor tip clearance
> Cover-to-inner rotor side clearance
> Cover-to-outer rotor side clearance

10   If any clearance is excessive, replace the entire oil pump assembly.

11   Pack the pump with petroleum jelly to prime it. Assemble the oil pump and tighten all fasteners to the torque listed in this Chapter's Specifications. Install the oil pressure regulator valve, spring and washer, then tighten the oil pressure regulator valve cap.

Installation

12   To install the pump, turn the flats in the rotor so they align with the flats on the crankshaft.

13   Install the pump-to-block bolts and tighten them to the torque listed in this Chapter's Specifications.

14   The remainder of installation is the reverse of removal.

---

15   Driveplate - removal and installation

---

1   Raise the vehicle and support it securely on jackstands, then remove the transaxle

14.9d Check the outer rotor-to-housing clearance

14.9e Check the clearance between the tips of the inner and outer rotors

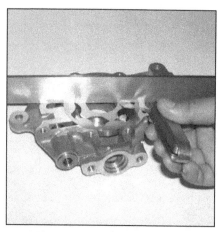

14.9f Using a straightedge and feeler gauge, check the side clearance between the surface of the oil pump and the inner and outer rotors

(see Chapter 7A).

Warning: *The engine must be supported from above with an engine hoist or an engine support fixture before working underneath the vehicle with the transaxle removed.*

2   Now would be a good time to check and replace the transaxle front pump seal.

3   Use paint or a center-punch to make alignment marks on the driveplate and crankshaft to ensure correct alignment during reinstallation.

4   Remove the bolts that secure the driveplate to the crankshaft. If the crankshaft turns, jam a large screwdriver or prybar through the driveplate to keep the crankshaft from turning, then remove the mounting bolts, bolt plate and driveplate.

5   Installation is the reverse of removal. Align the matching paint marks. Use thread locking compound on the bolt threads and tighten them in a criss-cross pattern to the torque listed in this Chapter's Specifications.

---

16   Rear main oil seal - replacement

1   All models use a one-piece rear main oil seal which is installed in a bolt-on housing. Replacing this seal requires removal of the transaxle and driveplate. The rear main seal and housing are serviced as one complete assembly. This procedure will require several special tools.

2   Remove the transaxle (see Chapter 7A) and the oil pan (see Section 13).

3   Remove the driveplate (see Section 15).

4   Remove the rear main oil seal housing bolts and the housing.

5   Lubricate the crankshaft with clean engine oil, then install the new oil seal housing onto the crankshaft.

Caution: *The new seal and housing comes equipped with an oil seal protector. Don't remove this protector before installation (it will be push out as the assembly is installed over the end of the crankshaft).*

6   Once the oil seal housing is attached and bolts slightly tight, install special alignment tools onto the oil pan rails.

7   While pressing the oil seal housing evenly, tighten the oil seal housing retainer bolts to the torque listed in this Chapter's Specifications. The housing must be flush with the plane of the oil pan mating surface of the engine to prevent oil leaks.

8   The remainder of installation is the reverse of removal.

---

17   Engine mounts - check and replacement

1   There are four powertrain mounts: the front mount attaches the transaxle to the crossmember; the left mount attaches the driver's side of the engine to the subframe; the right engine mount attaches the engine block to the passenger's side of the subframe; and a rear mount attaches the engine block to the subframe.

Check

2   During the check, the engine must be raised slightly to remove the weight from the mounts.

3   Raise the vehicle and support it securely on jackstands. Remove the front wheels and tires. Position two jacks, one under the crankshaft pulley and the other under the transaxle bellhousing. Place a block of wood between the floor jack head and the crankshaft pulley or bellhousing, then carefully raise the engine/transaxle just enough to take the weight off the mounts.

Warning: *DO NOT place any part of your body under the engine when it's supported only by a jack!*

4   Check the mounts to see if the rubber is cracked, hardened or separated from the metal plates. Sometimes the rubber will split right down the center.

5   Check for relative movement between the mount plates and the engine or subframe (use a large screwdriver or prybar to attempt to move the mounts). If movement is noted, lower the engine and tighten the mount fasteners.

6   Rubber preservative should be applied to the mounts to slow deterioration.

Replacement

Front mount

7   Raise the vehicle and support it securely on jackstands.

8   Place a block of wood between the head of the floor jack and the oil pan and support the engine.

9   Remove the front mount through-bolt.

10   Remove the crossmember-to-mount bolt from under the crossmember.

11   Remove the crossmember mounting bolts and remove the crossmember (see Chapter 10).

12   Remove the mount from the crossmember. If the mount bracket needs to be replaced, remove the bracket-to-block bolts and the bracket.

13   The remainder of installation is the reverse of removal.

Left mount

14   Disconnect the negative battery cable from the remote ground terminal (see Chapter 5), then raise the vehicle and support it securely on jackstands.

15   Remove the under-vehicle splash shield (see Section 9).

16   Remove the throttle body air inlet tube and the air filter housing (see Chapter 4).

17   Support the transaxle with a floor jack and wooden block placed between the jack and the transaxle.

18   Raise the transaxle enough to take the weight off of the mount.

19   Remove the mounting bolts and mount. Replace the left mount with a new one and tighten the bolts to the torque listed in this Chapter's Specifications.

20   The remainder of installation is the reverse of removal.

Right mount

21   Remove the coolant reservoir (see Chapter 3).

22   Raise the vehicle and support it securely on jackstands. Remove the under-vehicle splash shield (see Section 9).

23   Remove the heater tube-to-subframe bolts and move the heater tube out of the way.

24   Remove the three engine mount bracket-to-timing cover support bolts.

25   Remove the two engine mount bracket-to-mount bolts and remove the bracket.

26   Support the engine with a floor jack and wooden block placed between the jack and the engine.

27   Remove the two engine mount-to-body bolts.

28   Remove the body-to-mount bolts from underneath.

29   Remove the engine mount and replace it with the new one. Tighten the body-to-mount bolts to the torque listed in this Chapter's Specifications.

30   Lower the engine and tighten the mount-to-bracket bolts to the torque listed in this Chapter's Specifications.

31   The remainder of installation is the reverse of removal.

Rear mount

32   Remove the throttle body air inlet tube and the air filter housing (see Chapter 4).

33   Raise the vehicle and support it securely on jackstands.

34   Remove the three rear mount bracket-to-transaxle bolts.

35   Remove the bracket and mount through-bolt and separate the bracket from the mount.

36   Remove the mount-to-subframe mounting bolts and remove the mount assembly from the vehicle.

37   Replace the mount with a new one, and install the bracket and through-bolt to the mount.

38   Install the mount to the subframe and the transaxle, then tighten the bolts to the torque listed in this Chapter's Specifications.

39   Tighten the through-bolt to the torque listed in this Chapter's Specifications.

Notes

Chapter 2 Part C
3.5L V6 engine

Contents

	Section		Section
Camshafts - removal, inspection and installation	7	Oil pan - removal and installation	13
Crankshaft front oil seal - replacement	12	Oil pump - removal, inspection and installation	14
Crankshaft pulley - removal and installation	11	Rear main oil seal - replacement	16
Cylinder heads - removal and installation	10	Repair operations possible with the engine in the vehicle	2
Driveplate - removal and installation	15	Rocker arm assembly and hydraulic lash adjusters - removal,	
Engine mounts - check and replacement	17	inspection and installation	5
Exhaust manifold(s) - removal and installation	9	Timing belt - replacement	6
General information	1	Top Dead Center (TDC) for number one piston - locating	3
Intake manifolds - removal and installation	8	Valve cover(s) - removal and installation	4

Specifications

General

Firing order	1-2-3-4-5-6	
Bore and stroke	3.780 x 3.189 inches	96.012 x 81.001 mm
Displacement	214 cubic inches	
Cylinder numbers (front to rear)		
Right bank	1-3-5	
Left bank	2-4-6	
Compression pressure	See Chapter 2E	

Camshaft

Endplay	0.001 to 0.014 inch	0.0254 to 0.3556 mm
Lobe wear limit		
Standard	0.001 inch	0.0254 mm
Service limit	0.010 inch	0.254 mm
Camshaft journal diameter	1.6905 to 1.6913 inches	42.939 to 42.959 mm
Camshaft bore diameter (inside)	1.6944 to 1.6953 inches	43.038 to 43.061 mm
Camshaft bearing oil clearance		
Standard	0.003 to 0.0047 inch	0.0762 to 0.1194 mm
Service limit	0.0059 inch	0.1499 mm

Front of Vehicle

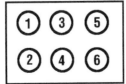

Cylinder identification diagram

30013-1-specs **HAYNES**

Oil pump

Cover warpage limit	0.001 inch	0.0254 mm
Inner and outer rotor thickness (minimum)	0.563 inch	14.3 mm
Outer rotor diameter (minimum)	3.149 inches	79.985 mm
Outer rotor-to-housing clearance (maximum)	0.015 inch	0.381 mm
Inner rotor-to-outer rotor lobe clearance	0.008 inch	0.203 mm
Oil pump housing-to-rotor side clearance	0.003 inch	0.076 mm

Torque specifications

Ft-lbs (unless otherwise indicated)   **Nm**

Note: One foot-pound (ft-lb) of torque is equivalent to 12 inch-pounds (in-lbs) of torque. Torque values below approximately 15 ft-lbs are expressed in inch-pounds, because most foot-pound torque wrenches are not accurate at these smaller values.

	Ft-lbs	Nm
Camshaft sprocket bolts*		
Step 1	75	102
Step 2	Tighten an additional 90-degrees	
Camshaft thrust plate bolts	21	28
Crankshaft pulley bolt	70	95
Cylinder head bolts (in sequence - **see illustration 10.28**)		
First step	45	61
Second step	65	88
Third step	65	88
Fourth step	Tighten an additional 90-degrees	
Drivebelt tensioner bolt	40	54
Driveplate-to-crankshaft bolts	70	95
Engine mounts		
Front mount (transaxle) bracket-to-crossmember bolts	74	100
Front mount through bolt	35	47
Front mount bracket-to-engine block bolts	35	47
Left mount bolts	45	61
Rear mount-to-subframe and engine bolts	37	50
Rear mount through bolt	35	47
Right mount-to-body bolts	55	75
Right mount bracket-to-mount nuts	55	75
Right mount-bracket-to-timing cover support bolts	50	68
Exhaust manifold-to-cylinder head bolts	17	23
Exhaust heat shield bolts	105 in-lbs	12
Exhaust pipe bolts	25	34
Heater supply tube mounting bolts	105 in-lbs	12
Intake manifold bolts		
Upper intake manifold bolts	105 in-lbs	12
Lower intake manifold bolts	21	28
Oil pan drain plug	20	27
Oil pan bolts		
Oil pan-to-engine block bolts	21	28
M6 bolts	105 in-lbs	12
M8 bolts	21	28
Oil pan-to-transaxle bolts (four)	40	54
Oil pump pick-up tube mounting bolt	21	28
Oil pump cover (plate) bolts (Torx no. 30)	105 in-lbs	12
Rear main oil seal retainer bolts	105 in-lbs	12
Rocker arm shaft bolts	23	31
Timing belt cover (front and rear) bolts		
M6	105 in-lbs	12
M8	21	28
M10	40	54
Timing belt tensioner	21	28
Timing belt tensioner pulley bolt	45	61
Throttle body bracket-to-throttle body	105 in-lbs	12
Throttle body bracket-to-cylinder head	21	28
Valve cover-to-cylinder head bolts	90 in-lbs	10

*Replace with new bolts

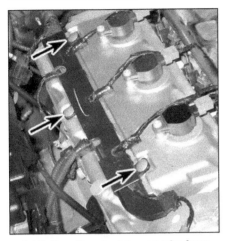

4.4 Pull up the wire harness tabs from the valve cover grommets using a trim panel tool and position the harness off to the side

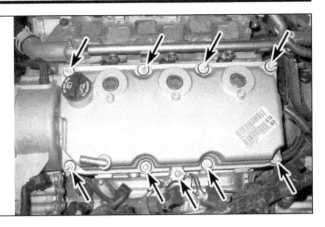

4.5 Location of the valve cover bolts and studs - front side shown, rear side similar

1   General information

1   This Part of Chapter 2 is devoted to in-vehicle repair procedures for the 3.5L V6 engine. These engines utilize an aluminum block with six cylinders arranged in a "V" shape at a 60-degree angle between the two banks. The overhead camshaft aluminum cylinder heads are equipped with replaceable valve guides and seats. Aluminum roller rockers mounted on two shafts in each cylinder head actuate the valves.

2   Information concerning engine removal and installation and engine block overhaul can be found in Chapter 2E. The following repair procedures are based on the assumption the engine is installed in the vehicle. If the engine has been removed from the vehicle and mounted on a stand, many of the steps outlined in this Part of Chapter 2 will not apply.

2   Repair operations possible with the engine in the vehicle

1   Many major repair operations can be accomplished without removing the engine from the vehicle.

2   Clean the engine compartment and the exterior of the engine with some type of degreaser before any work is done. It'll make the job easier and help keep dirt out of the internal areas of the engine.

3   Depending on the components involved, it may be helpful to remove the hood to improve access to the engine as repairs are performed (see Chapter 11 if necessary). Cover the fenders to prevent damage to the paint. Special pads are available, but an old bedspread or blanket will also work.

4   If vacuum, exhaust, oil or coolant leaks develop, indicating a need for gasket or seal replacement, the repairs can generally be done with the engine in the vehicle. The intake and exhaust manifold gaskets, timing belt cover gasket, oil pan gasket, crankshaft oil seals and cylinder head gaskets are all accessible with the engine in place.

5   Exterior engine components, such as the intake and exhaust manifolds, the oil pan, timing belt covers (and the oil pump), the water pump, the starter motor, the alternator and the fuel system components can be removed for repair with the engine in place.

6   Since the cylinder heads can be removed without pulling the engine, valve component servicing can also be accomplished with the engine in the vehicle. Replacement of the camshafts, timing belt and sprockets is also possible with the engine in the vehicle, although the cylinder head must be removed from the engine to remove the camshafts.

7   In extreme cases caused by a lack of necessary equipment, repair or replacement of piston rings, pistons, connecting rods and rod bearings is possible with the engine in the vehicle. However, this practice is not recommended because of the cleaning and preparation work that must be done to the components involved.

3   Top Dead Center (TDC) for number one piston - locating

This procedure is essentially the same as for the 2.7L V6 engine. Refer to Chapter 2B and follow the procedure outlined there.

4   Valve cover(s) - removal and installation

Removal

1   Disconnect the negative battery cable from the remote ground terminal (see Chapter 5).

2   If you're removing the rear valve cover, remove the upper intake manifold (see Section 8) and cover the lower intake manifold with rags to keep out dirt.

3   Remove the ignition coils (see Chapter 5).

4   Pull up the wire harness tabs from the valve cover grommets and position the harness off to the side **(see illustration)**.

5   Remove the valve cover bolts **(see illustration)**. Note the locations of the stud bolts so they can be reinstalled in their proper locations.

6   Detach the valve cover.

Note: *If the cover sticks to the cylinder head, use a block of wood and a hammer to dislodge it. If the cover still won't come loose, pry on it carefully, but don't distort the sealing flange.*

Installation

7   The mating surfaces of each cylinder head and valve cover must be perfectly clean when the covers are installed. Inspect the rubber gaskets; if they're in good condition and there were no oil leaks, they can be re-used.

8   Inspect the spark plug tube seals. Replace them if they're cracked or flattened, or if the rubber has hardened. Make sure the spark plug tube seals are in position before installing the valve cover **(see illustration)**.

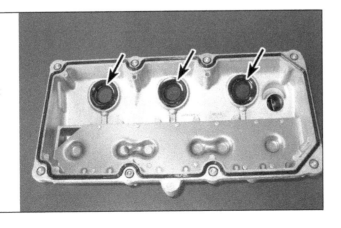

4.8 Install new spark plug tube seals before installing the valve cover

5.3 Unscrew the rocker arm shaft bolts and lift the rocker arm assembly from the cylinder head - loosen them in the reverse order of the tightening sequence

9   Clean the mounting bolt threads with a die if necessary to remove any corrosion and restore damaged threads. Use a tap to clean the threaded holes in the heads.
10   Place the valve cover and new gasket in position, then install the bolts. Tighten the bolts in several steps to the torque listed in this Chapter's Specifications.
11   The remainder of installation is the reverse of removal. Start the engine and check carefully for oil leaks.

---

5   Rocker arm assembly and hydraulic lash adjusters - removal, inspection and installation

Caution: *To prevent air ingestion into the lash adjusters, do not turn the rocker arm assemblies over. Keep them in an upright position.*
Caution: *When the rocker arm assemblies are placed on the bench, do not allow them to rest on the lash adjusters, as the plastic retainers and the lash adjusters may become damaged.*

Removal

1   Remove the valve cover(s) (see Section 4).
2   Mark the rocker arms and pedestals for identification before removing them. There are four per cylinder; two intake and two exhaust.
3   Loosen each rocker arm shaft bolt a little at a time, in the reverse of the tightening sequence **(see illustration 5.11)**, until they are all loose enough to be removed by hand. Remove the shaft and rockers as an assembly **(see illustration)**.
Caution: *Never allow the rocker arm assembly to be placed upside down or to rest on the lash adjusters as damage may occur.*

Inspection

4   Inspect each rocker arm for wear, cracks and other damage. This engine uses aluminum rocker arms with steel roller tips and

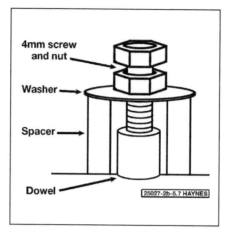

5.7 A 4 mm screw, nut, washer and spacer can be used to extract the dowel pins from the pedestals

hydraulic lash adjusters. Make sure the rollers turn freely and show no signs of wear; also check the lash adjuster contact surfaces for wear or damage. Make sure the lash adjuster retaining caps hold all of the lash adjusters securely in place on the rocker arms.
Caution: *DO NOT at anytime remove the lash adjuster from the rocker arms or damage to the lash adjuster and the rocker arm will occur.*
5   Check each rocker arm pivot area and shaft for wear, cracks and galling. If the rocker arms or shafts are worn or damaged, replace them with new ones.
6   Make sure the axle for the roller tip is not sticking out one side more than the other, and check for wear, cracks and galling on the hydraulic lash adjuster-to-valve stem tip contact surface.
7   To remove the rocker arms and pedestals from the shafts for inspection, the dowel pins from each pedestal must be pulled out. Use a 4 mm screw and nut, and a washer and spacer as a puller to extract the dowel pins from each pedestal **(see illustration)**.
Note: *New dowel pins must be used during installation.*

Installation

8   If the rockers arms or pedestals have been removed from the shafts for inspection,

lubricate the shafts and install the rocker arms and pedestals in their marked order. Keep the intake rockers on the intake shaft, and exhaust rockers on the exhaust shaft.
Caution: *The notches in the rocker arm shafts must face UP when assembled or damage to the rocker arms will occur.*
9   Reinstall new dowel pins in each pedestal (they press in until they bottom out in the pedestals).
10   The rocker arm assemblies should be installed, and the bolts tightened, with the valvetrain in a no load situation. To accomplish this, the camshafts must be placed in a neutral position. Remove the timing belt cover (see Section 6). Rotate the engine until the camshaft sprockets and the crankshaft sprocket are correctly aligned for TDC number 1 **(see illustrations 6.12a, 6.12b, 6.12c and 6.12d)**. Tighten the right side rocker arm assembly bolts, in the proper sequence **(see illustration 5.11)**, to the torque listed in this Chapter's Specifications.
11   Install the left rocker arm assembly and tighten the bolts, in the proper sequence **(see illustration)**, to the torque listed in this Chapter's Specifications.
Caution: *The notches in the left rocker arm shafts must face UP and towards the front of the engine. The notches in the right rocker arm shafts must face UP and towards the rear of the engine (firewall side). Improper oiling of the rocker arms and shafts will result if this procedure is not followed!*
12   The remainder of installation is the reverse of removal. Start the engine and check for oil leaks.

---

6   Timing belt - replacement

Caution: *Because this is an interference engine design, if the timing belt has broken, there will be damage to the valves (and possibly the pistons), and removal of the cylinder heads will be necessary.*
Note: *Because of work necessary to get at the timing belt and replace it, and because the water pump is driven by the timing belt, it is recommended that the water pump be replaced at the same time as the belt (see Chapter 3). The factory recommends replacing the timing belt at 100,000 miles.*

5.11 Rocker arm assembly bolt TIGHTENING sequence

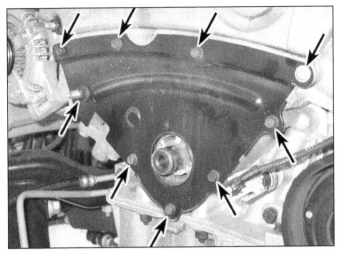

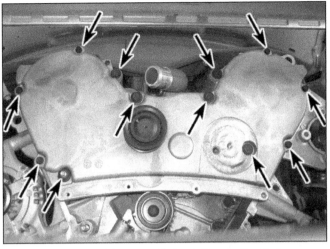

6.11a Location of the lower timing belt cover mounting fasteners

6.11b Location of the upper timing belt mounting fasteners

6.12a Rotate the crankshaft until the pointer on the crankshaft sprocket aligns with the TDC mark on the oil pump cover . . .

6.12b . . . the right camshaft sprocket timing mark aligns with the mark on the rear cover . . .

6.12c . . . and the left camshaft sprocket timing mark aligns with the mark on the rear cover

1   Disconnect the negative battery cable from the remote ground terminal (see Chapter 5).
2   Remove the cooling fan assembly (see Chapter 3) and the accessory drivebelt (see Chapter 1).
3   Remove the valve covers (see Section 4) and loosen the rocker arm shaft bolts (see Section 5).
4   Loosen the lug nuts on the right front wheel.
5   Raise the front of the vehicle and support it securely on jackstands, then remove the wheel.
6   Remove the right front wheel and inner fender liner (see Chapter 11).
7   Place a block of wood between a floor jack head and the oil pan, and support the engine.
8   Remove the right engine mount (see Section 17).
9   Remove the crankshaft pulley (see Section 11) and the drivebelt tensioner (see Chapter 1).
10   Remove the power steering pump mounting bolts and position the power steer-

ing pump off to the side without disconnecting the power steering system fluid lines (see Chapter 10).
11   Remove the lower timing belt cover and the upper timing belt cover from the cylinder heads **(see illustrations)**. If the same belt is to be reinstalled, mark the direction of rotation on the belt using white paint.

12   Temporarily install the crankshaft pulley bolt and turn the crankshaft with the bolt to align the timing marks on the crankshaft and camshaft sprockets. The crankshaft sprocket arrow should line up with the TDC indicator on the oil pump cover and the camshaft sprocket marks should line up with the marks on the top of the rear covers **(see illustrations)**.

6.12d Overall view of the crankshaft and camshaft sprocket alignment marks on the 3.5L V6 engine

6.13 Remove the timing belt tensioner bolts, then remove the tensioner and the timing belt

6.14 Compress the plunger on the tensioner in a vise until a drill bit or Allen wrench can be inserted through the hole to retain the plunger - orient the pin in a way that won't interfere with installing the tensioner with the pin in place

13   Unbolt the timing belt tensioner and remove the belt **(see illustration)**.

14   Slowly compress the timing belt tensioner in a vise until a drill bit or Allen wrench can be inserted through the hole to lock it in **(see illustration)**.

Note: *Total bleed down time of the tensioner should take approximately four to five minutes. DO NOT force tensioner to bleed down faster or damage to the tensioner may occur.*

15   Making sure that all timing marks are still aligned, start by installing the new belt at the crankshaft pulley, proceeding counterclockwise up to the left (driver's side) camshaft sprocket. After the belt is on the left camshaft sprocket, keep tension on the belt as it is fed under the water pump pulley, over the right camshaft sprocket, and past the tensioner. If the old timing belt is being reused, be sure the directional mark made in Step 11 is pointing in the proper direction.

16   Hold the tensioner pulley against the belt

and install the tensioner. Tighten the tensioner bolts to the torque listed in this Chapter's Specifications.

17   Remove the Allen key or drill bit from the tensioner, allowing it to tension the pulley on the belt. Tighten the rocker arm shaft bolts, in the proper sequence **(see illustration 5.11)**, to the torque listed in this Chapter's Specifications.

18   Rotate the crankshaft pulley through two complete revolutions to check that the timing marks still remain aligned. If not, repeat the belt installation process.

19   The remainder of installation is the reverse of removal.

20   Start the engine and check for leaks.

Note: *Noise may be present upon initial start up, due to air entering the timing chain tensioner. If this situation occurs, raise the idle to 1,600 - 2,000 rpm for ten minutes to purge the air from the tensioner. This noise should last no longer than 15 minutes after initial start up.*

7   Camshafts - removal, inspection and installation

Removal

1   Remove the valve cover (see Section 4), the timing belt (see Section 6) and the rocker arm assembly (see Section 5). If you're removing the camshaft from the rear cylinder head, remove the PCM (see Chapter 6).

2   Prevent the camshaft from turning by holding the sprocket with a 1-7/16 inch box end wrench, then remove the camshaft sprocket bolt and detach the sprocket from the camshaft. At the other end of the cylinder head, remove the camshaft thrust plate and O-ring seal and withdraw the camshaft from the cylinder head, being careful not to nick the cam bearings or journals. Also remove the camshaft oil seal.

Inspection

3   After the camshaft has been removed from the engine, cleaned with solvent and dried, inspect the bearing journals for uneven wear, pitting and evidence of seizure. If the camshaft journals are damaged, inspect the cylinder head also.

4   Measure the bearing journals with a micrometer to determine if they are excessively worn or out-of-round **(see illustration)**. Compare the measurements with this Chapter's Specifications. Use an inside micrometer to measure the journal diameters in the cylinder head. Compare the measurements with this Chapter's Specifications.

5   Measure the lobe height of each cam lobe on the intake camshaft and record your measurements **(see illustration)**. Compare the measurements for excessive variations. Standard measurements should be 0.001 inch (0.025 mm). If the lobe heights vary more than 0.010 inch (0.254 mm), replace the camshaft. Compare the lobe height measure-

7.4 Check the diameter of each camshaft bearing journal to pinpoint excessive wear and out-of-round conditions

7.5 Measure the camshaft lobe height (greatest dimension) with a micrometer

7.6 Check the cam lobes for pitting, excessive wear and scoring. If scoring is excessive, as shown here, replace the camshaft

7.7 Coat both the cam lobes and the journals with camshaft installation lube before installing the camshaft

8.6a Location of the front upper intake manifold mounting bracket mounting studs/bolts

ments on the exhaust camshaft and follow the same procedure. Do not compare intake camshaft lobe heights with exhaust camshaft lobe heights, as they are different. Only compare intake lobes with intake lobes and exhaust lobes with other exhaust lobes.

6 Check the camshaft lobes for heat discoloration, score marks, chipped areas, pitting and uneven wear **(see illustration)**. If the lobes are in good condition and if the lobe lift variation measurements recorded earlier are within the limits, the camshaft can be reused.

Installation

7 Lubricate the camshaft bearing journals and cam lobes with camshaft installation lube **(see illustration)**.

8 Insert the camshaft carefully into the cylinder head, then install a new seal at the front of the head. Using a seal driver or a deep socket with an outside diameter slightly smaller than that of the seal, tap the new seal in.

9 Install the camshaft cover at the back of the head with a new O-ring.

10 The remainder of installation is the reverse of removal (see Sections 4, 5 and 6).

8 Intake manifolds - removal and installation

Upper intake manifold

1 Relieve the fuel system pressure (see Chapter 4).

2 Disconnect the negative battery cable from the remote ground terminal (see Chapter 5).

3 Remove the air filter housing (see Chapter 4).

4 Label and disconnect the hoses and electrical connectors attached to the intake manifold and throttle body.

5 Remove the bolts securing the EGR pipe to the upper intake manifold (see Chapter 6).

6 Remove the brackets from the throttle body and the right side of the upper intake manifold **(see illustrations)**.

7 Loosen the upper intake manifold bolts, starting with the outer bolts and working to the inside bolts, and remove the upper intake manifold **(see illustration)**.

8 Clean the mounting surfaces of the lower intake manifold and the upper manifold.

8.6b Location of the rear upper intake manifold mounting bracket mounting studs/bolts

Inspect the rubber O-ring gasket for cracking, hardness or other signs of deterioration. If it's in good condition it can be reused.

9 Install the gasket into the groove in the upper intake manifold **(see illustration)**,

8.7 Lift the upper intake manifold from the engine

8.9 Install the upper intake manifold gasket, making sure it seats in the groove

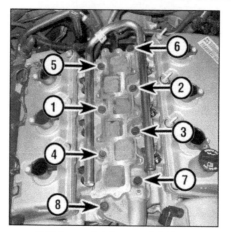

8.19 Apply a bead of RTV sealant around the water passages at the ends of the heads

8.22 Lower intake manifold bolt TIGHTENING sequence

then install the upper intake manifold onto the lower intake manifold. Tighten the bolts, starting with the center bolts and working to the outer bolts using a circular pattern, to the torque listed in this Chapter's Specifications. The remainder of installation is the reverse of removal.

Lower intake manifold

Warning: *Wait until the engine is completely cool before beginning this procedure.*

10 Drain the cooling system (see Chapter 1).

11 Remove the upper intake manifold (see Steps 1 through 7).

12 Label and detach any remaining hoses which would interfere with the removal of the lower intake manifold.

13 Remove the fuel rail and injectors from the lower intake manifold (see Chapter 4).

14 Remove the heater hoses and coolant hoses (see Chapter 1).

15 Loosen the remaining manifold mounting bolts/nuts in 1/4-turn increments until they can be removed by hand, and move the ignition coil capacitor out of the way.

16 The manifold will probably be stuck to

the cylinder heads and force may be required to break the gasket seal.

Caution: *Don't pry between the manifold and the heads or damage to the gasket sealing surfaces may occur, leading to vacuum leaks.*

17 Turn the manifold over, remove the heater supply tube and replace the O-ring with a new one.

18 Carefully use a scraper to remove all traces of old gasket material and sealant from the manifold and cylinder heads, then clean the mating surfaces with lacquer thinner or acetone.

19 Apply a bead of RTV sealant or equivalent around the front and rear cylinder head water passages before laying the new gaskets in place **(see illustration)**.

20 Install the intake manifold gaskets on the cylinder heads and apply RTV sealant to the manifold side of the gaskets, around the water passages.

21 Position the manifold on the engine, making sure the gaskets and manifold are aligned over the ports in the cylinder heads. Install the fuel rail and injectors onto the lower manifold, then install the manifold mounting bolts with the ignition coil capacitor.

22 Following the recommended tightening

sequence **(see illustration)**, tighten the nuts/bolts, in several steps, to the torque listed in this Chapter's Specifications.

23 The remainder of installation is the reverse of removal. Refill the cooling system (see Chapter 1), then run the engine and check for fuel, vacuum and coolant leaks.

9 Exhaust manifold(s) - removal and installation

Warning: *The engine must be completely cool before beginning this procedure.*

Removal

1 Disconnect the negative battery cable from the remote ground terminal (see Chapter 5).

2 Raise the front of the vehicle and support it securely on jackstands. Remove the engine splash shield fasteners and shield.

3 Disconnect the oxygen sensor electrical connectors and remove the sensors (see Chapter 6).

4 Remove the crossunder pipe bolts and detach the crossunder pipe from the exhaust manifolds.

5 On the rear manifold, remove the right front driveaxle (see Chapter 8).

6 Lower the vehicle, remove the exhaust manifold heat shield mounting bolts/nuts **(see illustration),** and remove the heat shields.

7 Remove the mounting bolts and detach the manifold from the cylinder head **(see illustration).**

Note: *Spray penetrating oil on the bolts before attempting to remove them.*

Installation

8 Clean the mating surfaces to remove all traces of old gasket material, then inspect the manifold for distortion and cracks. Warpage can be checked with a precision straightedge held against the mating flange. If a feeler gauge thicker than 0.030-inch can be inserted between the straightedge and flange surface, take the manifold to an automotive machine

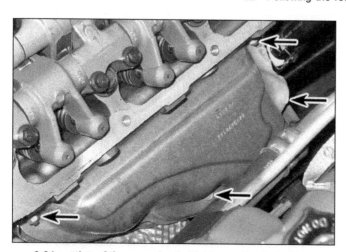

9.6 Location of the upper heat shield mounting bolts/nuts

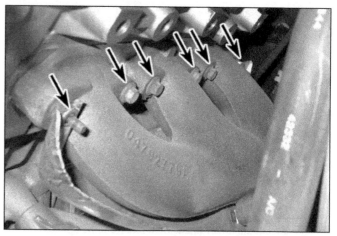

9.7 Location of the exhaust manifold mounting bolts/studs

10.16a Hold the camshaft sprocket with a wrench while loosening the bolt with a socket and breaker bar - these bolts are very tight

10.16b The camshaft sprocket bolts are not interchangeable - the left sprocket bolt is 10 inches while the right sprocket bolt is 8-3/8 inches

shop for resurfacing.

9   Place the exhaust manifold in position with a new gasket and install the mounting bolts finger tight.

Note: *Be sure to identify the exhaust manifold gaskets by the correct cylinder designation and the position of the exhaust ports on the gasket.*

10   Starting in the middle and working out toward the ends, tighten the mounting bolts in several increments, to the torque listed in this Chapter's Specifications.

11   The remainder of installation is the reverse of removal.

12   Start the engine and check for exhaust leaks between the manifold and cylinder head and between the manifold and exhaust pipe.

10   Cylinder heads - removal and installation

Warning: *The engine must be completely cool before beginning this procedure.*

Note: *Special tools are necessary to complete this procedure. Read through the entire procedure and obtain the special tools before beginning work.*

Removal

1   Relieve the fuel system pressure (see Chapter 4). Disconnect the negative battery cable from the remote ground terminal (see Chapter 5).

2   Drain the cooling system (see Chapter 1).

3   Remove the air filter housing (see Chapter 4).

4   Remove the coolant recovery bottle (see Chapter 3).

5   If you're removing the front bank cylinder head, remove the cooling fan and shroud assembly (see Chapter 3).

6   Remove the alternator (see Chapter 5).

7   Remove the upper intake manifold (see Section 8).

8   Remove the fuel rail (see Chapter 4) and the lower intake manifold (see Section 8).

9   Remove the ignition coils and the spark plugs (see Chapters 1 and 5, if necessary).

10   Detach the exhaust manifold from the cylinder head being removed (see Section 9).

11   Remove the power steering pump and set it aside without disconnecting the fluid lines (see Chapter 10).

12   Remove the valve cover(s) (see Section 4).

13   Remove the rocker arms and shafts (see Section 5).

14   Support the engine with a engine with an engine hoist (see Chapter 2E) and remove the crankshaft pulley.

15   Remove the timing belt covers, align the TDC marks and remove the timing belt (see Section 6).

Note: *If the belt is to be reused, mark its direction of rotation before removing it.*

16   Hold the camshaft sprocket hex with a wrench while using a socket and breaker bar to loosen the camshaft bolt, which is under considerable torque **(see illustration).**

Caution: *The two camshaft sprockets are not interchangeable, nor are the camshaft sprocket bolts* **(see illustrations)**. *The front sprocket bolt is 10 inches while the rear sprocket bolt is 8-3/8 inches. Always use NEW camshaft sprocket bolts.*

17   Remove the camshaft thrust plate from the rear of the head and slide the camshaft back approximately 3-1/2 inches. This will allow removal of the sprocket and bolt.

18   Remove the bolts securing the rear timing belt cover to the cylinder head. Don't disturb any of the other timing belt cover bolts.

19   Using the new head gasket, outline the cylinders and bolt pattern on a piece of cardboard. Be sure to indicate the front of the engine for reference. Punch holes at the bolt locations. This cardboard holder will be used to accurately store each cylinder head bolt after they are removed from the cylinder head.

10.16c The camshaft sprockets are also not interchangeable

20   Loosen each of the cylinder head mounting bolts 1/4-turn at a time until they can be removed by hand - work from bolt-to-bolt in the reverse of the tightening sequence **(see illustration 10.28)**. Store the bolts in the cardboard holder as they're removed.

21   Lift the head(s) off the engine. If resistance is felt, don't pry between the head and block, as damage to the mating surfaces will result. Recheck for head bolts that may have been overlooked, then use a hammer and block of wood to tap up on the head to break the gasket seal. Be careful because there are locating dowels in the block which position each head. As a last resort, pry each head up at the rear corner only and be careful not to damage anything. After removal, place the head on blocks of wood to prevent damage to the gasket surfaces.

22   If the camshaft is to be removed for examination or replacement, it must be done with the cylinder head off the engine, as it comes out the back of the cylinder head (see Section 7).

10.23 Use a putty knife or gasket scraper to remove gasket material from the cylinder head and block

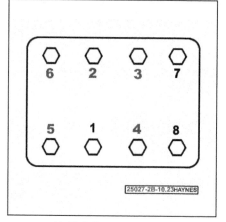

10.28 Cylinder head bolt TIGHTENING sequence

11.8 Insert a prybar or large screwdriver through the pulley to hold it while you loosen the crankshaft pulley bolt

Installation

23   The mating surfaces of each cylinder head and block must be perfectly clean when the head is installed. Use a gasket scraper to remove all traces of carbon and old gasket material **(see illustration)**, then clean the mating surfaces with lacquer thinner or acetone. If there's oil on the mating surfaces when the head is installed, the gasket may not seal correctly and leaks may develop. When working on the block, it's a good idea to cover the valley with shop rags to keep debris out of the engine. Use a shop rag or vacuum cleaner to remove any debris that falls into the cylinders.

24   Check the block and head mating surfaces for nicks, deep scratches and other damage. If damage is slight, it can be removed with a file; if it's excessive, machining may be the only alternative.

25   Use a tap of the correct size to chase the threads in the head bolt holes. Dirt, corrosion, sealant and damaged threads will affect torque readings. Check the cleaned head bolts for stretch by holding them next to a steel ruler. If all the threads don't touch the ruler, the bolt should be replaced.

26   Position the new gasket over the dowel pins in the block. Some gaskets are marked TOP or FRONT to ensure correct installation.

27   Make sure the camshaft is installed in the cylinder head, then carefully position the head on the block without disturbing the gasket.

28   Tighten the bolts in the recommended sequence **(see illustration)** to the torque listed in this Chapter's Specifications.

29   Install the rear timing belt cover-to-cylinder head bolts. Tighten the bolts to the torque listed in this Chapter's Specifications.

30   Slide the camshaft back into the cylinder head and install the camshaft sprocket and new bolt, then move the camshaft back into place. Hold the hex on the camshaft sprocket with a wrench and tighten the camshaft bolt to the torque listed in this Chapter's Specifica-

tions **(see illustration 10.16a)**.

31   Install the camshaft thrust plate at the back of the head, using a new O-ring. Tighten the bolts to the torque listed in this Chapter's Specifications.

32   The remainder of installation is the reverse of removal.

33   Change the oil and filter, refill the cooling system (see Chapter 1), run the engine and check for leaks.

11   Crankshaft pulley - removal and installation

1   Disconnect the negative battery cable from the remote ground terminal (see Chapter 5).

2   Loosen the lug nuts on the right front wheel, raise the front of the vehicle and support it securely on jackstands.

3   Remove the right front wheel and inner fender liner (see Chapter 11).

4   Remove the right engine mount (see Section 17).

5   Remove the accessory drivebelts (see Chapter 1).

6   Connect an engine hoist to the engine (see Chapter 2E) and support the engine.

7   Remove the lower crossmember (see Chapter 10).

8   Lower the engine enough to position a large screwdriver through the crankshaft pulley to keep the crankshaft from turning, and remove the pulley-to-crankshaft bolt **(see illustration)**.

9   Pull the crankshaft pulley off the crankshaft with a puller **(see illustration)**.

Caution: *The jaws of the puller must only contact the hub of the pulley - not the outer ring.*

Note: *A spacer that bears on the nose of the crankshaft or a long Allen-head bolt must be inserted into the crankshaft nose for the puller's tapered tip to push against. Don't allow the puller screw to bear against the crankshaft itself.*

10   To install the crankshaft pulley, slide the pulley onto the crankshaft as far as it will slide on, then use a vibration damper installation tool to press the pulley onto the crankshaft.

11   Install the crankshaft pulley retaining bolt and tighten it to the torque listed in this Chapter's Specifications.

12   The remainder of installation is the reverse of removal.

12   Crankshaft front oil seal - replacement

1   Remove the crankshaft pulley (see Section 11).

2   Remove the timing belt (see Section 6).

3   Use a two-bolt puller to remove the crankshaft sprocket, with the bolts threaded into the holes in the sprocket.

4   With a hammer and punch, tap the dowel pin out of the crankshaft. Drive the pin into the hollow, threaded area of the crankshaft snout, where it can be extracted with a small magnet.

11.9 Use a three jaw puller to remove the crankshaft pulley. A long Allen bolt must be inserted into the crankshaft for the tool to push against

12.5 Using a hook-type seal remover, pry out the old seal

12.6 Drive the new seal in to the same depth as the old one, using a seal driver or a socket with an outside diameter the same as that of the seal

5   Pry the old seal out with a hook-type seal tool, being very careful not to scratch the seal surface of the crankshaft **(see illustration)**. Note how the seal is installed - the new one must be installed to the same depth and facing the same way.

6   Lubricate the inner lip of the new seal with engine oil and drive it in with a seal driver or a large deep socket and hammer **(see illustration)**.

7   Installation is the reverse of removal. Tap the dowel pin back into the crankshaft, with 3/64-inch extending out.

13   Oil pan - removal and installation

Removal

1   Disconnect the negative battery cable from the remote ground terminal (see Chapter 5).

2   Remove the engine oil dipstick, then remove the oil dipstick tube mounting bolt and pull the tube out of the oil pan.

3   Raise the front of the vehicle and support it securely on jackstands. Apply the parking brake and block the rear wheels to keep the vehicle from rolling off the stands.

4   Connect an engine hoist to the engine (see Chapter 2E) and support the engine.

5   Remove the engine splash shield (see Section 9).

6   Drain the engine oil and remove the oil filter (see Chapter 1).

7   Raise the engine with the hoist just enough to allow the crossmember to be removed.

8   Remove the front crossmember mounting bolts and crossmember (see Chapter 10).

9   Loosen the exhaust manifold bolts (see Section 9), then remove the exhaust cross-over pipe nuts and pipe.

10   Remove the four oil pan-to-bellhousing bolts.

11   Remove the oil pan bolts, then carefully separate the oil pan from the block. Don't pry between the block and the pan or damage to

the sealing surfaces could occur and oil leaks may develop. Instead, tap the pan with a soft-face hammer to break the gasket seal.

Installation

12   Clean the pan with solvent and remove all old sealant and gasket material from the block and pan mating surfaces. Clean the mating surfaces with lacquer thinner or acetone and make sure the bolt holes in the block are clear. Check the oil pan flange for distortion, particularly around the bolt holes.

13   Apply a bead of RTV sealant to the parting lines between the engine block and the oil pump housing and then to the engine block and rear main oil seal retainer. Install the gasket on the block.

14   Place the oil pan in position on the block and install the bolts, tightening them finger-tight.

15   Lower the engine and install the engine mount nuts, tightening them to the torque listed in this Chapter's Specifications.

16   Install the oil pan-to-transaxle bolts and tighten them to approximately 12 in-lbs.

17   Tighten the oil pan fasteners to the torque listed in this Chapter's Specifications. Starting at the center, follow a criss-cross pattern and work up to the final torque in three steps. Don't forget the two small Allen bolts at the rear, accessible through the torque converter access opening.

18   Tighten the oil-pan-to-transaxle bolts to the torque listed in this Chapter's Specifications.

19   The remainder of installation is the reverse of removal.

20   Refill the engine with oil (see Chapter 1), replace the filter, run it until normal operating temperature is reached and check for leaks.

14   Oil pump - removal, inspection and installation

Caution: *Do not remove the oil pressure relief valve, spring and plunger from the oil pump. Replace the oil pump as one complete assembly if the oil pressure relief valve is damaged.*

Removal

1   Remove the crankshaft sprocket (see Section 11).

2   Remove the oil pan (see Section 13).

3   Remove the oil pump pick-up tube.

4   Remove the oil-pump-to-block bolts and pull the oil pump from the block.

Inspection

5   Unbolt the cover from the back of the oil pump housing.

6   Remove the inner and outer rotors from the oil pump body, noting their installed direction for reassembly.

7   Clean all parts thoroughly in solvent and carefully inspect the rotors, pump cover and oil pump housing for nicks, scratches or burrs. Replace the assembly if it is damaged.

8   Use a straightedge and measure the oil pump cover for warpage with a feeler gauge. If it's warped more than the limit listed in this Chapter's Specifications, the pump should be replaced.

9   Measure the thickness and the diameter of the outer rotor. If either measurement is less than the value listed in this Chapter's Specifications, the pump should be replaced **(see illustration)**.

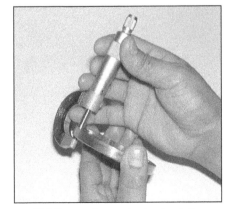

14.9 Use a micrometer to check the thickness of the outer rotor

14.11 Check the outer rotor-to-housing clearance

14.12 Check the clearance between the lobes of the inner and outer rotors

10   Measure the thickness of the inner rotor. If the measurement is less than the value listed in this Chapter's Specifications, the pump should be replaced.

11   Insert the outer rotor into the oil pump housing and, while holding the rotor against one side of the housing with your finger, measure the clearance at the opposite side between the rotor and housing **(see illustration)**. If the measurement is more than the maximum allowable clearance listed in this Chapter's Specifications, the pump should be replaced.

12   Install the inner rotor in the oil pump assembly and measure the clearance between the lobes on the inner and outer rotors **(see illustration)**. If the clearance is more than the value listed in this Chapter's Specifications, the pump should be replaced.

Note: *Install the inner rotor with the mark facing up.*

13   Position a straightedge across the face of the oil pump assembly **(see illustration)**. If the clearance between the pump surface and the rotors is greater than the limit listed in this Chapter's Specifications, the pump should be replaced.

Installation

14   Prime the oil pump by filling the housing with engine oil. Install the pump cover and tighten the bolts to the torque listed in this Chapter's Specifications.

15   To install the pump, place the gasket on the engine block mating surface, then turn the flats in the rotor so they align with the flats on the crankshaft, and install the oil pump housing to the engine block.

16   Install the pump-to-block bolts and tighten them to the torque listed in this Chapter's Specifications.

17   The remainder of installation is the reverse of removal.

15   Driveplate - removal and installation

This procedure is essentially the same as for the 2.7L V6 engine. Refer to Chapter 2B and follow the procedure outlined there, but use the bolt torque listed in this Chapter's Specifications.

16   Rear main oil seal - replacement

Note: *The rear main seal and housing are serviced as one complete assembly. This procedure will require several special tools. Read the procedure carefully before starting.*

This procedure is essentially the same as for the 2.7L V6 engine. Refer to Chapter 2B and follow the procedure outlined there, but use the bolt torque value listed in this Chapter's Specifications.

17   Engine mounts - check and replacement

1   There are four powertrain mounts: the front mount attaches the transaxle to the crossmember; the left mount attaches the driver's side of the engine to the subframe; the right engine mount attaches the engine block to the passenger's side of the subframe; and a rear mount attaches the engine block to the subframe.

Check

2   During the check, the engine must be raised slightly to remove the weight from the mounts.

3   Raise the vehicle and support it securely on jackstands. Remove the front wheels and tires. Position two jacks, one under the crankshaft pulley and the other under the transaxle bellhousing. Place a block of wood between a floor jack head and the crankshaft pulley or bellhousing, then carefully raise the engine/transaxle just enough to take the weight off the mounts.

Warning: *DO NOT place any part of your body under the engine when it's supported only by a jack!*

4   Check the mounts to see if the rubber is cracked, hardened or separated from the metal plates. Sometimes the rubber will split right down the center.

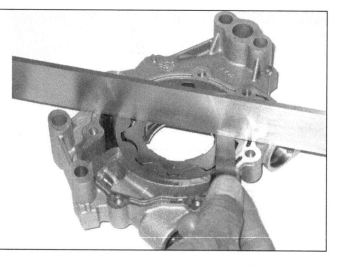

14.13 Using a straightedge and feeler gauge, check the side clearance between the surface of the oil pump housing and the rotors

5   Check for relative movement between the mount plates and the engine or subframe (use a large screwdriver or prybar to attempt to move the mounts). If movement is noted, lower the engine and tighten the mount fasteners.

6   Rubber preservative should be applied to the mounts to slow deterioration.

Replacement

Front mount

7   Raise the vehicle and support it securely on jackstands.

8   Remove the crossmember-to-mount bolts from under the crossmember.

9   Remove the front mount through bolt and remove the mount.

Note: *It may be necessary to support the transaxle, with a floor jack and wooden block placed between the jack and the transaxle, to remove the through bolt.*

10   The remainder of installation is the reverse of removal.

Left mount

11   Disconnect the negative battery cable from the remote ground terminal (see Chapter 5). Raise the vehicle and support it securely on jackstands.

12   Remove the middle engine splash shield

(see Section 9).

13   Remove the air filter housing (see Chapter 4).

14   Support the transaxle with a floor jack and wooden block placed between the jack and the transaxle.

15   Raise the transaxle enough to take the weight off of the mount.

16   Remove the mounting bolts and mount. Replace the left mount with a new one and tighten the bolts to the torque listed in this Chapter's Specifications.

17   The remainder of installation is the reverse of removal.

Right mount

18   Raise the vehicle and support it securely on jackstands.

19   Remove the power steering fluid reservoir (see Chapter 10) and set it out of the way.

20   Support the engine with a floor jack and wooden block placed between the jack and the engine.

21   Remove the three engine mount bracket-to-timing cover support bolts.

22   Remove the two engine mount bracket-to-mount nuts and remove the bracket.

23   Remove the two engine mount-to-body bolts.

24   Remove the engine mount and replace it with the new one. Tighten the mount-to-body bolts to the torque listed in this Chapter's Specifications.

25   Lower the engine and install the engine mount bracket nuts and bolts, then tighten them to the torque listed in this Chapter's Specifications.

Rear mount

26   Working from the top rear of the engine compartment, remove the mounting bracket-to-block bolts.

27   Raise the vehicle and support it securely on jackstands.

28   Remove the four mount-to-subframe mounting bolts and remove mount assembly form the vehicle.

29   Remove the bracket and mount through bolt and separate the bracket from the mount.

30   Replace the mount with a new one and install the bracket and through bolt to the mount.

31   Install the mount to the subframe, then install the bracket to the engine to the engine block and tighten the bolts to the torque listed in this Chapter's Specifications.

32   Tighten the through bolt to the torque listed in this Chapters Specifications.

Notes

Chapter 2 Part D
3.6L V6 engine

Contents

	Section			Section
Camshaft(s) – removal, inspection and installation	9	Oil pans - removal and installation		11
Crankshaft front oil seal - replacement	7	Oil pump - removal, inspection and installation		12
Crankshaft pulley - removal and installation	6	Rear main oil seal - replacement		15
Cylinder heads - removal and installation	10	Repair operations possible with the engine in the vehicle		2
Driveplate - removal and installation	14	Timing chain cover, chain and sprockets - removal,		
Engine mounts - check and replacement	16	inspection and installation		8
General information	1	Top Dead Center (TDC) for number one piston - locating		3
Intake manifolds - removal and installation	5	Valve covers - removal and installation		4
Oil cooler - removal and installation	13			

Specifications

General

Displacement	220 cubic inches	
Bore	3.779 inches	95.987 mm
Stroke	3.268 inches	83.007 mm
Compression ratio	10.2:1	
Cylinder numbers (drivebelt end-to-transaxle end)		
Rear bank	1-3-5	
Front bank (radiator side)	2-4-6	
Firing order	1-2-3-4-5-6	
Oil pressure		
At idle speed	5 psi (minimum)	34.5 kPa (minimum)
At 3,000 rpm	62 to 139 psi	427.5 to 958.4 kPa

Camshaft

Bore diameter		
Cam tower 1	1.2606 to 1.2615 inches	32.0192 to 32.0421 mm
Cam tower 2, 3, and 4	0.9457 to 0.9465 inch	24.021 to 24.041 mm
Bearing journal diameter		
No. 1	1.2589 to 1.2596 inches	31.976 to 31.994 mm
No. 2, 3, and 4	0.9440 to 0.9447 inch	23.978 to 23.995 mm
Bearing clearance		
No. 1	0.0001 to 0.0026 inch	0.0025 to 0.0660 mm
No. 2, 3, and 4	0.0009 to 0.0025 inch	0.0229 to 0.0635 mm
End play	0.003 to 0.01 inch	0.076 to 0.254 mm

Cylinder head

Gasket thickness (compressed)	0.019 to 0.024 inch	0.4826 to 0.6096 mm
Warpage limit	0.0035 inch	0.09 mm

**Front
of
Vehicle**

 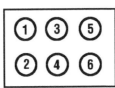

**Cylinder and coil
terminal locations**

30013-1-specs HAYNES

Torque specifications

Ft-lbs (unless otherwise indicated)   Nm

Note: *One foot-pound (ft-lb) of torque is equivalent to 12 inch-pounds (in-lbs) of torque. Torque values below approximately 15 ft-lbs are expressed in inch-pounds, since most foot-pound torque wrenches are not accurate at these smaller values.*

	Ft-lbs	Nm
Camshaft bearing cap bolts	84 in-lbs	9
Crankshaft pulley bolt		
Step 1	30	40
Step 2	Tighten an additional 105-degrees	
Cylinder head bolts* (in sequence - **see illustrations 10.29a and 10.29b**)		
2014 and earlier models		
Step 1	22	
Step 2	33	
Step 3	Tighten an additional 75 degrees	
Step 4	Tighten an additional 50 degrees	
Step 5	Loosen all in reverse of tightening sequence	
Step 6	22	
Step 7	33	
Step 8	Tighten an additional 70 degrees	
Step 9	Tighten an additional 70 degrees	
2015 and later models**		
Step 1	22	
Step 2	33	
Step 3	33	
Step 4	Tighten an additional 125 degrees	
Step 5	Loosen all in reverse of tightening sequence	
Step 6	22	
Step 7	33	
Step 8	33	
Step 9	Tighten an additional 130 degrees	
Drivebelt idler sprocket bolt	18	24
Driveplate-to-crankshaft bolts	70	95
Catalytic converter to cylinder head fasteners	27	37
Exhaust crossover bolts	21	28
Intake manifold (upper) retaining bolts		
2013 and earlier	71 in-lbs	8
2014 and later	89 in-lbs	10
Intake manifold (lower) retaining bolts		
2013 and earlier	71 in-lbs	8
2014 and later	106 in-lbs	12
Oil cooler		
Bolts	106 in-lbs	4
Screws	35 in-lbs	12
Oil pan drain plug	20	27
Oil pan		
Lower pan-to-upper pan nut/bolts	97 in-lbs	11
Upper pan-to-rear main seal housing (M6 bolts)	108 in-lbs	12
Upper pan-to-cylinder block (M8 bolts)	18	24
Upper pan-to-transaxle bolts	41	56
Oil pump pick-up tube mounting bolts	106 in-lbs	12
Oil pump cover (plate) screws	105 in-lbs	12
Oil pump-to-engine block fasteners	106 in-lbs	12
Rear main oil seal retainer bolts	105 in-lbs	12
Timing cover bolts		
M6 bolts	106 in-lbs	12
M8 bolts	18	24
M10 bolts	41	56
Timing gear splash shield bolts	35 in-lbs	4
Camshaft chain tensioner bolts (primary)	106 in-lbs	12
Camshaft chain guide bolts	106 in-lbs	12
Camshaft chain idler sprocket bolt	18	24
Camshaft chain tensioner bolts (secondary)	106 in-lbs	12
Camshaft oil control valve (sprocket bolt)	111	150
Variable valve timing solenoid bolt	35 in-lbs	4
Oil pump timing chain sprocket (T45)	18	24
Valve cover-to-cylinder head bolts	106 in-lbs	12
Water pump bolts	See Chapter 3	

* Use new bolts

** If using a new engine block, follow all nine Steps. If using the existing block, follow Steps 6 to 9.

1 General information

1 This Part of Chapter 2 is devoted to in-vehicle repair procedures for the 3.6L V6 engine.

2 The 3.6 liter engine utilizes Variable Valve Timing (VVT), Dual Overhead Camshafts (DOHC), four timing chains, an aluminum cylinder block, steel cylinder sleeves or liners with six cylinders arranged in a "V"-shape, with 60-degrees between the two banks. The 3.6 liter engine has a chain driven oil pump with a multi-stage pressure regulator to increase fuel economy. The exhaust manifolds are integral with the cylinder heads to make the engine lighter.

Caution: *This engine is of an interference design and severe engine damage will occur if the timing chain breaks.*

3 The cylinders are numbered from front to rear. The right bank is numbered 1, 3, 5 and the left bank is numbered 2, 4, 6. The firing order is 1–2–3–4–5–6.

4 Information concerning engine removal and installation can be found in Chapter 2E. The following repair procedures are based on the assumption that the engine is installed in the vehicle. If the engine has been removed from the vehicle and mounted on a stand, many of the steps outlined in this Part of Chapter 2 do not apply.

2 Repair operations possible with the engine in the vehicle

1 Many major repair operations can be done without removing the engine from the vehicle.

2 Clean the engine compartment and the exterior of the engine with degreaser before any work is done. It'll make the job easier and help keep dirt out of internal parts of the engine.

3 It may be helpful to remove the hood to improve engine access when repairs are performed (see Chapter 11). Cover the fenders to prevent damage to the paint. Special pads are available, but an old bedspread or blanket will also work.

4 If vacuum, exhaust, oil, or coolant leaks develop, indicating a need for gasket or seal replacement, the repairs can generally be done with the engine in the vehicle. The intake and exhaust manifold gaskets, timing chain cover gasket, oil pan gasket, crankshaft oil seals, and cylinder head gaskets are all accessible with the engine in the vehicle.

5 Exterior engine components, such as the intake and exhaust manifolds, the oil pan, the oil pump, the timing chain cover, the water pump, the starter motor, the alternator, and fuel system components can be removed for repair with the engine in the vehicle.

6 Cylinder heads can be removed without pulling the engine. Valve component servicing can also be done with the engine in the vehicle. Replacement of the timing chain and

4.4 Lift the insulator up and off of the retaining posts, then remove it from the front valve cover

sprockets is also possible with the engine in the vehicle, as is camshaft and valvetrain removal and installation.

7 Repair or replacement of piston rings, pistons, connecting rods, and rod bearings is possible with the engine in the vehicle, however, this practice is not recommended because of the cleaning and preparation work that must be done to the components.

3 Top Dead Center (TDC) for number one piston - locating

1 Top Dead Center (TDC) is the highest point in the cylinder that each piston reaches as it travels up the cylinder bore. Each piston reaches TDC on the compression stroke and again on the exhaust stroke, but TDC generally refers to piston position on the compression stroke.

2 Positioning the piston(s) at TDC is an essential part of certain procedures such as camshaft and timing chain/sprocket removal.

3 Before beginning this procedure, be sure to place the transaxle in Neutral and apply the parking brake or block the rear wheels. Disconnect the negative battery cable from the remote ground terminal or battery (see Chapter 5). Remove the ignition coils (see Chapter 5) and the spark plugs (see Chapter 1).

4 Install a compression pressure gauge in the number one spark plug hole (see Chapter 2E). It should be a gauge with a screw-in fitting and a hose at least six inches long.

5 Rotate the crankshaft using a socket and breaker bar on the crankshaft pulley bolt while observing for pressure on the compression gauge. The moment the gauge shows pressure, indicates that the number one cylinder has begun the compression stroke.

6 Once the compression stroke has begun, TDC for the compression stroke is reached by bringing the piston to the top of the cylinder.

Note: *If a compression gauge is not available, you can simply place a blunt object over the spark plug hole and listen for compression as the engine is rotated. Once compression at the No.1 spark plug hole is noted, the remainder of the Step is the same.*

7 This engine is not equipped with external components (crankshaft pulley, flywheel, timing hole, etc.) that are marked to identify the position of number 1 TDC. Therefore, the only method to double-check the location of TDC number 1 is to remove the valve cover to access the camshaft sprockets and alignment marks (see Section 8).

8 After the number one piston has been positioned at TDC on the compression stroke, TDC for any of the remaining cylinders can be located by turning the crankshaft 120-degrees and following the firing order (refer to this Chapter's Specifications). For example, rotating the engine 120-degrees past TDC number 1 will put the engine at TDC compression for cylinder number 2.

4 Valve covers - removal and installation

Removal

1 Disconnect the negative battery cable from the remote ground terminal or battery (see Chapter 5).

2 Remove the engine cover.

3 Remove the upper intake manifold (see Section 5).

Note: *Cover open ports on the intake to prevent debris from entering the engine.*

Caution: *Once the valve covers are removed, the magnetic timing wheels are exposed* **(see illustration 9.9)**. *The magnetic timing wheels on the camshafts must not come in contact with any type of magnet or magnetic field. If contact is made, the timing wheels will need to be replaced.*

4 Remove insulator from the front valve cover **(see illustration)**.

5 Before removing the variable valve timing solenoid connectors from the front of each valve cover, mark them appropriately so they can be reinstalled in their original locations.

6 Disconnect the wiring harness retainers from the valve cover and move the harnesses out of the way.

7 Remove the ignition coils on both sides of the engine (see Chapter 5).

8 Mark the Camshaft Position (CMP) sen-

4.10 Front valve cover mounting bolts

5.5 Loosen the resonator hose clamp, remove the attaching pin
and remove the resonator

sors to each valve cover so they can be reinstalled in their original locations, then remove
the sensor(s) (see Chapter 6).

9   Remove the PCV valve from the rear
cover (see Chapter 1).

10   Remove the valve cover fasteners **(see
illustration)** and remove the cover(s).

Caution: *If the cover is stuck to the cylinder
head, tap one end with a block of wood and
a hammer to jar it loose. If that doesn't work,
slip a flexible putty knife between the cylinder
head and cover to break the gasket seal. Don't
pry at the cover-to-cylinder head joint or damage to the sealing surfaces may occur (leading to future oil leaks).*

11   Remove and discard the valve cover
gasket, then remove the spark plug tube
seals.

Note: *The cover gaskets can be reused if they
are not damaged.*

Installation

12   The mating surfaces of each cylinder
head and valve cover must be perfectly clean
when the covers are installed. Use a gasket
scraper to remove all traces of sealant and

old gasket material, then clean the mating
surfaces with brake system cleaner. If there's
sealant or oil on the mating surfaces when the
cover is installed, oil leaks may develop.

13   Inspect the spark plug tube seals; if
damaged, carefully remove the seals using
an appropriate pry tool. Position the new seal
with the part number facing the valve cover,
then use a socket that contacts the outer edge
to drive the seal in place.

14   Apply a dab of RTV sealant at the joints
where the engine front cover meets the cylinder head.

15   Install the valve cover and bolts, then
tighten the bolts to the torque listed in this
Chapter's Specifications.

16   The remainder of installation is the
reverse of removal.

5   Intake manifolds - removal and installation

Warning: *Wait until the engine is completely
cool before beginning this procedure.*

Removal

1   If you will be removing the lower intake
manifold, relieve the fuel system pressure
(see Chapter 4).

2   Disconnect the negative battery cable
from the remote ground terminal or battery
(see Chapter 5).

3   If equipped, remove the engine cover.

Upper intake manifold

4   Remove the upper radiator hose retainer
from the upper intake manifold (see Chapter 1).

5   Remove the intake resonator **(see illustration)**.

6   Disconnect the wiring harness from the
MAP sensor and the Electronic Throttle Control (ETC) **(see illustration)**.

7   Disconnect the PCV valve hose (see
Chapter 1), vapor purge hose and brake
booster hoses.

8   Disconnect the wiring harness retainers
from the upper intake support bracket and
the retainer from the stud bolt **(see illustration)**.

9   Remove the nuts and stud bolt, then

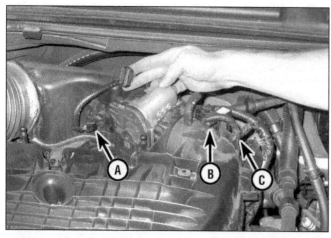

5.6 Disconnect the ETC connectors (A), the MAP sensor (B) and
the electrical harness (C) retainer

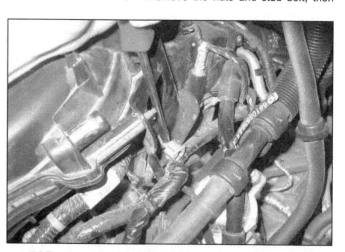

5.8 Pry the wiring harness retainer off of the bracket stud

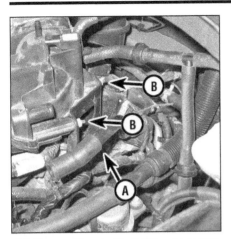

5.9 Remove the stud bolt (A) and bracket nuts (B), and remove the bracket

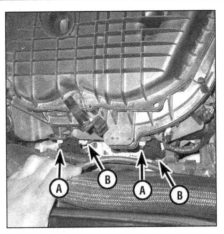

5.11 Pull the upper radiator hose back, remove the support bracket upper nuts (A), loosen the lower nuts (B) and remove the brackets from the upper intake manifold

5.12 Location of the upper intake manifold bolts

remove the upper intake manifold bracket **(see illustration)**.

10   Remove the nut from the bracket on the heater core return tube.

Note: *On 2015 and later models, there is no nut retaining the heater core return tube.*

11   Remove the support bracket-to-upper manifold nuts **(see illustration)**.

12   Loosen, but do not remove, the bolts on the manifold, and remove the upper intake manifold **(see illustration)**.

13   Discard the six upper-to-lower intake manifold seals, and cover the open intake ports to prevent debris from entering the engine.

14   If required, remove the insulator from the front valve cover **(see illustration 4.4)**.

Lower intake manifold

15   Remove the upper intake manifold (see Steps 4 through 14).

16   Disconnect the fuel line to the fuel rail (see Chapter 4).

17   Remove the fuel injectors and fuel rail (see Chapter 4).

Note: *The lower intake manifold can be removed with the injectors and fuel rail in place. Be careful not to damage the fuel injectors once the manifold is removed.*

18   Pry the wiring harness retainer from the end of the manifold and move the harness out of the way.

19   Remove the lower intake manifold bolts **(see illustration)**, and remove the manifold from the cylinder heads.

20   Discard the six manifold-to-cylinder head seals.

Installation

Lower intake manifold

Note: *The mating surfaces of the cylinder heads, cylinder block, and the intake manifold must be perfectly clean when the lower intake manifold is installed. Gasket removal solvents are available at most auto parts stores and may be helpful*

when removing old gasket material that's stuck to the cylinder heads, cylinder block and lower intake manifold (the lower intake manifold is made of aluminum - aggressive scrapping can cause damage). Be sure to follow the instructions printed on the solvent container.

21   Use a gasket scraper to remove all traces of sealant and old gasket material, then clean the mating surfaces with lacquer thinner or acetone. If there's old sealant or oil on the mating surfaces when the lower intake manifold is installed, oil or vacuum leaks may develop. Use a vacuum cleaner to remove gasket material that falls into the intake ports or the lifter valley.

22   If removed, install the fuel injectors and the fuel rail (see Chapter 4).

23   Install new intake manifold seals to the manifold.

Note: *Remove any rags or towels used in the manifold ports.*

24   Carefully lower the lower intake manifold into place **(see illustration)** and install the mounting bolts finger-tight.

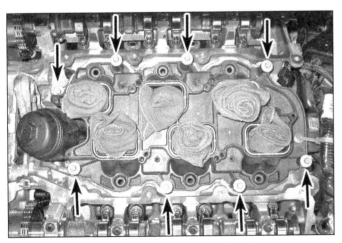

5.19 Lower intake manifold bolt locations

5.24 Install the manifold making sure the new intake seals do not fall out of the manifold

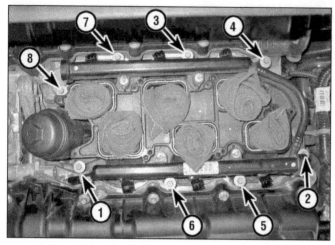

5.25 Lower intake manifold bolt TIGHTENING sequence

5.30 Upper intake manifold bolt TIGHTENING sequence

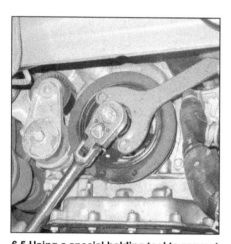

6.5 Using a special holding tool to prevent the crankshaft from turning, loosen, then remove the bolt

6.6 Slide the pulley from the end of the crankshaft; a puller shouldn't be required

7.2 Use a hook tool to pry the seal from the timing cover

25   Tighten the mounting bolts in steps, following the tightening sequence (see illustration), to the torque listed in this Chapter's Specifications.

26   Install the upper intake manifold. The remainder of installation is the reverse of removal.

Upper intake manifold

27   Check the condition of the rubber seals that are installed into each intake runner on the upper intake manifold. If they are damaged, replace the seals in the upper intake manifold.

28   Place the insulator on the mounting pins, if removed.

29   Install the upper intake manifold onto the lower intake manifold while pulling the bolts up.

Note: *The bolts are specially made for the composite material and should be turned slowly to prevent damage to the upper intake manifold.*

30   Tighten the mounting bolts in sequence (see illustration) to the torque listed in this Chapter's Specifications.

31   The remainder of installation is the reverse of removal.

32   Refill the cooling system, if drained (see Chapter 1), start the engine and check for leaks and proper operation.

6   Crankshaft pulley - removal and installation

Removal

1   Disconnect the negative battery cable from the remote ground terminal or battery (see Chapter 5).

2   Loosen the lug nuts on the right front wheel, raise the vehicle, and support it securely on jackstands.

3   Remove the right front wheel.

4   Remove the drivebelt (see Chapter 1).

5   The crankshaft pulley bolt is incredibly tight; using a breaker bar, socket and special tool #10198 or equivalent (see illustration), hold the balancer from turning while loosening the bolt.

6   Pull the crankshaft pulley off the crankshaft (see illustration).

Installation

7   Apply clean engine oil or multi-purpose grease to the seal contact surface of the balancer hub (if it isn't lubricated, the seal lip could be damaged and oil leakage would result).

8   Install the crankshaft pulley, aligning the keyway on the crankshaft with the slot in the pulley hub. Install the bolt and tighten it by hand.

9   Prevent the engine from rotating (see Step 5) then tighten the bolt to the torque listed in this Chapter's Specifications.

10   The remainder of installation is the reverse of removal.

7   Crankshaft front oil seal - replacement

1   Remove the crankshaft balancer (see Section 6).

2   Use a screwdriver or hook tool to carefully pry out the seal (see illustration).

Note: *Be careful not to damage the oil pump*

cover bore where the seal is seated, or the nose and sealing surface of the crankshaft.

3   Another procedure for removing the seal is to drill a small hole on each side of the seal and place a self-tapping screw in each hole **(see illustration)**. Use these screws as a means of pulling the seal out without having to pry on it.

4   If the seal is being replaced when the timing chain cover is removed, support the cover on top of two blocks of wood and drive the seal out from the backside with a hammer and punch.

Caution: *Be careful not to scratch, gouge or distort the area that the seal fits into or a leak will develop.*

5   Apply clean engine oil or multi-purpose grease to the outer edge of the new seal, then install it in the cover with the lip (spring side) facing IN. Drive the seal into place with a large socket and a hammer **(see illustration)**. Make sure the seal enters the bore squarely and stop when the front face is at the proper depth.

Note: *If a large socket isn't available, a piece of pipe will also work.*

6   Check the surface on the balancer hub that the oil seal rides on. If the surface has been grooved from long-time contact with the seal, the balancer will need to be replaced.

7   Lubricate the balancer hub with clean engine oil and install the crankshaft balancer (see Section 6).

8   The remainder of installation is the reverse of the removal.

---

8   Timing chain cover, chain and sprockets - removal, inspection and installation

---

Warning: *Wait until the engine is completely cool before beginning this procedure.*
Caution: *The timing system is complex, and severe engine damage will occur if you make any mistakes. Do not attempt this procedure unless you are highly experienced with this type of repair. If you are at all unsure of your abilities, be sure to consult an expert. Double-check all your work and be sure everything is correct before you attempt to start the engine.*
Caution: *Do not rotate the crankshaft or camshafts separately during this procedure (with the timing chains removed), as damage to the valves may occur.*
Note: *Several special tools are required to complete these procedures, so read through the entire Section and obtain the special tools before beginning work.*

Removal

Timing chain cover

1   Disconnect the negative battery cable from the remote ground terminal or battery (see Chapter 5).

2   Drain the engine coolant (see Chapter 1).

3   Loosen the right-front wheel lug nuts. Raise the vehicle and support it securely on

7.3 Another way of removing an old oil seal is to screw a self-tapping screw partially into the seal, then use pliers as a lever to pull it from the engine

jackstands. Drain the engine oil (see Chapter 1).

4   Remove the right-front wheel and drivebelt splash shield (see Chapter 1).

5   Remove the drivebelt, drivebelt tensioner and idler pulley (see Chapter 1).

6   Remove the thermostat housing and upper radiator hose, and disconnect the heater hose from the water pump (see Chapter 3).

7   Remove the heater core supply pipe fasteners from the rear cylinder head and move the pipe out of the way.

8   Remove the power steering pump and tie it out of the way with wire (see Chapter 10).

9   Remove the crankshaft pulley (see Section 6).

10   Remove the valve covers (see Section 4).

Caution: *Once the valve covers are removed, the magnetic timing wheels are exposed* **(see illustration 9.9)**. *The magnetic timing wheels on the camshafts must not come in contact with any type of magnet or magnetic field. If contact is made, the timing wheels will need to be replaced.*

7.5 Drive the seal squarely into the cover using a socket and hammer

11   Remove the upper and lower oil pans (see Section 11).

12   Once the oil pans are removed, temporarily install the engine mount crossmember and mount through-bolts. Place a floor jack under the engine (with a block of wood between the jack head and engine) and raise the engine slightly. Remove the right engine mount and bracket (see Section 16).

13   Remove the timing chain cover mounting bolts **(see illustration)**. There are seven indented prying points, one on top and three on each side; carefully pry the cover free of the engine block and cylinder heads. If it still sticks, slip a putty knife between the engine block and cover to break the bond (but be careful not to scratch the surfaces).

14   Remove and discard the coolant housing and water pump gaskets from the back side of the timing chain cover.

Timing chain

Warning: *When the timing chains are removed, do not rotate the camshafts or crankshaft; the valves and pistons can be damaged if contact is made.*

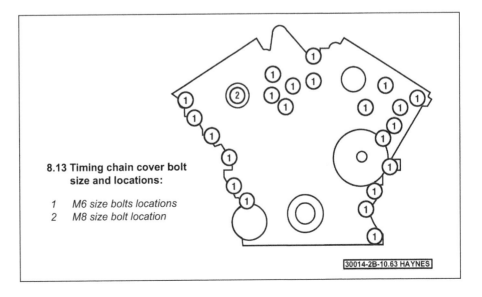

8.13 Timing chain cover bolt size and locations:

1   M6 size bolts locations
2   M8 size bolt location

30014-2B-10.63 HAYNES

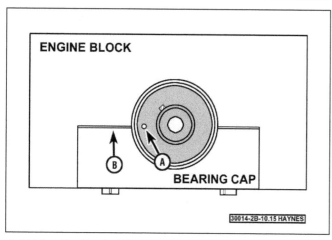

8.15 Align the dimple (A) on the crankshaft with the line (B) made where the engine block and bearing cap meets

8.17 Verity that there are 12 pins between the mark on each phaser

15   Temporarily install the crankshaft pulley bolt. Turn the crankshaft with the bolt to TDC number 1, on the exhaust stroke to align the timing marks on the crankshaft and camshaft sprockets. Rotate the engine clockwise only, until the mark on the crankshaft aligns with the line made where the engine block and bearing cap meets **(see illustration)**.

16   On the left (front) side camshaft phaser, the machined scribe lines should be facing away from each other, and the arrows should be pointing towards each other in a parallel line with the gasket surface of the cylinder head. On the right (rear) side camshaft phaser, the arrows should be facing away from each other and the machined scribe lines should be pointing towards each other in a parallel line with the gasket surface of the cylinder head **(see illustrations 9.10a and 9.10b)**. If, when you align the crankshaft mark with the bearing cap parting line, the camshaft marks are not in alignment as shown in **illustration 8.59**, rotate the engine one full revolution, realign the crankshaft mark, and verify that the camshaft marks are in proper alignment.

17   Verify the phaser marks are aligned with the plated links; if the plated links cannot be distinguished, make sure there are 12 pins between the two marks **(see illustration)**.

Note: *Use paint or a permanent marker to mark the direction of rotation on all chains before removing them so they can be installed in the same direction.*

18   Starting with the right side chain tensioner, press the tensioner plunger in until special tool #8514 or a 3 mm Allen wrench can be inserted through both small holes in the top and bottom of the tensioner body, holding the plunger in the compressed position.

19   Working on the left side chain tensioner, locate the access hole on the side of the tensioner. Working through the hole, lift and hold the pawl off of the rack of the plunger in the tensioner. Press the plunger in until special tool #8514 or a 3 mm Allen wrench can be inserted through both small holes in the top and bottom of the tensioner body, holding the

plunger in the compressed position.

20   Remove the timing gear splash shield fasteners, then remove the shield from the oil pump housing.

21   Remove the oil pump tensioner and sprocket (see Section 12), then remove the oil pump chain from the crankshaft gear.

Note: *The oil pump chain and sprocket do not have to be timed, but the chain should be marked to make sure it is installed in the same direction of rotation.*

22   Starting with the right side chain, slide camshaft phaser lock tool #10202-1 from the front, between the two camshaft phasers, towards the chain (with the tool number facing up).

Note: *It may be necessary to rotate the intake camshaft a few degrees using a wrench on the camshaft flat when installing the phaser lock tool.*

23   Using a large wrench on the camshaft flats and a socket and ratchet on the oil control valves, loosen, but do not remove, the oil control valves.

24   Remove the right side camshaft phaser lock tool, then unscrew the intake camshaft oil control valve from the center of the phaser.

25   Slide the intake camshaft phaser off of the end of the camshaft, then remove the right side timing chain.

Note: *If necessary, remove the exhaust camshaft oil control valve from the center of the phaser and remove the phaser.*

26   Working on the left side chain, slide camshaft phaser lock tool #10202-2 from the front, between the two camshaft phasers, towards the chain (with the tool number facing up).

Note: *It may be necessary to rotate the intake camshaft a few degrees using a wrench on the camshaft flat when installing the phaser lock tool.*

27   Using a large wrench on the camshaft flats and a socket and ratchet on the oil control valves, loosen, but do not remove, the oil control valves

28   Remove the left side camshaft phaser lock tool, then unscrew the exhaust camshaft oil control valve from the center of the phaser.

29   Slide the exhaust camshaft phaser off of the end of the camshaft, then remove the left side timing chain.

Note: *If necessary, remove the intake camshaft oil control valve from the center of the phaser and remove the phaser.*

30   Locate the primary chain tensioner to the side of the crankshaft chain and press the tensioner plunger in until special tool #8514 or a 3 mm Allen wrench can be inserted through the small hole in the side of the tensioner body, holding the plunger in the compressed position.

31   With the tensioner in the compressed position, remove the TORX (T30) mounting fasteners and the tensioner.

32   Remove the primary chain guide TORX (T30) mounting fasteners and the guide.

33   Remove the idler sprocket TORX (T45) mounting fastener and washer, then remove the idler sprocket, primary chain and crankshaft sprocket.

Note: *The chain should be marked to make sure it is installed in the same direction of rotation.*

34   If necessary, remove the chain tensioner (T30) fasteners and remove the tensioner(s), keeping the tensioners in the compressed position.

35   If necessary, remove the chain guide fasteners and guides for both chains.

Inspection

36   Inspect the timing chain dampener (guide) for cracks and wear and replace it, if necessary.

37   Clean the timing chain and sprockets with solvent and dry them with compressed air (if available).

Warning: *Wear eye protection when using compressed air.*

38   Inspect the components for wear and damage. Look for teeth that are deformed, chipped, pitted, and cracked.

39   The timing chain and sprockets should be replaced with new ones if the engine has high mileage, the chain has visible damage, or total freeplay midway between the sprockets exceeds one inch. Failure to replace a

worn timing chain and sprockets may result in erratic engine performance, loss of power, and decreased fuel mileage. Loose chains can jump timing. In the worst case, chain jumping or breakage will result in severe engine damage.

Installation

Caution: *Before starting the engine, carefully rotate the crankshaft by hand through at least two full revolutions (use a socket and breaker bar on the crankshaft pulley center bolt). If you feel anyresistance, STOP! There is something wrong - most likely, valves are contacting the pistons. You must find the problem before proceeding.*

40   Use a plastic gasket scraper to remove all traces of old gasket material and sealant from the cover, engine block and cylinder heads. The cover is made of aluminum, so be careful not to nick or gouge it. Only clean the gasket sealing surfaces with rubbing alcohol (isopropyl) - do not use any oil based fluids.

41   If removed, install the chain guides and tensioners (still in the compressed position).

42   Make sure the keyway is installed on the crankshaft and the dimple on the crankshaft is aligned with the line made where the engine block and bearing cap meets **(see illustration 8.15).**

43   Verify the camshafts are at TDC, with the alignment holes pointing up **(see illustration 9.29).**

44   Place the primary chain on the crankshaft sprocket, with the plated link of the primary chain aligned with the arrow on the bottom of the sprocket. Insert the idler sprocket into the chain, aligning the other plated link with the machined mark on the idler sprocket.

45   Using clean engine oil, coat the sprockets and chain. Install the assembly while keeping the marks aligned, then install the idler sprocket mounting fastener finger tight.

46   Check the alignment of the marks; the plated link on the idler sprocket should be on top (12 o'clock) and the machined mark on the crankshaft should be aligned with the line made where the engine block and bearing cap meet **(see illustration).** If the marks are all aligned, tighten the idler sprocket fastener to the torque listed in this Chapter's Specifications.

47   Install the primary chain guide and tensioner, then tighten the fasteners to the torque listed in this Chapter's Specifications. Remove the special tool from the tensioner plunger.

48   Starting with the left side chain, install the intake camshaft phaser and oil control valve, then tighten the valve finger tight.

49   Place the left side chain over the inside phaser and around the inside cogs of the idler sprocket so that the plate link of the chain is aligned with the machined arrow on the sprocket.

50   With the chain aligned at the idler sprocket, install the exhaust camshaft phaser so that the arrows are pointing towards each other and in a parallel line with the cylinder head gasket surface **(see illustration 9.10a),** then install the oil control valve finger tight.

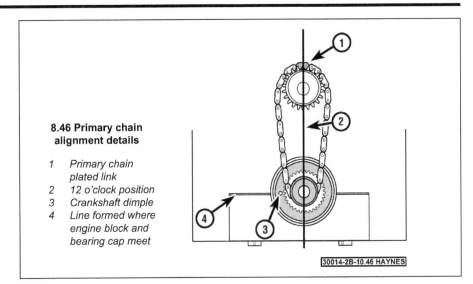

8.46 Primary chain alignment details

1   *Primary chain plated link*
2   *12 o'clock position*
3   *Crankshaft dimple*
4   *Line formed where engine block and bearing cap meet*

30014-2B-10.46 HAYNES

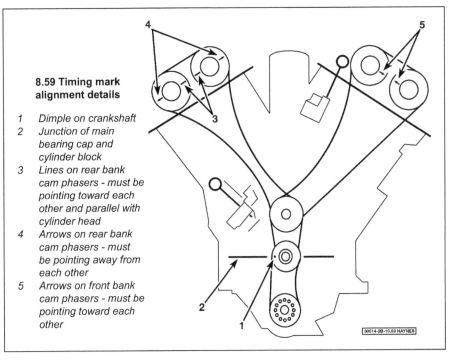

8.59 Timing mark alignment details

1   *Dimple on crankshaft*
2   *Junction of main bearing cap and cylinder block*
3   *Lines on rear bank cam phasers - must be pointing toward each other and parallel with cylinder head*
4   *Arrows on rear bank cam phasers - must be pointing away from each other*
5   *Arrows on front bank cam phasers - must be pointing toward each other*

30014-2B-10.59 HAYNES

51   Slide camshaft phaser lock tool #10202-2 from the front, between the two camshaft phasers towards the chain with the tool number facing up.

52   Using a large wrench on the camshaft flats and a socket and ratchet on the oil control valves, tighten both valves to the torque listed in this Chapter's Specifications.

53   Working on the right side (rear bank) chain, install the exhaust camshaft phaser and oil control valve, tightening the valve finger tight.

54   Place the right side chain over the intake phaser and around the outside cogs of the idler sprocket so that the plate link of the chain is aligned with the machined circle on the sprocket.

55   With the chain aligned at the idler sprocket, install the intake camshaft phaser so that the machined lines are pointing towards each other and in a parallel line with the gas-

ket surface of the cylinder head **(see illustration 9.10b),** then install the oil control valve finger tight.

56   Slide camshaft phaser lock tool #10202-1 from the front, between the two camshaft phasers, towards the chain (with the tool number facing up).

57   Using a large wrench on the camshaft flats and a socket and ratchet on the oil control valves, tighten both valves to the torque listed in this Chapter's Specifications.

58   Install the oil pump timing chain, tensioner, sprocket and splash shield (see Section 12).

Note: *There are no timing marks on the oil pump chain or sprocket.*

59   Verify all the marks are aligned **(see illustration),** then remove the special tool or Allen wrenches from the primary and secondary tensioners. Also remove the camshaft phaser lock tools.

9.9 Location of the magnetic timing wheels

60   Rotate the engine two complete turns using the machined mark on the crankshaft with the line made where the engine block and bearing caps meet as the reference. Verify all the marks are aligned and there are 12 pins between the phaser marks **(see illustration 8.17)**; if the marks are off, rotate the engine two more complete turns and check again.

61   Once the timing marks are correct, install the new coolant housing and water pump housing gaskets into the grooves on the back side of the timing cover

62   Apply a 1/8-inch wide by 1/16-inch high, bead of RTV sealant to the sealing surface of the cover, then install the cover on the alignment dowels.

63   Install the cover bolts **(see illustration 8.13)** and tighten them in a criss-cross pattern, in three steps, to the torque listed in this Chapter's Specifications.

64   The remainder of installation is the reverse of removal.

65   Add oil and coolant (see Chapter 1), start the engine and check for leaks.

---

9   Camshaft(s) – removal, inspection and installation

Caution: *The timing system is complex, and severe engine damage will occur if you make any mistakes. Do not attempt this procedure unless you are highly experienced with this type of repair. If you are at all unsure of your abilities, be sure to consult an expert. Double-check all your work and be sure everything is correct before you attempt to start the engine.*
Caution: *Once the valve covers are removed, the magnetic timing wheels are exposed. The magnetic timing wheels on the camshafts must not come in contact with any type of magnet or magnetic field. If contact is made, the timing wheels will need to be replaced.*
Note: *The timing chain for each camshaft can be removed from the camshafts individually, without removing all the timing chains, using the tools outlined in this Section. If the tools are not available, the timing chain cover and all chains will need to be removed before the camshafts can be removed (see Section 8).*

Removal

1   Disconnect the negative battery cable from the remote ground terminal or battery (see Chapter 5).

2   Loosen the lug nuts on the right front wheel, raise the front of the vehicle and support it securely on jackstands. Remove the right front wheel and the drivebelt splash shield.

3   Drain the engine oil and coolant, then remove the drivebelt (see Chapter 1).

4   Remove the air filter housing (see Chapter 4) and resonator **(see illustration 5.5)**.

5   Remove the intake manifolds (see Section 5).

6   Disconnect all wires and vacuum hoses from the cylinder heads. Label them to simplify reinstallation.

7   Disconnect the ignition coils and remove the spark plugs (see Chapter 1). Label the ignition coils to simplify reinstallation.

8   Remove the valve covers (see Section 4).

9   Once the valve covers are removed, the magnetic timing wheels are exposed **(see illustration)**. The magnetic timing wheels on the camshafts must not come in contact with any type of magnet or magnetic field. If contact is made, the timing wheels will need to be replaced.

10   Rotate the crankshaft clockwise and place the #1 piston at TDC on the exhaust stroke. On the left (front) side camshaft phaser, the machined scribe lines should be facing away from each other, and the arrows should be pointing towards each other in a parallel line with the gasket surface of the cylinder head. On the right (rear) side camshaft phaser, the arrows should be facing away from each other, and the machined scribe lines should be pointing towards each other in a parallel line with the gaskets surface of the cylinder head **(see illustrations)**.

11   Using a permanent marker or paint, mark the camshaft phasers to the timing chains for reinstallation.

9.10a With the engine at TDC #1, the left (front) camshaft phaser scribe marks (A) should be pointing away from each other, the arrow marks (B) should be pointing towards each other in a straight line and that line should be parallel with the cylinder head surface (C)

9.10b The right (rear) phaser scribe marks should be pointing towards each other in a straight line (and that line should be parallel with the cylinder head surface)

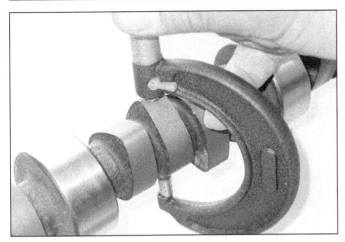

9.22 Use a micrometer to measure cam lobe height

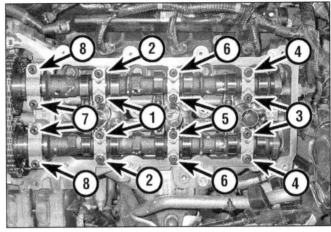

9.28 Camshaft bearing cap bolt TIGHTENING sequence – left (front) side shown, right side is identical

12   Working from the top of the timing chain cover, insert special tool #10200-3 down the side of the tensioner to the access hole on the side of the tensioner. Working through the small hole in the side of the tensioner, lift and hold the pawl off of the rack of the plunger in the tensioner. Slide chain holding tool #10200-1 between the cylinder head and the back side of the chain against the chain guide, forcing the rack and plunger back into the tensioner body.
Caution: *The chain holding tool must remain in place while the phasers are removed or the timing chain will fall off into the timing chain cover.*
13   Slide camshaft phaser lock tool #10202-1 (right side) or 10202-2 (left side), from the front, between the two camshaft phasers, towards the chain.
Note: *It may be necessary to rotate the intake camshaft a few degrees using a wrench on the camshaft flat when installing the phaser lock tool.*
14   Using a large wrench on the camshaft flats and a socket and ratchet on the oil control valves, loosen, then remove each of the oil control valves from the phaser end of the camshaft.
15   At the same time, carefully slide both the intake and exhaust phaser (with the phaser lock securely between them) forward until they are off the end of the camshafts.
Caution: *Do not remove the phaser lock or try to disassemble the phasers.*
16   Using the alignment holes in the camshaft as a reference point, slowly rotate both camshafts counterclockwise approximately 30-degrees Before Top Dead Center (BTDC). In this position, the camshafts are in a neutral or no load position.
Note: *The camshaft bearing caps are marked with a number and letter code; " 1I " is for the number one Intake camshaft bearing cap. The notch on the caps should always be installed towards the front.*
17   Loosen the camshaft bearing cap bolts in the reverse of the tightening sequence **(see illustration 9.28).**

18   Remove the camshaft bearing caps and carefully lift the camshafts from the cylinder head.
19   Mark the rocker arms so they can be installed in their original locations, then remove them.
20   Mark the hydraulic lash adjusters so they can be installed in their original locations, then remove them from the cylinder head.

Inspection
21   Check the camshaft bearing surfaces for pitting, score marks, galling, and abnormal wear. If the bearing surfaces are damaged, the cylinder head will have to be replaced.
22   Compare the camshaft lobe height by measuring each lobe with a micrometer **(see illustration)**. Measure each of the intake lobes and record the measurements and relative positions. Then measure each of the exhaust lobes and record the measurements and relative positions also. This will let you compare all of the intake lobes to one another and all of the exhaust lobes to one another. If the difference between the lobes exceeds 0.005 inch, the camshaft should be replaced. Do not compare intake lobe heights to exhaust lobe heights as lobe lift may be different. Only compare intake lobes to intake lobes and exhaust lobes to exhaust lobes for this comparison.
23   Check the rocker arms and shafts for abnormal wear, pits, galling, score marks, and rough spots. Don't attempt to restore rocker arms by grinding the pad surfaces. Replace defective parts.

Installation
Caution: *Before starting the engine, carefully rotate the crankshaft by hand through at least two full revolutions (use a socket and breaker bar on the crankshaft pulley center bolt). If you feel any resistance, STOP! There is something wrong - most likely, valves are contacting the pistons. You must find the problem before proceeding. Check your work and see if any updated repair information is available.*
24   Dip the hydraulic lash adjusters in clean engine oil and install them into their original locations.

25   Apply moly-base grease or engine assembly lube to the rocker arm contact points and rollers and install them into their original locations.
26   Lubricate the camshaft bearing journals and lobes with moly-base grease or engine assembly lube, then install them carefully in the cylinder head about 30-degrees before (counterclockwise of) TDC. Don't scratch the bearing surfaces with the cam lobes!
Caution: *Do not rotate the camshafts more than a few degrees to prevent the valves from contacting the pistons.*
27   Install the camshaft bearing caps, then install the mounting bolts and finger tighten them.
28   Tighten the bearing caps in sequence **(see illustration)** to the torque listed in this Chapter's Specifications.
29   Rotate the camshafts clockwise 30-degrees, and verify the alignment holes in the camshafts are in the 12 o'clock (pointing straight up) or neutral position **(see illustration)**.
30   Carefully slide both the intake and exhaust phaser (with the phaser lock tool

9.29 Locate the alignment holes on the camshafts and make sure they are in the neutral position (pointing straight up) – right (rear) side shown, left side is identical

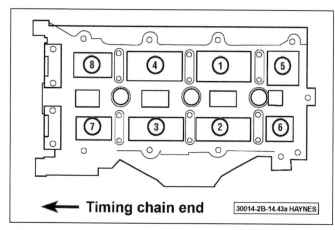
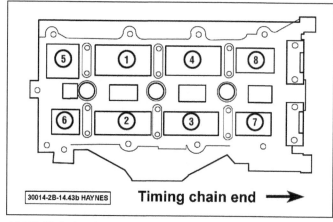

10.29a Left (front) side cylinder head bolt TIGHTENING sequence   **10.29b** Right (rear) side cylinder head bolt TIGHTENING sequence

securely between them) onto the camshafts and verify the marks are aligned.

31   Install the oil control valves into the camshaft phasers, then tighten them to the torque listed in this Chapter's Specifications. Remove the chain holding tool and release the tensioner plunger.
Caution: *Make sure to prevent the camshafts from turning by holding the camshaft with a large wrench on the camshaft flats.*
32   Slowly rotate the engine two complete turns (720-degrees) and verify the alignment marks are correct **(see illustrations 9.10a and 9.10b).**
33   The remainder of installation is the reverse of removal.

10   Cylinder heads - removal and installation

Warning: *Wait until the engine is completely cool before beginning this procedure.*

Removal

1   Relieve the fuel pressure (see Chapter 4), then disconnect the negative battery cable from the remote ground terminal or battery (see Chapter 5).
2   Loosen the lug nuts on the right front wheel, raise the front of the vehicle and support it securely on jackstands. Remove the right front wheel and the drivebelt splash shield.
3   Drain the engine oil and coolant, then remove the drivebelt (see Chapter 1).
4   Remove the air filter housing and resonator (see Chapter 4).
5   Remove the intake manifolds (see Section 5).
6   Disconnect all wires and vacuum hoses from the cylinder heads. Label them to simplify reinstallation.
7   Disconnect the ignition coils and remove the spark plugs (see Chapter 1).
8   Remove the catalytic converter(s) (see Chapter 6).
9   Remove the valve covers (see Section 4).

10   Remove the crankshaft balancer (see Section 6).
11   Remove the oil pans (see Section 11).
12   If you're working on the front cylinder head:

  • *Remove the alternator (see Chapter 5).*
  • *Remove the oil dipstick tube fastener and remove the tube from the oil pan.*
  • *Remove the air conditioning compressor (see Chapter 3)*
  • *Disconnect the main engine harness connectors at the rear of the cylinder and move the harness and retainers out of the way.*

13   If you're working on the rear cylinder head:

  • *Remove the power steering pump (see Chapter 10)*
  • *Remove the heater core tube fasteners and move the tube away from the cylinder head.*

14   Remove the timing chain cover (see Section 8).
15   Rotate the crankshaft clockwise and place the #1 piston at TDC on the exhaust stroke. When the crankshaft is at TDC, the dimple on the crankshaft will be in line with the line made where the bearing cap meets the engine block. The front cylinder bank cam phaser arrows should be pointing toward each other and be parallel with where the cylinder head and valve cover meets. The rear side cam phaser arrows should point away from each other and the lines on the phasers should be pointing towards each other **(see illustration 8.59).**
16   Remove the timing chain for the cylinder head or, if both cylinder heads are being removed, remove both chains (see Section 8).
17   Remove the oil control valves from the cam phasers (sprockets) (see Section 9).
18   Remove the timing chain tensioner and chain guides (see Section 8).
19   Remove the camshafts, rocker arms and lash adjusters (see Section 9).
Caution: *Once the valve covers are removed the magnetic timing wheels are exposed. The magnetic timing wheels on the camshafts must not come in contact with any type of*

magnet or magnetic field. If contact is made the timing wheels will need to be replaced (see Section 9).
Note: *Keep the rocker arms and lash adjusters in order so that they can be installed in their original locations.*
20   Loosen the cylinder head bolts in the reverse of the tightening sequence **(see illustrations 10.29a and 10.29b).**
21   Lift the cylinder head off the block. If resistance is felt, dislodge the cylinder head by striking it with a wood block and hammer. If prying is required, pry only on a casting protrusion - be very careful not to damage the cylinder head or block!
Caution: *Do not set the cylinder head on its gasket side; the sealing surface can be easily damaged.*
22   Have the cylinder head inspected and serviced by a qualified automotive machine shop.

Installation

23   The mating surfaces of each cylinder head and the engine block must be perfectly clean when the cylinder head is installed.
24   Carefully use a gasket scraper to remove all traces of carbon and old gasket material, then clean the mating surfaces with brake system cleaner. If there's oil on the mating surfaces when the cylinder head is installed, the gasket may not seal correctly and leaks may develop.
25   When working on the engine block, it's a good idea to cover the lifter valley with shop rags to keep debris out of the engine. Use a shop rag or vacuum cleaner to remove any debris that falls into the cylinders.
26   Check the engine block and cylinder head mating surfaces for nicks, deep scratches, and other damage. If damage is slight, it can be removed with a file; if it's excessive, machining may be the only alternative.
27   Position the new gasket over the dowel pins in the engine block. Some gaskets are marked TOP or FRONT to ensure correct installation.
28   Carefully position the cylinder head on the engine block without disturbing the gasket.
29   Install NEW cylinder head bolts and tighten them in the recommended sequence

(see illustrations) to the torque steps listed in this Chapter's Specifications.

Caution: *Do not use a torque wrench for steps requiring additional rotation or turns; apply a paint mark to the bolt head or use a torque-angle gauge (available at most automotive parts stores) and a socket and breaker bar.*

30 The remainder of installation is the reverse of removal.

31 Change the engine oil and filter (see Chapter 1).

32 Refill the cooling system (see Chapter 1). Start the engine and check for leaks and proper operation.

11 Oil pans - removal and installation

Removal

1 Disconnect the negative battery cable from the remote ground terminal or battery (see Chapter 5).

2 Raise the front of the vehicle and support it securely on jackstands. Apply the parking brake and block the rear wheels to keep it from rolling off the stands.

3 Drain the engine oil (see Chapter 1).

4 Remove the lower splash shield fasteners and remove the splash shield.

Lower oil pan

5 Remove the bolts and nuts, then carefully separate the lower oil pan from the upper oil pan **(see illustration)**. Don't pry between the upper pan and the lower pan or damage to the sealing surfaces could occur and oil leaks may develop. Tap the pan with a soft-face hammer to break the gasket seal. If it still sticks, slip a putty knife between the upper pan and lower pan to break the bond (but be careful not to scratch the surfaces).

Upper oil pan

6 Remove the dipstick tube bracket mounting bolt. Using a twisting motion, pull the dipstick tube out of the upper oil pan.

7 Remove the right side driveaxle (see Chapter 8).

8 Remove the lower oil pan (see Step 5).

9 Disconnect the exhaust crossunder pipe flange fasteners and remove the crossunder pipe.

10 Remove the engine mount crossmember (see Chapter 10).

11 Remove the coolant tube-to-upper pan fastener and move the tube back.

12 Remove the five upper oil pan-to-transaxle mounting bolts.

13 Remove the torque converter access plate, and the rubber plugs just below the plate.

14 Remove the two upper pan-to-rear main seal housing bolts (M6 size).

Caution: *The oil pan-to-rear main seal bolts are hard to see and can easily be missed. If they are not removed, the rear main seal housing will be severely damaged when the pan is lowered.*

15 Remove the nineteen upper oil pan bolts (M8 size) around the perimeter of the pan, then carefully separate the oil pan from the engine block. Use the two indented prying points on each side of the oil pan to carefully pry the pan free of the engine block. If it still sticks, slip a putty knife between the engine block and oil pan to break the bond (but be careful not to scratch the surfaces).

Installation

16 Clean the pan(s) with solvent and remove all old sealant and gasket material from the engine block and pan mating surfaces. Clean the mating surfaces with lacquer thinner or acetone and make sure the bolt holes in the engine block are clear. Check the oil pan flange(s) for distortion, particularly around the bolt holes. If necessary, place the pan(s) on a wood block and use a hammer to flatten and restore the gasket surface.

Upper oil pan

17 Apply a 1/8-inch wide by 1/16-inch high bead of RTV sealant to the sealing surface of the pan. Install the upper pan and the bolts, then tighten the bolts finger-tight.

18 Tighten the upper pan-to-transaxle bolts to the torque listed in this Chapter's Specifications.

19 Tighten the remaining bolts in a circular pattern, starting from the middle and working your way outwards, to the torque listed in this Chapter's Specifications.

20 The remainder of installation is the reverse of removal.

21 Refill the engine with oil (see Chapter 1). Start and run the engine until normal operating temperature is reached, then check for leaks.

Lower oil pan

22 Apply a 1/8-inch wide by 1/16-inch high bead of RTV sealant to the sealing surface of the pan. Install the lower pan and the bolts.

23 Tighten the bolts in a circular pattern, starting from the middle and working your way outwards, to the torque listed in this Chapter's Specifications.

24 The remainder of installation is the reverse of removal.

25 Refill the engine with oil (see Chapter 1). Start and run the engine until normal operating temperature is reached, then check for leaks.

12 Oil pump - removal, inspection and installation

Removal

1 Disconnect the negative battery cable from the remote ground terminal or battery (see Chapter 5).

2 Raise the front of the vehicle and support it securely on jackstands. Apply the park-

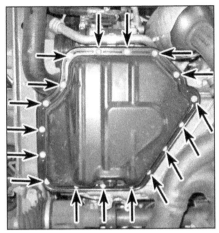

11.5 Lower oil pan fastener locations

ing brake and block the rear wheels to keep it from rolling off the stands.

3 Drain the engine oil (see Chapter 1).

4 Remove the lower splash shield fasteners and remove the splash shield.

5 Remove the lower and upper oil pans (see Section 11).

6 Remove the oil pump pick-up tube fastener, and remove the tube from the pump. Discard the pick-up tube O-ring.

7 Disconnect the oil pump solenoid electrical connector from the side of the engine.

8 Working from the side of the block, depress the oil pump solenoid electrical connector locking tab and push the connector into the block.

Note: *The connector will have to be maneuvered around the tensioner mounting bolt.*

9 Remove the oil pump timing gear splash shield bolts and remove the splash shield.

10 Press the oil pump chain tensioner away from the chain until a 3 mm Allen wrench can be inserted into the housing to hold the tensioner back.

11 Using a permanent marker or paint, make reference marks on the chain and oil pump gear.

12 Hold the oil pump gear from moving, then remove the T45 Torx mounting bolt and the oil pump gear.

13 Hold the tensioner and remove the Allen wrench, allowing the tensioner to release. Remove the spring from the dowel pin and slide the tensioner from the oil pump.

14 Remove the oil pump mounting bolts and remove the pump.

Inspection

Note: *The oil pump is not serviceable; if there is a problem, the pump assembly must be replaced.*

15 Clean all parts thoroughly in solvent and carefully inspect the rotors, pump cover, and timing chain cover for nicks, scratches, or burrs. Replace the assembly if it is damaged.

16 Use a straightedge and a feeler gauge to measure the oil pump cover for warpage

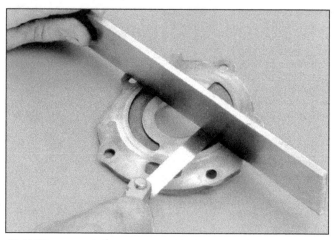

12.16 Place a straightedge across the oil pump cover and check it for warpage with a feeler gauge

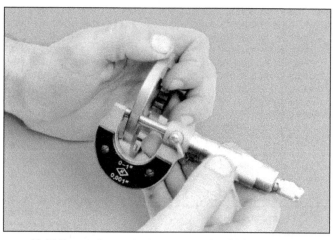

12.17 Use a micrometer to measure the thickness of the outer rotor

12.19 Check the outer rotor-to-housing clearance with a feeler gauge

12.20 Check the clearance between the lobes of the inner and outer rotors

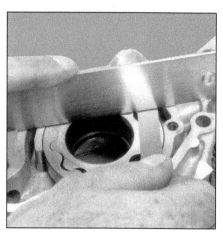

12.21 Using a straightedge and feeler gauge, check the clearance between the surface of the oil pump cover and the rotors

(see illustration). If it's warped more than the limit listed in this Chapter's Specifications, the pump should be replaced.

17   Measure the thickness of the outer rotor **(see illustration)**. If the thickness is less than the value listed in this Chapter's Specifications, the pump should be replaced.

18   Measure the thickness of the inner rotor. If the thickness is less than the value listed in this Chapter's Specifications, the pump should be replaced.

19   Insert the outer rotor into the oil pump housing and measure the clearance between the rotor and housing **(see illustration)**. If the measurement is more than the maximum allowable clearance listed in this Chapter's Specifications, the pump should be replaced.

20   Install the inner rotor in the oil pump assembly and measure the clearance between the lobes on the inner and outer rotors **(see illustration)**. If the clearance is more than the value listed in this Chapter's Specifications, the pump should be replaced.

21   Place a straightedge across the face of

the oil pump assembly **(see illustration)**. If the clearance between the pump surface and the rotors is greater than the limit listed in this Chapter's Specifications, the pump should be replaced.

Installation

22   Place the oil pump onto the engine block using the aligning dowels. Install the mounting bolts and tighten them to the torque listed in this Chapter's Specifications.

23   Slide the oil pump chain tensioner onto the pivot, then push the tensioner back against the spring. Insert a 3 mm Allen wrench into the tensioner to hold it in place.

24   Place the oil pump timing chain gear into the chain, center it onto the oil pump shaft and install the T45 mounting bolt. Tighten the bolt to the torque listed in this Chapter's Specifications.

Note: *Make sure the gear is facing the same way as when it was removed (see Step 11). There are no timing marks on the pump gear or chain, and no timing is necessary.*

25   Maneuver the oil pump solenoid into position and insert it through the block opening until it snaps in place.

26   Install the timing gear splash shield and bolts, then tighten the bolts to the torque listed in this Chapter's Specifications.

27   The remainder of installation is the reverse of removal.

28   Refill the engine with oil and change the oil filter (see Chapter 1).

---

13   Oil cooler - removal and installation

Warning: *Wait until the engine is completely cool before beginning this procedure.*

Removal

1   Relieve the fuel pressure (see Chapter 4), then disconnect the negative battery cable from the remote ground terminal or battery (see Chapter 5).

2   Drain the coolant (see Chapter 1).

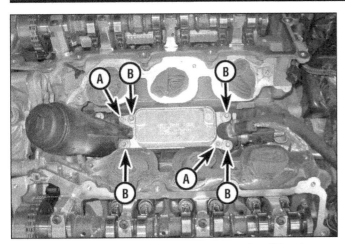

13.4 Remove the oil cooler mounting screws (A) and mounting bolts (B)

14.4 Mark the relative position of the driveplate to the crankshaft and, using an appropriate tool to hold the driveplate, remove the bolts

3   Remove the lower intake manifold (see Section 5).

4   Remove the oil cooler mounting fasteners **(see illustration)**.

5   Remove the oil cooler and discard the seals.

Installation

6   Install new seals to the oil cooler.

7   Place the oil cooler onto the block and install the two mounting screws **(see illustration 13.4)**.

8   Install the mounting bolts and tighten the screws and bolts to the torque listed in this Chapter's Specifications.

9   The remainder of installation is the reverse of removal.

10   Refill the cooling system (see Chapter 1). Run the engine until normal operating temperature is reached, and check for leaks.

14   Driveplate - removal and installation

Removal

1   Raise the vehicle and support it securely on jackstands.

2   Remove the transaxle (see Chapter 7A).

3   To ensure correct alignment during reinstallation, match-mark the driveplate and backing plate to the crankshaft so they can be reassembled in the same position.

4   Remove the bolts that hold the driveplate to the crankshaft **(see illustration)**. A special tool is available at most auto parts stores to hold the driveplate while loosening the bolts. If the tool is not available, wedge a screwdriver in the starter ring gear teeth to jam the driveplate.

5   Remove the driveplate from the crankshaft. The driveplate is fairly heavy; be sure to support it while removing the last bolt.

6   Clean the driveplate to remove grease

and oil. Inspect the driveplate for damage or other defects.

7   Clean and inspect the mating surfaces of the driveplate and the crankshaft.

8   If the crankshaft rear main seal is leaking, replace it before reinstalling the driveplate (see Section 15).

Installation

9   Position the driveplate and backing plate against the crankshaft. Align the previously applied match marks. Before installing the bolts, apply thread-locking compound to the threads.

10   Hold the driveplate with the holding tool, or wedge a screwdriver in the starter ring gear teeth to keep the driveplate from turning. Tighten the bolts to the torque listed in this Chapter's Specifications.

11   The remainder of installation is the reverse of removal.

15   Rear main oil seal - replacement

1   Remove the oil pans (see Section 11).

2   Remove the driveplate (see Section 14).

3   Unbolt the seal retainer from the engine block and slide the retainer and seal off the end of the crankshaft.

Note: *The rear main oil seal has been incorporated into the seal retainer and must be replaced as a unit.*

4   Clean the engine block, oil pan and crankshaft.

5   The new seal and retainer assembly comes with a plastic installation sleeve; make sure it's in place.

6   Apply a 1/4-inch bead of RTV sealant where the lower corners of the seal retainer meet the oil pan.

7   Place the assembly over the crankshaft and push it squarely into place, making sure the dowels in the seal retainer engage the locating holes in the engine block.

8   Install, but don't fully tighten, the seal retainer bolts.

9   Remove the plastic installation sleeve, then tighten the bolts, using an alternating pattern, to the torque listed in this Chapter's Specifications.

10   The remainder of installation is the reverse of removal.

16   Engine mounts - check and replacement

1   The engine mounting system on these models consists of four molded mounts. The right and left mounts support the engine/transaxle assembly while the front and rear mounts restrict torquing action of the powertrain.

2   Engine mounts seldom require attention, but broken or deteriorated mounts should be replaced immediately, or the added strain placed on driveline components may cause damage or accelerated wear.

Check

3   During the check, the engine must be raised slightly to remove the weight from the mounts.

4   Raise the vehicle and support it securely on jackstands, then position a jack under the engine oil pan. Place a large wood block between the jack head and the oil pan to prevent oil pan damage, then carefully raise the engine just enough to take the weight off the mounts.

Warning: *DO NOT place any part of your body under the engine when it's supported only by a jack!*

5   Check the mounts to see if the rubber is cracked, hardened or separated from the metal backing. Sometimes the rubber will split right down the center.

6   Check for relative movement between the mount plates and the engine or frame

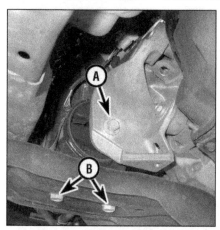

16.10 Front engine mount through bolt (A) and mount-to-subframe bolts (B)

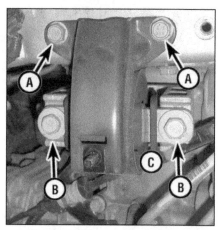

16.18 Remove the left side engine mount-to-body fasteners (A), remove the mount-to-transaxle bolts (B) and a hidden bolt at the bottom of the mount (C) then lift the mount from the engine compartment

(use a large screwdriver or pry bar to attempt to move the mounts). If movement is noted, lower the engine and tighten the mount fasteners.

7   Rubber preservative may be applied to the mounts to slow deterioration.

Replacement

Front mount

8   Raise the front of the vehicle and support it securely on jackstands.

9   Place a floor jack under the engine (with a wood block between the jack head and oil pan) and raise the engine slightly to relieve the weight from the mounts.

10   Remove the front engine mount through-bolt from the insulator and the engine mounting bracket **(see illustration)**.

11   Remove the front engine mount-to-crossmember bolts and remove the insulator assembly.

12   Install the new mount and tighten the bolts securely.

Left mount

13   Disconnect the negative battery cable from the remote ground terminal or battery (see Chapter 5).

14   Remove the air filter and housing (see Chapter 4).

Note: *On some models it may be necessary to remove the Powertrain Control Module (PCM) (see Chapter 6).*

15   Raise the front of the vehicle and support it securely on jackstands.

16   Place a floor jack under the engine (with a wood block between the jack head and oil pan) and raise the engine slightly to relieve the weight from the mounts.

17   Remove the engine mount crossmember (see Chapter 10).

18   Remove the transaxle mount fasteners and remove the mount **(see illustration)**.

19   Install the new mount and tighten the bolts securely.

Right mount

20   Raise the front of the vehicle and support it securely on jackstands. Remove the lower splash shield fasteners and remove the splash shield.

21   Remove the engine ground strap the mount bracket.

22   Place a floor jack under the engine (with a wood block between the jack head and oil pan) and raise the engine slightly to relieve the weight from the mounts.

23   Remove the two right engine mount insulator vertical fasteners from the frame rail and loosen the one horizontal fastener **(see illustration)**.

24   Remove the three bolts and two nuts from the top bracket, then remove the bracket.

25   Remove the mount fasteners and remove the mount.

26   Install the new mount and tighten the bolts securely.

Rear mount

27   Raise the front of the vehicle and support it securely on jackstands.

28   Place a floor jack under the engine (with a wood block between the jack head and oil pan) and raise the engine slightly to relieve the weight from the mounts.

29   Remove the rear mount heat shield and bracket mounting bolts **(see illustration)**.

30   Remove the insulator through-bolt from the mount and the rear mount bracket.

31   Remove the four mount fasteners and remove the mount.

32   Install the new mount and tighten the bolts securely.

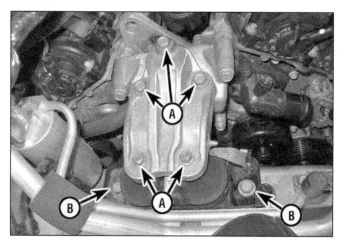

16.23 Remove the right side engine mount top bracket fasteners (A) and separate the top from the engine mount bracket then remove the mount fasteners (B) and the mount from the engine compartment

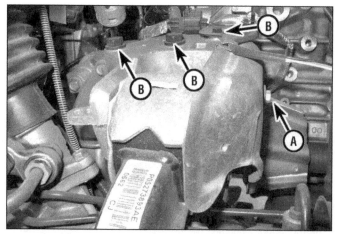

16.29 Remove the heat shield mounting bolt (A) and the heat shield to access the rear engine mount through bolt, then remove the bracket bolts (B) to access the insulator mounting bolts

Chapter 2 Part E
General engine overhaul procedures

Specifications

	Section		Section
Crankshaft - removal and installation	10	Engine removal - methods and precautions	6
Cylinder compression check	3	General information - engine overhaul	1
Engine - removal and installation	7	Initial start-up and break-in after overhaul	12
Engine overhaul - disassembly sequence	8	Oil pressure check	2
Engine overhaul - reassembly sequence	11	Pistons and connecting rods - removal and installation	9
Engine rebuilding alternatives	5	Vacuum gauge diagnostic checks	4

Specifications

General

Displacement
2.4L	146.5 cubic inches
2.7L	167 cubic inches
3.5L	214 cubic inches
3.6L	220 cubic inches

Bore
2.4L	3.465 inches
2.7L	3.386 inches
3.5L	3.780 inches
3.6L	3.779 inches

Stroke
2.4L (2014 and earlier)	3.819 inches
2.4L (2015 and later)	3.228 inches
2.7L	3.091 inches
3.5L	3.189 inches
3.6L	3.268 inches

Compression ratio
2.4L (2014 and earlier)	10.5:1
2.4L (2015 and later)	10.0:1
2.7L	9.67:1
3.5L	10:1
3.6L	10.2:1

Compression pressure .... 100 psi minimum and no more than 25% variance between cylinders

Oil pressure*

2.4L engine
At idle speed	4 psi (minimum)
At 3000 rpm	25 to 80 psi

2.7 and 3.5L engines
At idle speed	5 psi (minimum)
At 3000 rpm	45 to 105 psi

3.6L engine
At idle speed	5 psi (minimum)
At 600 to 1200 rpm	5 psi (warm) or 139 psi (cold)
At 1201 to 3500 rpm	30 psi (warm) or 139 psi (cold)
At 3501 to 6400 rpm	62 psi (warm) or 139 psi (cold)

*If the idle oil pressure test result was zero, don't perform the 3000 rpm or higher test

Torque specifications

Ft-lbs (unless otherwise indicated)

Note: *One foot-pound (ft-lb) of torque is equivalent to 12 inch-pounds (in-lbs) of torque. Torque values below approximately 15 ft-lbs are expressed in inch-pounds, because most foot-pound torque wrenches are not accurate at these smaller values.*

Connecting rod bearing cap bolts*
    2.4L engine (2014 and earlier)
        Step 1 15
        Step 2 .... Tighten an additional 1/4 turn (90-degrees)
    2.4L engine (2015 and later)
        Step 1 15
        Step 2 .... Tighten an additional 92-degrees
    2.7L and 3.5L engines
        Step 1 20
        Step 2 .... Tighten an additional 1/4 turn (90-degrees)
    3.6L engine
        Step 1 15
        Step 2 .... Tighten an additional 1/4 turn (90-degrees)
Crankshaft target wheel bolts
    2.4L (in sequence, **see illustration 10.7**).. 110 in-lbs
    3.6L .... 89 in-lbs
Driveplate-to-crankshaft bolts.. See Chapter 2A, 2B, 2C or 2D
Driveplate-to-torque converter bolts .... See Chapter 7A
Main bearing cap bolts*
    2.4L (in sequence, **see illustration 10.31a**)
        With main bolt markings b or 6
            Step 1.. 132 in-lbs
            Step 2.. 20
            Step 3.. Tighten an additional 1/8 turn (45-degrees)
        With main bolt markings M (and the default settings required
        for 2015 and later 2.4L engines)
            Step 1.. 132 in-lbs
            Step 2.. 33
            Step 3.. Tighten an additional 1/8 turn (45-degrees)
    2.7L and 3.5L (in sequence, **see illustrations 10.31b, 10.31c and 10.31d**)
        Step 1 (inner bolts) 15
        Step 2 (inner bolts) .... Tighten an additional 1/4 turn (90-degrees)
        Step 3 (outer bolts) 20
        Step 4 (outer bolts) .... Tighten an additional 1/4 turn (90-degrees)
        Step 5 (side bolts).. 21
    3.6L (in sequence, **see illustrations 10.31e, 10.31f and 10.31g**)
        Step 1 (inner bolts) M11.. 15
        Step 2 (inner bolts) .... Tighten an additional 1/4 turn (90-degrees)
        Step 3 (outer bolts and windage tray) M8.. 16
        Step 4 (outer bolts and windage tray).. Tighten an additional 1/4 turn (90-degrees)
        Step 5 (side bolts).. 22
Ladder frame (oil pan housing) bolts - 2.4L engines (in sequence, **see illustration 10.34**)
    Step 1.. 89 in-lbs
    Step 2.. 19

*Use new bolts

1.10a An engine block being bored. An engine rebuilder will use special machinery to recondition the cylinder bores

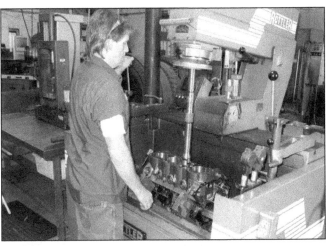

1.10b If the cylinders are bored, the machine shop will normally hone the engine on a machine like this

1   General information - engine overhaul

1   Included in this Part of Chapter 2 are general information and diagnostic testing procedures for determining the overall mechanical condition of your engine.

2   The information ranges from advice concerning preparation for an overhaul and the purchase of replacement parts and/or components to detailed, step-by-step procedures covering removal and installation.

3   The following Sections have been written to help you determine whether your engine needs to be overhauled and how to remove and install it once you've determined it needs to be rebuilt. For information concerning in-vehicle engine repair, see Chapter 2A, 2B, 2C or 2D.

4   The Specifications included in this Part are general in nature and include only those necessary for testing the oil pressure and checking the engine compression. Refer to Chapter 2A, 2B, 2C or 2D for additional engine Specifications.

5   It's not always easy to determine when, or if, an engine should be completely overhauled, because a number of factors must be considered.

6   High mileage is not necessarily an indication that an overhaul is needed, while low mileage doesn't preclude the need for an overhaul. Frequency of servicing is probably the most important consideration. An engine that's had regular and frequent oil and filter changes, as well as other required maintenance, will most likely give many thousands of miles of reliable service. Conversely, a neglected engine may require an overhaul very early in its service life.

7   Excessive oil consumption is an indication that piston rings, valve seals and/or valve guides are in need of attention. Make sure that oil leaks aren't responsible before deciding that the rings and/or guides are bad. Perform a cylinder compression check to deter-

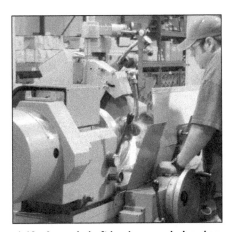

1.10c A crankshaft having a main bearing journal ground

mine the extent of the work required (see Section 3). Also check the vacuum readings under various conditions (see Section 4).

8   Check the oil pressure with a gauge installed in place of the oil pressure sending unit and compare it to this Chapter's Specifications (see Section 2). If it's extremely low, the bearings and/or oil pump are probably worn out.

9   Loss of power, rough running, knocking or metallic engine noises, excessive valve train noise and high fuel consumption rates may also point to the need for an overhaul, especially if they're all present at the same time. If a complete tune-up doesn't remedy the situation, major mechanical work is the only solution.

10   An engine overhaul involves restoring the internal parts to the specifications of a new engine. During an overhaul, the piston rings are replaced and the cylinder walls are reconditioned (rebored and/or honed) (see illustrations). If a rebore is done by an automotive machine shop, new oversize pistons will also be installed. The main bearings, connecting rod bearings and camshaft bearings are generally replaced with new ones and, if

1.11a A machinist checks for a bent connecting rod, using specialized equipment

necessary, the crankshaft may be reground to restore the journals (see illustration). Generally, the valves are serviced as well, since they're usually in less-than-perfect condition at this point. While the engine is being overhauled, other components, such as the starter and alternator, can be rebuilt as well. The end result should be similar to a new engine that will give many trouble free miles.

Note: Critical cooling system components such as the hoses, drivebelts, thermostat and water pump should be replaced with new parts when an engine is overhauled. The radiator should be checked carefully to ensure that it isn't clogged or leaking (see Chapter 3). If you purchase a rebuilt engine or short block, some rebuilders will not warranty their engines unless the radiator has been professionally flushed. Also, we don't recommend overhauling the oil pump - always install a new one when an engine is rebuilt.

11   Overhauling the internal components on today's engines is a difficult and time-consuming task which requires a significant amount of specialty tools and is best left to a professional engine rebuilder (see illustrations).

1.11b A bore gauge being used to check a clyinder bore

1.11c Uneven piston wear like this indicates a bent connecting rod

A competent engine rebuilder will handle the inspection of your old parts and offer advice concerning the reconditioning or replacement of the original engine, Never purchase parts or have machine work done on other components until the block has been thoroughly inspected by a professional machine shop. As a general rule, time is the primary cost of an overhaul, especially since the vehicle may be tied up for a minimum of two weeks or more. Be aware that some engine builders only have the capability to rebuild the engine you bring them while other rebuilders have a large inventory of rebuilt exchange engines in stock. Also be aware that many machine shops could take as much as two weeks time to completely rebuild your engine depending on shop workload. Sometimes it makes more sense to simply exchange your engine for another engine that's already rebuilt to save time.

2 Oil pressure check

1   Low engine oil pressure can be a sign of an engine in need of rebuilding. A low oil pressure indicator (often called an idiot light) is not a test of the oiling system. Such indicators only come on when the oil pressure is dangerously low. Even a factory oil pressure gauge in the instrument panel is only a relative indication, although much better for driver information than a warning light. A better test is with a mechanical (not electrical) oil pressure gauge.

2   Locate the oil pressure indicator sending unit on the engine block.

3   On 2014 and earlier 2.4L engines, the oil pressure sending unit is located behind the alternator and under the intake manifold (see illustration). To gain access to the sending unit, the air conditioning compressor may have to be unbolted and repositioned, without disconnecting the refrigerant lines (see Chapter 3). Be sure to reinstall the compressor and drivebelt before running the engine. On 2015 and later 2.4L engines, the oil pressure sending unit is at the right rear corner of the engine, just behind the drivebelt pulleys. On 2.7L engines, the oil pressure sending unit is located on the side of the engine block, directly under the exhaust manifold (see illustration) (refer to Chapter 2B for removal and installation of the exhaust manifolds. Also disable the fuel and ignition systems). On 3.5L engines, the

oil pressure sending unit is located on the front of the engine next to the crankshaft pulley (see illustration). On 3.6L engines, the oil pressure sending unit is located on the oil filter adapter below the intake manifolds. Refer to Chapter 2D for removal and installation of the manifolds.

4   Unscrew and remove the oil pressure sending unit, then screw in the hose for your oil pressure gauge. If necessary, install an adapter fitting. Use Teflon tape or thread sealant on the threads of the adapter and/or the fitting on the end of your gauge's hose.

5   Connect an accurate tachometer to the engine, according to the tachometer manufacturer's instructions.

6   Check the oil pressure with the engine running (normal operating temperature) at the specified engine speed, and compare it to this Chapter's Specifications. If it's extremely low, the bearings and/or oil pump are probably worn out.

3 Cylinder compression check

1   A compression check will tell you what mechanical condition the upper end of your engine (pistons, rings, valves, head gaskets) is in. Specifically, it can tell you if the compression is down due to leakage caused by worn piston rings, defective valves and seats or a blown head gasket.

Note: *The engine must be at normal operating temperature and the battery must be fully charged for this check.*

2   Begin by cleaning the area around the spark plugs before you remove them (compressed air should be used, if available). The idea is to prevent dirt from getting into the cylinders as the compression check is being done.

3   Remove all of the spark plugs from the engine (see Chapter 1).

4   Disable the fuel system by unplugging the fuel pump module electrical connector (see Chapter 4).

5   Install a compression gauge in the spark plug hole (see illustration).

6   Have an assistant hold the accelerator pedal to the floor and crank the engine over at least seven compression strokes while you

2.3a Location of the oil pressure sending unit - 2.4L engines

2.3b Location of the oil pressure sending unit - 2.7L engines

2.3c Location of the oil pressure sending unit - 3.5L engines

watch the gauge. The compression should build up quickly in a healthy engine. Low compression on the first stroke, followed by gradually increasing pressure on successive strokes, indicates worn piston rings. A low compression reading on the first stroke, which doesn't build up during successive strokes, indicates leaking valves or a blown head gasket (a cracked head could also be the cause). Deposits on the undersides of the valve heads can also cause low compression. Record the highest gauge reading obtained.

8   Repeat the procedure for the remaining cylinders and compare the results to this Chapter's Specifications.

9   Add some engine oil (about three squirts from a plunger-type oil can) to each cylinder, through the spark plug hole, and repeat the test.

10   If the compression increases after the oil is added, the piston rings are definitely worn. If the compression doesn't increase significantly, the leakage is occurring at the valves or head gasket. Leakage past the valves may be caused by burned valve seats and/or faces or warped, cracked or bent valves.

11   If two adjacent cylinders have equally low compression, there's a strong possibility that the head gasket between them is blown. The appearance of coolant in the combustion chambers or the crankcase would verify this condition.

12   If one cylinder is slightly lower than the others, and the engine has a slightly rough idle, a worn lobe on the camshaft could be the cause.

13   If the compression is unusually high, the combustion chambers are probably coated with carbon deposits. If that's the case, the cylinder head(s) should be removed and decarbonized.

14   If compression is way down or varies greatly between cylinders, it would be a good idea to have a leak-down test performed by an automotive repair shop. This test will pinpoint exactly where the leakage is occurring and how severe it is.

4   Vacuum gauge diagnostic checks

1   A vacuum gauge provides inexpensive but valuable information about what is going on in the engine. You can check for worn rings or cylinder walls, leaking head or intake manifold gaskets, restricted exhaust, stuck or burned valves, weak valve springs, improper ignition or valve timing and ignition problems.

2   Unfortunately, vacuum gauge readings are easy to misinterpret, so they should be used in conjunction with other tests to confirm the diagnosis.

3   Both the absolute readings and the rate of needle movement are important for accurate interpretation. Most gauges measure vacuum in inches of mercury (in-Hg). The following references to vacuum assume the diagnosis is being performed at sea level. As elevation increases (or atmospheric pressure decreases), the reading will decrease. For every 1,000 foot increase in elevation above

3.5 A compression gauge with a threaded fitting for the spark plug hole is preferred over the type that requires hand pressure to maintain the seal

4.4 A simple vacuum gauge can be handy in diagnosing engine condition and performance

approximately 2,000 feet, the gauge readings will decrease about one inch of mercury.

4   Connect the vacuum gauge directly to the intake manifold vacuum, not to ported (throttle body) vacuum **(see illustration)**. Be sure no hoses are left disconnected during the test or false readings will result.

5   Before you begin the test, allow the engine to warm up completely. Block the wheels and set the parking brake. With the transaxle in Park, start the engine and allow it to run at normal idle speed.

Warning: *Keep your hands and the vacuum gauge clear of the fans.*

6   Read the vacuum gauge; an average, healthy engine should normally produce about 17 to 22 in-Hg with a fairly steady needle **(see illustration)**. Refer to the following vacuum gauge readings and what they indicate about the engine's condition:

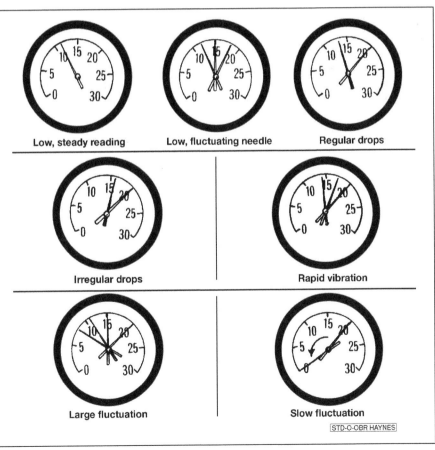

Low, steady reading   Low, fluctuating needle   Regular drops

Irregular drops   Rapid vibration

Large fluctuation   Slow fluctuation

STD-O-OBR HAYNES

4.6 Typical vacuum gauge readings

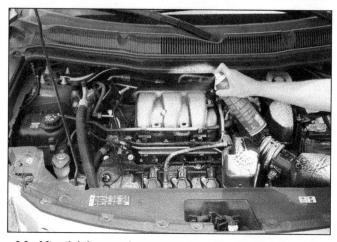

6.3a After tightly wrapping water-vulnerable components, use a spray cleaner on everything, with particular concentration on the greasiest areas, usually around the valve cover and lower edges of the block. If one section dries out, apply more cleaner

6.3b Depending on how dirty the engine is, let the cleaner soak in according to the directions and then hose off the grime and cleaner. Get the rinse water down into every area you can get at; then dry important components with a hair dryer or paper towels

7   A low, steady reading usually indicates a leaking gasket between the intake manifold and cylinder head(s) or throttle body, a leaky vacuum hose, late ignition timing or incorrect camshaft timing. Check ignition timing with a timing light and eliminate all other possible causes, utilizing the tests provided in this Chapter before you remove the timing chain cover to check the timing marks.

8   If the reading is three to eight inches below normal and it fluctuates at that low reading, suspect an intake manifold gasket leak at an intake port or a faulty fuel injector.

9   If the needle has regular drops of about two-to-four inches at a steady rate, the valves are probably leaking. Perform a compression check or leak-down test to confirm this.

10   An irregular drop or down-flick of the needle can be caused by a sticking valve or an ignition misfire. Perform a compression check or leak-down test and read the spark plugs.

11   A rapid vibration of about four in-Hg vibration at idle combined with exhaust smoke indicates worn valve guides. Perform a leakdown test to confirm this. If the rapid vibration occurs with an increase in engine speed, check for a leaking intake manifold gasket or head gasket, weak valve springs, burned valves or ignition misfire.

12   A slight fluctuation, say one inch up and down, may mean ignition problems. Check all the usual tune-up items and, if necessary, run the engine on an ignition analyzer.

13   If there is a large fluctuation, perform a compression or leak-down test to look for a weak or dead cylinder or a blown head gasket.

14   If the needle moves slowly through a wide range, check for a clogged PCV system, incorrect idle fuel mixture, throttle body or intake manifold gasket leaks.

15   Check for a slow return after revving the engine by quickly snapping the throttle open until the engine reaches about 2,500 rpm and let it shut. Normally the reading should drop to near zero, rise above normal idle reading (about 5 in-Hg over) and then return to the previous idle reading. If the vacuum returns slowly and doesn't peak when the throttle is snapped shut, the rings may be worn. If there is a long delay, look for a restricted exhaust system (often the muffler or catalytic converter). An easy way to check this is to temporarily disconnect the exhaust ahead of the suspected part and redo the test.

5   Engine rebuilding alternatives

1   The do-it-yourselfer is faced with a number of options when purchasing a rebuilt engine. The major considerations are cost, warranty, parts availability and the time required for the rebuilder to complete the project. The decision to replace the engine block, piston/connecting rod assemblies and crankshaft depends on the final inspection results of your engine. Only then can you make a cost effective decision whether to have your engine overhauled or simply purchase an exchange engine for your vehicle.

2   Some of the rebuilding alternatives include:

3   **Individual parts** - If the inspection procedures reveal that the engine block and most engine components are in reusable condition, purchasing individual parts and having a rebuilder rebuild your engine may be the most economical alternative. The block, crankshaft and piston/connecting rod assemblies should all be inspected carefully by a machine shop first.

4   **Short block** - A short block consists of an engine block with a crankshaft and piston/connecting rod assemblies already installed. All new bearings are incorporated and all clearances will be correct. The existing camshafts, valve train components, cylinder head and external parts can be bolted to the short block with little or no machine shop work necessary.

5   **Long block** - A long block consists of a short block plus an oil pump, oil pan, cylinder head, valve cover, camshaft and valve train components, timing sprockets and chain or gears and timing cover. All components are installed with new bearings, seals and gaskets incorporated throughout. The installation of manifolds and external parts is all that's necessary.

6   **Low mileage used engines** - Some companies now offer low mileage used engines which is a very cost effective way to get your vehicle up and running again. These engines often come from vehicles which have been in totaled in accidents or come from other countries which have a higher vehicle turnover rate. A low mileage used engine also usually has a similar warranty like the newly remanufactured engines.

7   Give careful thought to which alternative is best for you and discuss the situation with local automotive machine shops, auto parts dealers and experienced rebuilders before ordering or purchasing replacement parts.

6   Engine removal - methods and precautions

1   If you've decided that an engine must be removed for overhaul or major repair work, several preliminary steps should be taken. Read all removal and installation procedures carefully prior to committing to this job. These engines are removed by lowering the engine to the floor, along with the transaxle, and then raising the vehicle sufficiently to slide the assembly out; this will require a vehicle hoist as well as an engine hoist.

2   Locating a suitable place to work is extremely important. Adequate work space, along with storage space for the vehicle, will be needed. If a shop or garage isn't available, at the very least a flat, level, clean work surface made of concrete or asphalt is required.

3   Cleaning the engine compartment and engine before beginning the removal procedure will help keep tools clean and organized **(see illustrations)**.

4   An engine hoist will also be necessary. Make sure the hoist is rated in excess of the combined weight of the engine and transaxle. Safety is of primary importance, considering the potential hazards involved in removing the engine from the vehicle.

5   If you're a novice at engine removal, get at least one helper. One person cannot easily do all the things you need to do to remove a big heavy engine and transaxle assembly from the engine compartment. Also helpful is to seek advice and assistance from someone who's experienced in engine removal.

6   Plan the operation ahead of time. Arrange for or obtain all of the tools and equipment you'll need prior to beginning the job **(see illustration)**. Some of the equipment necessary to perform engine removal and installation safely and with relative ease are (in addition to a vehicle hoist and an engine hoist) a heavy duty floor jack (preferably fitted with a transmission jack head adapter), complete sets of wrenches and sockets as described in the front of this manual, wooden blocks, plenty of rags and cleaning solvent for mopping up spilled oil, coolant and gasoline.

7   Plan for the vehicle to be out of use for quite a while. A machine shop can do the work that is beyond the scope of the home mechanic. Machine shops often have a busy schedule, so before removing the engine, consult the shop for an estimate of how long it will take to rebuild or repair the components that may need work.

7   Engine - removal and installation

Warning: *Gasoline is extremely flammable, so take extra precautions when you work on any part of the fuel system. Don't smoke or allow open flames or bare light bulbs near the work area, and don't work in a garage where a gas-type appliance (such as a water heater or clothes dryer) is present. Since gasoline is carcinogenic, wear fuel-resistant gloves when there's a possibility of being exposed to fuel, and, if you spill any fuel on your skin, rinse it off immediately with soap and water. Mop up any spills immediately and do not store fuel-soaked rags where they could ignite. The fuel system is under constant pressure, so, if any fuel lines are to be disconnected, the fuel pressure in the system must be relieved first (see Chapter 4 for more information). When you perform any kind of work on the fuel system, wear safety glasses and have a Class B type fire extinguisher on hand.*

Warning: *The engine must be completely cool before beginning this procedure.*

Warning: *The air conditioning system is under high pressure. Do not loosen any hose fittings or remove any components until after the system has been discharged. Air conditioning refrigerant must be properly discharged into an EPA-approved recovery/recycling unit at a dealer service department or an automotive air conditioning repair facility. Always wear eye protection when disconnecting air conditioning system fittings.*

6.6 Get an engine stand sturdy enough to firmly support the engine while you're working on it. Stay away from three-wheeled models: they have a tendency to tip over more easily, so get a four-wheeled unit

Caution: *Engine removal on V6 models is a difficult job, especially for the do-it-yourself mechanic working at home. Because of the vehicle's design, the manufacturer states that on V6 engines, the engine and transaxle have to be removed as a unit from the bottom of the vehicle, not the top. With a floor jack and jackstands, the vehicle can't be raised high enough and supported safely enough for the engine/transaxle assembly to slide out from underneath. The manufacturer recommends that removal of the engine transaxle assembly only be performed on a frame-contact type vehicle hoist.*

Note: *On four-cylinder models, the engine can be separated from the transaxle and removed from the top of the engine compartment using a conventional engine hoist.*

Note: *Read through the entire Section before beginning this procedure.*

Removal

1   Have the air conditioning system discharged by an automotive air conditioning technician.

2   Park the vehicle on a frame-contact type vehicle hoist, then engage the arms of the hoist with the jacking points of the vehicle. Raise the hoist arms until they contact the vehicle, but not so much that the wheels come off the ground.

3   Remove the engine cover (see Chapter 1) then relieve the fuel system pressure (see Chapter 4).

4   Place protective covers on the fenders and cowl and remove the hood (see Chapter 11).

5   Remove the air filter housing (see Chapter 4).

6   Disconnect the negative battery cable from the remote ground terminal or battery (see Chapter 5).

7   Remove the lower splash shield from under the vehicle.

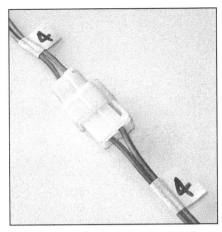

7.9 Label each wire before unplugging the connector

8   Clearly label and disconnect all vacuum lines, emissions hoses, wiring harness connectors, ground straps and fuel lines. Once all the harness connectors have been disconnected, move the engine harness out of the way.

9   Masking tape and/or a touch up paint applicator work well for marking items **(see illustration)**. Take instant photos or sketch the locations of components and brackets.

10   Loosen the front wheel lug nuts and the driveaxle/hub nuts (see Chapter 8), then raise the vehicle on the hoist.

Note: *Keep in mind that during this procedure you'll have to adjust the height of the vehicle to perform certain operations.*

11   Drain the cooling system and engine oil and remove the drivebelt (see Chapter 1).

12   Remove the cooling system reservoir (see Chapter 3).

13   Remove the alternator and its brackets (see Chapter 5).

14   Remove the power steering fluid reservoir and set it off to the side without disconnecting the fluid lines (see Chapter 10).

15   Remove the engine oil dipstick, and the tube mounting bolt, and remove the tube from the oil pan.

16   On 2.7L engines, remove the front bumper cover (see Chapter 11).

17   On V6 engines, remove the upper and lower intake manifolds (see Chapter 2D) and vacuum pump (see Chapter 9).

18   Lower the vehicle and detach the heater hoses at the firewall.

19   Detach the radiator hoses from the engine and, on 2.4L engines, remove the thermostat housing assembly (see Chapter 3).

20   On 2.4L engines, remove the power steering hose support bracket from the front engine mount.

21   On V6 engines, remove the cooling fan(s), shroud(s) and radiator (see Chapter 3).

Note: *Install new transmission fluid cooler lines on reassembly (see Chapter 7A).*

22   On V6 engines, remove the transaxle dipstick tube. Plug the opening with a suitable device.

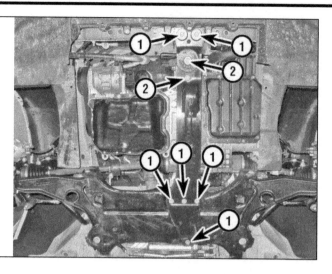

7.36 Longitudinal crossmember bolts (1) and transaxle front mounting bolt (2) locations

23 Disconnect the shift cable from the transaxle (see Chapter 7A). Also disconnect any wiring harness connectors from the transaxle.

24 Disconnect the upper air conditioning line from the condenser for additional clearance.

25 Disconnect the air conditioning lines at the compressor and the junction inside the engine compartment. Remove the air conditioning compressor (see Chapter 3) and then the compressor mounting bracket.

26 Remove the power steering pump and bracket (see Chapter 10).

27 Raise the vehicle on the hoist. Remove the front wheels.

28 On V6 models, remove the driveaxles (see Chapter 8).

29 Unplug the downstream oxygen sensor electrical connector.

30 Detach the exhaust pipe from the exhaust manifold(s) (see Chapter 2A, 2B, 2C or 2D) and crossunder pipe.

Note: *On 2.7L and 3.6L engines, the exhaust manifold and catalytic converters are combined and are referred to as "maniverters."*

31 Remove the power steering fluid cooler, if equipped.

Four-cylinder models

32 Remove the torque converter inspection cover and remove the torque converter bolts (see Chapter 7A).

33 Install a lifting chain strong enough to support the full weight of the engine. Using an engine hoist connected to the chain, slightly raise the engine and remove the front mount (see Chapter 2A).

34 Remove the transaxle-to-engine mounting bolts.

35 Carefully separate the engine from the transaxle and remove the engine from the vehicle. Place the engine on an engine stand and disconnect the chain and hoist.

V6 models

36 Remove the engine longitudinal crossmember bolts and remove the crossmember **(see illustration)**.

37 Mark the position of the driveplate and remove the torque converter bolts (see Chapter 7A).

38 Lower the vehicle.

39 Support the engine with a floor jack and block of wood. Remove the right (passenger's) side engine mount, including the portion that bolts to the engine. Using one of the mount-to-engine bolts, attach one end of an engine lifting sling or chain to the mount boss. Tighten the bolt securely. Attach the other end of the sling or chain to the other side of the engine, using one of the transaxle-to-engine bolts. Be sure the positioning of the chain or sling will support the engine and transaxle in a balanced manner.

Note: *The sling or chain must be long enough to allow the engine hoist to lower the engine/transaxle assembly to the ground, without letting the hoist arm contact the vehicle.*

40 Roll the hoist into position and attach the sling or chain to it. Take up the slack until there is slight tension on the hoist, then remove the jack from under the engine. Remember that the transaxle end of the engine will be heavier, so position the chain on the hoist so it balances the engine and the transaxle level with the vehicle.

Note: *Depending on the design of the engine hoist, it may be helpful to position the hoist from the side of the vehicle, so that when the engine/transaxle assembly is lowered, it will fit between the legs of the hoist.*

41 Recheck to be sure nothing except the remaining mount is still connecting the engine or transaxle to the vehicle. Disconnect and label anything still remaining.

42 Remove the driver's side transaxle mount (see Chapter 2A, 2B, 2C or 2D).

43 Slowly lower the engine/transaxle to the ground.

44 Once the engine/transaxle assembly is on the floor, disconnect the engine lifting hoist and raise the vehicle until it clears the assembly.

45 Reconnect the chain or sling and raise the engine and transaxle. Support the engine with blocks of wood or another floor jack, while leaving the sling or chain attached to the right-

side mounting boss. Support the transaxle with another floor jack, preferably one with a transmission jack head adapter. At this point the transaxle can be unbolted and removed from the engine. Be very careful to ensure that the components are supported securely so they won't topple off their supports during disconnection.

46 Reconnect the lifting chain to the engine, then raise the engine and attach it to an engine stand.

Installation

47 Installation is the reverse of removal, noting the following points:

- *Check the engine/transaxle mounts. If they're worn or damaged, replace them.*
- *Inspect the torque converter seal and bushing.*
- *Attach the transaxle to the engine (see Chapter 7A).*
- *Add coolant, oil, power steering and transmission fluids as needed (see Chapter 1).*
- *Run the engine and check for proper operation and leaks. Shut off the engine and recheck fluid levels.*

8   Engine overhaul - disassembly sequence

1 It's much easier to remove the external components if it's mounted on a portable engine stand. A stand can often be rented quite cheaply from an equipment rental yard. Before the engine is mounted on a stand, the driveplate should be removed from the engine.

2 If a stand isn't available, it's possible to remove the external engine components with it blocked up on the floor. Be extra careful not to tip or drop the engine when working without a stand.

3 If you're going to obtain a rebuilt engine, all external components must come off first, to be transferred to the replacement engine. These components include:

    Driveplate
    Ignition system components
    Emissions-related components
    Engine mounts and mount brackets
    Engine rear cover (spacer plate between driveplate and engine block)
    Intake/exhaust manifolds
    Fuel injection components
    Oil filter
    Thermostat and housing assembly
    Water pump

Note: *When removing the external components from the engine, pay close attention to details that may be helpful or important during installation. Note the installed position of gaskets, seals, spacers, pins, brackets, washers, bolts and other small items.*

4 If you're going to obtain a short block (assembled engine block, crankshaft, pistons and connecting rods), then remove the timing belt/timing chain, cylinder head, oil pan, oil pump pick-up tube, oil pump and water pump

9.1 Before you try to remove the pistons, use a ridge reamer to remove the raised material (ridge) from the top of the cylinders

9.3 Checking the connecting rod endplay (side clearance)

from your engine so that you can turn in your old short block to the rebuilder as a core. See Section 5 for additional information regarding the different possibilities to be considered.

9 Pistons and connecting rods - removal and installation

Removal

Note: *Prior to removing the piston/connecting rod assemblies, remove the cylinder head, oil pan, and the timing chain cover (see Chapter 2A, 2B, 2C or 2D).*

1 Use your fingernail to feel if a ridge has formed at the upper limit of ring travel (about 1/4-inch down from the top of each cylinder). If carbon deposits or cylinder wear have produced ridges, they must be completely removed with a special tool **(see illustration)**. Follow the manufacturer's instructions provided with the tool. Failure to remove the ridges before attempting to remove the piston/connecting rod assemblies may result in piston breakage.

2 After the cylinder ridges have been removed, turn the engine so the crankshaft is facing up. On 2.4L engines, remove the balance shaft module (see Chapter 2A) then remove the ladder frame mounting bolts and remove the ladder frame from the engine block. On V6 engines, remove the oil pump (see Chapter 2B, 2C or 2D), then remove the outer crankshaft main bearing cap bolts and the windage tray, if equipped.

3 Before the main bearing caps and connecting rods are removed, check the connecting rod endplay with feeler gauges. Slide them between the first connecting rod and the crankshaft throw until the play is removed **(see illustration)**. Repeat this procedure for each connecting rod. The endplay is equal to the thickness of the feeler gauge(s). Check with an automotive machine shop for the endplay service limit (a typical endplay limit should measure between 0.005 to 0.015 inch [0.127

to 0.396 mm]). If the play exceeds the service limit, new connecting rods will be required. If new rods (or a new crankshaft) are installed, the endplay may fall under the minimum allowable. If it does, the rods will have to be machined to restore it. If necessary, consult an automotive machine shop for advice.

4 Check the connecting rods and caps for identification marks. If they aren't plainly marked, use paint or a marker to clearly identify each rod and cap (1, 2, 3, etc., depending on the cylinder they're associated with) **(see illustration)**.

Caution: *Do not use a punch and hammer to mark the connecting rods or they may be damaged.*

5 Loosen each of the connecting rod cap bolts 1/2-turn at a time until they can be removed by hand.

Note: *New connecting rod cap bolts must be used when reassembling the engine, but save the old bolts for use when checking the connecting rod bearing oil clearance.*

6 Remove the number one connecting rod cap and bearing insert. Don't drop the bearing insert out of the cap.

7 Remove the bearing insert and push the connecting rod/piston assembly out through

the top of the engine. Use a wooden or plastic hammer handle to push on the upper bearing surface in the connecting rod. If resistance is felt, double-check to make sure that all of the ridge was removed from the cylinder.

8 Repeat the procedure for the remaining cylinders.

9 After removal, reassemble the connecting rod caps and bearing inserts in their respective connecting rods and install the cap bolts finger tight. Leaving the old bearing inserts in place until reassembly will help prevent the connecting rod bearing surfaces from being accidentally nicked or gouged.

10 The pistons and connecting rods are now ready for inspection and overhaul at an automotive machine shop.

Piston ring installation

11 Before installing the new piston rings, the ring end gaps must be checked. It's assumed that the piston ring side clearance has been checked and verified correct.

12 Lay out the piston/connecting rod assemblies and the new ring sets so the ring sets will be matched with the same piston and cylinder during the end gap measurement and engine assembly.

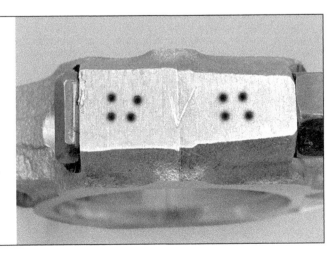

9.4 If the connecting rods or caps are not marked, use permanent ink or paint to mark the caps to the rods by cylinder number (for example, this would be number 4 cylinder connecting rod)

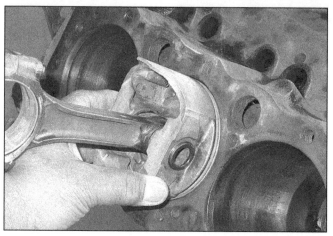

9.13 Install the piston ring into the cylinder then push it down into position using a piston so the ring will be square in the cylinder

9.14 With the ring square in the cylinder, measure the ring end gap with a feeler gauge

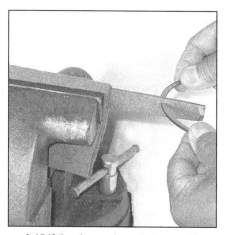

9.15 If the ring end gap is too small, clamp a file in a vise as shown and file the piston ring ends - be sure to remove all raised material

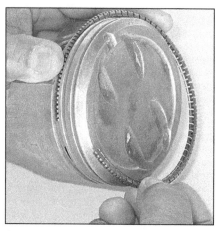

9.19a Installing the spacer/expander in the oil ring groove

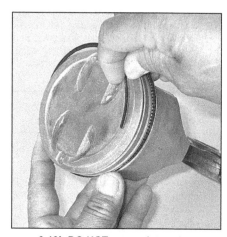

9.19b DO NOT use a piston ring installation tool when installing the oil control side rails

13 Insert the top (number one) ring into the first cylinder and square it up with the cylinder walls by pushing it in with the top of the piston **(see illustration)**. The ring should be near the bottom of the cylinder, at the lower limit of ring travel.

14 To measure the end gap, slip feeler gauges between the ends of the ring until a gauge equal to the gap width is found **(see illustration)**. The feeler gauge should slide between the ring ends with a slight amount of drag. A typical ring gap should fall between 0.010 and 0.020 inch [0.25 to 0.50 mm] for compression rings and up to 0.030 inch [0.76 mm] for the oil ring steel rails. If the gap is larger or smaller than specified, double-check to make sure you have the correct rings before proceeding.

15 If the gap is too small, it must be enlarged or the ring ends may come in contact with each other during engine operation, which can cause serious damage to the engine. If necessary, increase the end gaps by filing the ring ends very carefully with a fine file. Mount

the file in a vise equipped with soft jaws, slip the ring over the file with the ends contacting the file face and slowly move the ring to remove material from the ends. When performing this operation, file only by pushing the ring from the outside end of the file towards the vise **(see illustration)**.

16 Excess end gap isn't critical unless it's greater than 0.040 inch (1.01 mm). Again, double-check to make sure you have the correct ring type.

17 Repeat the procedure for each ring that will be installed in the first cylinder and for each ring in the remaining cylinders. Remember to keep rings, pistons and cylinders matched up.

18 Once the ring end gaps have been checked/corrected, the rings can be installed on the pistons.

19 The oil control ring (lowest one on the piston) is usually installed first. It's composed of three separate components. Slip the spacer/expander into the groove **(see illustration)**. If an anti-rotation tang is used, make sure it's

inserted into the drilled hole in the ring groove. Next, install the upper side rail in the same manner **(see illustration)**. Don't use a piston ring installation tool on the oil ring side rails, as they may be damaged. Instead, place one end of the side rail into the groove between the spacer/expander and the ring land, hold it firmly in place and slide a finger around the piston while pushing the rail into the groove. Finally, install the lower side rail.

20 After the three oil ring components have been installed, check to make sure that both the upper and lower side rails can be rotated smoothly inside the ring grooves.

21 The number two (middle) ring is installed next. It's usually stamped with a mark which must face up, toward the top of the piston. Do not mix up the top and middle rings, as they have different cross-sections.

Note: *Always follow the instructions printed on the ring package or box - different manufacturers may require different approaches.*

22 Use a piston ring installation tool and make sure the identification mark is facing

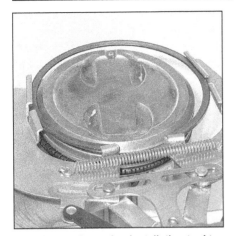

9.22 Use a piston ring installation tool to install the number 2 and the number 1 (top) rings - be sure the directional mark on the piston ring(s) is facing toward the top of the piston

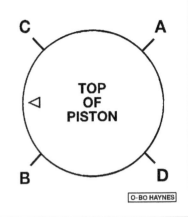

9.30 Position the piston ring end gaps as shown here before installing the piston/connecting rod assemblies into the engine

A *Top compression ring gap*
B *Second compression ring and oil ring spacer gap*
C *Upper oil ring gap*
D *Lower oil ring gap*

FRONT OF ENGINE

the top of the piston, then slip the ring into the middle groove on the piston **(see illustration)**. Don't expand the ring any more than necessary to slide it over the piston.

23 Install the number one (top) ring in the same manner. Make sure the mark is facing up. Be careful not to confuse the number one and number two rings.

24 Repeat the procedure for the remaining pistons and rings.

Installation

25 Before installing the piston/connecting rod assemblies, the cylinder walls must be perfectly clean, the top edge of each cylinder bore must be chamfered, and the crankshaft must be in place.

26 Remove the cap from the end of the number one connecting rod (refer to the marks made during removal). Remove the original bearing inserts and wipe the bearing surfaces of the connecting rod and cap with a clean, lint-free cloth. They must be kept spotlessly clean.

Connecting rod bearing oil clearance check

27 Clean the back side of the new upper bearing insert, then lay it in place in the connecting rod.

28 Make sure the tab on the bearing fits into the recess in the rod. Don't hammer the bearing insert into place and be very careful not to nick or gouge the bearing face. Don't lubricate the bearing at this time.

29 Clean the back side of the other bearing insert and install it in the rod cap. Again, make sure the tab on the bearing fits into the recess in the cap, and don't apply any lubricant. It's critically important that the mating surfaces of the bearing and connecting rod are perfectly clean and oil free when they're assembled.

30 Position the piston ring gaps at 90-degree intervals around the piston as shown **(see illustration)**.

31 Lubricate the piston and rings with clean engine oil and attach a piston ring compressor to the piston. Leave the skirt protruding about 1/4-inch to guide the piston into the cylinder. The rings must be compressed until they're flush with the piston.

32 Rotate the crankshaft until the number one connecting rod journal is at BDC (bottom dead center) and apply a liberal coat of engine oil to the cylinder walls.

33 With the directional stamp (arrow) on top of the piston facing the front (timing belt/timing chain end) of the engine, gently insert the piston/connecting rod assembly into the number one cylinder bore and rest the bottom edge of the ring compressor on the engine block. Install the pistons with the (arrow) mark facing toward the timing belt/timing chain end.

34 Tap the top edge of the ring compressor to make sure it's contacting the block around its entire circumference.

35 Gently tap on the top of the piston with the end of a wooden or plastic hammer handle **(see illustration)** while guiding the end of the connecting rod into place on the crankshaft journal. The piston rings may try to pop out of the ring compressor just before entering the cylinder bore, so keep some downward pressure on the ring compressor. Work slowly, and

if any resistance is felt as the piston enters the cylinder, stop immediately. Find out what's hanging up and fix it before proceeding. Do not force the piston into the cylinder - you might break a ring and/or the piston.

36 Once the piston/connecting rod assembly is installed, the connecting rod bearing oil clearance must be checked before the rod cap is permanently installed.

37 Cut a piece of the appropriate size Plastigage slightly shorter than the width of the connecting rod bearing and lay it in place on the number one connecting rod journal, parallel with the journal axis **(see illustration)**.

38 Clean the connecting rod cap bearing face and install the rod cap. Make sure the mating mark on the cap is on the same side as the mark on the connecting rod **(see illustration 9.4)**.

39 Install the old rod bolts, at this time, and tighten them to the torque listed in this Chapter's Specifications.

Note: *Use a thin-wall socket to avoid erroneous torque readings that can result if the socket is wedged between the rod cap and the bolt. If the socket tends to wedge itself between the bolt and the cap, lift up on it slightly until it no longer contacts the cap. DO NOT rotate the crankshaft at any time during this operation.*

9.35 Use a plastic or wooden hammer handle to push the piston into the cylinder

9.37 Place Plastigage on each connecting rod bearing journal parallel to the crankshaft centerline

9.41 Use the scale on the Plastigage package to determine the bearing oil clearance - be sure to measure the widest part of the Plastigage and use the correct scale; it comes with both standard and metric scales

10.1 Checking crankshaft endplay with a dial indicator

10.3 Checking crankshaft endplay with feeler gauges at the thrust bearing journal

40   Remove the bolts and detach the rod cap, being very careful not to disturb the Plastigage. Discard the cap bolts at this time as they cannot be reused.
Note: *You MUST use new connecting rod bolts.*
41   Compare the width of the crushed Plastigage to the scale printed on the Plastigage envelope to obtain the oil clearance **(see illustration)**. The connecting rod oil clearance is usually about 0.001 to 0.002 inch. Consult an automotive machine shop for the clearance specified for the rod bearings on your engine.
42   If the clearance is not as specified, the bearing inserts may be the wrong size (which means different ones will be required). Before deciding that different inserts are needed, make sure that no dirt or oil was between the bearing inserts and the connecting rod or cap when the clearance was measured. Also, recheck the journal diameter. If the Plastigage was wider at one end than the other, the journal may be tapered. If the clearance still exceeds the limit specified, the bearing will have to be replaced with an undersize bearing.
Caution: *When installing a new crankshaft, always use a standard size bearing.*

Final installation

43   Carefully scrape all traces of the Plastigage material off the rod journal and/or bearing face. Be very careful not to scratch the bearing - use your fingernail or the edge of a plastic card.
44   Make sure the bearing faces are perfectly clean, then apply a uniform layer of clean moly-base grease or engine assembly lube to both of them. You'll have to push the piston into the cylinder to expose the face of the bearing insert in the connecting rod.
Caution: *Install new connecting rod cap bolts. Do NOT reuse old bolts - they have stretched and cannot be reused.*
45   Slide the connecting rod back into place on the journal, install the rod cap, install the

new bolts and tighten them to the torque listed in this Chapter's Specifications. Again, work up to the torque in three steps.
46   Repeat the entire procedure for the remaining pistons/connecting rods.
47   The important points to remember are:
•   *Keep the back sides of the bearing inserts and the insides of the connecting rods and caps perfectly clean when assembling them.*
•   *Make sure you have the correct piston/rod assembly for each cylinder. The mark on the piston must face the front (timing belt/chain end) of the engine.*
•   *Lubricate the cylinder walls liberally with clean oil.*
•   *On 2.4L engines, apply a 1/8-inch bead of RTV engine sealant to the block, then place the ladder frame onto the block and tighten the bolts in a circular patter, starting from the center and working outwards, in two steps, to the torque listed in this Chapter's Specifications.*
48   After all the piston/connecting rod assemblies have been correctly installed, rotate the crankshaft a number of times by hand to check for any obvious binding.
49   As a final step, check the connecting rod endplay again.
50   Compare the measured endplay to the tolerance listed in this Chapter's Specifications to make sure it's acceptable. If it was correct before disassembly and the original crankshaft and rods were reinstalled, it should still be correct. If new rods or a new crankshaft were installed, the endplay may be inadequate. If so, the rods will have to be removed and taken to an automotive machine shop for resizing.

10   Crankshaft - removal and installation

Removal

Note: *The crankshaft can be removed only after the engine has been removed from the vehicle. It's assumed that the driveplate, crankshaft pul-*

ley, timing belt/timing chain, oil pan, oil pump body, oil filter, oil pump pick-up tube, windage tray and piston/connecting rod assemblies have already been removed. The rear main oil seal retainer must be unbolted and separated from the block before proceeding with crankshaft removal.
1   Before the crankshaft is removed, measure the endplay. Mount a dial indicator with the indicator in line with the crankshaft and just touching the end of the crankshaft as shown **(see illustration)**.
2   Pry the crankshaft all the way to the rear and zero the dial indicator. Next, pry the crankshaft to the front as far as possible and check the reading on the dial indicator. The distance traveled is the endplay. A typical crankshaft endplay will fall between 0.003 to 0.010 inch (0.076 to 0.254 mm). If it is greater than that, check the crankshaft thrust surfaces for wear after it's removed. If no wear is evident, new main bearings should correct the endplay.
3   If a dial indicator isn't available, feeler gauges can be used. Gently pry the crankshaft all the way to the front of the engine. Slip feeler gauges between the crankshaft and the front face of the thrust bearing or washer to determine the clearance **(see illustration)**.
4   On 2.4L engines, remove the ladder frame (oil pan housing) bolts and carefully separate the ladder frame from the engine block. **(see illustration 10.34)**.
Note: *New main bearing cap bolts must be used when reassembling the engine, but save the old bolts for use when checking the main bearing oil clearance.*
5   Loosen the main bearing cap bolts 1/4-turn at a time each, until they can be removed by hand, then gently tap the main bearing cap with a soft-face hammer around the perimeter of the assembly. Pull the main bearing cap straight up and off the cylinder block. Try not to drop the bearing inserts if they come out with the assembly.
6   Carefully lift the crankshaft out of the engine. It may be a good idea to have an assistant available, since the crankshaft is quite heavy and awkward to handle. With the bearing inserts in place inside the engine block

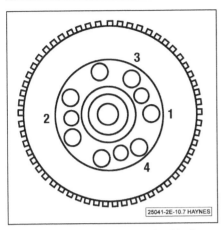

10.7 Crankshaft target wheel bolt tightening sequence (2.4L engines - 3.6L engines do not require bolts to be tightened in sequence)

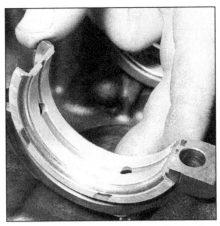

10.11 Installing a crankshaft main bearing onto the engine block main bearing saddle

10.18 Place the Plastigage onto the crankshaft bearing journal as shown

10.22 Use the scale on the Plastigage package to determine the bearing oil clearance - be sure to measure the widest part of the Plastigage and use the correct scale; it comes with both standard and metric scales

and main bearing caps, reinstall the bedplate/main bearing caps onto the engine block and tighten the bolts finger tight. Make sure you install the bedplate/main bearing caps with the arrow facing the front end of the engine.

Installation

7   Crankshaft installation is the first step in engine reassembly. It's assumed at this point that the engine block and crankshaft have been cleaned, inspected and repaired or reconditioned. On 2.4L and 3.6L engines, install the target wheel to the crankshaft and tighten the new bolts, in sequence, to the torque listed in this Chapter's Specifications **(see illustration)**, if removed.
8   Position the engine block with the bottom facing up.
9   Remove the mounting bolts and lift off the main bearing caps.
10   If they're still in place, remove the original bearing inserts from the block main bearing caps. Wipe the bearing surfaces of the block and bedplate with a clean, lint-free cloth. They must be kept spotlessly clean. This is critical for determining the correct bearing oil clearance.

Main bearing oil clearance check

11   Without mixing them up, clean the back sides of the new upper main bearing inserts (with grooves and oil holes) and lay one in each main bearing saddle in the block **(see illustration)**. Each upper bearing has an oil groove and oil hole in it.
Caution: *The oil holes in the block must line up with the oil holes in the upper bearing inserts.*
12   The thrust washers or thrust bearing inserts must be installed in the number 3 crankshaft journal on 2.4L and 2.7L engines or number 2 journal on 3.5L and 3.6L engines. Clean the back sides of the lower main bearing inserts and lay them in the corresponding location in the main bearing caps. Make sure the tab on the bearing insert fits into the recess in the block or bedplate or main bearing caps. The upper bearings with the oil

holes are installed into the engine block, while the lower bearings without the oil holes are installed in the main bearing caps.
Caution: *Do not hammer the bearing insert into place and don't nick or gouge the bearing faces. DO NOT apply any lubrication at this time.*
13   Clean the faces of the bearing inserts in the block and the crankshaft main bearing journals with a clean, lint-free cloth.
14   Check or clean the oil holes in the crankshaft, as any dirt here can go only one way - straight through the new bearings.
15   Once you're certain the crankshaft is clean, carefully lay it in position in the cylinder block.
16   Before the crankshaft can be permanently installed, the main bearing oil clearance must be checked.
17   Cut several strips of the appropriate size of Plastigage. They must be slightly shorter than the width of the main bearing journal.
18   Place one piece on each crankshaft main bearing journal, parallel with the journal axis as shown **(see illustration)**.
19   Clean the faces of the bearing inserts in the main bearing caps. Hold the bearing inserts in place and install the assembly onto the crankshaft and cylinder block. DO NOT disturb the Plastigage. Make sure you install the main bearing caps with the arrow facing the front (timing belt/timing chain end) of the engine.
Caution: *On 2.4L and 2.7 L engines the number 3 main bearing cap, on 3.5L and 3.6L engines the number 2 main bearing cap must be centered over the inner bolt holes of the block. If the bearing cap is not centered the crankshaft counterweights can contact the main bearing cap and cause severe engine damage.*
Caution: *On 2.4L engines, the manufacturer used three different head bolts. The bolts have different torque values and are not interchangeable. The bolts have an identifying mark on top of the bolt head - the letter "b", the letter "M," or the number "6" - the correct bolts must be reinstalled to prevent incorrect tightening torque.*

20   Apply clean engine oil to all bolt threads prior to installation, then install the old bolts finger-tight; do not install the side bolts at this time. Tighten all the bolts in the sequences shown **(see illustrations 10.31a through 10.31e)** progressing in steps, to the torque listed in this Chapter's Specifications. On V6 engines, only install and tighten the inner bolts. It is not necessary to install the main bearing cap outer bolts or side bolts at this time.
21   Remove the bolts in the reverse order of the tightening sequence and carefully lift the caps straight up and off the block. Do not disturb the Plastigage or rotate the crankshaft. If the cap(s) is difficult to remove, tap it gently from side-to-side with a soft-face hammer to loosen it.
22   Compare the width of the crushed Plastigage on each journal to the scale printed on the Plastigage envelope to determine the main bearing oil clearance **(see illustration)**. Check with an automotive machine shop for the crankshaft main bearing oil clearance limits.

ENGINE BEARING ANALYSIS

Debris

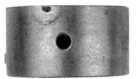

Babbitt bearing embedded with debris from machinings

Microscopic detail of debris

Microscopic detail of gouges

Overplated copper alloy bearing gouged by cast iron debris

Aluminum bearing embedded with glass beads

Microscopic detail of glass beads

Damaged lining caused by dirt left on the bearing back

Misassembly

Result of a lower half assembled as an upper - blocking the oil flow

Excessive oil clearance is indicated by a short contact arc

Polished and oil-stained backs are a result of a poor fit in the housing bore

Result of a wrong, reversed, or shifted cap

Overloading

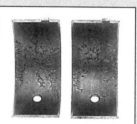

Damage from excessive idling which resulted in an oil film unable to support the load imposed

Damaged upper connecting rod bearings caused by engine lugging; the lower main bearings (not shown) were similarly affected

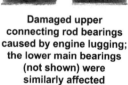

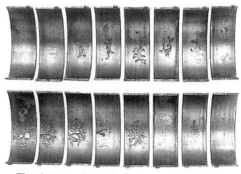

The damage shown in these upper and lower connecting rod bearings was caused by engine operation at a higher-than-rated speed under load

Misalignment

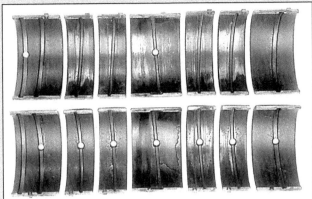

A warped crankshaft caused this pattern of severe wear in the center, diminishing toward the ends

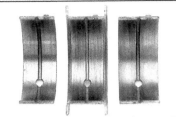

A poorly finished crankshaft caused the equally spaced scoring shown

A bent connecting rod led to the damage in the "V" pattern

A tapered housing bore caused the damage along one edge of this pair

Lubrication

Result of dry start: The bearings on the left, farthest from the oil pump, show more damage

Result of a low oil supply or oil starvation

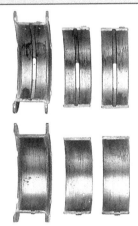

Severe wear as a result of inadequate oil clearance

Corrosion

Microscopic detail of corrosion

Corrosion is an acid attack on the bearing lining generally caused by inadequate maintenance, extremely hot or cold operation, or inferior oils or fuels

Microscopic detail of cavitation

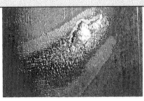

Example of cavitation - a surface erosion caused by pressure changes in the oil film

Damage from excessive thrust or insufficient axial clearance

Bearing affected by oil dilution caused by excessive blow-by or a rich mixture

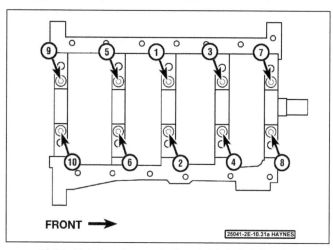

10.31a Main bearing cap bolt and side bolt tightening sequence - 2.4L engines

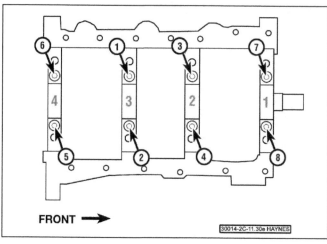

10.31b Main bearing cap inner bolt tightening sequence - 2.7L and 3.5L engines

23 If the clearance is not as specified, the bearing inserts may be the wrong size (which means different ones will be required). Before deciding if different inserts are needed, make sure that no dirt or oil was between the bearing inserts and the cap assembly or block when the clearance was measured. If the Plastigage was wider at one end than the other, the crankshaft journal may be tapered. If the clearance still exceeds the limit specified, the bearing insert(s) will have to be replaced with an undersize bearing insert(s). **Caution:** *When installing a new crankshaft, always install a standard bearing insert set.*
24 Carefully scrape all traces of the Plastigage material off the main bearing journals and/or the bearing insert faces. Be sure to remove all residue from the oil holes. Use your fingernail or the edge of a plastic card - don't nick or scratch the bearing faces.

Final installation
25 Carefully lift the crankshaft out of the cylinder block.

26 Clean the bearing insert faces in the cylinder block, then apply a thin, uniform layer of moly-base grease or engine assembly lube to each of the bearing surfaces. Coat the thrust faces as well as the journal face of the thrust bearing.
27 Make sure the crankshaft journals are clean, then lay the crankshaft back in place in the cylinder block.
28 Clean the bearing insert faces and then apply the same lubricant to them.
29 Install the main bearing caps onto the designated journals.
30 Prior to installation, apply clean engine oil to the **NEW** bolt threads, wiping off any excess, then install all bolts finger-tight.
31 Tighten the cap bolts, in sequence **(see illustrations)**, to the torque listed in this Chapter's Specifications.
32 Recheck crankshaft endplay with a feeler gauge or a dial indicator. The endplay should be correct if the crankshaft thrust faces aren't worn or damaged and if new bearings have been installed.

33 Rotate the crankshaft a number of times by hand to check for any obvious binding. It should rotate with a running torque of 50 inlbs or less. If the running torque is too high, correct the problem at this time.
34 On 2.4L engines, apply a continuous 1/8 inch bead of RTV sealant to the engine block surface then carefully place the ladder frame on to the block. Install the the mounting bolts hand tight. Tighten the bolts in the sequence shown **(see illustration)** to the torque listed in this Chapter's Specifications.
35 Install the new rear main oil seal assembly (see Chapter 2A, 2B, 2C or 2D).
36 Install the oil pump, pick up tube or balance shaft module (see Chapter 2A, 2B, 2C or 2D).

11 Engine overhaul - reassembly sequence

1 Before beginning engine reassembly, make sure you have all the necessary new parts, gaskets and seals as well as the follow-

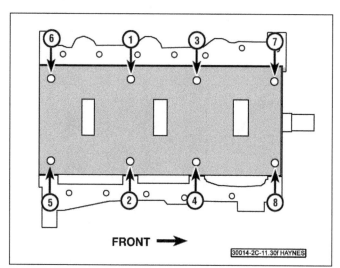

10.31c Main bearing cap outer bolt and windage tray tightening sequence - 2.7L and 3.5L engines

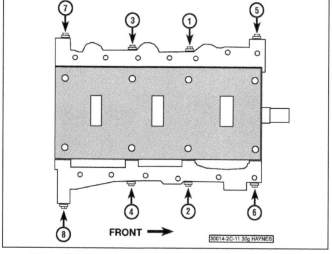

10.31d Main bearing cap side bolt tightening sequence - 2.7L and 3.5L engines

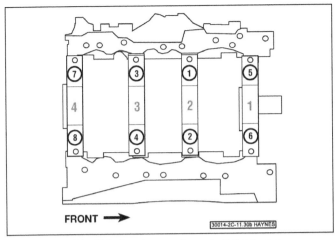

10.31e Main bearing cap inner bolt tightening sequence - 3.6L engines

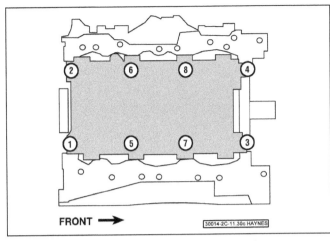

10.31f Main bearing cap outer bolt and windage tray tightening sequence - 3.6L engines

ing items on hand:

Common hand tools
A 1/2-inch drive torque wrench
New engine oil
Gasket sealant
Thread locking compound

2   If you obtained a short block, it will be necessary to install the cylinder head, the oil pump and pick-up tube, the oil pan, the water pump, the timing belt/timing chain and timing cover, and the valve cover (see Chapter 2A, 2B, 2C or 2D) and balance shaft module on 2.4L models (see Chapter 2A). In order to save time and avoid problems, the external components must be installed in the following general order:

Thermostat and housing cover
Water pump
Intake and exhaust manifolds
Fuel injection components
Emissions control components
Spark plug wires and spark plugs
Ignition coils
Oil filter
Engine mounts and mount brackets
Driveplate

12 Initial start-up and break-in after overhaul

Warning: *Have a fire extinguisher handy when starting the engine for the first time.*

1   Once the engine has been installed in the vehicle, double-check the engine oil and coolant levels.

2   With the spark plugs out of the engine and the ignition system and fuel pump disabled (see Section 3, Step 5), crank the engine until oil pressure registers on the gauge or the light goes out.

3   Install the spark plugs, hook up the plug wires or install the coils, and restore the ignition system and fuel pump functions.

4   Start the engine. It may take a few moments for the fuel system to build up pressure, but the engine should start without a great deal of effort.

5   After the engine starts, it should be allowed to warm up to normal operating temperature. While the engine is warming up,

make a thorough check for fuel, oil and coolant leaks.

6   Shut the engine off and recheck the engine oil and coolant levels.

7   Drive the vehicle to an area with minimum traffic, accelerate from 30 to 50 mph, then allow the vehicle to slow to 30 mph with the throttle closed. Repeat the procedure 10 or 12 times. This will load the piston rings and cause them to seat properly against the cylinder walls. Check again for oil and coolant leaks.

8   Drive the vehicle gently for the first 500 miles (no sustained high speeds) and keep a constant check on the oil level. It is not unusual for an engine to use oil during the break-in period.

9   At approximately 500 to 600 miles, change the oil and filter.

10   For the next few hundred miles, drive the vehicle normally. Do not pamper it or abuse it.

11   After 2000 miles, change the oil and filter again and consider the engine broken in.

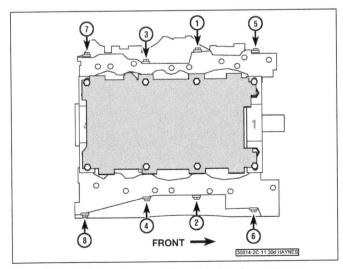

10.31g Main bearing cap side bolt tightening sequence - 3.6L engines

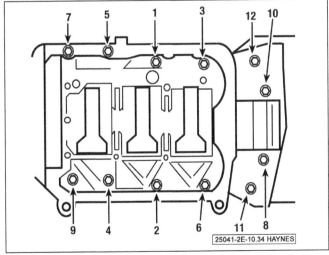

10.34 Ladder frame (oil pan housing) mounting bolt tightening sequence - 2.4L engines only

Notes

Chapter 3
Cooling, heating and air conditioning systems

Specifications

	Section		Section
Air conditioning and heating system - check and maintenance.....	3	General information	1
Air conditioning compressor - removal and installation	12	Heater core - replacement...	10
Air conditioning condenser - removal and installation	14	Heater/air conditioner control assembly - removal	
Air conditioning receiver-drier - removal and installation..............	13	and installation	11
Blower motor resistor/power module and blower motor		Orifice tube (2014 and earlier models) - removal and installation ...	15
assembly - replacement	9	Radiator - removal and installation..	7
Coolant reservoir/expansion tank - removal and installation	6	Thermostat - replacement	4
Engine cooling fans - check and replacement..............................	5	Troubleshooting..	2
Expansion valve (2015 and later models) - replacement	16	Water pump - replacement	8

Contents

General

Radiator cap pressure rating	14 to 18 psi
Expansion tank cap pressure rating (2015 and later models)...............	21 psi
Thermostat rating (opening temperature)..	195 degrees F
Cooling system capacity...	See Chapter 1
Refrigerant type	
2014 and earlier models..	R-134a
2015 and later models..	R-1234yf
Refrigerant capacity*	
2010 and earlier models..	1.00 lb
2011 and later models..	1.25 lbs
2015 and later models...	1.13 lbs

Check the refrigerant capacity listed on the underhood HVAC label; if the charge capacity listed on the label differs from that shown here, assume the label is correct.

Torque specifications   Ft-lbs (unless otherwise indicated)

Note: *One foot-pound (ft-lb) of torque is equivalent to 12 inch-pounds (in-lbs) of torque. Torque values below approximately 15 ft-lbs are expressed in inch-pounds, since most foot-pound torque wrenches are not accurate at these smaller values.*

Thermostat housing bolts	
2.4L engine (2014 and earlier models)	
Primary thermostat (2010 and earlier models)	79 in-lbs
Primary thermostat (2011 to 2014 models).................................	159 in-lbs
Secondary thermostat..	159 in-lbs
2.4L engine (2015 and later)...	89 in-lbs
2.7L engine	105 in-lbs
3.5L engine	105 in-lbs
3.6L engine	106 in-lbs
Water inlet tube bolts (2014 and earlier models)..................................	17.5
Water pump mounting bolts	
2.4L engine (2014 and earlier)	18
2.4L engine (2015 and later)..	89 in-lbs
2.7L engine	105 in-lbs
3.5L engine	108 in-lbs
3.6L engine	
2014 and earlier (in sequence – **see illustration 8.35**).................	106 in-lbs
2015 and later	
Smaller bolts (M6)........................	106 in-lbs
Large bolt (M8)...	17
Water pump pulley bolts (2.4L engine)...	80 in-lbs
Idler pulley bolt (3.6L engine)	18

2.2 The cooling system pressure tester is connected in place of the pressure cap, then pumped up to pressurize the system

2.5a The combustion leak detector consists of a bulb, syringe and test fluid

1   General information

Warning: *Do not allow antifreeze to come in contact with your skin or painted surfaces of the vehicle. Rinse off spills immediately with plenty of water. Antifreeze is highly toxic if ingested. Never leave antifreeze lying around in an open container or in puddles on the floor; children and pets are attracted by it's sweet smell and may drink it. Check with local authorities about disposing of used antifreeze. Many communities have collection centers which will see that antifreeze is disposed of safely. Never dump used antifreeze on the ground or pour it into drains.*

Engine cooling system

1   All modern vehicles employ a pressurized engine cooling system with thermostatically controlled coolant circulation. The cooling system consists of a radiator, an expansion tank or coolant reservoir, a pressure cap (located on the expansion tank or cooling system filler neck), a thermostat, a cooling fan, and a water pump.
2   The water pump circulates coolant through the engine. The coolant flows around each cylinder and around the intake and exhaust ports, near the spark plug areas and in close proximity to the exhaust valve guides.
3   A thermostat controls engine coolant temperature. During warm up, the closed thermostat prevents coolant from circulating through the radiator. As the engine nears normal operating temperature, the thermostat opens and allows hot coolant to travel through the radiator, where it's cooled before returning to the engine.

Heating system

4   The heating system consists of a blower fan and heater core located in a housing under the dash, the hoses connecting the heater core to the engine cooling system and the heater/air conditioning control head on the dashboard. Hot engine coolant is circulated through the heater core. When the heater mode is activated, a flap door in the housing opens to expose the heater core to the passenger compartment through air ducts. A fan switch on the control head activates the blower motor, which forces air through the core, heating the air.

Air conditioning system

5   The air conditioning system consists of a condenser mounted in front of the radiator, an evaporator mounted adjacent to the heater core, a compressor mounted on the engine, a receiver-drier or accumulator and the plumbing connecting all of the above components.
6   A blower fan forces the warmer air of the passenger compartment through the evaporator core (sort of a radiator-in-reverse), transferring the heat from the air to the refrigerant. The liquid refrigerant boils off into low pressure vapor, taking the heat with it when it leaves the evaporator.

2   Troubleshooting

Coolant leaks

1   A coolant leak can develop anywhere in the cooling system, but the most common causes are:

A loose or weak hose clamp
A defective hose
A faulty pressure cap
A damaged radiator
A bad heater core
A faulty water pump
A leaking gasket at any joint that carries coolant

2   Coolant leaks aren't always easy to find. Sometimes they can only be detected when the cooling system is under pressure. Here's where a cooling system pressure tester comes in handy. After the engine has cooled completely, the tester is attached in place of the pressure cap, then pumped up to the pressure value equal to that of the pressure cap rating **(see illustration)**. Now, leaks that only exist when the engine is fully warmed up will become apparent. The tester can be left connected to locate a nagging slow leak.

Coolant level drops, but no external leaks

3   If you find it necessary to keep adding coolant, but there are no external leaks, the probable causes include:

a) A blown head gasket
b) A leaking intake manifold gasket (only on engines that have coolant passages in the manifold), or a cracked cylinder head or cylinder block

4   Any of the above problems will also usually result in contamination of the engine oil, which will cause it to take on a milkshake-like appearance. A bad head gasket or cracked head or block can also result in engine oil contaminating the cooling system.
5   Combustion leak detectors (also known as block testers) are available at most auto parts stores. These work by detecting exhaust gases in the cooling system, which indicates a compression leak from a cylinder into the coolant. The tester consists of a large bulb-type syringe and bottle of test fluid **(see illustration)**. A measured amount of the fluid is

2.5b Place the tester over the cooling system filler neck and use the bulb to draw a sample into the tester

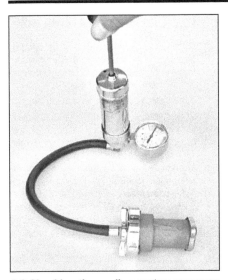

2.8 Checking the cooling system pressure cap with a cooling system pressure tester

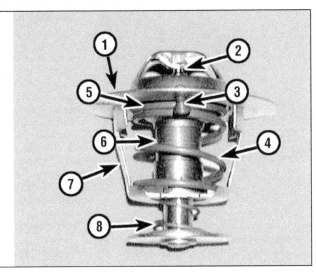

2.10 Typical thermostat

1   *Flange*
2   *Piston*
3   *Jiggle valve*
4   *Main coil spring*
5   *Valve seat*
6   *Valve*
7   *Frame*
8   *Secondary coil spring*

added to the syringe. The syringe is placed over the cooling system filler neck and, with the engine running, the bulb is squeezed and a sample of the gases present in the cooling system are drawn up through the test fluid **(see illustration)**. If any combustion gases are present in the sample taken, the test fluid will change color.

6   If the test indicates combustion gas is present in the cooling system, you can be sure that the engine has a blown head gasket or a crack in the cylinder head or block, and will require disassembly to repair.

Pressure cap

Warning: *Wait until the engine is completely cool before beginning this check.*

7   The cooling system is sealed by a spring-loaded cap, which raises the boiling point of the coolant. If the cap's seal or spring are worn out, the coolant can boil and escape past the cap. With the engine completely cool, remove the cap and check the seal; if it's cracked, hardened or deteriorated in any way, replace it with a new one.

8   Even if the seal is good, the spring might not be; this can be checked with a cooling system pressure tester **(see illustration)**. If the cap can't hold a pressure within approximately 1-1/2 lbs of its rated pressure (which is marked on the cap), replace it with a new one.

9   The cap is also equipped with a vacuum relief spring. When the engine cools off, a vacuum is created in the cooling system. The vacuum relief spring allows air back into the system, which will equalize the pressure and prevent damage to the radiator (the radiator tanks could collapse if the vacuum is great enough). If, after turning the engine off and allowing it to cool down you notice any of the cooling system hoses collapsing, replace the pressure cap with a new one.

Thermostat

10   Before assuming the thermostat **(see illustration)** is responsible for a cooling system problem, check the coolant level (see Chapter 1), drivebelt tension (see Chapter 1) and temperature gauge (or light) operation.

11   If the engine takes a long time to warm up (as indicated by the temperature gauge or heater operation), the thermostat is probably stuck open. Replace the thermostat with a new one.

12   If the engine runs hot or overheats, a thorough test of the thermostat should be performed.

13   Definitive testing of the thermostat can only be made when it is removed from the vehicle. If the thermostat is stuck in the open position at room temperature, it is faulty and must be replaced.

Caution: *Do not drive the vehicle without a thermostat. The computer may stay in open loop and emissions and fuel economy will suffer.*

14   To test a thermostat, suspend the (closed) thermostat on a length of string or wire in a pot of cold water.

15   Heat the water on a stove while observing the thermostat. The thermostat should fully open before the water boils.

16   If the thermostat doesn't open and close as specified, or sticks in any position, replace it.

Cooling fan

Electric cooling fan

17   If the engine is overheating and the cooling fan is not coming on when the engine temperature rises to an excessive level, unplug the fan motor electrical connector(s) and connect the motor directly to the battery with fused jumper wires. If the fan motor doesn't come on, replace the motor.

18   If the radiator fan motor is okay, but it isn't coming on when the engine gets hot, the fan relay might be defective. A relay is used to control a circuit by turning it on and off in response

to a control decision by the Powertrain Control Module (PCM). These control circuits are fairly complex, and checking them should be left to a qualified automotive technician. Sometimes, the control system can be fixed by simply identifying and replacing a bad relay.

19   Locate the fan relays in the engine compartment fuse/relay box.

20   Test the relay (see Chapter 12).

21   If the relay is okay, check all wiring and connections to the fan motor. Refer to the wiring diagrams at the end of Chapter 12. If no obvious problems are found, the problem could be the Engine Coolant Temperature (ECT) sensor or the Powertrain Control Module (PCM). Have the cooling fan system and circuit diagnosed by a dealer service department or repair shop with the proper diagnostic equipment.

Belt-driven cooling fan

22   Disconnect the negative battery cable from the remote ground terminal (see Chapter 5) and rock the fan back and forth by hand to check for excessive bearing play.

23   With the engine cold (and not running), turn the fan blades by hand. The fan should turn freely.

24   Visually inspect for substantial fluid leakage from the clutch assembly. If problems are noted, replace the clutch assembly.

25   With the engine completely warmed up, turn off the ignition switch and disconnect the negative battery cable from the remote ground terminal. Turn the fan by hand. Some drag should be evident. If the fan turns easily, replace the fan clutch.

Water pump

26   A failure in the water pump can cause serious engine damage due to overheating.

Drivebelt-driven water pump

27   There are two ways to check the operation of the water pump while it's installed on the engine. If the pump is found to be defec-

2.28 The water pump weep hole is generally located on the underside of the pump

tive, it should be replaced with a new or rebuilt unit.

28   Water pumps are equipped with weep (or vent) holes **(see illustration)**. If a failure occurs in the pump seal, coolant will leak from the hole.

29   If the water pump shaft bearings fail, there may be a howling sound at the pump while it's running. Shaft wear can be felt with the drivebelt removed if the water pump pulley is rocked up and down (with the engine off). Don't mistake drivebelt slippage, which causes a squealing sound, for water pump bearing failure.

Timing chain or timing belt-driven water pump

30   Water pumps driven by the timing chain or timing belt are located underneath the timing chain or timing belt cover.

31   Checking the water pump is limited because of where it is located. However, some basic checks can be made before deciding to remove the water pump. If the pump is found to be defective, it should be replaced with a new or rebuilt unit.

32   One sign that the water pump may be failing is that the heater (climate control) may not work well. Warm the engine to normal operating temperature, confirm that the coolant level is correct, then run the heater and check for hot air coming from the ducts.

33   Check for noises coming from the water pump area. If the water pump impeller shaft or bearings are failing, there may be a howling sound at the pump while the engine is running.

Note: *Be careful not to mistake drivebelt noise (squealing) for water pump bearing or shaft failure.*

34   It you suspect water pump failure due to noise, wear can be confirmed by feeling for play at the pump shaft. This can be done by rocking the drive sprocket on the pump shaft up and down. To do this you will need to remove the tension on the timing chain or belt as well as access the water pump.

All water pumps

35   In rare cases or on high-mileage vehicles, another sign of water pump failure may be the presence of coolant in the engine oil. This condition will adversely affect the engine in varying degrees.

Note: *Finding coolant in the engine oil could indicate other serious issues besides a failed water pump, such as a blown head gasket or a cracked cylinder head or block.*

36   Even a pump that exhibits no outward signs of a problem, such as noise or leakage, can still be due for replacement. Removal for close examination is the only sure way to tell. Sometimes the fins on the back of the impeller can corrode to the point that cooling efficiency is diminished significantly.

Heater system

37   Little can go wrong with a heater. If the fan motor will run at all speeds, the electrical part of the system is okay. The three basic heater problems fall into the following general categories:

a) Not enough heat
b) Heat all the time
c) No heat

38   If there's not enough heat, the control valve or door is stuck in a partially open position, the coolant coming from the engine isn't hot enough, or the heater core is restricted. If the coolant isn't hot enough, the thermostat in the engine cooling system is stuck open, allowing coolant to pass through the engine so rapidly that it doesn't heat up quickly enough. If the vehicle is equipped with a temperature gauge instead of a warning light, watch to see if the engine temperature rises to the normal operating range after driving for a reasonable distance.

39   If there's heat all the time, the control valve or the door is stuck wide open.

40   If there's no heat, coolant is probably not reaching the heater core, or the heater core is plugged. The likely cause is a collapsed or plugged hose, core, or a frozen heater control valve. If the heater is the type that flows coolant all the time, the cause is a stuck door or a broken or kinked control cable.

Air conditioning system

41   If the cool air output is inadequate:

a) Inspect the condenser coils and fins to make sure they're clear
b) Check the compressor clutch for slippage
c) Check the blower motor for proper operation
d) Inspect the blower discharge passage for obstructions
e) Check the system air intake filter for clogging

42   If the system provides intermittent cooling air:

a) Check the circuit breaker, blower switch and blower motor for a malfunction
b) Make sure the compressor clutch isn't slipping
c) Inspect the plenum door to make sure it's operating properly

d) Inspect the evaporator to make sure it isn't clogged
e) If the unit is icing up, it may be caused by excessive moisture in the system, incorrect super heat switch adjustment, or low thermostat adjustment

43   If the system provides no cooling air:

a) Inspect the compressor drivebelt; make sure it isn't loose or broken
b) Make sure the compressor clutch engages; if it doesn't, check for a blown fuse
c) Inspect the wire harness for broken or disconnected wires
d) If the compressor clutch doesn't engage, bridge the terminals of the AC pressure switch(es) with a jumper wire; if the clutch now engages, and the system is properly charged, the pressure switch is bad
e) Make sure the blower motor is not disconnected or burned out
f) Make sure the compressor isn't partially or completely seized
g) Inspect the refrigerant lines for leaks
h) Check the components for leaks
i) Inspect the receiver-drier/accumulator or expansion valve/tube for clogged screens

44   If the system is noisy:

a) Look for loose panels in the passenger compartment
b) Inspect the compressor drivebelt; it may be loose or worn
c) Check the compressor mounting bolts; they should be tight
d) Listen carefully to the compressor; it may be worn out
e) Listen to the idler pulley and bearing, and the clutch; either may be defective
f) The winding in the compressor clutch coil or solenoid may be defective
g) The compressor oil level may be low
h) The blower motor fan bushing or the motor itself may be worn out
i) If there is an excessive charge in the system, you'll hear a rumbling noise in the high pressure line, a thumping noise in the compressor, or see bubbles or cloudiness in the sight glass
j) If there is a low charge in the system, you might hear hissing in the evaporator case at the expansion valve, or see bubbles or cloudiness in the sight glass

---

3   Air conditioning and heating system - check and maintenance

---

Air conditioning system

Warning: *The air conditioning system is under high pressure. Do not loosen any hose fittings or remove any components until after the system has been discharged. Air conditioning refrigerant should be properly discharged into an EPA-approved recovery/recycling unit at a dealer service department or an automotive air conditioning repair facility. Always wear eye protection when disconnecting air conditioning system fittings.*

3.9 Insert a thermometer in the center vent, turn on the air conditioning system and wait for it to cool down;depending on the humidity, the output air should be 35 to 40 degrees cooler than the ambient air temperature

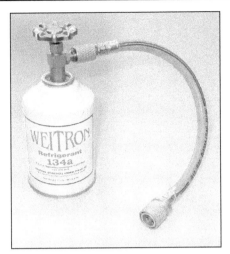

3.11 Typical automotive air conditioning charging kit

3.13 The low-side charging port is located near the right strut tower

Caution: *All models covered by this manual use environmentally friendly R-134a or R-1234yf. This refrigerant (and its appropriate refrigerant oils) are not compatible with R-12 refrigerant system components and must never be mixed or the components will be damaged.*
Caution: *When replacing entire components, additional refrigerant oil should be added equal to the amount that is removed with the component being replaced. Be sure to read the can before adding any oil to the system, to make sure it is compatible with the R-134a or R-1234yf system.*

1   The following maintenance checks should be performed on a regular basis to ensure that the air conditioning continues to operate at peak efficiency.

a) *Inspect the condition of the compressor drivebelt. If it is worn or deteriorated, replace it (see Chapter 1).*

b) *Check the drivebelt tension (see Chapter 1).*

c) *Inspect the system hoses. Look for cracks, bubbles, hardening and deterioration. Inspect the hoses and all fittings for oil bubbles or seepage. If there is any evidence of wear, damage, or leakage, replace the hose(s).*

d) *Inspect the condenser fins for leaves, bugs and any other foreign material that may have embedded itself in the fins. Use a fin comb or compressed air to remove debris from the condenser.*

e) *Make sure the system has the correct refrigerant charge.*

f) *If you hear water sloshing around in the dash area or have water dripping on the carpet, check the evaporator housing drain tube, and insert a piece of wire into the opening to check for blockage.*

2   It's a good idea to operate the system for about ten minutes at least once a month. This is particularly important during the winter months because long term non-use can

cause hardening, and subsequent failure, of the seals. Note that using the Defrost function operates the compressor.

3   If the air conditioning system is not working properly, proceed to Step 6 and perform the general checks outlined below.

4   Because of the complexity of the air conditioning system and the special equipment necessary to service it, in-depth troubleshooting and repairs beyond checking the refrigerant charge and the compressor clutch operation are not included in this manual. However, simple checks and component replacement procedures are provided in this Chapter. For more complete information on the air conditioning system, refer to the *Haynes Automotive Heating and Air Conditioning Manual.*

5   The most common cause of poor cooling is simply a low system refrigerant charge. If a noticeable drop in system cooling ability occurs, one of the following quick checks will help you determine if the refrigerant level is low.

Checking the refrigerant charge

6   Warm the engine up to normal operating temperature.

7   Place the air conditioning temperature selector at the coldest setting and put the blower at the highest setting.

8   After the system reaches operating temperature, feel the larger pipe exiting the evaporator at the firewall. The outlet pipe should be cold (the tubing that leads back to the compressor). If the evaporator outlet pipe is warm, the system probably needs a charge.

9   Insert a thermometer in the center air distribution duct **(see illustration)** while operating the air conditioning system at its maximum setting - the temperature of the output air should be 35 to 40 degrees F below the ambient air temperature (down to approximately 40 degrees F). If the ambient (outside) air temperature is very high,

say 110 degrees F, the duct air temperature may be as high as 60 degrees F, but generally the air conditioning is 35 to 40 degrees F cooler than the ambient air.

10   Further inspection or testing of the system requires special tools and techniques and is beyond the scope of the home mechanic.

Adding refrigerant

Caution: *Make sure any refrigerant, refrigerant oil or replacement component you purchase is designated as compatible with your system.*

11   Purchase an air conditioning charging kit at an auto parts store **(see illustration).** A charging kit includes a can of refrigerant, a tap valve and a short section of hose that can be attached between the tap valve and the system low side service valve.
Caution: *Never add more than one can of refrigerant to the system. If more refrigerant than that is required, the system should be evacuated and leak tested.*

12   Back off the valve handle on the charging kit and screw the kit onto the refrigerant can, making sure first that the O-ring or rubber seal inside the threaded portion of the kit is in place.
Warning: *Wear protective eyewear when dealing with pressurized refrigerant cans.*

13   Remove the dust cap from the low-side charging port and attach the hose's quick-connect fitting to the port **(see illustration).**
Warning: *DO NOT hook the charging kit hose to the system high side! The fittings on the charging kit are designed to fit only on the low side of the system.*

14   Warm up the engine and turn On the air conditioning. Keep the charging kit hose away from the fan and other moving parts.
Note: *The charging process requires the compressor to be running. If the clutch cycles off, you can put the air conditioning switch on High and leave the car doors open to keep the clutch on and compressor working.The compressor can be kept on during the charg-*

ing by removing the connector from the pressure switch and bridging it with a paper clip or jumper wire during the procedure.

15   Turn the valve handle on the kit until the stem pierces the can, then back the handle out to release the refrigerant. You should be able to hear the rush of gas. Keep the can upright at all times, but shake it occasionally. Allow stabilization time between each addition.

Note: *The charging process will go faster if you wrap the can with a hot-water-soaked rag to keep the can from freezing up.*

16   If you have an accurate thermometer, you can place it in the center air conditioning duct inside the vehicle and keep track of the output air temperature. A charged system that is working properly should cool down to approximately 40 degrees F. If the ambient (outside) air temperature is very high, say 110 degrees F, the duct air temperature may be as high as 60 degrees F, but generally the air conditioning is 35 to 40 degrees F cooler than the ambient air.

17   When the can is empty, turn the valve handle to the closed position and release the connection from the low-side port. Reinstall the dust cap.

18   Remove the charging kit from the can and store the kit for future use with the piercing valve in the UP position, to prevent inadvertently piercing the can on the next use.

Heating systems

19   If the carpet under the heater core is damp, or if antifreeze vapor or steam is coming through the vents, the heater core is leaking. Remove it (see Section 10) and install a new unit (most radiator shops will not repair a leaking heater core).

20   If the air coming out of the heater vents isn't hot, the problem could stem from any of the following causes:

a) *The thermostat is stuck open, preventing the engine coolant from warming up enough to carry heat to the heater core. Replace the thermostat (see Section 4).*

b) *There is a blockage in the system, preventing the flow of coolant through the heater core. Feel both heater hoses at the firewall. They should be hot. If one of them is cold, there is an obstruction in one of the hoses or in the heater core, or the heater control valve is shut. Detach the hoses and back flush the heater core with a water hose. If the heater core is clear but circulation is impeded, remove the two hoses and flush them out with a water hose.*

c) *If flushing fails to remove the blockage from the heater core, the core must be replaced (see Section 10).*

Eliminating air conditioning odors

21   Unpleasant odors that often develop in air conditioning systems are caused by the growth of a fungus, usually on the surface of the evaporator core. The warm, humid envi-

ronment there is a perfect breeding ground for mildew to develop.

22   The evaporator core on most vehicles is difficult to access, and factory dealerships have a lengthy, expensive process for eliminating the fungus by opening up the evaporator case and using a powerful disinfectant and rinse on the core until the fungus is gone. You can service your own system at home, but it takes something much stronger than basic household germ-killers or deodorizers.

23   Aerosol disinfectants for automotive air conditioning systems are available in most auto parts stores, but remember when shopping for them that the most effective treatments are also the most expensive. The basic procedure for using these sprays is to start by running the system in the RECIRC mode for ten minutes with the blower on its highest speed. Use the highest heat mode to dry out the system and keep the compressor from engaging by disconnecting the wiring connector at the compressor.

24   The disinfectant can usually comes with a long spray hose. Insert the nozzle into an intake port inside the cabin filter housing, and spray according to the manufacturer's recommendations. Try to cover the whole surface of the evaporator core, by aiming the spray up, down and sideways. Follow the manufacturer's recommendations for the length of spray and waiting time between applications.

25   Once the evaporator has been cleaned, the best way to prevent the mildew from coming back again is to make sure your evaporator housing drain tube is clear.

Automatic heating and air conditioning systems

26   Some vehicles are equipped with an optional automatic climate control system. This system has its own computer that receives inputs from various sensors in the heating and air conditioning system. This computer, like the PCM, has self-diagnostic capabilities to help pinpoint problems or faults within the system. Vehicles equipped with automatic heating and

4.4 The thermostat housing cover and mounting bolts (2.4L shown)

4.5 Water inlet removed, primary thermostat exposed (2.4L shown)

air conditioning systems are very complex and considered beyond the scope of the home mechanic. Vehicles equipped with automatic heating and air conditioning systems should be taken to a dealer service department or other qualified facility for repair.

---

4   Thermostat - replacement

Warning: *Do not remove the radiator cap, drain the coolant or replace the thermostat until the engine has cooled completely.*

1   Disconnect the negative battery cable from the remote ground terminal or battery (see Chapter 5).

2   Drain the cooling system (see Chapter 1). If the coolant is relatively new or in good condition, save it and reuse it. Read the **Warning** in Section 2.

2.4L four cylinder engine (2014 and earlier)

Primary thermostat

3   Remove the air filter and intake hose assembly (see Chapter 4).

4   Follow the coolant hose to the thermostat housing inlet **(see illustration)**. Detach the hose from the fitting. If it's stuck, grasp it near the engine end with a pair of adjustable pliers and twist it to break the seal, then pull it off. If the hose is old or deteriorated, cut it off and install a new one.

Note: *If the outer surface of the large fitting that mates with the hose is deteriorated (corroded, pitted, etc.), it may be damaged further by hose removal. If it is, the thermostat housing cover will have to be replaced.*

Note: *If the hose has recently been replaced and the fitting is known to be in good condition, the thermostat can be serviced without removing the hose from the housing cover.*

5   Remove thermostat housing fasteners and cover. If the cover is stuck, tap it with a soft-face hammer to jar it loose. Be prepared for some coolant to spill as the seal is broken. Remove the thermostat assembly **(see illustration)**.

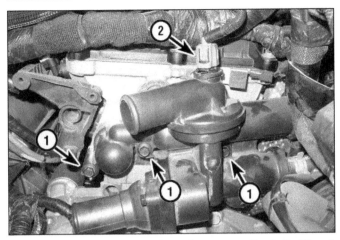

4.8 Secondary thermostat coolant adapter mounting bolts (1) and electrical connector (2)

4.12 Remove the O-ring from the water pipe

Secondary thermostat

6   Remove the air filter and intake hose assembly (see Chapter 4).

7   Remove the coolant hose from the rear of the coolant adapter on the thermostat.

8   Disconnect the electrical connector at the top of the housing **(see illustration)**.

9   Disconnect the radiator hoses from the front and top of the coolant adapter.

10   Loosen and remove the mounting bolts from the coolant adapter **(see illustration 4.8)**.

11   Carefully remove the coolant adapter and secondary thermostat assembly from the water pump inlet tube.

12   Carefully remove the O-ring from the inlet located on the engine block **(see illustration)**. Avoid gouging the metal, or leaks may result.

2.7L V6 engine

13   Raise the vehicle and place it securely on jackstands. Remove the left front wheel. Remove the accessory drivebelt (see Chapter 1).

14   Remove the air conditioning compressor mounting bolts. Remove the air conditioning compressor.

15   Remove the bolts holding the heater line to the thermostat housing and move the heater line aside.

16   Loosen and remove the hose from the thermostat housing. If it's stuck, grasp it near the end with a pair of adjustable pliers and twist to break the seal, then pull it off. If the hose is old or deteriorated, cut it off and install a new one.

Note: *Note the position of the thermostat and where the small valve is located, because it must be installed in the same position. Also note how the thermostat is installed (which end is facing up, or out, and the position of the air bleed jiggle valve, if equipped).*

3.5L V6 engine

17   Loosen and remove the upper radiator hose from the thermostat housing. If it's stuck, grasp it near the end with a pair of adjustable pliers and twist to break the seal, then pull it

4.18 Thermostat housing cover fasteners - 3.5L engine

off. If the hose is old or deteriorated, cut it off and install a new one.

18   Remove the thermostat housing cover fasteners and remove the cover **(see illustration)**.

19   Remove the thermostat and seal.

3.6L V6 engine

Note: *On 3.6L engines, the thermostat is an integral part of the housing and must be replaced as an assembly.*

20   Loosen and remove the coolant recovery bottle mounting bolts and set the bottle aside.

21   Follow the upper radiator hose to the engine to locate the thermostat housing cover **(see illustration)**.

22   Loosen the hose clamp, then detach the hose from the fitting. If it's stuck, grasp it near the end with a pair of adjustable pliers and twist it to break the seal, then pull it off. If the hose is old or deteriorated, cut it off and install a new one.

Note: *If the outer surface of the large fitting that mates with the hose is deteriorated (corroded, pitted, etc.), it may be damaged further by hose removal. If it is, the thermostat housing cover will have to be replaced.*

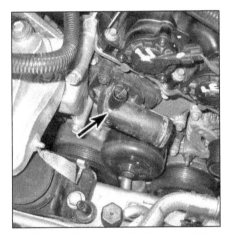

4.21 On 3.6L engines, the thermostat is located at the left end of the engine, near the coolant recovery tank

Note: *If the hose has recently been replaced and the fitting is known to be in good condition, the thermostat can be serviced without removing the hose from the housing cover.*

23   Remove the thermostat housing cover fasteners and cover. If the cover is stuck, tap it with a soft-face hammer to jar it loose. Be prepared for some coolant to spill as the seal is broken.

2.4L four cylinder engine (2015 and later)

Note: *On these models, the thermostat in an integral part of the thermostat housing and must be replaced as an assembly.*

24   Remove the engine top cover, then raise and support the front of the vehicle securely on jackstands. Remove the engine under cover.

25   Remove the battery and battery tray (see Chapter 5).

26   Remove the fasteners securing the fuse box, then position the fuse box aside.

27   Follow the upper radiator hose to the thermostat housing at the engine. Disengage

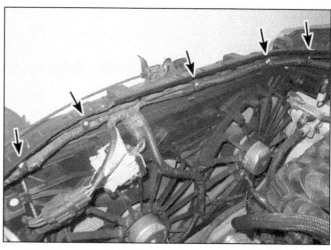

5.6 Fan electrical connector and harness retaining pin

5.7 Detach the wiring harness from the radiator support

the wiring harness(es) from the radiator hose brackets.

28   Loosen the hose clamp securing the upper radiator hose. If the clamp is the non-reusable type, replace it with a screw-type clamp when installing. Disconnect the upper radiator hose from the thermostat housing.

29   Remove the brake vacuum pump (see Chapter 9).

30   Release the transaxle shift cable from the bracket (see Chapter 7) and position it aside.

31   Disconnect the engine coolant temperature sensor electrical connector (see Chapter 6). Disconnect the expansion tank hose quick-connect fitting from the thermostat housing.

32   Remove the intake manifold-to-thermostat housing support bracket bolts, but only loosen the nuts so the bracket can be repositioned as the housing is removed.

33   Disconnect the remaining coolant hoses from the thermostat housing, noting their locations for installation. Replace any hose clamps that required cutting with screw-type hose clamps.

34   Remove the remaining thermostat hous-

ing mounting bolts, noting the lower bracket which requires being pried aside slightly.

35   Guide the thermostat housing out of the engine compartment.

36   Remove the O-ring from the housing. When installing the thermostat housing on these models, the lower metal bracket must be pried away (a little more) to allow the housing to be seated on the engine - make sure this bracket is retained by the mounting bolt when reassembling.

All models

37   Take note of how the thermostat and gasket or O-ring are installed.

38   Remove all traces of the old gasket from the mating surfaces and clean them thoroughly.

39   Install a new gasket, seal or O-ring as applicable. Make sure that it is oriented in the same way as the original.

Note: *It is standard practice to use a thin layer of RTV sealant when installing flat replacement gaskets. However, if the gasket is designed with a raised crushable sealing surface (not flat), or if it is an O-ring, no RTV sealant is necessary.*

40   The remainder of installation is the reverse of removal. Make sure that the replacement thermostat is installed in the same direction and position as the one removed. Tighten the thermostat housing cover bolts to the torque listed in this Chapter's Specifications.

41   Reattach the hose(s) to the housing cover and fitting(s), then tighten the hose clamp(s) securely.

42   Reconnect the battery (see Chapter 5).

43   Refill the cooling system (see Chapter 1).

44   Start the engine and allow it to reach normal operating temperature, then check for leaks and proper thermostat operation (as described in Section 2).

---

5   Engine cooling fans - check and replacement

Warning: *To avoid possible injury or damage, DO NOT operate the engine with a damaged fan. Do not attempt to repair fan blades - replace a damaged fan with a new one.*

Warning: *The electric fans can start at any time; keep hands, clothes and tools away from the fan until the battery is disconnected to avoid possible injury or damage.*

Check

1   If the engine is overheating and the cooling fan is not coming on when the engine temperature rises to an excessive level, see Section 2. Check the fan relays in the underhood fuse/relay box.

2   If the relays are okay, check all wiring and connections to the fan motor. Refer to the wiring diagrams at the end of Chapter 12. If no obvious problems are found, the problem could be the Engine Coolant Temperature (ECT) sensor or the Powertrain Control Module (PCM). Have the cooling fan system and circuit diagnosed by a dealer service department or repair shop with the proper diagnostic equipment.

5.8 Squeeze the clamp and slide it back on the hose, then disconnect the hose from the radiator

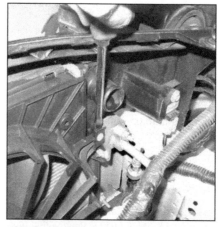

5.10 Remove the fan assembly-to-radiator upper clips and fasteners (some models may have mounting bolts instead)

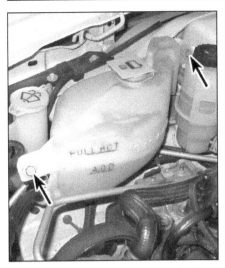

6.2 Coolant reservoir mounting bolts

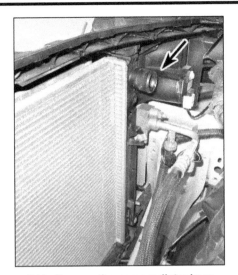

7.11a Remove the upper radiator hose from the radiator

7.11b Remove the lower radiator hose from the radiator

Replacement

Warning: *Wait until the engine is completely cool before beginning this procedure.*

3   Disconnect the negative battery cable from the remote ground terminal or battery (see Chapter 5).

4   Set the parking brake, raise the front of the vehicle and support it securely on jackstands.

5   Drain the cooling system (see Chapter 1).

6   Disconnect the cooling fan electrical connector and harness retaining pins **(see illustration)**.

7   On 2011 and later models, remove the fasteners holding the wire loom to the upper radiator core support **(see illustration)**.

8   Remove the upper radiator hose from the radiator **(see illustration)**.

9   On 2011 and later models, remove the oil dipstick tube to make room for fan removal.

10   Remove the upper clips and fasteners holding the fan assembly to the radiator **(see illustration)**. On later models, remove the fan shroud mounting bolts.

11   Remove the cooling fan assembly by pulling it straight up and out of the engine compartment.

12   Installation is the reverse of removal. Place the fan assembly back into the retaining clips for the side and bottom.

13   Refill the cooling system (see Chapter 1).

6   Coolant reservoir/expansion tank - removal and installation

Warning: *Wait until the engine is completely cool before beginning this procedure.*

Non-pressurized coolant reservoir

1   Place a drain pan under the reservoir, then detach the reservoir hose from the radiator outlet under the radiator cap.

2   Remove the coolant reservoir mounting bolts **(see illustration)**.

Pressurized coolant reservoir (expansion tank)

3   Drain the cooling system (see Chapter 1) to a point below the coolant reservoir container level. If the coolant is relatively new or in good condition, save it and reuse it. Read the **Warning** in Section 1.

4   Remove the power steering reservoir screw (if equipped).

5   Disconnect the return hose on top of the coolant reservoir container.

6   Remove the coolant reservoir mounting bolts.

7   Disconnect the hose(s) from the bottom of the coolant reservoir container.

All coolant reservoir types

8   Lift the reservoir straight up and out.

9   Installation is the reverse of removal. While the reservoir is off the vehicle, it should be cleaned with soapy water and a brush to remove any deposits inside. Inspect it for damage and replace it if necessary. Fill the reservoir with the proper type and amount of coolant (see Chapter 1).

7   Radiator - removal and installation

Warning: *Wait until the engine is completely cool before beginning this procedure.*

Removal

1   Have the air conditioning system discharged and the refrigerant recovered by an air conditioning technician.

2   Disconnect the negative battery cable from the remote ground terminal or battery (see Chapter 5).

3   Set the parking brake, raise the front of the vehicle and support it securely on jackstands.

4   Remove the engine cover.

5   Remove the air filter and intake hose assembly (see Chapter 4).

6   Drain the cooling system (see Chapter 1).

7   Remove the front bumper cover (see Chapter 11).

8   On 2010 and earlier models, disconnect the electrical horn connectors. Loosen the fastener and remove the horn assembly from the upper radiator core support.

9   On 2011 and later models, remove the under-vehicle splash shield.

10   On 2015 and later models, perform the following to remove the large multi-purpose bracket to the front of the radiator/condenser:

a)   *With the front bumper cover removed, remove the bumper reinforcement.*

b)   *On the support brackets located to the rear of the headlights - loosen the lower-rear bolt, then remove the four top bracket bolts (both sides).*

c)   *Remove the headlight housings (see Chapter 12).*

d)   *Disconnect the horn electrical connectors and separate the wiring harness clips from along the front face of the multi-purpose bracket.*

e)   *Disconnect the hood latch cable (see Chapter 11).*

f)   *Remove the plastic baffle and also the clip fasteners from the right and left sides of the bracket.*

g)   *Remove the clip fasteners along the lower end of the bracket.*

h)   *Using a marker or paint, mark the relationship of the multi-purpose bracket to the upper radiator support to allow for correct alignment during installation.*

i)   *Remove the bolts along the top of the bracket, then remove the bracket.*

11   Disconnect the upper and lower radiator hoses from the radiator **(see illustrations)**.

7.12 Disconnect the electrical connectors for the radiator fan

7.14 Disconnect the refrigerant line(s) attached to the air conditioning condenser (on 2015 and later models, the refrigerant line connections are located at the front side of the condenser)

7.15 Disconnect the transmission cooler lines from the condenser and move them aside (2014 and earlier models)

12   Disconnect the electrical connectors for the radiator fan **(see illustration)**.

13   On 2011 and later models, remove the engine cooling fan assembly (see Section 5).

14   Disconnect the A/C line(s) attached to the air conditioning condenser **(see illustration)**. On 2015 and later models, remove the bolt and detach the transmission cooler bracket from the radiator, without disconnecting the cooler lines.

15   On 2014 and earlier models, disconnect the transmission cooler lines from the condenser and move aside **(see illustration)**. On 2015 and later models, remove the bolt and detach the transmission cooler bracket from the radiator, without disconnecting the cooler lines.

16   On 2010 and earlier models, the radiator is held to the support frame place by plastic tabs on top of the radiator. The tabs must be compressed to release the radiator. Compress the plastic tabs and then tilt the radiator and air conditioner condenser assembly away from the radiator support.

17   On 2011 and later models, the radiator is held to the support frame by Torx screws on top of the radiator. Loosen and remove the screws. Remove the radiator by tilting the top of it rearward to release it from the mounting bracket.

18   On 2014 and earlier models, lift the radiator and air conditioning condenser assembly from the vehicle. Carefully separate the condenser from the radiator by removing the fasteners and disconnecting the condenser retaining tabs.

19   On 2015 and later models, remove the condenser-to-radiator mounting bolts. Tilt the condenser slightly forward to allow the radiator to pass through and be lifted out from the front of the engine.

Note: *The cooling fan assembly can be removed with or without the radiator.*

20   Check the radiator for leaks and damage. If it needs repair, have a radiator shop or dealer service department perform the work, as special techniques are required.

21   Bugs and dirt can be removed from the radiator by spraying it from the back side with a garden hose. The radiator should be flushed out with a garden hose before reinstallation.

22   Check the radiator rubber mounts for deterioration and replace them if necessary.

Installation

23   Installation is the reverse of removal. Make sure the A/C condenser is properly attached to the radiator before seating the radiator into the lower rubber mounts.

Note: *Be sure that the flexible air seals on each side of the radiator are in the correct position while installing the radiator.*

24   After installation, fill the cooling system with the proper mixture of antifreeze and water (see Chapter 1).

25   Start the engine and check for leaks. Allow the engine to reach normal operating temperature, indicated by the upper radiator hose becoming hot. Recheck the coolant level and add more if required.

26   Have the air conditioning system recharged by the shop that discharged it.

8   Water pump - replacement

Warning: *Wait until the engine is completely cool before beginning this procedure.*

1   Disconnect the negative battery cable from the remote ground terminal or battery (see Chapter 5).

2   Set the parking brake, raise the front of the vehicle and support it securely on jackstands.

3   Drain the cooling system (see Chapter 1).

8.6 Water pump pulley bolts

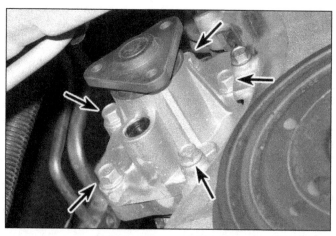

8.8 Water pump mounting bolts

8.12 Timing chain guide bolts (A) and water pump mounting bolts (B) - 2.7L engine

8.14 Water pump mounting bolts - 3.5L engine

2.4L four cylinder engine

4   Remove the accessory drivebelt (see Chapter 1).
5   Remove the accessory drivebelt splash shield.
6   Remove the water pump pulley bolts **(see illustration)**.
7   Remove the water pump pulley.
8   Remove the water pump mounting bolts **(see illustration)** and remove the pump.

2.7L V6 engine

9   Remove the radiator fan assembly (see Section 5).
10   Remove the accessory drivebelt (see Chapter 1).
11   Remove the timing chain and guides.
12   Remove the bolts holding the water pump to the engine block, and remove the water pump **(see illustration)**.

3.5L V6 engine

13   Remove the timing belt (see Chapter 1).
Note: *The water pump is equipped with a weep hole. It is normal for a small amount of coolant to leak from this area. If there is a*

steady leak, or large amount of coolant on the weep hole, then the water pump may have a failure. Ensure that the water pump is at fault before removal.
14   Remove the water pump mounting bolts. Note the various bolt lengths and locations as they're removed to ensure correct installation **(see illustration)**.
15   Remove the water pump.

3.6L V6 engine

16   Remove the coolant reservoir tank and set it aside (see Section 6).
17   Remove the right side engine mount (see Chapter 2D).
18   Remove the engine mount block from the water pump.
19   Remove the lower engine cover.
20   Raise the front of the vehicle and support it securely on jackstands, then remove the right front wheel.
21   Remove the inner fender splash shield (see Chapter 11).
22   Remove the drivebelt (see Chapter 1).
23   Remove the idler pulley **(see illustration)**.

24   Loosen the air conditioning compressor mount bolts. Remove compressor and set it aside.
25   Disconnect the bypass and lower radiator hoses from the water pump **(see illustration)**.
26   Remove the 10 water pump mounting bolts; 4 of the water pump mount bolts are mounted directly into the timing chain cover. Note the various bolt lengths and locations as they're removed to ensure correct installation.
27   Remove the water pump and pulley.

Installation

28   Clean the bolt threads and threaded holes in the timing case cover to remove any corrosion and sealant. Do not gouge or scratch the mating surface on the timing case cover.
29   Compare the new pump to the old one to make sure they're identical. If the old pump is being reused, check the impeller blades on the backside for corrosion. If any fins are missing or badly corroded, replace the pump with a new one.

8.23 Remove the fastener securing the idler pulley

8.25 Disconnect the bypass (A) and lower radiator (B) hoses

8.32 In this example, the O-ring is completely seated in its groove (2.7L water pump shown)

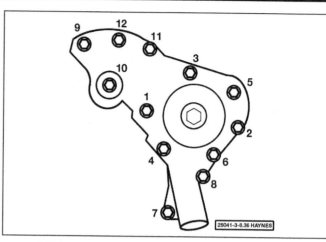

8.35 3.6L V6 water pump bolt TIGHTENING sequence (2014 and earlier)

30   Remove all traces of old gasket material or the O-ring seal from the engine and water pump (if the same one is to be installed).

31   Clean the engine and water pump mating surfaces with lacquer thinner or acetone.

32   On water pumps equipped with an O-ring seal, apply a thin coat of RTV sealant to the new water pump O-ring and position it in the groove on the back of the water pump **(see illustration)**. **Caution:** *Make sure the O-ring is correctly seated in the water pump groove to avoid a coolant leak.*

33   Install a new seal, then carefully mate the pump to the engine.

34   Install the water pump mounting bolts and tighten them to the torque listed in this Chapter's Specifications.

35   On 2014 and earlier 3.6L V6 engines, the bolts must be torqued in a specific sequence **(see illustration)**. 2015 and later V6 models have a slightly different mounting bolt arrangement, and do not require the bolts to be tightened in a specific sequence, but tightened in a criss-cross pattern to the torque value listed in this Chapter's Specifications. On all other models, thread the bolts finger tight first, then tighten them a little at a time to the torque listed in this Chapter's Specifications. Turn the water pump by hand to make sure it rotates freely. **Caution:** *Don't overtighten the mounting bolts; doing so will damage the pump.*

36   The remainder of installation is the reverse of removal. Refill and bleed the cooling system (see Chapter 1). Run the engine and check for leaks and proper operation.

9   Blower motor resistor/power module and blower motor assembly - replacement

Warning: *The models covered by this manual are equipped with Supplemental Restraint Systems (SRS), more commonly known as airbags. Always disable the airbag system before working in the vicinity of any airbag system component to avoid the possibility of accidental deployment of the airbag, which could cause personal injury (see Chapter 12).*

1   Disconnect the negative battery cable from the remote ground terminal or battery (see Chapter 5).

Blower motor assembly

Note: *The blower motor and blower wheel are balanced to each other at the factory and can only be replaced as an assembly.*

2   The blower motor assembly is located on the passenger's side of the HVAC housing. Remove the insulation panel from underneath the glove box **(see illustration)**.

3   Unlock the connector and disconnect the wire harness connector from the blower motor.

4   Remove the 3 mounting screws holding the blower motor and wire bracket **(see illustration)**. Remove the blower motor assembly.

5   Installation is the reverse of removal.

Blower motor power module

Note: *Models equipped with automatic temperature control utilize a power module instead of a blower motor resistor. They are similar in the way they are mounted and connected.*

6   Disconnect the electrical connectors for the blower motor power module, located under the glove box **(see illustration 9.4)**.

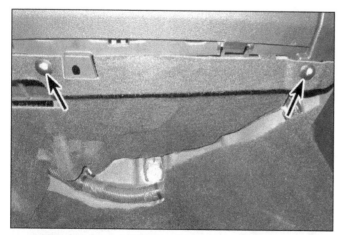

9.2 Remove the insulation panel underneath the glove box

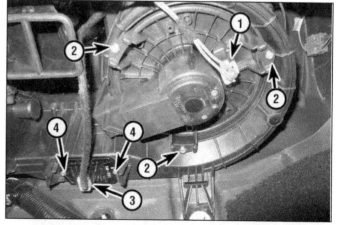

9.4 Blower motor electrical connector (1), blower motor mounting fasteners (2), power module electrical connector (3) and fasteners (4)

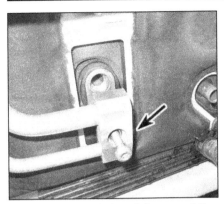

10.6 Remove the bolt and disconnect the lines to the air conditioning evaporator

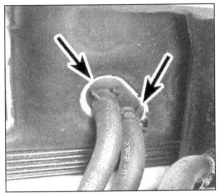

10.7 Remove the clamps and disconnect the heater hoses from the heater core

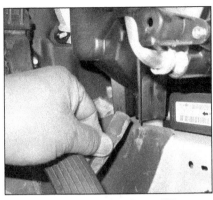

10.17 Disconnect the air conditioner condensation tube

7   Remove the mounting fasteners from the blower to the heating/air conditioner housing and remove the power module from the heater/air conditioning housing.

8   Installation is the reverse of removal.

Blower motor resistor (all except 2011 and later Avenger models)

Warning: *If the vehicle has been recently driven, the blower motor resistors may be hot. In order to avoid personal injury, wait at least five minutes for the blower motor resistors to cool before servicing.*

Caution: *Do not operate the blower motor without the blower motor resistor, it may damage the vehicle.*

9   Disconnect the electrical connectors from the blower motor resistor.

10   Remove the mounting fasteners from the blower motor resistor to the heating/air conditioner housing and remove the blower motor resistor.

11   Installation is the reverse of removal.

10   Heater core - replacement

Warning: *The air conditioning system is under high pressure. DO NOT loosen any fittings or remove any components until after the system has been discharged. Air conditioning refrigerant must be properly discharged into an EPA-approved container at a dealer service department or an automotive air conditioning repair facility. Always wear eye protection when disconnecting air conditioning system fittings.*

Warning: *Wait until the engine is completely cool before beginning this procedure.*

Warning: *The models covered by this manual are equipped with Supplemental Restraint Systems (SRS), more commonly known as airbags. Always disable the airbag system before working in the vicinity of any airbag system component to avoid the possibility of accidental deployment of the airbag, which could cause personal injury (see Chapter 12).*

Note: *Removing interior trim requires the use of special upholstery removal tools used to pry apart fastening clips without damaging them.*

10.18a Disconnect the driver's side rear distribution air duct from the heating/air conditioning housing

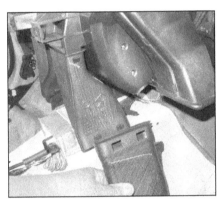

10.18b Disconnect the passenger's side rear distribution air duct from the heating/air conditioning housing

Be sure to use these tools when removing interior components.

1   Have the system discharged and the refrigerant recovered by an air conditioning technician.

2   Disconnect the negative battery cable from the remote ground terminal or battery (see Chapter 5).

3   Drain the cooling system (see Chapter 1).

4   On the driver's side of the firewall, remove the uppermost nut that holds the heat shield.

5   Remove the two nuts holding the heat shield to studs on the dash panel. The heat shield can then be rotated and tilted to aid in removal.

6   Remove the bolt securing the refrigerant lines to the air conditioning evaporator. Disconnect the lines and remove the seals **(see illustration)**.

Note: *To prevent the entry of moisture or debris, place plugs in the lines, or use tape to cover the openings.*

7   Place towels underneath the heater core tube fittings. Remove the clamps and disconnect the heater hoses from the heater core **(see illustration)**.

Note: *To prevent spillage, place plugs in the heater core tubes, or use tape to cover the openings.*

8   Remove the bolts holding the heating/air conditioning housing to the instrument panel.

9   Remove the front sill trim on both the passenger's and driver's sides. Remove the front seats (see Chapter 11).

10   If equipped, loosen fasteners and remove the passenger side amplifier.

11   Remove the B pillar trim panels and cowl trim panel on the driver's side.

12   Remove the center console (see Chapter 11).

13   Pull back the carpet to expose the rear seat ducts.

14   Remove the instrument panel (see Chapter 11).

Note: *On 2015 and later models, the heater core/HVAC housing is removed as an assembly with the instrument panel (see Chapter 11). Once removed, the heater core can be separated from the panel by simply disengaging the wiring harness and removing the four mounting fasteners.*

15   Disconnect the shift interlock cable connected to the floor duct and move aside.

16   Remove the screws holding the left front floor duct to the heating/air conditioning housing and remove the duct.

17   Remove the air conditioner condensation tube **(see illustration)**.

18   Remove the fasteners holding the left and right rear distribution ducts. Disconnect the ducts from the heating and air conditioning housing **(see illustrations)**.

10.19 Remove the nut holding the heating/air conditioning housing to the passenger's side dash panel

10.20 Remove the heating/air conditioner housing through the passenger side of the vehicle

10.21 Place the heating/air conditioner housing on a work bench

10.23 Unclip the fastener and remove the flange from the heating/air conditioning housing

19   Remove the nut holding the heater/air conditioner housing to the passenger's side dash panel **(see illustration)**.
20   Pull the heater/air conditioning unit rear-

ward. Remove the heating/air conditioning housing through the passenger compartment **(see illustration)**.
Note: *During removal of the heating/air conditioning housing, ensure that the interior of the car is properly covered with towels to catch any fluids that may come off the housing components.*
21   Place the heater/air conditioning housing on a work bench **(see illustration)**.
22   Remove the foam seal from the connection tube flange holding the heater core outlets.
23   Un-clip the fasteners securing the flange and remove the flange from the heating/air conditioning housing **(see illustration)**.
24   Remove the fasteners securing the retaining brackets. Remove the heater core line retaining brackets **(see illustrations)**.
25   Carefully remove the heater core from the heater/air conditioning housing **(see illustration)**.
26   Installation is the reverse of removal. Use new seals for the heater core fittings and refill the cooling system (see Chapter 1).
Note: *If a new heater core is being installed, the cooling system must be flushed (see Chapter 1).*

11   Heater/air conditioner control assembly - removal and installation

Warning: *The models covered by this manual are equipped with Supplemental Restraint Systems (SRS), more commonly known as airbags. Always disable the airbag system before working in the vicinity of any airbag system component to avoid the possibility of accidental deployment of the airbag, which could cause personal injury (see Chapter 12).*
Note: *The heater and air conditioner control assembly on 2015 and later models looks a bit different than what is shown in the accompanying photos, but the principle of replacement is identical.*
1   Disconnect the negative battery cable from the remote ground terminal or battery (see Chapter 5).
2   Carefully pry the heater/air conditioning control bezel assembly from the instrument panel **(see illustration)**.
Note: *On 2010 and earlier models, do not at-*

10.24a Remove the fasteners holding the retaining bracket

10.24b Remove the bracket holding the heater core lines

10.25 Carefully remove the heater core from the heater/air conditioning housing

11.2 Pry the control panel from the instrument panel

11.3 Disconnect all electrical connectors from the back of the heating/air conditioning control assembly

tempt to pry the heater/air conditioning bezel from the top, or the retaining clips may become damaged.

3   Disconnect all electrical connectors from the back of the heater/air conditioning control assembly **(see illustration)**.

4   Remove the mounting fasteners at each corner **(see illustration)**, then remove the control assembly from the bezel.

5   Installation is the reverse of removal. If the heater/air conditioning control assembly is being replaced, calibration/diagnostic tests will be necessary. This will require a specialized scan tool; take the vehicle to a dealer service department or other qualified repair shop to have this service performed.

12   Air conditioning compressor - removal and installation

Warning: *The air conditioning system is under high pressure. DO NOT loosen any fittings or remove any components until after the system has been discharged. Air conditioning refriger-*

ant must be properly discharged into an EPA-approved container at a dealer service department or an automotive air conditioning repair facility. Always wear eye protection when disconnecting air conditioning system fittings.

Caution: *When replacing entire components, additional refrigerant oil must be added equal to the amount that is removed with the component being replaced. Read the label on the oil container to verify that it is compatible with the R-134a or R-1234yf system before adding any of it to the system.*

Note: *The receiver-drier should always be replaced when the compressor is replaced (see Section 13).*

Removal

1   Have the system discharged and the refrigerant recovered by an air conditioning technician.

2   Disconnect the negative battery cable from the remote ground terminal or battery (see Chapter 5).

3   Raise the front of the vehicle and support it securely on jackstands.

4   Remove the front wheels and the front wheel well splash shields (see Chapter 11)

5   Remove the accessory drivebelt (see Chapter 1).

6   Disconnect the wire harness connected to the air conditioning compressor **(see illustration)**.

7   Remove the fasteners securing the air conditioning lines to the compressor and detach the refrigerant lines from the compressor **(see illustration 12.6)**.

8   Remove the refrigerant lines from the compressor. Plug all open fittings to prevent entry of dirt and moisture.

Note: *On V6 models, if more access is necessary for compressor removal, you can remove the alternator (see Chapter 5).*

Note: *The rearmost refrigerant line must be removed before the rearmost mounting bolt can be accessed.*

9   Remove the compressor mounting fasteners **(see illustration 12.6)** and lower the compressor from the vehicle.

10   If a new compressor is being installed, pour out the oil from the old compressor into a graduated container and add that amount of new refrigerant oil to the new compressor. Also follow any directions included with the new compressor.

Installation

Caution: *On clutch type compressors, ensure the clutch coil wiring is not damaged during installation. Be careful not to damage the contact surface of the pulley.*

11   Installation is the reverse of removal. Install new O-rings onto the line fittings and lightly coat them with the correct refrigerant oil.

Note: *Only use O-rings that are designed specifically for A/C system applications.*

12   Have the system evacuated, recharged and leak tested by the shop that discharged it.

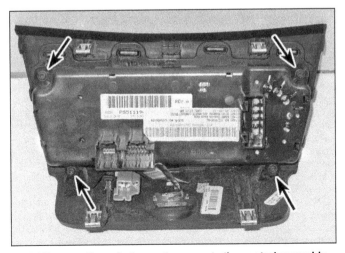

11.4 Remove these fasteners to separate the control assembly from the bezel

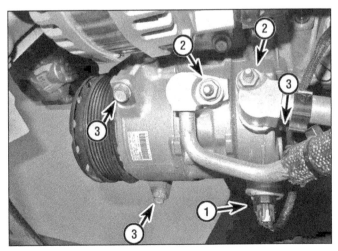

12.6 Air conditioning compressor details

1	Electrical connector	3	Compressor mounting
2	Refrigerant line fitting nuts		bolts

13.5 Receiver-drier details:

1   *Refrigerant line fitting nuts*
2   *Receiver-drier*

3   *Receiver-drier mounting*
    fastener

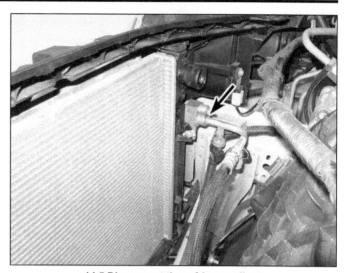

14.5 Disconnect the refrigerant lines

13   Air conditioning receiver-drier - removal and installation

Warning: *The air conditioning system is under high pressure. DO NOT loosen any fittings or remove any components until after the system has been discharged. Air conditioning refrigerant must be properly discharged into an EPA-approved container at a dealer service department or an automotive air conditioning repair facility. Always wear eye protection when disconnecting air conditioning system fittings.*
Caution: *When replacing entire components, additional refrigerant oil must be added equal to the amount that is removed with the component being replaced. Read the label on the oil container to verify that it is compatible with the R-134a or R-1234yf system before adding any of it to the system.*
Note: *On some models, the receiver-drier is contained in the side of the air conditioning condenser, and must be replaced as a unit (see Section 14).*

Removal

1   Have the system discharged and the refrigerant recovered by an air conditioning technician.
2   Disconnect the negative battery cable from the remote ground terminal or battery (see Chapter 5).
3   Loosen the right-front wheel lug nuts. Set the parking brake, raise the front of the vehicle and support it securely on jackstands. Remove the right front wheel.
4   Remove the inner fender splash shield retainers and pull out the splash shield far enough to gain access to the receiver-drier.
Note: *The front bumper cover (see Chapter 11) and windshield washer reservoir (see Section 6) may be removed for more access to the receiver-drier.*

5   Remove both line fittings from the receiver-drier **(see illustration).**
Note: *Plug all openings immediately to prevent contamination.*
6   Remove the mounting fastener retaining the receiver-drier to the mounting bracket.
7   Loosen the mounting strap and remove the receiver-drier.

Installation

8   Installation is the reverse of removal. Install new O-rings onto the line fittings and lightly coat them with the correct refrigerant oil.
Note: *Only use O-rings that are designed specifically for A/C system applications.*
9   If you are replacing the receiver-drier with a new unit, add 0.7-ounce (20 ml) of refrigerant oil to the replacement.
10   Have the system evacuated, recharged and leak tested by the shop that discharged it.

14   Air conditioning condenser - removal and installation

Warning: *The air conditioning system is under high pressure. DO NOT loosen any fittings or remove any components until after the system has been discharged. Air conditioning refrigerant must be properly discharged into an EPA-approved container at a dealer service department or an automotive air conditioning repair facility. Always wear eye protection when disconnecting air conditioning system fittings.*
Caution: *When replacing entire components, additional refrigerant oil must be added equal to the amount that is removed with the component being replaced. Read the label on the oil container to verify that it is compatible with the R-134a or R-1234yf system before adding any of it to the system.*
Note: *If the condenser is being replaced because of damage (cracked or punctured), the*

receiver-drier should also be replaced (see Section 13).
Note: *For 2015 and later models, the condenser replacement procedure is very similar to the radiator replacement procedure (see Section 7) - with the only difference being that the transaxle oil cooler needs to be unbolted and separated from the condenser (without disconnecting the oil cooler lines).*

Removal

1   Have the system discharged and the refrigerant recovered by an air conditioning technician.
2   Disconnect the negative battery cable from the remote ground terminal or battery (see Chapter 5).
3   Remove the radiator support brace (see Section 7) and tilt the radiator rearward.
4   Remove the nut retaining the air conditioning lines at the block to the right of the air conditioning condenser.
5   Remove the retaining bolts and disconnect the refrigerant lines **(see illustration).** Plug all open fittings to prevent entry of dirt and moisture.
6   On 2008 and later models, remove the air conditioning condenser connecting line-to-radiator mounting bolts. Remove the bolts holding the connecting line to the air conditioning condenser, and remove the connecting line.
Note: *On 2008 and later models, the connecting line has an O-ring that must be replaced upon installation.*
7   Remove the mounting bolts securing the lower condenser bracket to the radiator.
8   Disconnect the automatic transaxle cooler lines from the left side of the condenser.
9   Carefully pull straight up to release the condenser from the lower clips, then remove the condenser from the vehicle.

Installation

10   Installation is the reverse of removal. Make certain to fully seat the condenser into the mounting clips and retainers. Install new O-rings onto the line fittings and lightly coat them with refrigerant oil.

Note: *Only use O-rings that are designed specifically for A/C system applications.*

11   If you are replacing the condenser with a new unit, add 0.7-ounce (20 ml) of refrigerant oil to the replacement.

12   Have the system evacuated, recharged and leak tested by the shop that discharged it.

15   Orifice tube (2014 and earlier models) - removal and installation

Warning: *The air conditioning system is under high pressure. DO NOT loosen any fittings or remove any components until after the system has been discharged. Air conditioning refrigerant must be properly discharged into an EPA-approved container at a dealer service department or an automotive air conditioning repair facility. Always wear eye protection when disconnecting air conditioning system fittings.*

Note: *The orifice tube is located inside the liquid line of the air conditioning system. The liquid line runs between the condenser and the evaporator. If the orifice tube is part of the air conditioning liquid line. If the orifice needs to be replaced, the entire liquid line needs to be replaced.*

Removal

1   Have the air conditioning system refrigerant discharged and recovered by an air conditioning technician.

2   Disconnect the negative battery cable from the remote ground terminal (see Chapter 5).

3   Set the parking brake, raise the front of the vehicle and support it securely on jackstands.

4   Remove the air conditioning condenser (see Section 14).

5   Reach through the bottom right of the front bumper towards the air conditioning condenser. Remove the liquid line-to-condenser fastener.

6   Disconnect the liquid line from the condenser.

Note: *Plug all openings immediately to prevent contamination.*

7   Lower the vehicle and remove the heat shield-to-dash panel fastener.

8   Remove the two fasteners behind the engine which retain the heat shield to the dash panel. Rotate and tilt the heat shield to free it and remove from vehicle.

9   Remove the power steering fluid reservoir and coolant reservoir mounting fasteners and move the reservoirs aside.

Note: *Neither the power steering or the coolant are required to be drained for this step.*

10   Remove the refrigerant line bracket-to-right front strut tower fastener.

11   Remove the fastener securing the liquid line and suction lines to the air conditioning evaporator.

12   Disconnect the liquid and suction lines from the evaporator. Remove the liquid line assembly from the vehicle.

Note: *Plug all openings immediately to prevent contamination.*

Installation

13   Installation is the reverse of removal. Install new O-rings and seals onto the line fittings and lightly coat them with the correct refrigerant oil.

Note: *Only use O-rings that are designed specifically for A/C system applications.*

14   Have the system evacuated, recharged and leak tested by the shop that discharged it.

16   Expansion valve (2015 and later models) - replacement

Warning: *The air conditioning system is under high pressure. DO NOT loosen any fittings or remove any components until after the system has been discharged. Air conditioning refrigerant must be properly discharged into an EPA-approved container at a dealer service department or an automotive air conditioning repair facility. Always wear eye protection when disconnecting air conditioning system fittings.*

Note: *The expansion valve is located in the firewall.*

1   Have the air conditioning system discharged by an air conditioning technician.

2   Disconnect the cable from the negative battery terminal (see Chapter 5).

3   Remove the nut, then separate the refrigerant lines from the expansion valve at the firewall **(see illustration 10.6)**. Tape off the ports to prevent contamination of the system while it is disassembled.

4   Remove the two expansion valve bolts and pull it from the evaporator in the firewall. If necessary, transfer the foam piece to the new expansion valve.

5   Installation is the reverse of removal. DO NOT reuse the old seals - replace and lubricate them with the correct refrigerant oil before installing.

6   Have the system evacuated, leak tested and recharged by the shop that discharged it.

Notes

Chapter 4
Fuel and exhaust systems

Contents

	Section			Section
Air filter housing - removal and installation	10		Fuel pump/fuel pressure regulator/fuel level sending unit assembly - removal and installation	8
Exhaust system servicing - general information	6		Fuel rail and injectors - removal and installation	12
Fuel level sending unit - replacement	9		Fuel tank - removal and installation	7
Fuel lines and fittings - general information and disconnection	5		General information	1
Fuel pressure - check	4		Throttle body - removal and installation	11
Fuel pressure relief procedure	3		Troubleshooting	2

Specifications

Fuel system

Fuel system pressure (all models)... 58 psi +/- 5 psi

Torque specifications

Ft-lbs (unless otherwise indicated)

Note: One foot-pound (ft-lb) of torque is equivalent to 12 inch-pounds (in-lbs) of torque. Torque values below approximately 15 foot-pounds are expressed in inch-pounds, because most foot-pound torque wrenches are not accurate at these smaller values.

Fuel rail bolts
    2.4L engine (2014 and earlier) 20
    2.4L engine (2015 and later).. 62 in-lbs
    2.7L V6 engine .... 100 in-lbs
    3.5L V6 engine 20.5
    3.6L V6 engine .... 62 in-lbs
Throttle body mounting bolts
    2.4L engine
        2007 models .... 79 in-lbs
        2008 to 2014 models .... 65 in-lbs
        2015 and later models .... 71 in-lbs
    2.7L V6 engine
        2007 models .... 120 in-lbs
        2008 and later models .... 50 in-lbs
    3.5L V6 engine
        2007 models .... 120 in-lbs
        2008 and later models .... 50 in-lbs
    3.6L V6 engine .... 80 in-lbs

2.2 The underhood fuse block is located at the left side of the engine compartment

2.10 An automotive stethoscope is used to listen to the fuel injectors in operation (2.4L engine shown)

1   General information

Fuel system warnings

Gasoline is extremely flammable and repairing fuel system components can be dangerous. Consider your automotive repair knowledge and experience before attempting repairs which may be better suited for a professional mechanic.

a) *Don't smoke or allow open flames or bare light bulbs near the work area*
b) *Don't work in a garage with a gas-type appliance (water heater, clothes dryer)*
c) *Use fuel-resistant gloves. If any fuel spills on your skin, wash it off immediately with soap and water*
d) *Clean up spills immediately*
e) *Do not store fuel-soaked rags where they could ignite*
f) *Prior to disconnecting any fuel line, you must relieve the fuel pressure (see Section 3)*
g) *Wear safety glasses*
h) *Have a proper fire extinguisher on hand*

Fuel system

1   The fuel system consists of the fuel tank, electric fuel pump/fuel level sending unit (located in the fuel tank), fuel rail and fuel injectors. The fuel injection system is a multi-port system; multi-port fuel injection uses timed impulses to inject the fuel directly into the intake port of each cylinder. The Powertrain Control Module (PCM) controls the injectors. The PCM monitors various engine parameters and delivers the exact amount of fuel required into the intake ports.

2   Fuel is circulated from the fuel pump to the fuel rail through fuel lines running along the underside of the vehicle. Various sections of the fuel line are either rigid metal or nylon, or flexible fuel hose. The various sections of the fuel hose are connected either by quick-connect fittings or threaded metal fittings.

Exhaust system

3   The exhaust system consists of the exhaust manifold(s), catalytic converter(s), muffler(s), tailpipe and all connecting pipes, flanges and clamps. The catalytic converters are an emission control device added to the exhaust system to reduce pollutants.

2   Troubleshooting

Fuel pump

1   The fuel pump is located inside the fuel tank. Sit inside the vehicle with the windows closed, turn the ignition key to ON (not START) and listen for the sound of the fuel pump as it's briefly activated. You will only hear the sound for a second or two, but that sound tells you that the pump is working. Alternatively, have an assistant listen at the fuel filler cap.

2   If the pump does not come on, check all of the fuses in the underhood fuse block **(see illustration)**.
Note: *There is no replaceable fuel pump relay; it is actually just a circuit incorporated into the Totally Integrated Power Module, which is part of the underhood fuse/relay block.*

3   If the fuses are okay, check the wiring back to the fuel pump. If the wiring is okay, the fuel pump module or Totally Integrated Power Module is probably defective. If the pump runs continuously with the ignition key in the ON position, the Totally Integrated Power Module or Powertrain Control Module (PCM) is probably defective. Have the circuit checked by a professional mechanic.

Fuel injection system

Note: *The following procedure is based on the assumption that the fuel pump is working and the fuel pressure is adequate (see Section 4).*
4   Check all electrical connectors that are related to the system. Check the ground wire connections for tightness.

5   Verify that the battery is fully charged (see Chapter 5).
6   Inspect the air filter element (see Chapter 1).
7   Check all fuses related to the fuel system (see Chapter 12).
8   Check the air induction system between the throttle body and the intake manifold for air leaks. Also inspect the condition of all vacuum hoses connected to the intake manifold and to the throttle body.
9   Remove the air intake duct from the throttle body and look for dirt, carbon, varnish, or other residue in the throttle body, particularly around the throttle plate. If it's dirty, clean it with carb cleaner, a toothbrush and a clean shop towel.
10   With the engine running, place an automotive stethoscope against each injector, one at a time, and listen for a clicking sound that indicates operation **(see illustration)**.
Warning: *Stay clear of the drivebelt and any rotating or hot components.*
Note: *This check will not be possible on some models, as the upper intake manifold (plenum) is in the way.*
11   If you can hear the injectors operating, but the engine is misfiring, the electrical circuits are functioning correctly, but the injectors might be dirty or clogged. Try a commercial injector cleaning product (available at auto parts stores). If cleaning the injectors doesn't help, replace the injectors.
12   If an injector is not operating (it makes no sound), disconnect the injector electrical connector and measure the resistance across the injector terminals with an ohmmeter. Compare this measurement to the other injectors. If the resistance of the non-operational injector is quite different from the other injectors, replace it.
13   If the injector is not operating, but the resistance reading is within the range of resistance of the other injectors, the PCM or the circuit between the PCM and the injector might be faulty.

3.3 Pull off this cover for access to the fuel pump module

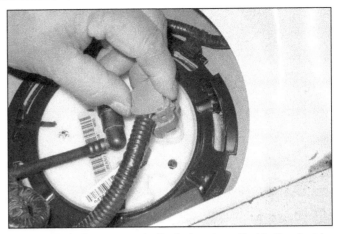

3.4 To relieve the pressure in the fuel system, disable the electric fuel pump by unplugging the electrical connector

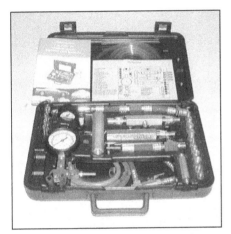

4.2 This typical fuel pressure testing kit contains all the necessary fittings and adapters, along with the fuel pressure gauge, to test most automotive fuel systems

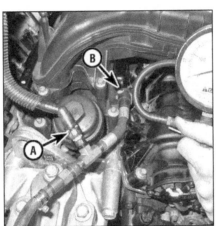

4.4a The fuel pressure gauge is installed between the fuel feed line (A) and fuel rail (B) (3.6L engine)

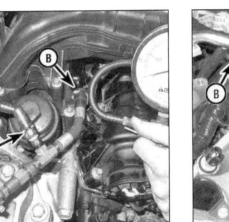

4.4b Fuel gauge connection details (3.5L V6 engine shown, other engines similar; intake manifold removed for clarity)

A   Fuel feed line fitting
B   Fuel rail
C   Tee fitting
D   Quick-connect fitting attached to fuel rail

4.4c Fuel gauge connection details (2.4L engine)

1   Fuel feed line
2   Tee fitting
3   Quick-connect fitting attached to fuel rail

3   Fuel pressure relief procedure

Warning: *Gasoline is extremely flammable. See **Fuel system warnings** in Section 1.*
1   Remove the fuel filler cap to relieve any pressure built-up in the fuel tank.
2   Remove the lower rear seat cushion.
3   Remove the fuel pump module cover **(see illustration)**.
Note: *On 2015 and later models, the pump access cover is retained by bolts.*
4   Unplug the electrical connector from the fuel pump module **(see illustration)**.
5   Start the engine; it should run momentarily then stall. Crank the engine several more times to ensure the fuel system has been completely relieved. Disconnect the negative battery cable from the remote ground terminal or battery before working on the fuel system.
6   It's a good idea to cover any fuel connection to be disassembled with rags to absorb the residual fuel that may leak out. Properly dispose of the rags.

7   After servicing the fuel system, diagnostic trouble codes may have been stored in the PCM's memory due to disconnecting the fuel pump circuit (see Chapter 6).

4   Fuel pressure - check

Warning: *Gasoline is extremely flammable. See **Fuel system warnings** in Section 1.*
Note: *The following procedure assumes that the fuel pump is receiving voltage and runs.*
1   Relieve the fuel system pressure (see Section 3).
2   In addition to a fuel pressure gauge capable of reading fuel pressure up to 70 psi, you'll need a hose and an adapter suitable for tee-ing into the fuel system at the quick-connect fitting between the fuel delivery hose and the fuel rail **(see illustration)**.
3   Disconnect the quick-connect fitting at the connection between the fuel delivery hose and the fuel rail (if you're unfamiliar with quick-connect fittings, refer to Section 5).
4   Tee-in the fuel pressure gauge between the fuel delivery hose and the fuel rail **(see illustrations)**.

Disconnecting Fuel Line Fittings

Two-tab type fitting; depress both tabs with your fingers, then pull the fuel line and the fitting apart

On this type of fitting, depress the two buttons on opposite sides of the fitting, then pull it off the fuel line

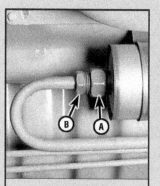

Threaded fuel line fitting; hold the stationary portion of the line or component (A) while loosening the tube nut (B) with a flare-nut wrench

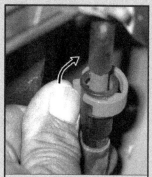

Plastic collar-type fitting; rotate the outer part of the fitting

Metal collar quick-connect fitting; pull the end of the retainer off the fuel line and disengage the other end from the female side of the fitting . . .

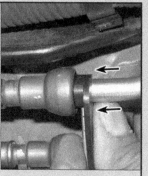

. . . insert a fuel line separator tool into the female side of the fitting, push it into the fitting and pull the fuel line off the pipe

Some fittings are secured by lock tabs. Release the lock tab (A) and rotate it to the fully-opened position, squeeze the two smaller lock tabs (B) . . .

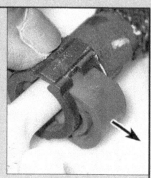

. . . then push the retainer out and pull the fuel line off the pipe

Spring-lock coupling; remove the safety cover, install a coupling release tool and close the tool around the coupling . . .

. . . push the tool into the fitting, then pull the two lines apart

Hairpin clip type fitting: push the legs of the retainer clip together, then push the clip down all the way until it stops and pull the fuel line off the pipe

5   Start the engine and allow it to idle. Note the gauge reading as soon as the pressure stabilizes, and compare it with the pressure listed in this Chapter's Specifications.

6   If the fuel pressure is not within specifications, check the following:

- *If the pressure is lower than specified, check for a restriction in the fuel system (kinked fuel line, plugged fuel pump inlet strainer or clogged fuel filter). If no restrictions are found, replace the fuel pump module (see Section 8).*
- *If the fuel pressure is higher than specified, replace the fuel pump module (see Section 8).*

7   Turn off the engine. Fuel pressure should not fall more than 8 psi over five minutes. If it does, the problem could be a leaky fuel injector, fuel line leak, or faulty fuel pump module.

8   Relieve the fuel system pressure, then disconnect the fuel pressure gauge. Reconnect the fuel line and wipe up any spilled gasoline.

6.1 Inspect the exhaust system rubber hangers for damage. This is a typical exhaust hanger for the intermediate exhaust pipe and muffler

5   Fuel lines and fittings - general information and disconnection

Warning: *Gasoline is extremely flammable. See **Fuel system warnings** in Section 1.*

1   Relieve the fuel pressure before servicing fuel lines or fittings (see Section 3), then disconnect the cable from the remote grounnd terminal or negative battery terminal (see Chapter 5) before proceeding.

2   The fuel supply line connects the fuel pump in the fuel tank to the fuel rail on the engine. The Evaporative Emission (EVAP) system lines connect the fuel tank to the EVAP canister and connect the canister to the intake manifold.

3   Whenever you're working under the vehicle, be sure to inspect all fuel and evaporative emission lines for leaks, kinks, dents and other damage. Always replace a damaged fuel or EVAP line immediately.

4   If you find signs of dirt in the lines during disassembly, disconnect all lines and blow them out with compressed air. Inspect the fuel strainer on the fuel pump pick-up unit for damage and deterioration.

Steel tubing

5   It is critical that the fuel lines be replaced with lines of equivalent type and specification.

6   Some steel fuel lines have threaded fittings. When loosening these fittings, hold the stationary fitting with a wrench while turning the tube nut.

Plastic tubing

Warning: *When removing or installing plastic fuel line tubing, be careful not to bend or twist it too much, which can damage it. Also, plastic fuel tubing is NOT heat resistant, so keep it away from excessive heat.*

7   When replacing fuel system plastic tubing, use only original equipment replacement plastic tubing.

Flexible hoses

8   When replacing fuel system flexible hoses, use only original equipment replacements.

9   Don't route fuel hoses (or metal lines) within four inches of the exhaust system or within ten inches of the catalytic converter. Make sure that no rubber hoses are installed directly against the vehicle, particularly in places where there is any vibration. If allowed to touch some vibrating part of the vehicle, a hose can easily become chafed and it might start leaking. A good rule of thumb is to maintain a minimum of 1/4-inch clearance around a hose (or metal line) to prevent contact with the vehicle underbody.

6   Exhaust system servicing - general information

Warning: *Allow exhaust system components to cool before inspection or repair. Also, when working under the vehicle, make sure it is securely supported on jackstands.*

1   The exhaust system consists of the exhaust manifolds, catalytic converter, muffler, tailpipe and all connecting pipes, flanges and clamps. The exhaust system is isolated from the vehicle body and from chassis components by a series of rubber hangers. Periodically inspect these hangers for cracks or other signs of deterioration, replacing them as necessary **(see illustration)**.

2   Conduct regular inspections of the exhaust system to keep it safe and quiet. Look for any damaged or bent parts, open seams, holes, loose connections, excessive corrosion or other defects which could allow exhaust fumes to enter the vehicle. Do not repair deteriorated exhaust system components; replace them with new parts.

3   If the exhaust system components are extremely corroded, or rusted together, a cutting torch is the most convenient tool for removal. Consult a properly-equipped repair shop. If a cutting torch is not available, you can

use a hacksaw, or if you have compressed air, there are special pneumatic cutting chisels that can also be used. Wear safety goggles to protect your eyes from metal chips and wear work gloves to protect your hands.

4   Here are some simple guidelines to follow when repairing the exhaust system:

- a) *Work from the back to the front when removing exhaust system components.*
- b) *Apply penetrating oil to the exhaust system component fasteners to make them easier to remove.*
- c) *Use new gaskets, hangers and clamps.*
- d) *Apply anti-seize compound to the threads of all exhaust system fasteners during reassembly.*
- e) *Be sure to allow sufficient clearance between newly installed parts and all points on the underbody to avoid overheating the floor pan and possibly damaging the interior carpet and insulation. Pay particularly close attention to the catalytic converter and heat shield.*

7   Fuel tank - removal and installation

Warning: *Gasoline is extremely flammable. See **Fuel system warnings** in Section 1.*
Note: *The following procedure is much easier to perform if the fuel tank is empty. If the fuel tank isn't empty or nearly empty, you can siphon fuel from the tank with a siphon kit, available at most auto parts stores. NEVER start the siphoning action with your mouth!*

1   Relieve the fuel system pressure (see Section 3).

2   Disconnect the negative battery cable from the remote ground terminal or battery (see Chapter 5).

3   Raise the rear of the vehicle and support it securely on jackstands.

4   If you're working on an AWD model, remove the driveshaft and the rear differential electronically controlled clutch (see Chapter 8).

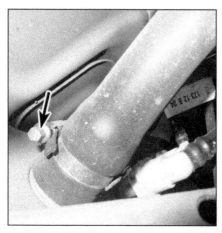

7.5 Loosen the clamp and detach the filler hose from the fuel tank

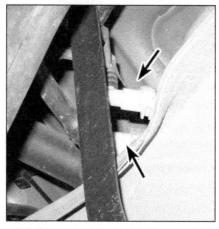

7.7a Disconnect the fuel delivery quick disconnect line and EVAP line

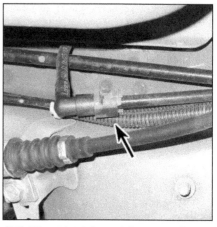

7.7b Disconnect the quick connect fitting from the fuel supply line

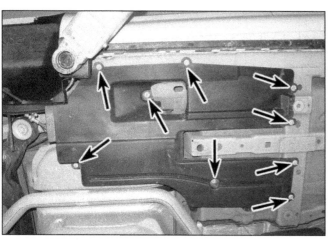

7.8a Right-side close-out panel fasteners

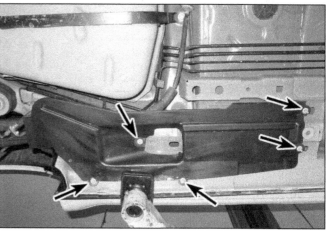

7.8b Left-side close-out panel fasteners

5   Loosen the hose clamp and disconnect the fuel filler hose from the fuel tank **(see illustration)**.

6   Unplug the electrical connector in the fuel pump module harness **(see illustration 3.5)**.

7   Disconnect the fuel line and EVAP line quick-connect fittings, both of which are located at the front of the fuel tank **(see illustrations)**. If you're unfamiliar with fuel line

quick-connect fittings, refer to Section 5.

8   Remove the close-out panels at each side of the fuel tank **(see illustrations)**.

9   On convertible models, remove the two front bolts from the cross braces and move the braces out from under that fuel tank.

10   Remove the exhaust pipe from below the fuel tank (see Section 6).

11   Support the fuel tank with a transmission

jack or with a floor jack. If you're using a floor jack, put a piece of plywood between the jack head and the tank to protect the tank.

12   Remove the fuel tank support strap bolts and carefully lower the fuel tank from the vehicle, making sure no hoses or wiring harness are still attached **(see illustration)**.

13   Installation is the reverse of removal. Tighten the fuel tank strap bolts securely.

14   Start the engine and check for leaks at any fuel line connectors that were disconnected.

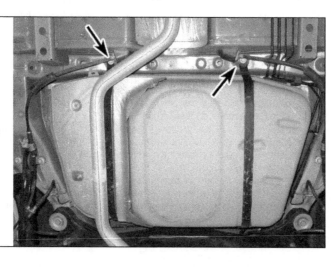
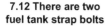

7.12 There are two fuel tank strap bolts

8   Fuel pump/fuel pressure regulator/fuel level sending unit assembly - removal and installation

Warning: *Gasoline is extremely flammable. See* **Fuel system warnings** *in Section 1.*
Note: *All wheel drive (AWD) models have a saddle fuel tank configuration and have main and auxiliary fuel pumps. Each fuel pump has a sending unit. The auxiliary fuel pump contains the fuel pump reservoir, fuel pressure regulator and fuel filter.*

8.3 Remove the fuel pump module access cover (2014 and earlier models shown - 2015 and later models have bolts retaining the cover)

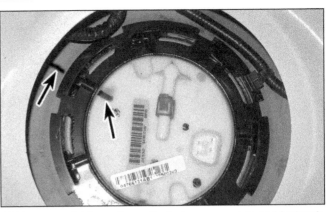

8.6 Arrows point to marks on the fuel pump module and fuel tank that will help you install the assembly in the correct orientation

1   Relieve the fuel system pressure (see Section 3).

2   Disconnect the negative battery cable from the remote ground terminal or battery (see Chapter 5).

3   Remove the rear seat cushion and fuel pump module cover (on AWD models remove main and auxiliary fuel pump module covers) **(see illustration)**.

4   Disconnect the fuel delivery line quick-connect fitting (and on 2015 and later models, also disconnect the fuel vapor line) from the fuel pump/fuel pressure regulator/fuel level sending unit assembly (see Section 5).

5   Disconnect the electrical connector from the fuel pump/fuel pressure regulator/fuel level sending unit assembly.

6   Before removing the fuel pump/fuel pressure regulator/fuel level sending unit assembly, make alignment marks on the assembly and the fuel tank (if marks don't already exist), to ensure the assembly will be correctly realigned when it's installed again **(see illustration)**.

7   Using adjustable-joint pliers, loosen the lock ring that secures the fuel pump/fuel pressure regulator/fuel level sending unit assembly **(see illustration)**.

8   Carefully lift the fuel pump/fuel pressure

regulator/fuel level sending unit assembly from the fuel tank **(see illustration)**. Angle the module so that you don't bend the float arm of the fuel level sending unit or damage the fuel pump inlet strainer.

Note: *If you're removing the auxiliary pump on an AWD model, raise the fuel pump module and disconnect the internal fuel line.*

9   While the pump is removed, inspect the pump inlet strainer. Make sure that it's not clogged or damaged. If the inlet strainer is dirty, try washing it in clean solvent. If it's still clogged, replace it.

10   Installation is the reverse of removal. Line up the marks, and tighten the fuel pump module lock ring securely.

9   Fuel level sending unit - replacement

Warning: *Gasoline is extremely flammable. See **Fuel system warnings** in Section 1.*

Note: *This procedure only applies to 2014 and earlier models. On 2015 and later models, the fuel level sending unit is not serviceable and must be replaced as an assembly with the fuel pump module.*

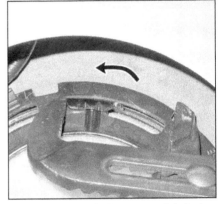

8.7 Loosen the fuel pump module lock ring using adjustable-joint pliers (ring must be turned counterclockwise)

1   Remove the fuel pump/fuel pressure regulator/fuel level sending unit assembly (see Section 8).

2   Disconnect the fuel level sending unit electrical connector **(see illustration)**.

Caution: *Do not touch the terminals or the sending unit resistor card. Do not pull the sending unit by the wires or float arm.*

8.8 The fuel pump module is a tight fit and must be removed carefully to avoid damaging the sending unit float arm.

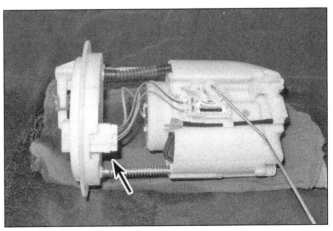

9.2 Depress the tab and disconnect the electrical connector from the underside of the fuel pump module

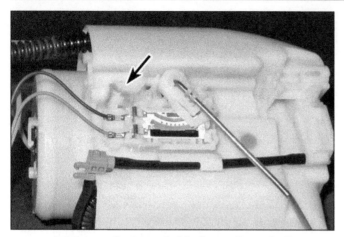

9.3 Release the tab and slide the sending unit off the fuel pump module

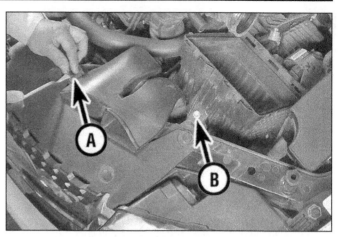

10.24 Fresh air intake duct push fastener (A) and housing mounting bolt (B)

11.4a Throttle body mounting bolts - 2.4L engine

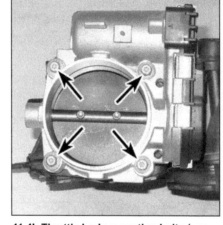

11.4b Throttle body mounting bolts (upper intake manifold removed for clarity) - 3.6L engine

3   Press the fuel level sending unit snap tab to the left while gently pushing up on the bottom of the sending unit to remove it **(see illustration)**.
4   Using a narrow electrical connector terminal tool, dislodge the sending unit terminals from the connector body, then install the terminals of the new sending unit into the correct positions in the connector.
5   The remainder of installation is the reverse of removal.

---

10   Air filter housing - removal and installation

Air intake duct

2.4L engines (2014 and earlier)

1   Loosen the clamp and detach the air intake hose from the air filter housing.
2   On PZEV models, disconnect the air injection hose from the intake duct.
3   Loosen the clamp retaining the duct to the throttle body, then remove the duct.
4   Installation is the reverse of removal.

2.4L engines (2015 and later)

5   Remove the engine cover by pulling it straight up. At the rear of the engine (next to the brake fluid reservoir), disconnect the IAT sensor from the intake tube, then loosen the hose clamp and disconnect the intake tube from the throttle body.
6   Remove the bolt that secures the intake tube to the valve cover, then loosen the hose clamp and disconnect the intake tube from the air filter housing. Remove the intake tube from across the top of the engine.
7   Installation is the reverse of removal.

2.7L and 3.5L engines

8   Disconnect the inlet air temperature sensor and fresh air hose from the air filter housing.
9   Loosen the hose clamp at the throttle body and remove the air inlet hose.
10   Disengage the wire harness connector from the starter cable support bracket.
11   Remove the air filter housing hold down fastener and pull the housing straight up off the locating pins.
12   Installation is the reverse of removal.

3.6L engines (2014 and earlier)

13   Remove the engine cover and disconnect the fresh air hose.
14   Remove the push pin and two bolts from the air filter body.
15   Loosen the hose clamp at the resonator and the remove air filter body.
16   Installation is the reverse of removal.

3.6L engines (2015 and later)

17   Remove the engine cover and disconnect the Intake Air Temperature (IAT) sensor electrical connector, located near the throttle body.
18   Loosen the hose clamps, then remove the intake flex tube from between the air filter housing and intake resonator.
19   Detach the fuel vapor line retaining clip from the resonator, remove the resonator-to-intake manifold mounting pushpin, then loosen the resonator hose clamp at the throttle body.
20   Remove the purge control solenoid (see Chapter 6), then pull the intake resonator straight up to release it from the grommet and off of the throttle body.

Air filter housing

2.4L (2014 and earlier), 2.7L and 3.5L engines

Caution: *Do not use a screwdriver or excessive force when unlocking the air filter housing cover tabs.*

21   Unlock the air filter housing cover tabs and lift off the cover.
22   Pull the cover forward to disengage the rear tabs.
23   Remove the air filter element.
24   Remove the plastic fastener from the fresh air intake duct (as applicable) and the filter housing mounting bolt, then pull the housing off the mounting stud(s) **(see illustration)**.
25   Inspect the rubber mounting grommets on the underside of the air filter housing. If the grommets are cracked, dried out, torn or otherwise damaged, replace them.
26   Installation is the reverse of removal.

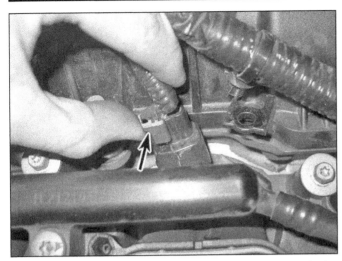

12.5 Slide the connector lock up and disconnect the electrical connector from the fuel injector

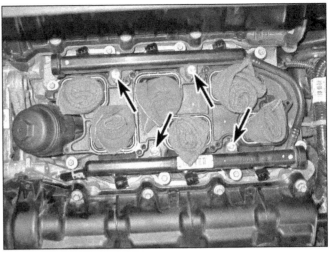

12.6a Fuel rail mounting bolts - 3.6L engine shown

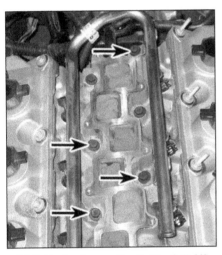

12.6b Fuel rail mounting bolts - 3.5L V6 engine shown, 2.7L V6 similar

12.6c Fuel rail mounting bolts - 2.4L four-cylinder engine

3.6L and 2015 and later 2.4L engines

27 Disconnect the PCV hose from the air filter housing, if equipped.

28 Loosen the intake duct hose clamp, then disconnect the duct from the housing.

29 On 2014 and earlier 3.6L engines, remove the plastic fastener from the fresh air inlet duct, then remove the two housing mounting bolts.

30 On 2015 and later models, pull the housing straight up to release it from the mounting grommets.

31 Installation is the reverse of removal.

---

11 Throttle body - removal and installation

Warning: *Wait until the engine is completely cool before beginning this procedure.*

1 Disconnect the negative battery cable from the remote ground terminal or battery (see Chapter 5). Remove the engine cover.

2 Detach the air intake duct or resonator, as applicable, from the throttle body.

3 Disconnect the throttle body electrical connector. If equipped, remove the throttle body support bracket.

4 Remove the throttle body mounting fasteners and detach the throttle body from the upper intake manifold **(see illustrations)**.

5 Remove the throttle body gasket and inspect it for cracks, tears and deterioration. If it isn't in perfect condition, replace it.

6 Make sure that the gasket mating surfaces of the throttle body and the intake manifold are clean.

7 Installation is the reverse of removal. Tighten the throttle body bolts to the torque listed in this Chapter's Specifications.

---

12 Fuel rail and injectors - removal and installation

Warning: *Gasoline is extremely flammable. See Fuel system warnings in Section 1.*

Warning: *Wait until the engine is completely cool before beginning this procedure.*

1 Relieve the fuel system pressure (see Section 3).

2 Disconnect the negative battery cable from the remote ground terminal or battery (see Chapter 5).

3 If you're working on a V6 engine, remove the upper intake manifold (see Chapter 2B, 2C or 2D). If you're working on a 2015 and later 2.4L engine, remove the air intake tube/resonator (see Section 10).

4 Disconnect the fuel delivery line quick-connect fitting and disconnect the fuel delivery line from the fuel rail (if you're unfamiliar with quick-connect fittings, see Section 5).

5 Disconnect the fuel injector electrical connectors **(see illustration)**. Detach the injector wiring harness mounting clips from the fuel rail (if equipped) and set the harness aside.

6 Remove the fuel rail mounting bolts **(see illustrations)**.

12.7a Using a screwdriver or pliers, remove the injector retaining clip . . .

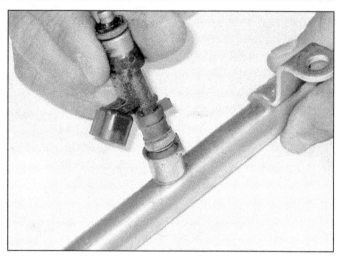

12.7b . . . and withdraw the injector from the fuel rail

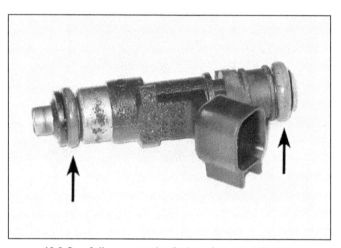

12.8 Carefully remove the O-rings from the injectors

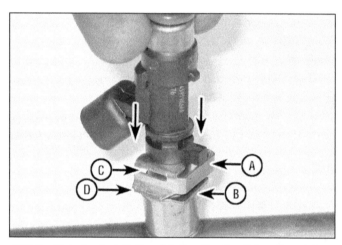

12.9 Fuel injector installation details: align the flat on the retainer (A) with the flat on the flange (B), and engage the slots in the retainer (C) with the semi-circular flanges (D) when injector is pushed into fuel rail pipe (3.5L V6 engine shown)

7   Carefully pull up on the fuel rail to disengage the injectors from their respective bores in the intake manifold, then remove the fuel rail and injectors as a single assembly. The injectors might initially stick in their bores, but they'll pull free when sufficient force is applied. Detach the injectors from the fuel rail **(see illustrations).**

8   Remove the O-rings from each injector **(see illustration)** and discard them. Install new O-rings and coat them with some clean engine oil to facilitate installation of the injectors.
9   Make sure that the injector is square to the bore of the mounting pipe and push it down into the mounting pipe until it's fully seated **(see illustration).**

10   Installation is otherwise the reverse of removal. Tighten the fuel rail retaining bolts to the torque listed in this Chapter's Specifications.
11   Start the engine and check for leaks at the quick-connect fitting that connects the fuel supply hose to the fuel rail. Also look for leaks at the upper end of each injector, where it's installed into the fuel rail.

Chapter 5
Engine electrical systems

Contents

	Section		Section
Alternator - removal and installation	7	General information and precautions	1
Battery - disconnection	3	Ignition coils - removal and installation	6
Battery - removal and installation	4	Starter motor - removal and installation	8
Battery cables - replacement	5	Troubleshooting	2

Specifications

Charging system
Charging voltage .... 13.5 to 14.5 volts

Torque specifications   **Ft-lbs**
Alternator mounting bolts
    2.4L engine (2014 and earlier) 40
    2.4L engine (2015 and later) 18
    2.7L engines 20
    3.5L engines 19
Starter mounting bolts 40

1 General information and precautions

General information
Ignition system
1   The electronic ignition system consists of the Crankshaft Position (CKP) sensor, the Camshaft Position (CMP) sensor, the Knock Sensor (KS), the Powertrain Control Module (PCM), the ignition switch, the battery, the individual ignition coils, and the spark plugs. For more information on the CKP, CMP and KS sensors, as well as the PCM, refer to Chapter 6.

Charging system
2   The charging system includes the alternator, the Powertrain Control Module (PCM), which incorporates the Electronic Voltage Regulator (EVR), the Body Control Module (BCM), a charge indicator light on the dash, the battery, a fuse or fusible link and the wiring connecting all of these components. The charging system supplies electrical power for the ignition system, the lights, the radio, etc. The alternator is driven by the drivebelt.

Starting system
3   The starting system consists of the battery, the ignition switch, the starter relay, the Powertrain Control Module (PCM), the Body Control Module (BCM), the Transmission Range (TR) switch, the starter motor and solenoid assembly, and the wiring connecting all of the components.

Precautions
4   Always observe the following precautions when working on the electrical system:
a) *Be extremely careful when servicing engine electrical components. They are easily damaged if checked, connected or handled improperly.*
b) *Never leave the ignition switched on for long periods of time when the engine is not running.*

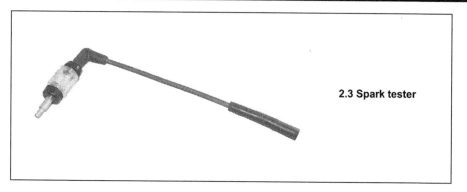

2.3 Spark tester

c) *Never disconnect the battery cables while the engine is running.*

d) *Maintain correct polarity when connecting battery cables from another vehicle during jump starting - see the "Booster battery (jump) starting" Section at the front of this manual.*

e) *Always disconnect the cable from the negative battery terminal before working on the electrical system, but read the battery disconnection procedure first (see Section 3).*

5   It's also a good idea to review the safety-related information regarding the engine electrical systems located in the *Safety first!* Section at the front of this manual before beginning any operation included in this Chapter.

2   Troubleshooting

Ignition system

1   If a malfunction occurs in the ignition system, do not immediately assume that any particular part is causing the problem. First, check the following items:

a) *Make sure that the cable clamps at the battery terminals are clean and tight.*

b) *Test the condition of the battery (see Steps 14 through 18). If it doesn't pass all the tests, replace it.*

c) *Check the ignition coil or coil pack connections.*

d) *Check any relevant fuses in the engine compartment fuse and relay box (see Chapter 12). If they're burned, determine the cause and repair the circuit.*

Check

Warning: *Because of the high voltage generated by the ignition system, use extreme care when performing a procedure involving ignition components.*

Warning: *If you're working on a model with a 3.6L V6 engine, relieve the fuel system pressure by disconnecting the fuel pump module electrical connector (see Chapter 4).*

Note: *The ignition system components on these vehicles are difficult to diagnose. In the event of ignition system failure that you can't diagnose, have the vehicle tested at a dealer service department or other qualified auto repair facility.*

Note: *You'll need a spark tester for the following test. Spark testers are available at most*

auto supply stores.

2   If the engine turns over but won't start, verify that there is sufficient ignition voltage to fire the spark plugs as follows.

3   On models with a coil-over-plug type ignition system, remove a coil and install the tester between the boot at the lower end of the coil and the spark plug **(see illustration)**.

4   Crank the engine and note whether or not the tester flashes.

Caution: *Do NOT crank the engine or allow it to run for more than five seconds; running the engine for more than five seconds may set a Diagnostic Trouble Code (DTC) for a cylinder misfire.*

5   If the tester flashes during cranking, the coil is delivering sufficient voltage to the spark plug to fire it. Repeat this test for each cylinder to verify that the other coils are OK.

6   If the tester doesn't flash, remove a coil from another cylinder and swap it for the one being tested. If the tester now flashes, you know that the original coil is bad. If the tester still doesn't flash, the PCM or wiring harness is probably defective. Have the PCM checked out by a dealer service department or other qualified repair shop (testing the PCM is beyond the scope of the do-it-yourselfer because it requires expensive special tools).

7   If the tester flashes during cranking but a misfire code (related to the cylinder being tested) has been stored, the spark plug could be fouled or defective.

Charging system

8   If a malfunction occurs in the charging system, do not automatically assume the alternator is causing the problem. First check the following items:

a) *Check the drivebelt tension and condition, as described in Chapter 1. Replace it if it's worn or deteriorated.*

b) *Make sure the alternator mounting bolts are tight.*

c) *Inspect the alternator wiring harness and the connectors at the alternator and voltage regulator. They must be in good condition, tight and have no corrosion.*

d) *Check the fusible link (if equipped) or main fuse in the underhood fuse/relay box. If it is burned, determine the cause, repair the circuit and replace the link or fuse (the vehicle will not start and/or the accessories will not work if the fusible link or main fuse is blown).*

e) *Start the engine and check the alternator for abnormal noises (a shrieking or squealing sound indicates a bad bearing).*

f) *Check the battery. Make sure it's fully charged and in good condition (one bad cell in a battery can cause overcharging by the alternator).*

g) *Disconnect the battery cables (negative first, then positive). Inspect the battery posts and the cable clamps for corrosion. Clean them thoroughly if necessary (see Chapter 1). Reconnect the cables (positive first, negative last).*

Alternator - check

9   Use a voltmeter to check the battery voltage with the engine off. It should be at least 12.6 volts **(see illustration 2.15)**.

10   Start the engine and check the battery voltage again. It should now be approximately 13.5 to 15 volts.

11   If the voltage reading is more or less than the specified charging voltage, the alternator might be defective, the Electronic Voltage Regulator (EVR) within the Powertrain Control Module (PCM) might be defective, or there might be a problem in the circuitry between the alternator, PCM or battery. Check for the presence of trouble codes related to the system (see Chapter 6). If no codes are found, remove the alternator and have it bench tested (most auto parts stores will do this for you); if it checks out ok, inspect all connectors and wiring in the circuit. If no problems are found, have the PCM checked by a dealer service department or other qualified repair shop.

12   The charging system (battery) light on the instrument cluster lights up when the ignition key is turned to ON, but it should go out when the engine starts.

13   If the charging system light stays on after the engine has been started, there is a problem with the charging system.

Battery - check

14   Check the battery state of charge. Visually inspect the indicator eye on the top of the battery (if equipped with one); if the indicator eye is black in color, charge the battery as described in Chapter 1. Next perform an open circuit voltage test using a digital voltmeter.

Note: *The battery's surface charge must be removed before accurate voltage measurements can be made. Turn on the high beams for ten seconds, then turn them off and let the vehicle stand for two minutes.*

15   With the engine and all accessories Off, touch the negative probe of the voltmeter to the negative terminal of the battery and the positive probe to the positive terminal of battery **(see illustration)**. The battery voltage should be 12.6 volts or slightly above. If the battery is less than the specified voltage, charge the battery before proceeding to the next test. Do not proceed with the battery load test unless the battery charge is correct.

16   Disconnect the negative battery cable, then the positive cable from the battery.

2.15 To test the open circuit voltage of the battery, connect the black probe of the voltmeter to the negative terminal and the red probe to the positive terminal of the battery; a fully charged battery should be at least 12.6 volts

2.17 Connect a battery load tester to the battery and check the battery condition under load following the tool manufacturer's instructions

17   Perform a battery load test. An accurate check of the battery condition can only be performed with a load tester **(see illustration)**. This test evaluates the ability of the battery to operate the starter and other accessories during periods of high current draw. Connect the load tester to the battery terminals. Load test the battery according to the tool manufacturer's instructions. This tool increases the load demand (current draw) on the battery.

18   Maintain the load on the battery for 15 seconds and observe that the battery voltage does not drop below 9.6 volts. If the battery condition is weak or defective, the tool will indicate this condition immediately.

Note: *Cold temperatures will cause the minimum voltage reading to drop slightly. Follow the chart given in the manufacturer's instructions to compensate for cold climates. Minimum load voltage for freezing temperatures (32 degrees F) should be approximately 9.1 volts.*

Starting system

The starter rotates, but the engine doesn't

19   Remove the starter (see Section 8). Check the overrunning clutch and bench test the starter to make sure the drive mechanism extends fully for proper engagement with the flywheel ring gear. If it doesn't, replace the starter.

20   Check the flywheel ring gear for missing teeth and other damage. With the ignition turned off, rotate the flywheel so you can check the entire ring gear.

The starter is noisy

21   If the solenoid is making a chattering noise, first check the battery (see Steps 14 through 18). If the battery is okay, check the cables and connections.

22   If you hear a grinding, crashing metallic sound when you turn the key to Start, check for loose starter mounting bolts. If they're tight, remove the starter and inspect the teeth on the starter pinion gear and flywheel ring gear. Look for missing or damaged teeth.

23   If the starter sounds fine when you first turn the key to Start, but then stops rotating the engine and emits a zinging sound, the problem is probably a defective starter drive that's not staying engaged with the ring gear. Replace the starter.

The starter rotates slowly

24   Check the battery (see Steps 14 through 18).

25   If the battery is okay, verify all connections (at the battery, the starter solenoid and motor) are clean, corrosion-free and tight. Make sure the cables aren't frayed or damaged.

26   Check that the starter mounting bolts are tight so it grounds properly. Also check the pinion gear and flywheel ring gear for evidence of a mechanical bind (galling, deformed gear teeth or other damage).

The starter does not rotate at all

27   Check the battery (see Steps 14 through 18).

28   If the battery is okay, verify all connections (at the battery, the starter solenoid and motor) are clean, corrosion-free and tight. Make sure the cables aren't frayed or damaged.

29   Check all of the fuses in the underhood fuse/relay box.

30   Check that the starter mounting bolts are tight so it grounds properly.

31   Check for voltage at the starter solenoid "S" terminal when the ignition key is turned to the start position. If voltage is present, replace the starter/solenoid assembly. If no voltage is present, the problem could be the starter relay, the Transmission Range (TR) switch (see Chapter 6), or with an electrical connector somewhere in the circuit (see the wiring diagrams - Chapter 12). Also, on many modern vehicles, the Powertrain Control Module

(PCM) and the Body Control Module (BCM) control the voltage signal to the starter solenoid; on such vehicles a special scan tool is required for diagnosis.

3   Battery - disconnection

Warning: *Always disconnect the cable from the negative battery terminal FIRST and hook it up LAST or the battery may be shorted by the tool being used to loosen the cable clamps.*

Warning: *Hydrogen gas is produced by the battery, so keep open flames and lighted cigarettes away from it at all times. Always wear eye protection when working around the battery. Rinse off spilled electrolyte immediately with large amounts of water.*

1   Some systems on the vehicle require battery power to be available at all times, either to maintain continuous operation (alarm system, power door locks, etc.), or to maintain control unit memory (radio station presets, Powertrain Control Module and other control units). When the battery is disconnected, the power that maintains these systems is cut. So, before you disconnect the battery, please note that on a vehicle with power door locks, it's a wise precaution to remove the key from the ignition and to keep it with you, so that it does not get locked inside if the power door locks should engage accidentally when the battery is reconnected!

2   Devices known as "memory-savers" can be used to avoid some of these problems. Precise details vary according to the device used. The typical memory saver is plugged into the cigarette lighter and is connected to a spare battery. Then the vehicle battery can be disconnected from the electrical system. The memory saver will provide sufficient current to maintain audio unit security codes, PCM memory, etc. and will provide power to always hot circuits such as the clock and radio memory circuits.

3.3 Remote negative battery terminal properly disconnected and isolated (2014 and earlier models)

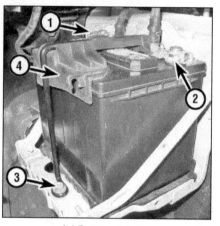

4.4 Battery details:

1   *Negative cable*
2   *Positive cable*
3   *Battery hold-down clamp bolt*
4   *Battery hold-down clamp*

---

4   Battery - removal and installation

2014 and earlier models

1   Disconnect and isolate the remote battery cable; there is an isolation hole designed to retain the cable away from the battery post in the terminal casing **(see illustratrion 3.3)**.
2   Loosen the left front wheel lug nuts, raise the front of the vehicle and support it securely on jackstands, then remove the wheel.
3   Remove the push pins and the inner-fender splash shield from the left front wheel well (see Chapter 11).
4   Remove and isolate the negative battery cable before doing the same to the positive battery cable **(see illustration)**.
5   Remove the battery retaining bracket nut, then remove the bracket.
6   Remove the battery from the wheelwell. Be careful.

2015 and later models

7   Open the hood, disconnect the intelligent Battery Sensor (IBS) connector, then loosen the nut and disconnect the cable from the negative battery terminal (see Section 3).
8   Loosen the nut and disconnect the cable from the positive battery terminal.
9   Remove the battery cover by pulling it straight up.
10   Remove the battery hold-down clamp.
11   Lift the battery up and out of the engine compartment.
Note: *Battery straps and handlers are available at most auto parts stores for reasonable prices. They make it easier to remove and carry the battery.*
12   If you are replacing the battery, make sure you get one that's identical, with the same dimensions, amperage rating, cold cranking rating, etc.
13   Installation is the reverse of removal. Always connect the positive cable first and the negative cable last.

6.4 Disconnect the ignition coil electrical connector

Battery tray

14   Remove the battery (see previous Steps).

2014 and earlier models

15   Remove the four battery tray mounting bolts retaining the tray to the left front frame rail.
16   Remove the battery tray through the wheel opening.
17   Thoroughly wash the battery tray in clean water, then dry it with compressed air.
18   Installation is the reverse of removal.

2015 and later models

19   Loosen the left-front wheel lug nuts, then raise the front of the vehicle and support it on jackstands. Remove the left-front wheel and inner fender splash shield (see Chapter 11). Also remove the engine under cover.
20   Remove the fuse box assembly mounting bolts and position it to the side.
21   Remove the Powertrain Control Module (PCM) mounting bolts, then position it to the side (see Chapter 6). If necessary, also remove the PCM mount for better access.
22   Detach any wiring harness clips from the battery tray, then remove the battery tray mounting bolts - accessed by working from the engine compartment, fenderwell, and from below the vehicle.
23   Remove the battery tray, rotating it as necessary. Installation is the reverse of removal.

---

5   Battery cables - replacement

1   When removing the cables, always disconnect the cable from the remote ground terminal first and hook it up last, or you might accidentally short out the battery with the tool you're using to loosen the cable clamps. Even if you're only replacing the cable for the positive terminal, Aways disconnect the negative cable from the battery first.

Warning: *Some memory savers deliver a considerable amount of current in order to keep vehicle systems operational after the main battery is disconnected. If you're using a memory saver, make sure that the circuit concerned is actually open before servicing it.*
Warning: *If you're going to work near any of the airbag system components, the battery MUST be disconnected and a memory saver must NOT be used. If a memory saver is used, power will be supplied to the airbag, which means that it could accidentally deploy and cause serious personal injury.*
3   The battery on 2014 and earlier models is mounted behind the lower front portion of the left front fender, in front of the inner fender splash shield. There are remote positive and negative terminals in the engine compartment, at the left strut tower. To disconnect the battery for service procedures requiring power to be cut from the vehicle, loosen the cable end nut of the remote negative terminal, disconnect the cable from the remote ground terminal, then attach the plastic insulator portion of the cable to the stud and reinstall the nut **(see illustration)**.
4   The battery on 2015 and later models is located in the engine compartment. Unlike 2014 and earlier models, the battery cables are accessible after opening the hood, which eliminates the need for a remote ground cable/terminal. To disconnect the battery for service procedures requiring power to be cut from the vehicle, first disconnect the Intelligent Battery Sensor (IBS) electrical connector from the negative cable (if equipped). Loosen the negative battery cable clamp nut securing the cable to the battery post, then lift the cable off the negative terminal and secure it aside to prevent accidental reconnection.
Note: *Whenever the battery is disconnected, the Intelligent Battery Sensor (IBS) electrical connector on 2015 and later models must be disconnected prior to disconnecting the negative cable from the battery post.*

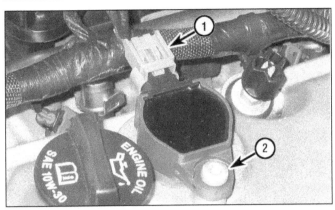

6.9 Disconnect the electrical connector (1) and remove the coil mounting bolt (2)

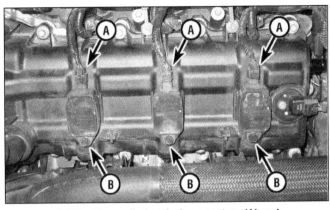

6.15 Ignition coil electrical connectors (A) and mounting fasteners (B)

2   Disconnect the old cables from the battery, then trace each of them to their opposite ends and disconnect them. Be sure to note the routing of each cable before disconnecting it to ensure correct installation.

3   If you are replacing any of the old cables, take them with you when buying new cables. It is vitally important that you replace the cables with identical parts.

4   Clean the threads of the solenoid or ground connection with a wire brush to remove rust and corrosion. Apply a light coat of battery terminal corrosion inhibitor or petroleum jelly to the threads to prevent future corrosion.

5   Attach the cable to the solenoid or ground connection and tighten the mounting nut/bolt securely.

6   Before connecting a new cable to the battery, make sure that it reaches the battery post without having to be stretched.

7   Connect the cable to the positive battery terminal first, *then* connect the ground cable to the negative battery terminal.

6   Ignition coils - removal and installation

1   If available, spray compressed air around the tops of the coils to ensure debris will not fall into the spark plug tube upon removal.
Warning: *Wear eye protection when using compressed air.*

2   Remove the engine cover(s) as necessary to access the components.

3   Disconnect the negative battery cable from the remote ground terminal or battery (see Section 3).

2.4L engine

4   Disconnect the electrical connector from the ignition coil **(see illustration)**.
Note: *Slide back the lock on the connector to allow it to be released.*

5   Remove the bolt attaching the ignition coil to the valve cover.

6   Grasp the ignition coil firmly and pull off the spark plug using a twisting motion.

7   Installation is the reverse of removal.

7.6 Loosen, but do not remove the lower alternator bolt (2014 and earlier 4-cylinder engines)

2.7L / 3.5L V6 engine

8   Remove the upper intake manifold to access the coils (see Chapter 2B or 2C).

9   Disconnect the electrical connector from the ignition coil **(see illustration)**.
Note: *If equipped, slide back the lock on the connector to allow it to be released.*

10   Remove the bolt attaching the ignition coil to the valve cover.

11   Grasp the ignition coil firmly and pull off the spark plug using a twisting motion.

12   Installation is the reverse of removal. Tighten the ignition coil bolts securely.

3.6L V6 engine

13   To remove an ignition coil from the rear cylinder bank (cylinder 1, 3 or 5), first remove the resonator.

14   To remove an ignition coil from the front cylinder bank (cylinder 2, 4, or 6), remove the upper intake manifold and insulator (see Chapter 2D).

15   Disconnect the electrical connector from the ignition coil **(see illustration)**.
Note: *If equipped, slide back the lock on the connector to allow it to be released.*

16   Remove the bolts attaching the ignition coil to the valve cover.

17   Grasp the coil firmly and pull it off the spark plug using a twisting motion.

7.7 Relocate the air conditioning compressor out of the way to allow the alternator to be removed from the vehicle (2014 and earlier 4-cylinder engines)

18   Installation is the reverse of removal. Tighten the ignition coil bolts securely.

7   Alternator - removal and installation

1   Disconnect the cable from the negative terminal of the battery or the remote ground terminal (see Section 3).

2.4L engine

2   Loosen the right front wheel lug nuts, raise the front of the vehicle and support it securely on jackstands, then remove the wheel.

3   Remove the underbody air dam, and the drivebelt splash shield (see Chapter 1).

4   Remove the drivebelt (see Chapter 1).

5   On 2014 and earlier models, remove the drivebelt idler pulley.

6   On 2014 and earlier models, loosen the lower alternator bolt **(see illustration).**

7   On 2014 and earlier models, unbolt and relocate the air conditioning compressor out of the way enough to allow the alternator to be removed **(see illustration).**
Warning: *Do not disconnect the refrigerant lines.*
Caution: *Secure the compressor with bailing wire or similar. DO NOT let the compressor hang by the refrigerant lines or electrical harness to prevent damage to the lines or wiring.*

7.8a Unplug the field connector from the alternator . . .

7.8b . . . then remove the nut that secures the B+ wire terminal to the stud on the back of the alternator (2.4L engine)

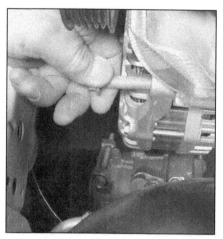

7.9a Remove the upper alternator bolt (2014 and earlier 2.4L engine)

7.9b Remove the lower alternator bolt (2014 and earlier 2.4L engine)

8   Unplug the field connector from the alternator, then remove the nut that secures the B+ wire terminal to the stud on the back of the alternator **(see illustrations)**.

9   Remove the upper and lower alternator mounting bolts **(see illustrations)**. On 2015 and later models, there is one upper mounting bolt and two lower bolts.

10   On 2014 and earlier models, position the refrigerant line between the B+ lug and center of the alternator, rotate the alternator with the pulley facing down, then slide the alternator out from the bottom of the vehicle **(see illustration)**. On 2015 and later models, carefully guide the alternator out of the engine compartment.

11   Installation is the reverse of removal. Tighten the alternator bolts to the torque listed in this Chapter's Specifications.

2.7L / 2014 and earlier models 3.6L V6 engines

12   Remove the air filter housing (see Chapter 4).

13   On 3.6L engines, remove the engine oil dipstick tube.

14   Remove the radiator cooling fan (see Chapter 3).

15   Remove the drivebelt (see Chapter 1).

16   Open the protective cover, remove the nut that secures the B+ wire terminal to the stud on the back of the alternator, and disconnect the field wire electrical connector from the alternator **(see illustration)**.

17   Remove the two upper alternator mounting bolts and one lower bolt **(see illustrations)**.

18   Remove the alternator without damaging the radiator.

19   Installation is the reverse of removal. Tighten the alternator bolts to the torque listed in this Chapter's Specifications.

3.5L V6 engine

20   Remove the engine cover.

21   Open the protective cover, remove the nut that secures the B+ wire terminal to the stud on the back of the alternator, and disconnect the field wire electrical connector from the alternator **(see illustration)**.

7.10 Remove the alternator (2014 and earlier 2.4L engine)

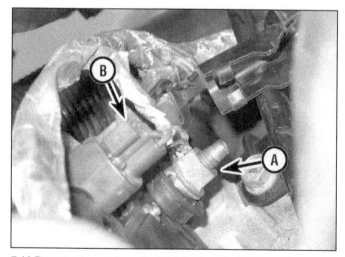

7.16 Remove the nut from the B+ stud terminal (A) and disconnect the battery cable from the stud, then disconnect the field wire electrical connector (B) from the alternator (3.6L V6 engines shown, 2.7L similar)

7.17a Remove lower alternator mounting bolt/nut . . .

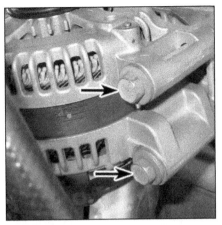

7.17b . . . then remove the upper alternator mounting bolts (2.7L and 3.6L V6 engines)

7.21 Remove the nut from the battery (B+) output terminal stud (A) and disconnect the battery cable from the stud, then disconnect the electrical connector (B) for the field terminals (3.5L V6 engines)

22   Remove the drivebelt (see Chapter 1).
23   Remove the two mounting bolts at the bottom of the alternator.
24   Remove the alternator from the vehicle.
25   Installation is the reverse of removal. Tighten the alternator bolts to the torque listed in this Chapter's Specifications.

2015 and later 3.6L V6 engine

Warning: *Gasoline is extremely flammable. See* Fuel System Warnings *in Chapter 4.*
26   Before beginning, release the fuel system pressure (see Chapter 4), then disconnect the cable from the negative battery terminal (see Section 3).
27   Remove the air filter housing (see Chapter 4).
28   Loosen the right-front wheel lug nuts, then raise the vehicle and support it securely on jackstands. Remove the right-front wheel, then remove the engine under cover.
29   Remove the drivebelt (see Chapter 1).
30   Unbolt and remove the fresh air inlet tube.
31   Remove the air conditioning compressor mounting bolts (see Chapter 3), then position the compressor aside and support it with a length of wire.
Warning: *DO NOT disconnect the air conditioning lines from the compressor, which are under constant high pressure!*
32   Disconnect the field wire electrical connector from the alternator.
33   Remove the protective cover and cable securing nut, then disconnect the battery positive (B+) cable from the alternator.
34   Lower the vehicle.
35   Using a floor jack with a block of wood placed between the jack head and engine oil pan, support the engine. Remove the right side engine mount-to-engine mount bracket bolts (see Chapter 2).
36   Disconnect the fuel line quick-connect fitting from the fuel rail (see Chapter 4).
37   Remove the lower and upper alternator mounting bolts, then carefully guide the alternator out from the fenderwell (between the upper and lower frame rails and around the radiator hoses).

8.5 Remove the starter mounting bolts (2014 and earlier 2.4L engine)

38   Installation is the reverse of removal. Tighten the alternator mounting bolts to the torque listed in this Chapter's specifications.

8   Starter motor - removal and installation

1   Disconnect the cable from the negative terminal of the battery or the remote ground terminal (see Section 3).

2.4L engine (2014 and earlier)

2   Remove the air filter housing and duct (see Chapter 4).
3   Disconnect the throttle body electrical connector, then remove the throttle body bolts.
4   Remove the throttle body (see Chapter 4).
5   Remove starter mounting bolts (see **illustration**).
6   Push the starter under the intake manifold, position the starter to face the nose downwards, then pull the starter up and out enough to access the electrical connections **(see illustration)**.

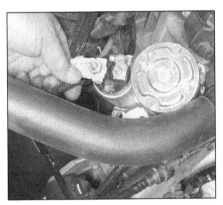

8.6 Position the starter nose-down so the electrical connectors can be removed (2014 and earlier 2.4L engine)

7   Disconnect the starter wiring and remove the starter.
8   Installation is the reverse of removal. Tighten the starter mounting bolts to the torque listed in this Chapter's Specifications.

2.7L / 3.5L V6 engine

9   Raise the front of the vehicle and place it securely on jackstands.
10   Support the engine with an engine support fixture across the top of the motor or a floor jack with a block of wood placed under the transaxle.
11   On 3.5L engines, remove the three starter wiring harness heat shield nuts and remove the heat shield if equipped.
12   Remove the engine splash shield.
13   Remove the front engine mount (see Chapter 2B or 2C).
14   Remove the longitudinal crossmember (see Chapter 2E , **illustration 7.36**).
15   On 3.5L engines, remove the front exhaust heat shield and oxygen sensor if necessary for access.
16   Remove the front engine mount bracket

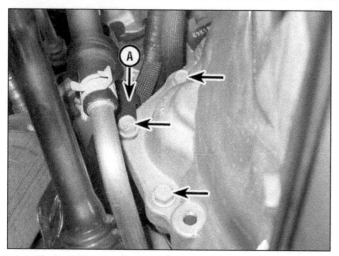

8.17 Remove the starter bolt attaching the ground wire (A), then remove the two remaining starter bolts (3.5L shown, 2.7L similar)

8.18 Depress this release tab (A) and disconnect the electrical connector from the solenoid then remove the nut that secures the battery cable to the stud terminal on the solenoid (B)to disconnect the B+ cable (3.5L shown, 2.7L similar)

bolts attaching the bracket to the engine and remove the bracket.

17   Remove the starter bolt securing the ground wire **(see illustration)**.

18   Remove the nut that secures the battery (B+) cable to the stud terminal on the solenoid, disconnect the B+ cable, then disconnect the solenoid electrical connector from the solenoid **(see illustration)**.

19   Remove the two remaining starter bolts.

20   Tip the rear of the starter upwards to clear the exhaust and remove the starter from the vehicle.

21   Remove the shield from the starter (if equipped).

22   Installation is the reverse of removal. Tighten the starter mounting bolts to the torque listed in this Chapter's Specifications.

3.6L V6 engine

23   Raise the front of the vehicle and support it securely on jackstands.

24   Remove the catalytic converter (see Chapter 6).

25   On 2014 and earlier models, remove the front engine mount (see Chapter 2D).

26   Remove the nut that secures the battery (B+) cable to the stud terminal on the solenoid, disconnect the B+ cable, then disconnect the solenoid electrical connector from the solenoid.

27   Remove the starter mounting bolts, then remove the starter assembly.

28   Installation is the reverse of removal. Tighten the starter mounting bolts to the torque listed in this Chapter's Specifications.

2.4L engine (2015 and later)

29   Remove the fuse box/PDC (Power Distribution Center) from the engine (see Chapter 12).

30   Remove the battery and battery tray (see Section 4).

31   Release the fuse box wiring harness

clips from along the engine block (which should enable greater access to the starter motor upper mounting bolt).

32   Loosen the starter upper mounting bolt.

33   Raise the front of the vehicle and support it securely on jackstands.

34   Remove the driveaxle that blocks access to the starter motor (see Chapter 8).

35   Remove the battery positive cable nut, then disconnect the battery positive cable from the starter motor.

36   Remove the starter motor lower mounting bolt, then remove it from the transaxle and rotate it down enough to disconnect the starter solenoid electrical connector. Be sure to release the locking tab of the connector before disconnecting it.

37   Carefully guide the starter out from under the vehicle.

38   Installation is the reverse of removal. Tighten the starter motor mounting bolts to the torque listed in this Chapter's specifications.

Chapter 6
Emissions and engine control systems

Contents

	Section
Accelerator Pedal Position (APP) sensor - replacement	4
Air Injection system (2.4L PZEV engines) - component replacement	22
Camshaft Position (CMP) sensor - replacement	5
Catalytic converter - replacement	18
Crankshaft Position (CKP) sensor - replacement	6
Engine Coolant Temperature (ECT) sensor - replacement	9
Evaporative Emissions Control (EVAP) system - component replacement	19
Exhaust Gas Recirculation (EGR) system - component replacement	20
General information	1
Intake Air Temperature (IAT) sensor - replacement	10
Knock sensor - replacement	11
Manifold Absolute Pressure (MAP) sensor - replacement	12

	Section
Manifold tuning valve solenoids (2.7L and 3.5L V6 engines) - removal and installation	7
Obtaining and clearing Diagnostic Trouble Codes (DTCs)	3
Oil temperature sensor - removal and installation	23
On Board Diagnosis (OBD) system	2
Oxygen sensors - general information and replacement	13
Positive Crankcase Ventilation (PCV) system	21
Powertrain Control Module (PCM) - replacement	17
Throttle Position (TP) sensor - replacement	14
Transaxle speed sensors - replacement	16
Transmission Range (TR) and transmission temperature sensors - replacement	15
Variable valve timing solenoids (2.4L and 3.6L engines) - removal and installation	8

Specifications

Torque specifications

Ft-lbs (unless otherwise indicated)

Note: *One foot-pound (ft-lb) of torque is equivalent to 12 inch-pounds (in-lbs) of torque. Torque values below approximately 15 ft-lbs are expressed in inch-pounds, because most foot-pound torque wrenches are not accurate at these smaller values.*

Accelerator pedal position (APP) sensor	
2014 and earlier models	144 in-lbs
2015 and later models	53 in-lbs
Air pump heat shield/bracket mount bolts	71 in-lbs
Air injection system check valve bolts	105 in-lbs
Positive Crankcase Ventilation (PCV) valve	
2.4L engine (2014 and earlier)	44 in-lbs
2.4L engine (2015 and later)	27 in-lbs
2.7L and 3.5L engines	35 in-lbs

Torque specifications (continued)

Ft-lbs (unless otherwise indicated)

Note: *One foot-pound (ft-lb) of torque is equivalent to 12 inch-pounds (in-lbs) of torque. Torque values below approximately 15 ft-lbs are expressed in inch-pounds, because most foot-pound torque wrenches are not accurate at these smaller values.*

Camshaft/Crankshaft Position (CMP/CKP) sensor mounting bolt
    2.4L and 2.7L engines... 80 in-lbs
    3.5L engines... 105 in-lbs
    3.6L engines... 80 in-lbs
Engine Coolant Temperature (ECT) sensor
    2.4L engines (2014 and earlier) .... 168 in-lbs
    2.4L engine (2015 and later)
        Step 1 .... 62 in-lbs
        Step 2 .... Two complete turns (720-degrees)
    2.7L and 3.5L engines.. 20
    3.6L engines (2014 and earlier) .... 96 in-lbs
    3.6L engines (2015 and later)... 22
Exhaust Gas Recirculation (EGR) valve-to-cylinder head bolts,
    2.7L and 3.5L engines.. 22
EGR tube flange-to-EGR valve bolts, 2.7L and 3.5L engines .... 106 in-lbs
EGR tube to exhaust manifold, 2.7L engines.................................... 51.5
EGR tube to intake manifold, 2.7L and 3.5L.................................... 53 in-lbs
Oil temperature sensor
    2.4L engine (2014 and earlier).. 160 in-lbs
    2.4L engines (2015 and later) 35
    3.6L engine (2014 and earlier).. 177 in-lbs
    3.6L engines (2015 and later)…….................................... 22
Oxygen sensors.. 30
Transaxle speed sensors
    41TE and 40TE transaxle (input and output)................... 20
    62TE transaxle (input, output and transfer shaft)........... 105 in-lbs
Knock sensors
    2.4L engine (2014 and earlier).. 168 in-lbs
    2.4L engines (2015 and later) .... 180 in-lbs
    2.7L and 3.5L engines.. 132 in-lbs
    3.6L engine (2014 and earlier).. 96 in-lbs
    3.6L engines (2015 and later) .... 180 in-lbs
Manifold Absolute Pressure (MAP) sensor
    (2014 and earlier 2.4L engine).. 40 in-lbs

1 General information

1   To prevent pollution of the atmosphere from incompletely burned and evaporating gases, and to maintain good driveability and fuel economy, a number of emission control systems are incorporated. They include the:

Catalytic converter

2   A catalytic converter is an emission control device in the exhaust system that reduces certain pollutants in the exhaust gas stream. There are two types of converters: oxidation converters and reduction converters.

3   Oxidation converters contain a monolithic substrate (a ceramic honeycomb) coated with the semi-precious metals platinum and palladium. An oxidation catalyst reduces unburned hydrocarbons (HC) and carbon monoxide (CO) by adding oxygen to the exhaust stream as it passes through the substrate, which, in the presence of high temperature and the catalyst materials, converts the HC and CO to water vapor (H_2O) and carbon dioxide (CO_2).

4   Reduction converters contain a monolithic substrate coated with platinum and rhodium. A reduction catalyst reduces oxides of nitrogen (NOx) by removing oxygen, which in the presence of high temperature and the catalyst material produces nitrogen (N) and carbon dioxide (CO_2).

5   Catalytic converters that combine both types of catalysts in one assembly are known as "three-way catalysts" or TWCs. A TWC can reduce all three pollutants.

Evaporative Emissions Control (EVAP) system

6   The Evaporative Emissions Control (EVAP) system prevents fuel system vapors (which contain unburned hydrocarbons) from escaping into the atmosphere. On warm days, vapors trapped inside the fuel tank expand until the pressure reaches a certain threshold. Then the fuel vapors are routed from the fuel tank through the fuel vapor vent valve and the fuel vapor control valve to the EVAP canister, where they're stored temporarily until the next time the vehicle is operated. When the conditions are right (engine warmed up, vehicle up to speed, moderate or heavy load on the engine, etc.) the PCM opens the canister purge valve, which allows fuel vapors to be drawn from the canister into the intake manifold. Once in the intake manifold, the fuel vapors mix with incoming air before being drawn through the intake ports into the combustion chambers where they're burned up with the rest of the air/fuel mixture. The EVAP system is complex and virtually impossible to troubleshoot without the right tools and training.

Exhaust Gas Recirculation (EGR) system

7   The EGR system reduces oxides of nitrogen by recirculating exhaust gases from the exhaust manifold, through the EGR valve

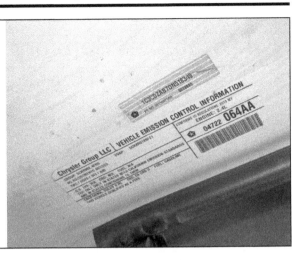

1.13 The Vehicle Emission Control Information (VECI) label is affixed to the underside of the hood

and intake manifold, then back to the combustion chambers, where it mixes with the incoming air/fuel mixture before being consumed. These recirculated exhaust gases dilute the incoming air/fuel mixture, which cools the combustion chambers, thereby reducing NOx emissions.

8   The EGR system consists of the Powertrain Control Module (PCM), the EGR valve, the EGR valve position sensor and various other information sensors that the PCM uses to determine when to open the EGR valve. The degree to which the EGR valve is opened is referred to as "EGR valve lift." The PCM is programmed to produce the ideal EGR valve lift for varying operating conditions. The EGR valve position sensor, which is an integral part of the EGR valve, detects the amount of EGR valve lift and sends this information to the PCM. The PCM then compares it with the appropriate EGR valve lift for the operating conditions. The PCM increases current flow to the EGR valve to increase valve lift and reduces the current to reduce the amount of lift. If EGR flow is inappropriate to the operating conditions (idle, cold engine, etc.) the PCM simply cuts the current to the EGR valve and the valve closes.

Secondary Air Injection (AIR) system

9   Some models are equipped with a secondary air injection (AIR) system. The secondary air injection system is used to reduce tailpipe emissions on initial engine start-up. The system uses an electric motor/pump assembly, relay, vacuum valve/solenoid, air shut-off valve, check valves and tubing to inject fresh air directly into the exhaust manifolds. The fresh air (oxygen) reacts with the exhaust gas in the catalytic converter to reduce HC and CO levels. The air pump and solenoid are controlled by the PCM through the AIR relay. During initial start-up, the PCM energizes the AIR relay, the relay supplies battery voltage to the air pump and the vacuum valve/solenoid, engine vacuum is applied to the air shut-off valve which opens

and allows air to flow through the tubing into the exhaust manifolds. The PCM will operate the air pump until closed loop operation is reached (approximately four minutes). During normal operation, the check valves prevent exhaust backflow into the system.

Powertrain Control Module (PCM)

10   The Powertrain Control Module (PCM) is the brain of the engine management system. It also controls a wide variety of other vehicle systems. In order to program the new PCM, the dealer needs the vehicle as well as the new PCM. If you're planning to replace the PCM with a new one, there is no point in trying to do so at home because you won't be able to program it yourself.

Positive Crankcase Ventilation (PCV) system

11   The Positive Crankcase Ventilation (PCV) system reduces hydrocarbon emissions by scavenging crankcase vapors, which are rich in unburned hydrocarbons. A PCV valve or orifice regulates the flow of gases into the intake manifold in proportion to the amount of intake vacuum available.

12   The PCV system generally consists of the fresh air inlet hose, the PCV valve or orifice and the crankcase ventilation hose (or PCV hose). The fresh air inlet hose connects the air intake duct to a pipe on the valve cover. The crankcase ventilation hose (or PCV hose) connects the PCV valve or orifice in the valve cover to the intake manifold.

Vehicle Emission Control Information (VECI) label

13   This label (**see illustration**), located on the underside of the hood, indicates what emission control systems the vehicle is equipped with and for what market the vehicle is certified (California, Federal, etc.), as well as any tune-up specifications and adjustments that may be needed.

Information Sensors

Accelerator Pedal Position (APP) sensor - as you press the accelerator pedal, the APP sensor alters its voltage signal to the PCM in proportion to the angle of the pedal, and the PCM commands a motor inside the throttle body to open or close the throttle plate accordingly

Camshaft Position (CMP) sensor - produces a signal that the PCM uses to identify the number 1 cylinder and to time the firing sequence of the fuel injectors

Crankshaft Position (CKP) sensor - produces a signal that the PCM uses to calculate engine speed and crankshaft position, which enables it to synchronize ignition timing with fuel injector timing, and to detect misfires

Engine Coolant Temperature (ECT) sensor - a thermistor (temperature-sensitive variable resistor) that sends a voltage signal to the PCM, which uses this data to determine the temperature of the engine coolant

Fuel tank pressure sensor - measures the fuel tank pressure and controls fuel tank pressure by signaling the EVAP system to purge the fuel tank vapors when the pressure becomes excessive

Intake Air Temperature (IAT) sensor - monitors the temperature of the air entering the engine and sends a signal to the PCM to determine injector pulse-width (the duration of each injector's on-time) and to adjust spark timing (to prevent spark knock)

Knock sensor - a piezoelectric crystal that oscillates in proportion to engine vibration which produces a voltage output that is monitored by the PCM. This retards the ignition timing when the oscillation exceeds a certain threshold

Manifold Absolute Pressure (MAP) sensor - monitors the pressure or vacuum inside the intake manifold. The PCM uses this data to determine engine load so that it can alter the ignition advance and fuel enrichment

Mass Air Flow (MAF) sensor - measures the amount of intake air drawn into the engine. It uses a hot-wire sensing element to measure the amount of air entering the engine

Oxygen sensors - generates a small variable voltage signal in proportion to the difference between the oxygen content in the exhaust stream and the oxygen content in the ambient air. The PCM uses this information to maintain the proper air/fuel ratio. A second oxygen sensor monitors the efficiency of the catalytic converter

Throttle Position (TP) sensor - a potentiometer that generates a voltage signal that varies in relation to the opening angle of the throttle plate inside the throttle body. Works with the PCM and other sensors to calculate injector pulse width (the duration of each injector's on-time)

Photos courtesy of Wells Manufacturing, except APP and MAF sensors.

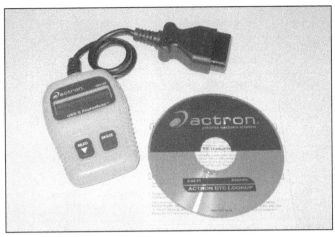

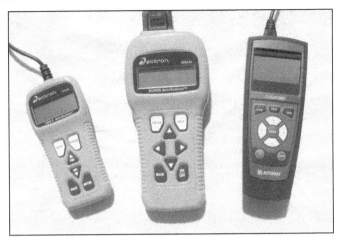

2.4a Simple code readers are an economical way to extract trouble codes when the CHECK ENGINE light comes on

2.4b Hand-held scan tools like these can extract computer codes and also perform diagnostics

2 On Board Diagnosis (OBD) system

General description

1 All models are equipped with the second generation OBD-II system. This system consists of an on-board computer known as the Powertrain Control Module (PCM), and information sensors, which monitor various functions of the engine and send data to the PCM. This system incorporates a series of diagnostic monitors that detect and identify fuel injection and emissions control system faults and store the information in the computer memory. This system also tests sensors and output actuators, diagnoses drive cycles, freezes data and clears codes.

2 The PCM is the brain of the electronically controlled fuel and emissions system. It receives data from a number of sensors and other electronic components (switches, relays, etc.). Based on the information it receives, the PCM generates output signals to control various relays, solenoids (fuel injectors) and other actuators. The PCM is specifically calibrated to optimize the emissions, fuel economy and driveability of the vehicle.

3 It isn't a good idea to attempt diagnosis or replacement of the PCM or emission control components at home while the vehicle is under warranty. Because of a federally-mandated warranty which covers the emissions system components and because any owner-induced damage to the PCM, the sensors and/or the control devices may void this warranty, take the vehicle to a dealer service department if the PCM or a system component malfunctions.

Scan tool information

4 Because extracting the Diagnostic Trouble Codes (DTCs) from an engine management system is now the first step in troubleshooting many computer-controlled systems and components, a code reader, at the very least, will be required **(see illustration)**. More

powerful scan tools can also perform many of the diagnostics once associated with expensive factory scan tools **(see illustration)**. If you're planning to obtain a generic scan tool for your vehicle, make sure that it's compatible with OBD-II systems. If you don't plan to purchase a code reader or scan tool and don't have access to one, you can have the codes extracted by a dealer service department or an independent repair shop.

Note: *Some auto parts stores even provide this service.*

3 Obtaining and clearing Diagnostic Trouble Codes (DTCs)

1 All models covered by this manual are equipped with on-board diagnostics. When the PCM recognizes a malfunction in a monitored emission or engine control system, component or circuit, it turns on the Malfunction Indicator Light (MIL) on the dash. The PCM will continue to display the MIL until the problem is fixed and the Diagnostic Trouble Code (DTC) is cleared from the PCM's memory. You'll need a scan tool to access any DTCs stored in the PCM.

2 Before outputting any DTCs stored in the PCM, thoroughly inspect ALL electrical connectors and hoses. Make sure that all electrical connections are tight, clean and free of corrosion. And make sure that all hoses are correctly connected, fit tightly and are in good condition (no cracks or tears).

Accessing the DTCs

3 The Diagnostic Trouble Codes (DTCs) can only be accessed with a code reader or scan tool. Professional scan tools are expensive, but relatively inexpensive generic code readers or scan tools **(see illustrations 2.4a and 2.4b)** are available at most auto parts stores. Simply plug the connector of the scan tool into the diagnostic connector **(see illustration)**. Then follow the instructions included with the scan tool to extract the DTCs.

4 Once you have outputted all of the

stored DTCs, look them up on the accompanying DTC chart.

5 After troubleshooting the source of each DTC, make any necessary repairs or replace the defective component(s).

Clearing the DTCs

6 Clear the DTCs with the code reader or scan tool in accordance with the instructions provided by the tool's manufacturer.

Diagnostic Trouble Codes

7 The accompanying tables are a list of the Diagnostic Trouble Codes (DTCs) that can be accessed by a do-it-yourselfer working at home (there are many, many more DTCs available to professional mechanics with proprietary scan tools and software, but those codes cannot be accessed by a generic scan tool). If, after you have checked and repaired the connectors, wire harness and vacuum hoses (if applicable) for an emission-related system, component or circuit, the problem persists, have the vehicle checked by a dealer service department or other qualified repair shop.

3.3 The 16-pin Data Link Connector (DLC) is located under the left side of the dash

OBD-II trouble codes

Code	Probable cause
P0016	Crankshaft/camshaft timing misalignment
P0031	Upstream oxygen sensor (cylinder bank no. 1), heater circuit low voltage
P0032	Upstream oxygen sensor heater (cylinder bank no. 1), heater circuit high voltage
P0037	Downstream oxygen sensor (cylinder bank no. 1), heater circuit low voltage
P0038	Downstream oxygen sensor (cylinder bank no. 1), heater circuit high voltage
P0068	Manifold pressure/throttle position correlation - high-flow/vacuum leak
P0070	Ambient temperature sensor stuck
P0071	Ambient temperature sensor performance
P0072	Ambient temperature sensor, low voltage
P0073	Ambient temperature sensor, high voltage
P0107	Manifold Absolute Pressure (MAP) sensor, low voltage
P0108	Manifold Absolute Pressure (MAP) sensor, high voltage
P0110	Intake Air Temperature (IAT) sensor, stuck
P0111	Intake Air Temperature (IAT) sensor performance
P0112	Intake Air Temperature (IAT) sensor, low voltage
P0113	Intake Air Temperature (IAT) sensor, high voltage
P0116	Engine Coolant Temperature (ECT) sensor performance
P0117	Engine Coolant Temperature (ECT) sensor, low voltage
P0118	Engine Coolant Temperature (ECT) sensor, high voltage
P0121	Throttle Position (TP) sensor performance
P0122	Throttle Position (TP) sensor, low voltage
P0123	Throttle Position (TP) sensor, high voltage
P0125	Insufficient coolant temperature for closed-loop control; closed-loop temperature not reached
P0128	Thermostat rationality
P0129	Barometric pressure out-of-range (low)
P0131	Upstream oxygen sensor (cylinder bank no. 1), low voltage or shorted to ground
P0132	Upstream oxygen sensor (cylinder bank no. 1), high voltage or shorted to voltage
P0133	Upstream oxygen sensor (cylinder bank no. 1), slow response
P0134	Upstream oxygen sensor (cylinder bank no. 1), sensor remains at center (not switching)

Code	Probable cause
P0135	Upstream oxygen sensor (cylinder bank no. 1) heater failure
P0137	Downstream oxygen sensor (cylinder bank no. 1), low voltage or shorted to ground
P0138	Downstream oxygen sensor (cylinder bank no. 1), high voltage or shorted to voltage
P0139	Downstream oxygen sensor (cylinder bank no. 1), slow response
P0140	Downstream oxygen sensor (cylinder bank no. 1), sensor remains at center (not switching)
P0141	Downstream oxygen sensor (cylinder bank no. 1), heater failure
P0171	Fuel control system too lean (cylinder bank no. 1)
P0172	Fuel control system too rich (cylinder bank no. 1)
P0201	Injector circuit malfunction - cylinder no. 1
P0202	Injector circuit malfunction - cylinder no. 2
P0203	Injector circuit malfunction - cylinder no. 3
P0204	Injector circuit malfunction - cylinder no. 4
P0205	Injector circuit malfunction - cylinder no. 5
P0206	Injector circuit malfunction - cylinder no. 6
P0300	Multiple cylinder misfire detected
P0301	Cylinder no. 1 misfire detected
P0302	Cylinder no. 2 misfire detected
P0303	Cylinder no. 3 misfire detected
P0304	Cylinder no. 4 misfire detected
P0305	Cylinder no. 5 misfire detected
P0306	Cylinder no. 6 misfire detected
P0315	No crank sensor learned
P0320	No crankshaft reference signal at Powertrain Control Module (PCM)
P0325	Knock sensor circuit malfunction
P0335	Crankshaft Position (CKP) sensor circuit
P0339	Crankshaft Position (CKP) sensor intermittent
P0340	Camshaft Position (CMP) sensor circuit
P0344	Camshaft Position (CMP) sensor intermittent
P0351	Ignition coil no. 1, primary circuit
P0352	Ignition coil no. 2, primary circuit
P0353	Ignition coil no. 3, primary circuit
P0354	Ignition coil no. 4, primary circuit

Code	Probable cause
P0401	Exhaust Gas Recirculation (EGR) system failure
P0403	Exhaust Gas Recirculation (EGR) solenoid circuit
P0404	Exhaust Gas Recirculation (EGR) sensor performance
P0405	Exhaust Gas Recirculation (EGR) sensor low voltage
P0406	Exhaust Gas Recirculation (EGR) sensor high voltage
P0420	Catalytic converter efficiency below threshold (upstream catalyst, cylinder bank no. 1)
P0432	Catalyst system efficiency below threshold (cylinder bank no. 2)
P0440	General Evaporative Emission Control (EVAP) system failure
P0441	Evaporative Emission Control (EVAP) system, incorrect purge flow
P0442	Evaporative Emission Control (EVAP) system, medium leak (0.040-inch) detected
P0443	Evaporative Emission Control (EVAP) system, purge solenoid circuit malfunction
P0452	Natural Vacuum Leak Detector (NVLD) pressure sensor circuit, low voltage
P0453	Natural Vacuum Leak Detector (NVLD) pressure sensor circuit, high input
P0455	Evaporative Emission Control (EVAP) system, large leak detected
P0456	Evaporative Emission Control (EVAP) system, small leak (0.020-inch) detected
P0460	Fuel level sending unit, no change as vehicle is operated
P0461	Fuel level sensor circuit, range or performance problem
P0462	Fuel level sending unit or sensor circuit, low voltage
P0463	Fuel level sending unit or sensor circuit, high voltage
P0480	Low-speed fan control relay circuit malfunction
P0498	Natural Vacuum Leak Detector (NVLD) canister vent valve solenoid circuit, low voltage
P0499	Natural Vacuum Leak Detector (NVLD) canister vent valve solenoid circuit, high voltage
P0500	No vehicle speed signal (four-speed automatic transaxles)
P0501	Vehicle speed sensor, range or performance problem
P0506	Idle speed control system, rpm lower than expected
P0507	Idle speed control system, rpm higher than expected
P0508	Idle Air Control (IAC) valve circuit, low voltage
P0509	Idle Air Control (IAC) valve circuit, high voltage
P0513	Invalid SKIM key (engine immobilizer problem)

Code	Probable cause
P0516	Battery temperature sensor, low voltage
P0517	Battery temperature sensor, high voltage
P0519	Idle speed performance
P0522	Engine oil pressure sensor/switch circuit, low voltage
P0532	Air conditioning refrigerant pressure sensor, low voltage
P0533	Air conditioning refrigerant pressure sensor, high voltage
P0551	Power Steering Pressure (PSP) switch circuit, range or performance problem
P0562	Battery voltage low
P0563	Battery voltage high
P0579	Speed control switch circuit, range or performance problem
P0580	Speed control switch circuit, low voltage
P0581	Speed control switch circuit, high voltage
P0582	Speed control vacuum solenoid circuit
P0586	Speed control vent solenoid circuit
P0594	Speed control servo power circuit
P0600	Serial communication link malfunction
P0601	Powertrain Control Module (PCM), internal controller failure
P0622	Alternator field control circuit malfunction or field not switching correctly
P0627	Fuel pump relay circuit
P0630	Vehicle Identification Number (VIN) not programmed in Powertrain Control Module (PCM)
P0632	Odometer not programmed in Powertrain Control Module (PCM)
P0633	SKIM key not programmed in Powertrain Control Module (PCM)
P0645	Air conditioning clutch relay circuit
P0685	Automatic Shutdown (ASD) relay control circuit
P0688	Automatic Shutdown (ASD) relay sense circuit, low voltage
P0700	Electronic Automatic Transaxle (EATX) control system malfunction or DTC present
P0703	Brake switch circuit malfunction
P0833	Clutch released switch circuit
P0850	Park/Neutral switch malfunction
P0856	Traction control torque request circuit

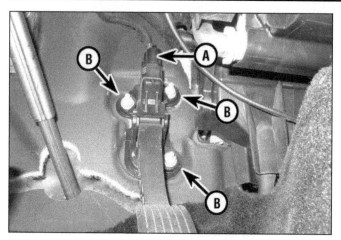

4.2 APP sensor electrical connector (A) and mounting fasteners (B)

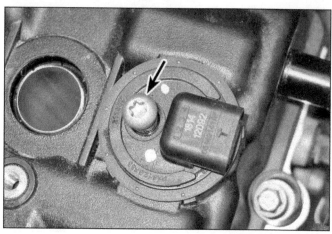

5.5a Remove the CMP sensor (3.6L models) mounting bolt . . .

5.5b . . . then pull the sensor straight up and out of the valve cover (3.6L models)

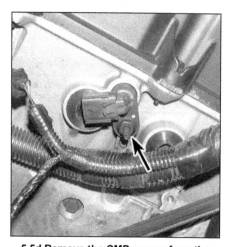

5.5c To disconnect the electrical connector from the CMP (2.7L V6 engine) depress the release tab and pull the connector off. Before removing the sensor, carefully pry off the harness clip (A) from the sensor mounting stud, if applicable

5.5d Remove the CMP sensor from the cylinder head on a 2.7L V6; remove this nut, then pull the sensor straight out

4   Accelerator Pedal Position (APP) sensor - replacement

Note: *Disconnect the negative battery cable from the remote ground terminal or battery before beginning (see Chapter 5).*

1   If necessary, remove the knee bolster trim panel and the knee bolster (see Chapter 11).

2   Disconnect the electrical connector from the upper end of the APP sensor assembly **(see illustration)**.

3   Remove the accelerator pedal/APP sensor assembly mounting nuts and remove the assembly.

4   Installation is the reverse of removal. Tighten the mounting fasteners to the torque listed in this Chapter's Specifications.

5   Camshaft Position (CMP) sensor - replacement

Caution: *After the sensor has been removed, do not insert any magnetic tools into the hole*

in the valve cover. Doing so could damage the magnetic timing wheels on the ends of the camshafts.

Note: *On 2.7L and 3.5L engines, the CMP sensor is located in the cylinder head near the camshaft.*

Note: *On 2014 and earlier 2.4L and 3.6L engines, there are two CMP sensors. On the 2.4L engine they are located at the left end of the cylinder head, at the front (intake) and rear (exhaust), just below the valve cover. On the 3.6L engine they are located at the left end of the valve covers.*

Note: *2015 and later 2.4L engines have only one CMP sensor, which is located on the front-left (driver's side) of the cylinder head.*

1   Disconnect the negative battery cable from the remote ground terminal or battery (see Chapter 5).

2   On all but 2015 and later 2.4L models, remove the air filter housing (see Chapter 4).

3   If you're removing the sensor from the front cylinder bank (bank 2), remove the upper intake manifold (see Chapter 2A, 2B, 2C or 2D).

4   Disconnect the electrical connector from the sensor.

5   Unscrew the sensor mounting bolt and pull the sensor from the valve cover or cylinder head **(see illustrations)**.

6   If you're going to reinstall the same sensor, check the O-ring for damage. If it's OK, it can be reused.

7   Apply a film of clean engine oil to the O-ring, then insert the sensor into the valve cover and install the mounting bolt, tightening it securely.

8   Installation is the reverse of removal.

6   Crankshaft Position (CKP) sensor - replacement

2.4L, 2.7L and 3.5L engines

Note: *The CKP sensor is located at either the rear of the cylinder block, near the transmission*

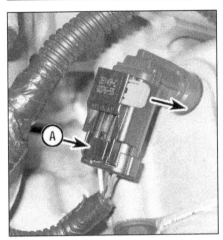

5.5e To remove the CMP sensor connector on a 3.5L V6, push the lock in the direction of the arrow, then depress the release tab (A) and pull off the connector

5.5f To detach the CMP sensor from the timing belt cover on a 3.5L V6, remove the sensor mounting bolt

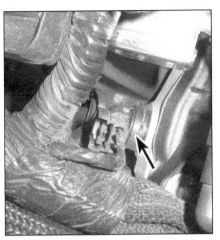

5.5g The intake CMP sensor on the 2.4L engine is located at the left front end of the cylinder head (2014 and earlier)

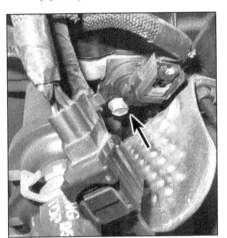

5.5h The exhaust CMP sensor on the 2.4L engine is located at the left rear end of the cylinder head (2014 and earlier)

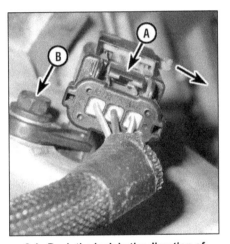

6.4a Push the lock in the direction of the arrow, depress the release tab (A), disconnect the electrical connector and remove the sensor mounting bolt (B) – 2.7L V6 engine

6.4b Location of the CKP on a 2.4L engine (2014 and earlier)

(2014 and earlier 2.4L models), on the transaxle housing (2.7L models), on the driver's side of the vehicle, above the differential housing (3.5L models), or on the front-left end of the cylinder block (2015 and later 2.4L models).

1   Disconnect the negative battery cable from the remote ground terminal or battery (see Chapter 5).
2   On all but 2015 and later models, raise the vehicle and support it securely on jackstands.
3   Remove the heat shield (if equipped) from the sensor.
4   Disconnect the electrical connector from the sensor **(see illustrations)**.
5   Remove the sensor mounting bolt and remove the sensor.
6   If you're going to reinstall the same sensor, check the O-ring for damage.
7   Apply a film of clean engine oil to the O-ring and insert the sensor into the cylinder block.
8   Install the mounting bolt and tighten it to the torque listed in this Chapter's Specifications.

3.6L engine

Note: *The CKP sensor is located at the left rear side of the cylinder block and is accessed from underneath.*
9   Disconnect the negative battery cable from the remote ground terminal or battery (see Chapter 5).
10   Raise the vehicle and support it securely on jackstands.
11   Pull back the heat shield (if equipped) from the sensor.
12   Disconnect the electrical connector from the sensor **(see illustration)**.
13   Unscrew the mounting bolt and remove the sensor from the engine block.
14   If you're going to reinstall the same sensor, check the O-ring for damage. If it's OK, it can be reused.
15   Apply a film of clean engine oil to the O-ring, then insert the sensor into the bell-housing.

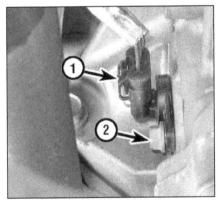

6.12 CKP sensor electrical connector retaining tab (1) and mounting bolt (2) (3.6L engine)

16   Install the mounting bolt and tighten it to the torque listed in this Chapter's Specifications.
Caution: *Before tightening the CKP sensor mounting bolt, make sure that the sensor is flush to the cylinder block.*

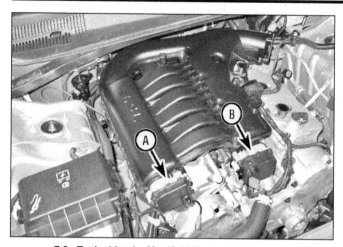

7.3a Typical Intake Manifold Tuning (IMT) system
(3.5L V6 engine shown, 2.7L engine similar):

A   *Manifold tuning valve actuator (used on 2.7L and 3.5L engines)*
B   *Short runner valve actuator (used only on 3.5L engines)*

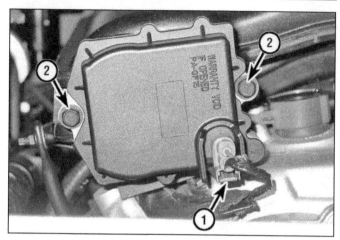

7.3b To remove the intake manifold tuning valve actuator from
the engine, depress the release tab (1), disconnect the electrical
connector and remove the two mounting bolts (2)
(3.5L engine shown, 2.7L similar)

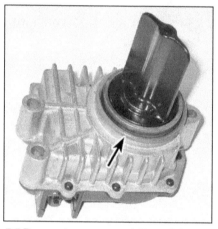

7.5 Be sure to remove and discard the old
actuator O-ring from the intake manifold
tuning valve actuator (shown) or short
runner valve actuator. Use a new O-ring
whether you're installing the old actuator
or a new unit

7   Manifold tuning valve solenoids (2.7L and 3.5L V6 engines) - removal and installation

Warning: *Wait until the engine has cooled completely before beginning this procedure.*

1   Disconnect the negative battery cable from the remote ground terminal (see Chapter 5).

2   Remove engine cover.

3   Disconnect the electrical connector from the intake manifold tuning valve **(see illustrations)**.

4   Unscrew the mounting bolts and remove the manifold tuning valve actuator from the intake manifold.

5   Remove the old O-ring from the manifold tuning valve and discard it **(see illustration)**. Install a new O-ring to the valve.

6   Remove the mounting bolts from the short runner valve actuator and remove the actuator (3.5L engine only) **(see illustration)**.

7   Installation is the reverse of removal.

8   Variable valve timing solenoids (2014 and earlier 2.4L and 3.6L engines) - removal and installation

Warning: *Wait until the engine has cooled completely before beginning this procedure.*

Note: *The variable valve timing solenoids are located at the right end of the exhaust and intake valve covers.*

1   Disconnect the negative battery cable from the remote ground terminal or battery (see Chapter 5).

2   Disconnect the solenoid valve electrical connector.

3   If you're removing more than one solenoid, mark their positions (LI for left intake, LE for left exhaust. etc) **(see illustration)**.

4   Unscrew the solenoid valve mounting bolt(s) and remove the variable valve timing solenoid **(see illustrations)**.

5   Remove the solenoid valve by pulling it straight out of the manifold. Inspect the O-ring for damage. If it's cracked, torn or

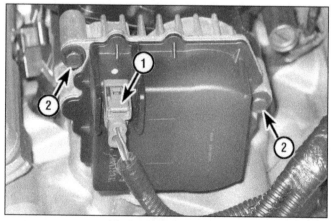

7.6 To remove the short runner valve actuator from a 3.5L engine,
depress the release tab (1), disconnect the electrical connector
and remove the two mounting bolts (2)

8.3 Mark the positions of the variable valve timing solenoids
(3.6L engine shown)

8.4a Remove the Torx screws and twist the solenoid out of the valve cover (3.6L engine shown)

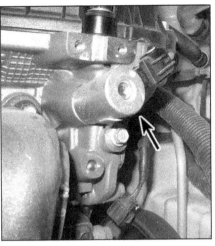

8.4b Variable valve timing solenoid - exhaust (2.4L engine)

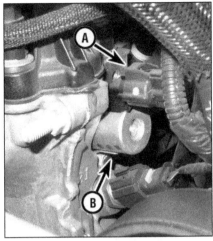

8.4c Variable valve timing solenoid connector (A) and mounting bolt (B) - intake (2.4L engine)

9.3a Slide the red lock (A) up, then depress the retaining tab (B) and pull the connector off (3.6L V6 engine shown)

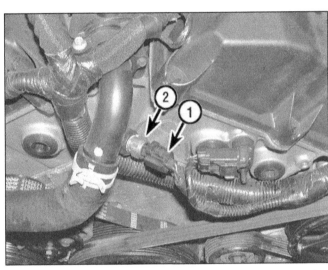

9.3b Depress the release tab (1) and disconnect the electrical connector, then unscrew the ECT sensor (2) (2.7L V6 engine)

otherwise deteriorated, replace it.

6   Lubricate the sensor O-ring with clean engine oil. Installation is the reverse of removal.

9   Engine Coolant Temperature (ECT) sensor - replacement

Warning: *Wait until the engine has cooled completely before beginning this procedure.*

Caution: *Handle the Engine Coolant Temperature (ECT) sensor with care. Damage to the ECT sensor will affect the operation of the entire fuel injection system.*

Note: *2011 and earlier 2.4L engines have two ECT sensors: Sensor 1 is located at the left end of the engine on the coolant adapter housing; Sensor 2 is located on the right front side of the engine block, above the oil*

pressure sensor. *2012 and later 2.4L models only have one ECT sensor, located at the left end of the engine on the coolant adapter housing.*

Note: *On 2.7L engines, the ECT sensor is located at the right end of the engine on the coolant outlet housing.*

Note: *On 3.5L engines, the ECT sensor is located at the left end of the engine, near the throttle body and EGR valve.*

Note: *On 3.6L engines, the ECT sensor is lthreaded into the left front end of the front-bank cylinder head.*

1   Disconnect the negative battery cable from the remote ground terminal or battery (see Chapter 5).

2   Drain the engine coolant to a point lower than that of the sensor (see Chapter 1).

3   Disconnect the electrical connector from the sensor, then unscrew the sensor **(see illustrations)**.

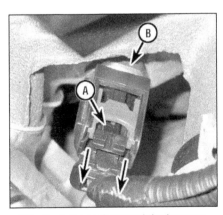

9.3c Push the lock toward the harness, depress the release tab (A) and disconnect the electrical connector, then unscrew the ECT sensor (B) from the lower intake manifold (3.5L shown)

9.3d On 2.4L engines, depress the tab and disconnect the electrical connector . . .

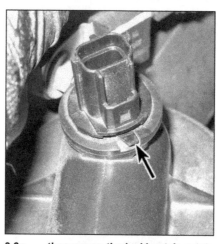

9.3e . . . then pry up the locking tab, rotate the sensor counterclockwise and pull it out of the coolant adapter housing

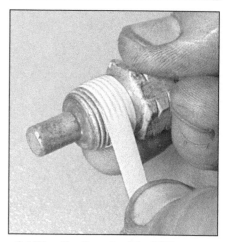

9.4 Wrap the threads of the ECT sensor with Teflon tape to prevent coolant from leaking past the threads

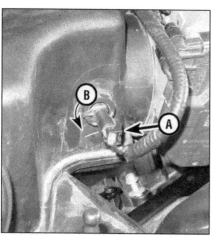

10.1a Depress the tab (A) and disconnect the electrical connector, then twist the IAT sensor 1/4-turn counterclockwise (B) and pull it from the resonator or air filter housing (3.6L engine)

10.1b To disconnect the IAT sensor electrical connector, push the lock toward the harness, then depress the release tab (A) and pull off the connector (3.5L engine)

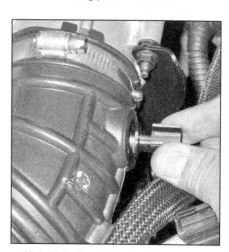

10.1c Note the orientation of the sensor (it must be installed in the exact same orientation), then grasp the sensor firmly and pull it straight out of the air intake duct (3.5L engine)

10.1d The IAT sensor on 2.4L models is located in the air filter housing . . .

10.1e . . . to remove it, twist it counterclockwise and pull it from the housing. When fully installed, the tab on the sensor must point towards the the housing

4   Before installing a threaded ECT sensor, wrap the threads of the sensor with Teflon tape to prevent coolant leakage **(see illustration)**.

5   Installation is otherwise the reverse of removal. Tighten threaded ECT sensors to the torque listed in this Chapter's Specifications.

6   Refill the cooling system (see Chapter 1).

10   Intake Air Temperature (IAT) sensor - replacement

Caution: *The IAT sensor must be installed with the correct orientation to the air duct or housing in order to function properly.*

Note: *The IAT sensor is located on the air filter housing or air inlet ducting.*

1   Disconnect the electrical connector from the IAT sensor **(see illustrations)**.

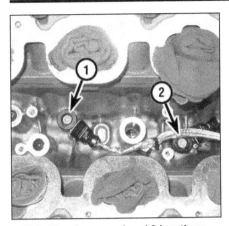

11.6a Knock sensor 1 and 2 locations -
3.6L engine

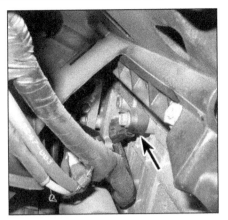

11.6b The knock sensor on the 2.4L
engine is located on the side of the
cylinder block in front of the starter

12.2a To disconnect the electrical
connector, pull out the lock, then
depress the release tab (A)
and detach the connector
(3.5L V6 shown, 2.7L V6 similar)

12.2b To remove the MAP sensor from the
intake manifold, rotate it counterclockwise
and pull it straight up and out of the
manifold (3.5L V6 shown, 2.7L V6 similar)

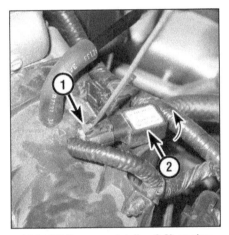

12.2c MAP sensor details - 3.6L engine

1   Connector lock (slide out, then
    depress and unplug connector)
2   MAP sensor (rotate counterclockwise,
    then pull out)

12.2d On the 2.4L engine, the MAP sensor
is located at the front of the engine in the
intake manifold, just above the alternator,
and is retained by a screw

2   Remove the sensor by turning it 1/4-turn
counterclockwise and pulling it out.
3   Inspect the condition of the sensor
O-ring. If it is cracked, torn or otherwise deteriorated, replace it.
4   Installation is the reverse of removal.

---

11   Knock sensor - replacement

Warning: *Wait for the engine to cool completely before performing this procedure.*
Note: *On 2014 and earlier 2.4L engines, the
knock sensor is located on the side of the cylinder block in front of the starter, under the intake manifold. On 2015 and later models, it is
located at the front of the engine block, under
the exhaust manifold.*
Note: *On V6 engines, the knock sensors are
located in the valley between the cylinder
heads. 2.7L and 3.5L engines have one knock
sensor; 3.6L engines have two.*

Note: *On 3.6L engines, the sensor nearest the
front of the engine is Sensor no. 1 and the sensor nearest the rear of the engine is Sensor no.
2. Due to the work involved, both knock sensors should be replaced at the same time.*
1   On V6 engines, relieve the fuel system
pressure (see Chapter 4) and drain the cooling system (see Chapter 1).
2   Disconnect the negative battery cable
from the remote ground terminal or battery (see
Chapter 5).
3   On 2007 2.7L engines, remove the cylinder head from the rear cylinder bank (see
Chapter 2B).
4   On 2008 and later 2.7L engines and
all 3.5L and 3.6L engines, remove the lower
intake manifold (see Chapter 2B, 2C, 2D). On
3.6L engines, also remove the engine oil filter
housing.
5   On 2.4L engines, remove the intake
manifold (2014 and earlier) or the exhaust
manifold (2015 and later) (see Chapter 2A).

6   Disconnect the electrical connector(s)
from the knock sensor(s) **(see illustrations).**
7   Remove the mounting bolt(s) and detach
the sensor(s) from the engine block.
8   Installation is the reverse of removal.
Tighten the knock sensor to the torque listed
in this Chapter's Specifications.
9   On V6 engines, refill the cooling system.

---

12   Manifold Absolute Pressure
(MAP) sensor - replacement

Note: *The MAP sensor is located on the upper
intake manifold.*
1   On 3.6L engines, remove the engine
cover by pulling it up off the ballstuds.
2   Disconnect the electrical connector from
the MAP sensor **(see illustrations).** On V6
models, twist the sensor to remove it from the
manifold. On 2.4L engines, remove the screw
and pull the sensor from the manifold.

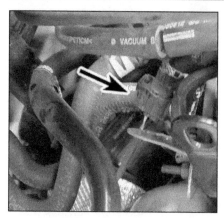

13.4 Upstream oxygen sensor connector (2.4L engine)

13.5 Disconnect the upstream oxygen sensor connector from the engine wiring harness

13.6 The upstream oxygen sensor is located in the exhaust pipe near the exhaust manifold (2.4L engine)

3   Inspect the MAP sensor O-ring for cracks, tears and deterioration; if it's damaged, replace it
4   Installation is the reverse of removal.

13 Oxygen sensors - general information and replacement

1   Be particularly careful when servicing an oxygen sensor:

a) *Oxygen sensors have a permanently attached pigtail and an electrical connector that cannot be removed. Damaging or removing the pigtail or electrical connector will render the sensor useless.*

b) *Keep grease, dirt and other contaminants away from the electrical connector and the louvered end of the sensor.*

c) *Do not use cleaning solvents of any kind on an oxygen sensor.*

d) *Oxygen sensors are extremely delicate. Do not drop a sensor or handle it roughly.*

e) *Make sure that the silicone boot on the sensor is installed in the correct position. Otherwise, the boot might melt and it might prevent the sensor from operating correctly.*

Note: *Because it is installed in the exhaust manifold, catalytic converter or pipe, all of which contract when cool, an oxygen sensor might be very difficult to loosen when the engine is cold. Rather than risk damage to the sensor, start and run the engine for a minute or two, then shut it off. Be careful not to burn yourself during the following procedure.*
Note: *Use an oxygen sensor socket, if available, for removal and installation of oxygen sensors.*
Note: *On 2.4L engines, the upstream sensor is located near the exhaust manifold. On all other engines, the left and right upstream sensors are located at the top of the catalytic converters.*
Note: *The downstream sensors are located on the side of each catalytic converter.*

2   Disconnect the negative battery cable from the remote ground terminal or battery (see Chapter 5).

2.4L engine (2014 and earlier)
Upstream oxygen sensor

3   Remove the engine exhaust manifold cover.
4   Disconnect the oxygen sensor wire harness mounting clips from the engine or body, if equipped **(see illustration)**.

5   Disconnect the oxygen sensor connector from the engine wiring harness **(see illustration)**.
6   Remove the oxygen sensor from the exhaust pipe **(see illustration)**.
7   Installation is the reverse of removal.

Downstream oxygen sensor
8   Disconnect the oxygen sensor connector mounting clips from the engine or body, if equipped.
9   Disconnect the oxygen sensor connector from the engine wiring harness **(see illustration)**.
10   Raise and support vehicle.
11   Remove the oxygen sensor from the catalytic converter **(see illustration)**.
12   Installation is the reverse of removal.

V6 engines
13   Raise the vehicle and place it securely on jackstands.
14   Trace the electrical lead from the oxygen sensor to the electrical connector and disconnect it.
15   Using a special oxygen sensor socket, if available, unscrew the sensor **(see illustrations)**.
16   After removing the old sensor, clean the threads of the sensor bore.

13.9 Downstream oxygen sensor connector is located near the upstream sensor (2.4L engine)

13.11 Downstream oxygen sensor is located in the side of the catalytic converter (2.4L engine)

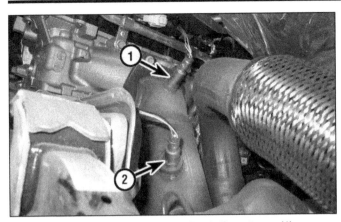

13.15a Front cylinder bank (bank 2) upstream (1) and downstream (2) oxygen sensors (3.6L engine shown)

13.15b Unscrew the downstream oxygen sensor. An oxygen sensor socket is being used to unscrew the downstream sensor, but there's plenty of room to work here, so you could also use a wrench on this sensor

17   If you're going to install the old sensor, apply anti-seize compound to the threads of the sensor to facilitate future removal. If you're going to install a new oxygen sensor, it's not necessary to apply anti-seize compound to the threads. The threads on new sensors already have anti-seize compound on them.

18   Install the oxygen sensor and tighten it to the torque listed in this Chapter's Specifications.

19   Installation is otherwise the reverse of removal.

2.4L engine (2015 and later)

Note: *Both oxygen sensor (upstream and downstream) wiring harness connectors are mounted in the same location, and the sensors themselves are located near each other on the exhaust manifold and catalytic converter at the front of the engine.*

20   Trace the electrical lead from the upstream or downstream oxygen sensor to the electrical connector and disconnect it. Also detach the wiring harness from any clips.

21   If equipped, remove the wiring harness metal protector piece from the oxygen sensor.

22   Using a special oxygen sensor socket, if available, unscrew the sensor **(see illustration 13.15b)**.

23   If you're going to install the old sensor, apply anti-seize compound to the threads of the sensor to facilitate future removal. If you're going to install a new oxygen sensor, it's not necessary to apply anti-seize compound to the threads. The threads on new sensors already have anti-seize compound on them.

24   Install the oxygen sensor and tighten it to the torque listed in this Chapter's Specifications.

25   Installation is otherwise the reverse of removal.

14   Throttle Position (TP) sensor - replacement

The TP sensor is an integral component of the electronic throttle body, and is not sepa-

rately serviceable. If you need to replace the TP sensor, you must replace the throttle body (see Chapter 4).

15   Transmission Range (TR) and transmission temperature sensors - replacement

The TR sensor and transmission temperature sensor (which is an integral part of the TR sensor) are located on the automatic transaxle valve body. In order to replace the TR sensor/transmission temperature sensor, you must remove the valve body, which is beyond the scope of the home mechanic.

16   Transaxle speed sensors - replacement

Note: *On 948TE/9HP48 (9-speed) transaxles, the speed sensors are an integral part of the automatic transaxle valve body. In order to replace either of the speed sensors, you must remove the valve body.*

40TE and 41TE transaxles

Note: *Transaxle speed sensors are located on the front side of the transaxle. The sensor closest to the engine is the input sensor, and the sensor to the far left (driver's side) is the output speed sensor.*

1   Disconnect the negative battery cable from the remote ground terminal or battery (see Chapter 5). Remove the air filter housing (see Chapter 4).

Input speed sensor

2   Disconnect the electrical connector from the sensor **(see illustration)**.

3   Unscrew the sensor from the case

4   If you're going to install the same sensor, remove the old O-ring from the sensor and install a new one.

5   Installation is the reverse of removal. Tighten the sensor to the torque listed in this Chapter's Specifications.

Output speed sensor

6   Disconnect the electrical connector from the sensor **(see illustration)**.

7   Unscrew the sensor from the case.

8   If you're going to install the same sensor,

16.2 Input speed sensor (40TE shown) 2.4L engine, sensor located near engine

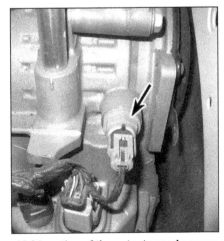

16.6 Location of the output speed sensor (40TE shown)

16.12 Location of the transaxle input speed sensor (62TE transaxle)

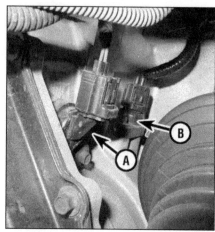

16.16 Location of the transaxle output speed sensor (A) and transfer shaft speed sensor (B) (62TE transaxle)

remove the old O-ring from the sensor and install a new one.

9   Installation is the reverse of removal. Tighten the sensor to the torque listed in this Chapter's Specifications.

62TE transaxle

10   Disconnect the negative battery cable from the remote ground terminal or battery (see Chapter 5).

11   Remove the air filter housing (see Chapter 4).

Input speed sensor

Note: *This sensor is located on the top of the transaxle.*

12   Disconnect the electrical connector from the input speed sensor **(see illustration)**.

13   Remove the retaining bolt and remove the sensor from the case.

14   If you're going to install the same sensor, remove the old O-ring from the sensor and install a new one.

15   Installation is the reverse of removal. Tighten the sensor to the torque listed in this Chapter's Specifications.

Output speed sensor/transfer shaft speed sensor

Note: *These sensors are located on the left rear side of the transaxle. On some models they may be more easily accessed by removing the left front wheel and inner fender splash shield.*

16   Disconnect the electrical connector from the sensor **(see illustration)**.

17   Remove the retaining bolt and remove the sensor from the case.

18   If you're going to install the same sensor, remove the old O-ring from the sensor and install a new one.

19   Installation is the reverse of removal. Tighten the sensor to the torque listed in this Chapter's Specifications. Tighten the wheel lug nuts to the torque listed in the Chapter 1 Specifications.

17   Powertrain Control Module (PCM) - replacement

Caution: *To avoid electrostatic discharge damage to the PCM, handle the PCM only by its case. Do not touch the electrical terminals during removal and installation. If available, ground yourself to the vehicle with an anti-static ground strap, available at computer supply stores.*

Note: *This procedure applies only to disconnecting, removing and installing the PCM that is already installed in your vehicle. If, however, you need to replace the PCM, it must be programmed with new software and calibrations, and information from the old PCM must be transferred to the new one. This will require the use of a special scan tool, so you will not be able to replace the PCM at home.*

Note: *The PCM is located in the left-rear corner of the engine compartment.*

1   Disconnect the negative battery cable from the remote ground terminal or battery (see Chapter 5).

2   If required, remove the air filter housing cover.

3   Unlock the electrical connectors and disconnect them from the PCM **(see illustration)**.

4   Remove the PCM mounting bolts and remove the PCM.

5   Installation is the reverse of removal.

18   Catalytic converter - replacement

Warning: *Wait until the engine has cooled completely before beginning this procedure.*

Warning: *Disconnect the negative cable from the remote ground terminal or battery before beginning (see Chapter 5).*

1   Raise the vehicle and place it securely on jackstands. Remove the engine under cover

2   Before trying to loosen the nuts and bolts at the flange(s) and the clamp bolt and nut

17.3 PCM mounting details

1   *Connector latches (flip up to release connectors)*
2   *Mounting bolts*

18.5 Catalytic converter-to-exhaust manifold flange nuts (2.4 engine)

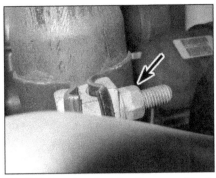

18.55 Exhaust under-pipe-to-rear exhaust pipe clamp (3.6L models)

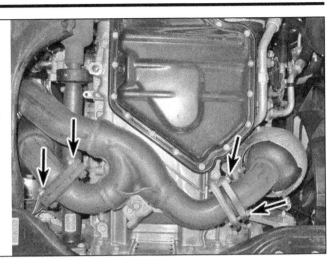

18.56 Exhaust under-pipe-to-catalytic converter fasteners (3.6L models)

behind the converter (on models so equipped), spray them with penetrating oil and wait the specified amount of time (see the instructions on the can) for the penetrant to loosen things up.

2.4L engines

3   On 2014 and earlier models, remove the oxygen sensor from the converter (see Section 13). On 2015 and later models, disconnect both oxygen sensor electrical connectors (see Section 13).
4   On 2014 and earlier models, remove the rear portion of the exhaust system. On 2015 and later models, disconnect the exhaust pipe from the catalytic converter by removing the two flange bolts. Also remove the catalytic converter support bracket bolt.
5   Remove the flange nuts at the exhaust manifold and detach the converter from the exhaust manifold **(see illustration)**.
6   Before installing the converter, coat the threads of the exhaust manifold flange nuts and bolts, and the clamp bolt with anti-seize compound. Tighten the fasteners securely.
7   Installation is otherwise the reverse of removal.

2.7L engines

Note: *The catalytic converters (front and rear) are incorporated into the exhaust manifolds.*

Front exhaust manifold/catalytic converter

8   Remove the engine cover.
9   Disconnect the upstream and downstream oxygen sensor engine wiring harness connectors.
10   Remove the engine oil level indicator tube.
11   Remove the front upper converter heat shield fasteners and remove the heat shield.
12   Remove the upstream and downstream oxygen sensors (see Section 13).
13   Remove the under-vehicle splash shield.
14   Remove the front catalytic converter-to-crossover fasteners.
15   Lower the vehicle and remove the front exhaust manifold/converter.
16   Before installing the converter, coat the threads of the exhaust manifold flange nuts and bolts, and the clamp bolt with anti-seize compound. Tighten the fasteners securely.
17   Installation is otherwise the reverse of removal.

Rear exhaust manifold/catalytic converter

18   Remove the engine cover.
19   Loosen the lower EGR tube flange nut, but don't remove it.
20   Raise the vehicle and support it securely on jackstands.
21   Remove the under-vehicle splash shield.
22   Remove the exhaust system (see Chapter 4).
23   Remove the crossunder pipe.
24   Lower the vehichle and remove the lower EGR tube.
25   Disconnect the rear upstream and downstream oxygen sensor engine wiring harness connector.
26   Remove the rear exhaust manifold heat shield.
27   Raise the vehicle and support it securely on jackstands.
28   Remove the upstream and downstream oxygen sensors (see Section 13).
29   Remove the rear catalytic converter fasteners and converter.
30   Before installing the converter, coat the threads of the exhaust manifold flange nuts and bolts, and the clamp bolt with anti-seize compound. Tighten the fasteners securely.
31   Installation is otherwise the reverse of removal.

3.5L engines

Note: *The catalytic converters (front and rear) are incorporated into the exhaust manifolds.*

Front exhaust manifold/catalytic converter

32   Remove the under-vehicle splash shield.
33   Remove the three fasteners and remove the upper heat shield.
34   Remove the under-vehicle splash shield.
35   Remove the front exhaust cross-under pipe bolts.
36   Disconnect and remove the front downstream oxygen sensor.
37   Loosen the oil level indicator tube retaining bolt and position it out of the way.
38   Disconnect and remove the front upstream oxygen sensor.

39   Remove the catalytic converter retaining bolts and remove the converter.
40   Before installing the converter, coat the threads of the exhaust manifold flange nuts and bolts, and the clamp bolt with anti-seize compound. Tighten the fasteners securely.
41   Installation is otherwise the reverse of removal.

Rear exhaust manifold/catalytic converter

42   Remove the right front driveaxle (see Chapter 8).
43   Remove the retainer nuts at the extension pipe.
44   Remove the retainer from the bracket to the transaxle bellhousing.
45   Remove the center crossmember.
46   Remove the exhuast manifold crossunder pipe.
47   Remove the front extension pipe.
48   Disconnect and remove the downstream oxygen sensor.
49   Remove the lower heat shield retainers and upper heat shield nuts.
50   Disconnect the upstream oxygen sensor and remove the heat shields.
51   Remove the exhaust/catalytic converter bolts and remove the converter.
52   Before installing the converter, coat the threads of the exhaust manifold flange nuts and bolts, and the clamp bolt with anti-seize compound. Tighten the fasteners securely.
53   Installation is otherwise the reverse of removal.

3.6L engines

54   If you're removing the rear catalytic converter, remove the right front wheel and the right driveaxle and intermediate shaft (see Chapter 8).
55   If you're removing the rear catalytic converter, loosen the exhaust under-pipe-to-rear exhaust pipe clamp **(see illustration)**, then separate the exhaust pipe from the under-pipe. Support the rear portion of the exhaust system with a floor jack.
56   Remove the exhaust under-pipe-to-catalytic converter fasteners **(see illustration)**. If you're removing the rear converter, remove

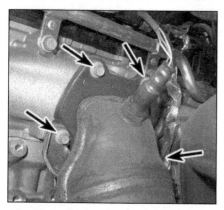

18.57 Catalytic converter-to-cylinder head bolts (one not visible here)

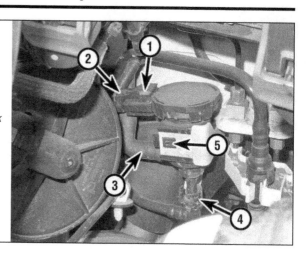

19.2 Canister purge solenoid details (typical)

1 *Electrical connector lock*
2 *Electrical connector release tab*
3 *Purge hose*
4 *Fuel tank vapor hose*
5 *Solenoid retaining tab*

the under-pipe fasteners from both converters and remove the under-pipe completely.
57 Remove the converter-to-cylinder head bolts **(see illustration)** and guide the converter out from the bottom of the vehicle.
58 Before installing the converter, coat the threads of the exhaust manifold flange nuts and bolts, and the clamp bolt with anti-seize compound. Tighten the fasteners securely.
59 Installation is otherwise the reverse of removal.

19 Evaporative Emissions Control (EVAP) system - component replacement

EVAP canister purge solenoid
Note: *The EVAP canister purge solenoid is located on the left side of the firewall, below the power brake booster.*
1 Remove the Powertrain Control Module (see Section 17).
2 Clearly label the EVAP hoses to ensure correct reassembly **(see illustration)**, then disconnect them from the solenoid.

3 Slide out the red lock, depress the tab and disconnect the electrical connector from the EVAP canister purge solenoid.
4 Depress the retaining tab and detach the purge solenoid from the mounting bracket.
5 Installation is the reverse of removal.

EVAP canister
Note: *The EVAP canister is located behind the left rear wheel opening.*
6 Raise the vehicle and place it securely on jackstands.
7 Disconnect the electrical connector from the EVAP canister.
8 Disconnect the hoses from the EVAP canister **(see illustration)**.
9 Remove the mounting fasteners from the mounting bracket and remove the EVAP canister assembly **(see illustration)**.
10 Installation is the reverse of removal.

20 Exhaust Gas Recirculation (EGR) system - component replacement

Warning: *Make sure that the engine is cool before removing any EGR component.*

Note: *This Section applies to 2.7L and 3.5L models only; 2.4L and 3.6L engines are not equipped with an EGR system.*
Note: *The EGR valve is bolted to the rear end of the rear-bank cylinder head. The lower EGR pipe connects the exhaust manifold to the EGR valve. The upper EGR pipe connects the EGR valve to the intake manifold.*
1 Remove the EVAP purge solenoid valve (see Section 19).
2 Disconnect the electrical connector from the EGR valve **(see illustration)** and set it aside.

Lower EGR tube
3 Remove the lower EGR tube nut from the exhaust manifold.
4 Remove the lower EGR tube bolts from the EGR valve and remove the lower EGR tube.
5 Installation is the reverse of removal.

Upper EGR tube
Note: *The photos accompanying this Section depict the EGR system on a 3.5L V6 model. Except for the lower EGR pipe (see above), the systems on 2.7L and 3.5L engines are virtually identical.*

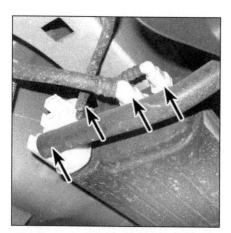

19.8 Remove all the hoses from the EVAP canister, taking care not to damage the plastic fittings

19.9 EVAP canister mounting bolt

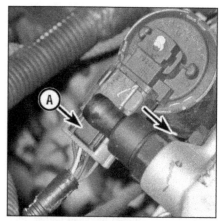

20.2 To disconnect the electrical connector from the EGR valve, slide the lock sideways, then depress the release tab (A) and pull off the connector

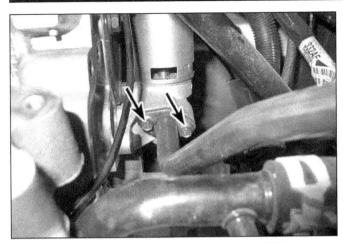

20.6 To disconnect the lower end of the upper EGR pipe from the EGR valve, remove these two bolts (3.5L V6 shown, 2.7L V6 similar)

20.7a Grasp the EGR pipe firmly and pull it out

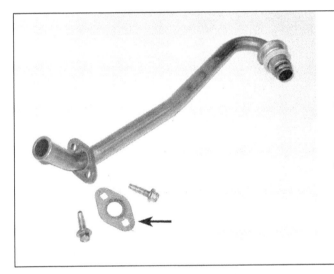

20.7b Always replace the gasket between the flange at the lower end of the upper EGR pipe, and the mounting surface on the EGR valve

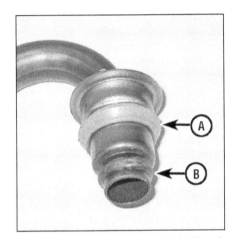

20.8 Always replace the dust seal (A) and O-ring (B) on the upper (intake manifold) end of the upper EGR pipe (3.5L V6 shown, 2.7L V6 similar)

6   Remove the two bolts that secure the lower end of the upper EGR pipe to the EGR valve **(see illustration)**.

7   Disconnect the upper end of the upper EGR pipe from the intake manifold **(see illustration)** and remove the upper EGR pipe. Remove the old gasket between upper pipe and EGR valve **(see illustration)**.

8   Remove the old rubber silicone seals from the EGR pipe **(see illustration)**. The smaller black seal is the actual O-ring; the larger blue seal is a dust seal. To ensure a positive seal, you should always replace both of these seals whenever you disconnect the EGR pipe.

9   Installation is the reverse of removal.

EGR valve

10   Disconnect the upper EGR pipe and, on 2.7L V6 engines, the lower EGR pipe, from the EGR valve.

11   Remove the EGR mounting bolts **(see illustration)** and remove the EGR valve from the right cylinder head.

12   Installation is the reverse of removal.

21   Positive Crankcase Ventilation (PCV) system

1   The Positive Crankcase Ventilation (PCV) system reduces hydrocarbon emissions by scavenging crankcase vapors. It does this by circulating fresh air from the air filter housing through the crankcase, where it mixes with blow-by gases, before being drawn through a PCV valve into the intake manifold.

2   The PCV system consists of the PCV valve and two hoses. The fresh air inlet hose connects the air filter housing to the valve cover. The crankcase ventilation hose (or PCV hose) connects the rear valve cover to the intake manifold. The PCV valve is located at the valve cover end of the crankcase ventilation hose that connects to the intake manifold.

3   To maintain idle quality, the PCV valve restricts the flow when the intake manifold vacuum is high. If abnormal operating conditions (such as piston ring problems) arise, the system is designed to allow excessive amounts of blow-by gases to flow back through the crankcase vent tube into the air cleaner to be con-

sumed by normal combustion.

4   Checking and replacement of the PCV valve is covered in Chapter 1.

20.11 To detach the EGR valve from the cylinder head, remove these two bolts (3.5L V6 shown, 2.7L V6 similar)

22.9 Push the ends of the fitting together to detach the air hose from the check valve

22.10 Check valve mounting bolts

22 Air injection system (2.4L PZEV engines) - component replacement

1   Disconnect the negative battery cable from the remote ground terminal or battery (see Chapter 5). Remove the engine cover.

Air injection pump

Note: *The air pump is located behind the left end of the engine, above the rear portion of the transaxle case.*

2   Remove the Powertrain Control Module (PCM) (see Section 17). Also remove the PCM mounting bracket.

3   Detach the tubes from the air pump by squeezing the collars of the quick-connect fittings.

4   Raise the vehicle and support it securely on jackstands.

5   Remove the bolt securing the air pump relay to the pump, then reposition the heat shield.

6   Lower the vehicle and remove the heat shield upper mounting bolt. Detach the shield.

7   Disconnect the electrical connector from the air pump, remove the three mounting bolts and detach the air pump from its bracket.

8   Installation is the reverse of the removal procedure.

Check valve

9   Disconnect the air hose from the check valve **(see illustration)**.

10   Remove the check valve mounting bolts and detach the valve from the exhaust manifold pipe **(see illustration)**.

11   Clean the exhaust manifold pipe flange (and the check valve flange, if the same one is to be reinstalled).

12   Installation is the reverse of removal. Be sure to use a new gasket and tighten the bolts to the torque listed in this Chapter's Specifications.

23 Oil temperature sensor - removal and installation

2.4L engine

Note: *The oil temperature sensor is located at the left front corner of the cylinder head, behind the power steering pump (if equipped).*

1   Remove the engine cover.

2   Depress the tab and disconnect the electrical connector from the sensor **(see illustration)**.

3   Unscrew the sensor from the cylinder head.

4   If necessary for clearance, remove the drivebelt (see Chapter 1), loosen the power steering pump mounting bolts (see Chapter 10) and pivot the pump forward to provide extra room.

5   Before installing the new sensor, wrap the threads with Teflon sealing tape or coat the threads with thread sealant.

6   Thread the sensor into the cylinder head, then tighten it to the torque listed in this Chapter's Specifications. Reconnect the electrical connector.

3.6L engine

Note: *The oil temperature sensor is located under the lower intake manifold, at the left end of the engine, threaded into the oil filter housing.*

7   Relieve the fuel system pressure (see Chapter 4).

23.2 Location of the oil temperature sensor (2.4L engine) (2014 and earlier 2.4L model shown, 2015 and later 2.4L model similar)

8   Disconnect the negative battery cable from the remote ground terminal (see Chapter 5).

9   Remove the upper and lower intake manifolds (see Chapter 2D).

10   Disconnect the electrical connector from the sensor, then unscrew the sensor from the oil filter housing.

11   Before installing the new sensor, wrap the threads with Teflon sealing tape or coat the threads with thread sealant.

12   Thread the sensor into the cylinder head, then tighten it to the torque listed in this Chapter's Specifications. Reconnect the electrical connector.

13   The remainder of installation is the reverse of the removal procedure.

Chapter 7 Part A
Automatic transaxle

Contents

	Section
Automatic transaxle - removal and installation	8
Automatic transaxle overhaul - general information	9
Brake Transmission Shift Interlock (BTSI) system - description, check, replacement and adjustment	5
Diagnosis - general	2

	Section
Driveaxle oil seals - replacement	3
General information	1
Shift cable - removal, installation and adjustment	4
Shift lever - removal and installation	6
Transaxle oil cooler - removal and installation	7

Specifications

General

Lubricant type and capacity	See Chapter 1

Torque specifications
Ft-lbs (unless otherwise indicated)

Note: *One foot-pound (ft-lb) of torque is equivalent to 12 inch-pounds (in-lbs) of torque. Torque values below approximately 15 ft-lbs are expressed in inch-pounds, because most foot-pound torque wrenches are not accurate at these smaller values.*

40TE / 41TE

Rear mount through-bolt	55
Shift cable adjustment lever bolt	16.5
Torque converter-to-driveplate bolts	65
Transaxle-to-engine bolts	55

62TE

Left transaxle mount bolts	
To bracket	75
To transaxle	75
Lower transaxle bracket-to-engine bolts	52
Lower transaxle mount bolts	52
Shift cable adjustment lever bolt	16.5
Stamped transaxle mounting brace	
Brace-to-bellhousing bolts	52
Brace-to-upper transaxle mount	37
Torque converter-to-driveplate bolts	65
Transaxle-to-engine bolts	52

948TE/9HP48 (9-speed)

Transaxle mounts	
Pivot bracket-to-transaxle	118
Transaxle bracket-to-transaxle	
M6 bolts	108 in-lbs
M12 bolts	77
Torque converter-to-driveplate bolts	30
Transaxle bellhousing-to-engine bolts (M10)	37

1   General information

1   These models are equipped with the 40TE (4-speed), 41TE (four-speed), 62TE (6-speed), or the 948TE/9HP48 (9-speed) automatic transaxle. The automatic transaxle and the differential are housed in a compact, lightweight, two-piece aluminum alloy housing.

2   These models are equipped with a Transmission Control Module (TCM) which is the brain of the transaxle. The TCM monitors engine and transaxle operating parameters through numerous sensors, then generates output signals to various relays and solenoids to regulate hydraulic pressures, optimize drivability, provide efficient torque management and maintain maximum fuel economy. All models incorporate the TCM into the PCM. The TCM is part of the On-Board Diagnostic system (OBD-II). For more information, see Chapter 6.

3   Because of the complexity of the automatic transaxles and the specialized equipment necessary to perform most service operations, this Chapter contains only those procedures related to general diagnosis, adjustment and removal and installation procedures.

4   If the transaxle requires major repair work, it should be left to a dealer service department or an automotive or transmission repair shop. Once properly diagnosed you can, however, remove and install the transaxle yourself and save the expense, even if the repair work is done by a transmission shop.

2   Diagnosis - general

1   Automatic transaxle malfunctions may be caused by five general conditions:

a) *Poor engine performance*
b) *Improper adjustment*
c) *Hydraulic malfunctions*
d) *Mechanical malfunctions*
e) *Malfunctions in the computer or its signal network*

2   Diagnosis of these problems should always begin with a check of the easily repaired items: fluid level and condition (see Chapter 1), shift cable adjustment and shift lever installation. Next, perform a road test to determine if the problem has been corrected or if more diagnosis is necessary. If the problem persists after the preliminary tests and corrections are completed, additional diagnosis should be performed by a dealer service department or other qualified transmission repair shop. Refer to *Troubleshooting* at the beginning of this manual for information on symptoms of transaxle problems.

Preliminary checks

3   Drive the vehicle to warm the transaxle to normal operating temperature.

4   Check the fluid level as described in Chapter 1:

a) *If the fluid level is unusually low, add enough fluid to bring the level within the designated area of the dipstick, then check for external leaks (see following).*

b) *If the fluid level is abnormally high, drain off the excess, then check the drained fluid for contamination by coolant. The presence of engine coolant in the automatic transaxle fluid indicates that a failure has occurred in the internal radiator oil cooler walls that separate the coolant from the transaxle fluid (see Chapter 3).*

c) *If the fluid is foaming, drain it and refill the transaxle, then check for coolant in the fluid, or a high fluid level.*

5   Check the engine idle speed.

Note: *If the engine is malfunctioning, do not proceed with the preliminary checks until it has been repaired and runs normally.*

6   Check and adjust the shift cable, if necessary (see Section 4).

7   If hard shifting is experienced, inspect the shift cable under the center console and at the manual lever on the transaxle (see Section 4).

Fluid leak diagnosis

8   Most fluid leaks are easy to locate visually. Repair usually consists of replacing a seal or gasket. If a leak is difficult to find, the following procedure may help.

9   Identify the fluid. Make sure it's transaxle fluid and not engine oil or brake fluid (automatic transaxle fluid is a deep red color).

10   Try to pinpoint the source of the leak. Drive the vehicle several miles, then park it over a large sheet of cardboard. After a minute or two, you should be able to locate the leak by determining the source of the fluid dripping onto the cardboard.

11   Make a careful visual inspection of the suspected component and the area immediately around it. Pay particular attention to gasket mating surfaces. A mirror is often helpful for finding leaks in areas that are hard to see.

12   If the leak still cannot be found, clean the suspected area thoroughly with a degreaser or solvent, then dry it thoroughly.

13   Drive the vehicle for several miles at normal operating temperature and varying speeds. After driving the vehicle, visually inspect the suspected component again.

14   Once the leak has been located, the cause must be determined before it can be properly repaired. If a gasket is replaced but the sealing flange is bent, the new gasket will not stop the leak. The bent flange must be straightened.

15   Before attempting to repair a leak, check to make sure that the following conditions are corrected or they may cause another leak.

Note: *Some of the following conditions cannot be fixed without highly specialized tools and expertise. Such problems must be referred to a qualified transmission shop or a dealer service department.*

Gasket leaks

16   Check the pan periodically. Make sure the bolts are tight, no bolts are missing, the gasket is in good condition and the pan is flat (dents in the pan may indicate damage to the valve body inside).

17   If the pan gasket is leaking, the fluid level or the fluid pressure may be too high, the vent may be plugged, the pan bolts may be too tight, the pan sealing flange may be warped, the sealing surface of the transaxle housing may be damaged, the gasket may be damaged or the transaxle casting may be cracked or porous. If sealant instead of gasket material has been used to form a seal between the pan and the transaxle housing, it may be the wrong type of sealant.

Seal leaks

18   If a transaxle seal is leaking, the fluid level may be too high, the vent may be plugged, the seal bore may be damaged, the seal itself may be damaged or improperly installed, the surface of the shaft protruding through the seal may be damaged or a loose bearing may be causing excessive shaft movement.

19   Make sure the dipstick tube seal is in good condition and the tube is properly seated. Periodically check the area around the sensors for leakage. If transaxle fluid is evident, check the seals for damage.

Case leaks

20   If the case itself appears to be leaking, the casting is porous and will have to be repaired or replaced.

21   Make sure the oil cooler hose fittings are tight and in good condition.

Fluid comes out vent pipe or fill tube

22   If this condition occurs, the possible causes are: the transaxle is overfilled; there is coolant in the fluid; the dipstick is incorrect; the vent is plugged or the drain-back holes are plugged.

3   Driveaxle oil seals - replacement

1   The driveaxle oil seals are located on the sides of the transaxle, where the inner ends of the driveaxles are splined into the differential side gears. If you suspect that a driveaxle oil seal is leaking, raise the vehicle and support it securely on jackstands. If the seal is leaking, you'll see lubricant on the side of the transaxle, below the seal.

2   Remove the driveaxle or driveaxle/intermediate shaft (see Chapter 8).

3   Using a screwdriver or prybar, carefully pry the oil seal out of the transaxle bore **(see illustration)**.

4   If the oil seal cannot be removed with a screwdriver or prybar, a special oil seal removal tool (available at auto parts stores) will be required.

5   Using a seal installer, install the new oil seal. Drive it into the bore squarely until it bottoms **(see illustration)**.

3.3 Using a large screwdriver or prybar, carefully pry the oil seal out of the transaxle (if you can't remove the oil seal with a screwdriver or prybar, you may need to obtain a special seal removal tool, available at most auto parts stores) – left side seal on 41TE transaxle shown, 40TE and 62TE similar

3.5 Using a seal installer, large section of pipe or a large deep socket as a drift, drive the new seal squarely into the bore and make sure that it's completely seated; lubricate the lip of the new seal with multi-purpose grease (left side seal on 41TE transaxle shown, 40TE and 62TE similar)

4.3 Pry the shift cable from the manual lever using a trim panel tool or flat-bladed screwdriver

4.4 Pry up the clip, then pull the shift cable up off the bracket on the transaxle

6   Install the driveaxle or intermediate shaft/driveaxle (see Chapter 8).
7   Check the fluid level (see Chapter 1) and adjust as necessary.

4   Shift cable - removal, installation and adjustment

Warning: *The models covered by this manual are equipped with Supplemental Restraint Systems (SRS), more commonly known as airbags. Always disarm the airbag system before working in the vicinity of any airbag system component to avoid the possibility of accidental deployment of the airbag, which could cause personal injury (see Chapter 12). Do not use a memory saving device to preserve the PCM's memory when working on or near airbag system components.*

Warning: *Do not attempt this procedure until the vehicle has cooled completely. The exhaust system components must be cold to avoid physical harm.*

Removal and installation

1   Shift the vehicle into Park.
2   Disconnect the negative battery cable from the remote ground terminal or battery (see Chapter 5).
3   Working in the engine compartment, disconnect the shift cable from the shift lever **(see illustration).**
4   Disconnect the shift cable from the bracket **(see illustration).**
5   Remove the center console (see Chapter 11).
6   Disconnect the shift cable housing at the shifter assembly **(see illustrations).**
7   Raise the vehicle and support it securely on jackstands.
8   Working under the vehicle, remove the

grommet from the floorpan and pull shift cable out of opening.
9   Carefully unfasten the cable from the retainers under the vehicle.
10   Remove the cable from the vehicle.
11   Installation is the reverse of removal.

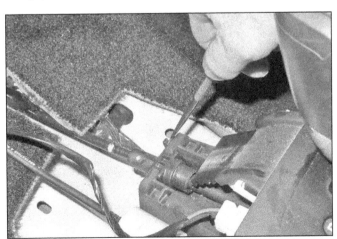

4.6a Remove the clip retaining the shift cable housing from the shift lever

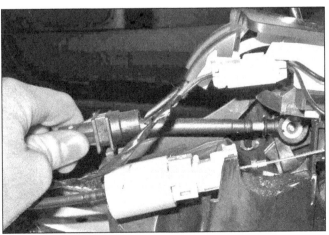

4.6b Remove the cable from the bracket

4.14 Loosen the shift cable adjustment bolt (A) and slide the cable (B) to adjust

Feed the shift cable into the passenger compartment and ensure the grommet is installed properly. Lubricate the grommet if necessary to ensure proper installation.

Adjustment

12   Park the vehicle on a flat surface and set the parking brake.

13   Place the shift lever in the Park position. Remove the key from the ignition.

14   Loosen the shift cable adjustment lever bolt at the transaxle shift lever **(see illustration)**.

15   Make sure the shift lever at the transaxle is in the Park position by pulling it forward all the way. The parking pawl must be engaged when adjusting the cable. If applied, release the parking brake, then rock the vehicle back and forth to ensure that the parking pawl is fully engaged.

16   Tighten the shift cable adjustment lever bolt to the torque listed in this Chapter's Specifications.

17   Check the shift lever for proper operation. It should operate smoothly without binding. The engine should start only in the Park or Neutral positions.

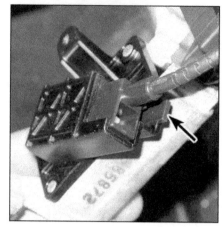

5.12 Depress the retaining tab and disconnect the BTSI cable from the lock cylinder housing

18   Shift the transaxle into all gear positions to make sure the cable is functioning properly. Readjust if necessary.

---

5   Brake Transmission Shift Interlock (BTSI) system - description, check, replacement and adjustment

---

Description

1   The Brake Transmission Shift Interlock (BTSI) system prevents the shift lever from being moved out of PARK unless the brake pedal is depressed. The BTSI system also prevents the ignition key from being turned to the LOCK or ACCESSORY position unless the shift lever is fully locked into the PARK position.

Check

2   Verify that the ignition key can be removed only when the shift lever is in the PARK position.

3   When the shift lever is in the PARK posi-

tion, you should be able to rotate the ignition key from OFF to LOCK. But when the shift lever is in any gear position other than PARK (including NEUTRAL), you should not be able to rotate the ignition key to the LOCK position.

4   You should be able to move the shift lever out of the PARK position when the ignition key is turned to the OFF position.

5   You should not be able to move the shift lever out of the PARK position when the ignition key is turned to the RUN or START position until you depress the brake pedal.

6   With the shifter in any gear selection other than Park, you should not be able to turn the key back to the ACC or LOCK position.

7   Once in gear, with the ignition key in the RUN position, you should be able to move the shift lever between gears, or put it into NEUTRAL or PARK, without depressing the brake pedal.

8   If the BTSI system doesn't operate as described, the cable requires adjustment as outlined below.

Component replacement

BTSI cable

9   Insert the key into the ignition switch and turn it to the ACC position.

10   Disconnect the negative battery cable from the remote ground terminal or battery (see Chapter 5).

11   Remove the driver's side knee bolster and hush panel (see Chapter 11).

12   Detach the BTSI cable from the ignition lock cylinder housing **(see illustration)**.

13   Remove the center floor console from the vehicle (see Chapter 11).

14   Remove the HVAC ducts as necessary to allow removal of the cable.

15   Release the retaining tab on the BTSI cable **(see illustration)**.

16   Disconnect the BTSI cable from the shifter assembly **(see illustration)**.

17   Remove the cable from the vehicle.

18   Installation is the reverse of removal. Adjust the BTSI cable (see Steps 19 through 24).

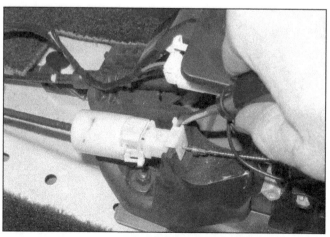

5.15 Use a screwdriver to release the retaining tab on the BTSI cable

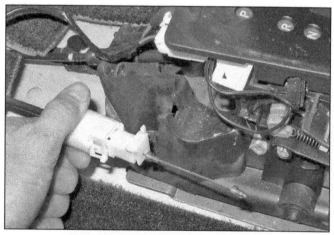

5.16 Disconnect the cable from the shifter base

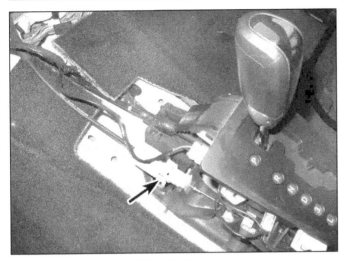

5.22 BTSI cable locking clip

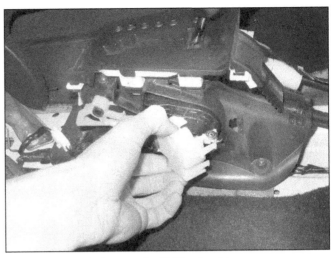

5.27 Remove the BTSI solenoid cover

BTSI cable adjustment

19   Remove the center floor console (see Chapter 11).

20   Place the shifter assembly in the Park position.

21   Insert the key into the ignition and turn it to the Lock position. Remove the key from the ignition switch.

22   NEW CABLE: Remove the lock pin and the cable will adjust automatically. Once adjusted, push the cable locking clip **(see illustration)** down to secure the adjustment.

23   USED/EXISTING CABLE: Pull up on the cable locking clip **(see illustration 5.22)** and the cable will adjust automatically. Once adjusted, push the clip down to secure the adjustment.

24   Verify correct operation as outlined above and readjust as necessary.

BTSI solenoid

25   Disconnect the negative battery cable from the remote ground terminal or battery (see Chapter 5).

26   Remove the center floor console (see Chapter 11).

27   Working on the passenger's side of the shifter assembly, remove the BTSI solenoid cover **(see illustration)**.

28   Disconnect the BTSI solenoid electrical connector, located at the rear of the shifter, and remove the solenoid from the shifter assembly **(see illustration)**.

Note: *Disconnect the BTSI cable from the shifter assembly if necessary (see Steps 15 and 16).*

29   Installation is the reverse of removal.

30   Verify correct operation as outlined above and adjust the BTSI cable if necessary (see Steps 19 through 24).

Steering column BTSI mechanism

31   Remove the upper and lower steering column covers (see Chapter 11).

32   Disconnect the BTSI cable from the steering column BTSI mechanism (see Step 12).

33   Remove the 2 screws attaching the steering column BTSI mechanism.

34   Remove the steering column BTSI mechanism.

35   Installation is the reverse of removal. Adjust the BTSI cable as necessary (see Steps 19 through 24).

6   Shift lever - removal and installation

1   Disconnect the negative battery cable from the remote ground terminal or battery (see Chapter 5).

2   Remove the center floor console (see Chapter 11).

3   Disconnect the electrical connectors to the shifter.

4   Detach the shift cable (see Section 4) and the BTSI cable (see Section 5) from the shift lever.

5   Remove the shifter asssembly mounting bolts **(see illustration)** and detach the shifter from the floor.

6   Installation is the reverse of removal.

7   Adjust the BTSI cable (see Section 5) and the shift cable (see Section 4).

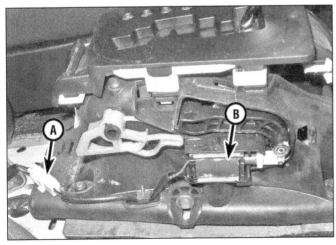

5.28 Disconnect the BTSI solenoid connector (A) and remove the solenoid (B) from the shifter base

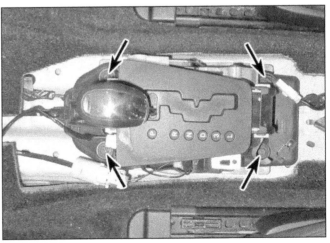

6.5 Shift lever mounting bolts

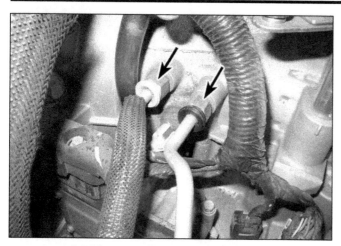

8.6 Disconnect the transaxle cooler lines

8.7 Unplug the transaxle electrical connectors

---

7   Transaxle oil cooler - removal and installation

The transaxle oil cooler and air conditioning condenser are serviced as an assembly. See Chapter 3 for the condenser removal procedure.

---

8   Automatic transaxle - removal and installation

Removal

40TE/41TE transaxles

1   Disconnect the negative battery cable from the remote ground terminal or battery (see Chapter 5).
2   Loosen the front wheel lug nuts and the driveaxle/hub nuts.
Note: *Depending on the type of wheels installed on the vehicle and the thickness of the socket you are using, you may have to loosen the driveaxle/hub nuts after the wheels have been removed (see Chapter 8).*
3   Remove the engine cover and air filter from the vehicle (see Chapter 4).

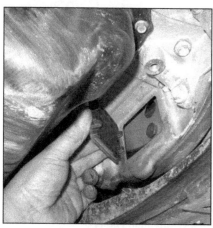

8.24 Torque converter cover removed

4   Remove the engine oil dipstick and plug the hole to prevent contamination.
5   Drain the transaxle fluid (see Chapter 1).
6   Disconnect the transaxle cooler lines from the transaxle **(see illustration)**.
7   Unplug the electrical connectors from the transaxle **(see illustration)**.
8   Working in the engine compartment, disconnect the shift cable from the shift lever and the bracket (see Section 4).
9   Disconnect the Crankshaft Position (CKP) sensor connector (see Chapter 6).
10   Remove the throttle body brace.
11   Remove the oxygen sensor harness bracket from the transaxle.
12   Remove the bolts attaching the rear mount bracket to the transaxle case.
13   Remove the upper starter brace bolt at the engine block (see Chapter 5).
14   Raise the vehicle and support it securely on jackstands. Remove the front wheels.
15   Remove the lower splash shields (see Chapter 1).
16   Remove the front driveaxles (see Chapter 8).
17   On V6 engines, remove the exhaust crossover pipe (see Chapter 4). Support the rear exhaust pipe with wire.
18   Remove the bolts attaching the rear mount bracket and mount to the tranaxle and crossmember and remove the rear mount bracket and mount.
19   Support the engine from above with a hoist or engine support fixture, or place a jack and a block of wood under the transaxle or engine oil pan.
20   Remove the front mount through bolt.
21   Remove the longitudinal crossmember (see Chapter 2E , **illustration 7.36**)
22   Remove the bolts attaching the front mount support and mount to the engine and the transaxle and remove the mount.
23   Remove the starter (see Chapter 5).
24   Remove the torque converter cover **(see illustration)**.
25   Mark the relationship of the torque converter to the driveplate so they can be installed in the same position.

26   Wedge a screwdriver between the teeth on the driveplate and the opening to prevent the engine from rotating, then remove the torque converter-to-driveplate bolts.
27   After all the bolts are removed, push the torque converter into the bellhousing so it doesn't stay with the engine when the transaxle is removed.
28   Support the transaxle with a transmission jack, if available, or use a floor jack. Secure the transaxle to the jack using straps or chains so it doesn't fall off during removal.
29   Lower the engine/transaxle slightly and remove the engine-to-transaxle bolts.
30   Make a final check that all connectors, harnesses and hoses have been disconnected from the transaxle, then move the transaxle jack toward the side of the vehicle until the transaxle is clear of the engine locating dowels. Make sure you keep the transaxle level as you do this.

62TE transaxle

31   Disconnect the negative battery cable from the remote ground terminal or battery (see Chapter 5).
32   Loosen the front wheel lug nuts and the driveaxle/hub nuts (see Chapter 8), then raise the vehicle and support it securely on jackstands. Remove the wheels.
Note: *Depending on the type of wheels installed on the vehicle and the thickness of the socket you are using, you may have to loosen the driveaxle/hub nuts after the wheels have been removed (see Chapter 8).*
33   Remove the air filter housing and air intake duct assembly (see Chapter 4).
34   Remove the coolant reservoir (see Chapter 3).
35   Detach the transaxle oil cooler lines and plug them to prevent fluid from spilling **(see illustration 8.6)**.
36   Remove the heater hose bracket bolts and move the hoses back. Clearly label, and unplug, all electrical connectors from the transaxle.
37   Disconnect the shift cable from the manual lever and bracket (see Section 4).

38 Disconnect and remove the Crankshaft Position (CKP) sensor (see Chapter 6).
39 Drain the transaxle fluid (see Chapter 1).
40 Disconnect the input and output speed sensors and the range sensor at the transaxle (see Chapter 6).
41 Remove the bolts attaching the stamped transaxle mounting brace.
42 Remove the starter motor (see Chapter 5).
43 Remove the 6 upper transaxle-to-engine bolts - 4 on the transaxle side and 2 on the engine side.
44 Support the engine from above with a hoist or engine support fixture, or place a jack and a block of wood under the transaxle.
45 Remove the left transaxle mount and bracket and disconnect the ground cable (if equipped).
46 Remove the 3 bolts attaching the left mount to the bracket.
47 Remove the lower splash shields.
48 Remove both driveaxles (see Chapter 8).
49 On AWD models, remove the transfer case (see Chapter 7B).
50 Unbolt the lower transaxle mount from the subframe and bracket and remove.
51 Remove the 4 bolts attaching the lower transaxle mount bracket to the engine and remove the bracket.
52 Remove the torque converter cover **(see illustration 8.24)**.
53 Mark the relationship of the torque converter to the driveplate so they can be installed in the same position.
54 Remove the torque converter-to-driveplate bolts. After all the bolts are removed, push the torque converter into the bellhousing so it doesn't stay with the engine when the transaxle is removed.
55 Disconnect the oxygen sensor connector located near the differential of the transaxle.
56 Support the transaxle with a transmission jack, if available, or use a floor jack. Secure the transaxle to the jack using straps or chains so it doesn't fall off during removal.
57 Remove the transaxle upper mount-to-bracket bolts (see Chapter 2A, 2B, 2C, or 2D). Remove the upper transaxle-to-engine bolts.
58 Remove the front transaxle/engine bracket and the rear engine mount and bracket (see Chapter 2A, 2B, 2C or 2D).
59 Remove the lower transaxle-to-engine bolts.
60 Make a final check that all connectors, harnesses and hoses have been disconnected from the transaxle, then move the transaxle jack toward the side of the vehicle until the transaxle is clear of the engine locating dowels. Make sure you keep the transaxle level as you do this.

948TE/9HP48 (9-speed)

Note: *You will need an above-engine support fixture to accomplish this procedure.*

61 Disconnect the cable from the negative battery terminal (see Chapter 5).

62 Loosen the front wheel lug nuts and the driveaxle/hub nuts (see Chapter 8), then raise the vehicle and support it securely on jackstands. Remove the wheels.
63 Remove the battery and battery tray (see Chapter 5). Remove the fuse box and fuse box tray (see Chapter 12).
64 Remove the PCM (see Chapter 6).
65 Detach the transaxle oil cooler quick-connect lines at the transaxle end and plug them to prevent fluid from spilling.
66 Disconnect the large TCM connector from the transaxle (located in front of the transaxle mount).
67 Disconnect the shift/manual park release cable from the manual lever and bracket (see Section 4).
68 Remove the retaining nuts, then separate the wiring harness bracket and ground cable from the upper side of the transaxle.
69 Drain the transaxle fluid (see Chapter 1).
70 Disconnect the transaxle solenoid wiring harness connector at the front, by rotating the connector lock counterclockwise (viewed from above).
71 Remove the starter motor (see Chapter 5).
72 Remove the upper transaxle-to-engine bellhousing bolts, leaving the four lower bellhousing bolts in place.
72 Attach an engine support fixture bracket to the left end of the cylinder head, then support the engine from above with a hoist or engine support fixture.
73 Using a transmission jack, raise the transaxle slightly to relieve tension, then remove the left transaxle mount and bracket(s).
74 Disconnect any remaining wiring harness connectors, ground cables, and detach any harness clips from the transaxle.
75 Remove the fenderwell and lower splash shields, then remove both driveaxles (see Chapter 8).
76 Disconnect the exhaust pipe from the catalytic converter and position/hang it aside.
77 On AWD models, remove the transfer case (see Chapter 7B).
78 Remove the torque converter cover **(see illustration 8.24)**.
79 Mark the relationship of the torque converter to the driveplate so they can be installed in the same position.
80 Remove the torque converter-to-driveplate bolts. Rotate the engine by hand (turning by crankshaft pulley bolt) to access all the bolts through the cover opening.
81 Support the transaxle with a transmission jack, if available, or use a floor jack with a suitable and wide enough head. Secure the transaxle to the jack using straps or chains so it doesn't fall off during removal.
82 Remove the rear engine/transaxle mount.
83 Remove the left side vehicle load beam frame (see Chapter 10).
84 Remove the four lower transaxle-to-engine bellhousing bolts.
85 Carefully pry the transaxle away from the engine, making sure not to gouge the mating surfaces.

86 Make a final check that all connectors, harnesses and hoses have been disconnected from the transaxle, then move the transaxle jack toward the side of the vehicle until the transaxle is clear of the engine locating dowels. Make sure you keep the transaxle level as you do this.

Installation

87 Installation is the reverse of removal, noting the following points:
Note: *Wedge a screwdriver between the teeth on the driveplate and the opening to prevent the engine from rotating while tightening the torque converter bolts.*

a) *As the torque converter is reinstalled, ensure that the drive tangs at the center of the torque converter hub engage with the recesses in the automatic transaxle fluid pump inner gear. This can be confirmed by turning the torque converter while pushing it towards the transaxle. If it isn't fully engaged, it will clunk into place.*
b) *When installing the transaxle, make sure the match marks you made on the torque converter and driveplate line up.*
c) *Install all of the driveplate-to-torque converter bolts before tightening any of them.*
d) *Tighten the driveplate-to-converter bolts to the torque listed in this Chapter's Specifications.*
e) *Tighten the transaxle mounting bolts to the torque listed in this Chapter's Specifications.*
f) *Tighten the driveaxle/hub nuts to the torque listed in the Chapter 8 Specifications.*
g) *Tighten the wheel lug nuts to the torque listed in the Chapter 1 Specifications.*
h) *Fill the transaxle with the correct type and amount of fluid (see Chapter 1).*
i) *Adjust the shift cable (see Section 4).*

---

9 Automatic transaxle overhaul - general information

---

1 In the event of a problem occurring, it will be necessary to establish whether the fault is electrical, mechanical or hydraulic in nature, before repair work can be contemplated. Diagnosis requires detailed knowledge of the transaxle's operation and construction, as well as access to specialized test equipment, and so is deemed to be beyond the scope of this manual. It is therefore essential that problems with the automatic transaxle are referred to a dealer service department or other qualified repair facility for assessment.
2 Note that a faulty transaxle should not be removed before the vehicle has been diagnosed by a knowledgeable technician equipped with the proper tools, as troubleshooting must be performed with the transaxle installed in the vehicle.

Notes

Chapter 7 Part B
Transfer case

Contents

	Section
Driveaxle oil seal (right side) - replacement	3
General information	1
Input shaft oil seal - replacement	4

	Section
Pinion seal - replacement	2
Transfer case - removal and installation	5

Specifications

Transfer case fluid type .... See Chapter 1

Torque specifications

Ft-lbs (unless otherwise indicated)

Note: *One foot-pound (ft-lb) of torque is equivalent to 12 inch-pounds (in-lbs) of torque. Torque values below approximately 15 ft-lbs are expressed in inch-pounds, because most foot-pound torque wrenches are not accurate at these smaller values.*

Subframe	(See Chapter 10)
2014 and earlier models	
Drain/fill plugs	26
Lower heat shield bolt	22
Pinion yoke nut	75
Transmission bracket bolts	40
Transfer case-to-transaxle bolts	
Lower	40
Upper	22
2015 and later models	
Fill plug	30
Heat shield bolts	120 in-lbs
Transfer case-to-transaxle flange	35
Mounting bracket-to-engine block	32
Mounting bracket-to-transfer case	16
Transfer case damper	19

1 General information

1   Due to the complexity of the transfer case covered in this manual and the need for specialized equipment to perform most service operations, this Chapter contains only removal and installation procedures.

2   If the transfer case requires major repair work, it should be taken to a dealer service department or an automotive or transmission repair shop. You can, however, remove and install the transfer case yourself and save the expense of that labor, even if the repair work is done by a transmission shop.

2 Pinion seal - replacement

1   Raise the vehicle and support it securely on jackstands.

2   Use chalk or a scribe to index the relationship of the driveshaft to the pinion yoke at the transfer case.

Caution: *DO NOT allow the driveshaft to hang from the front, rear or center support bearing - always suppport the driveshaft. Damage to the joints, boots and/or center support bearing may occur, resulting in vibration.*

3   Remove the front Constant Velocity (CV) joint bolts and washers from the transfer case.

Support the front of the driveshaft.

4   Remove the transfer case pinion yoke nut.

5   Using a puller, remove the transfer case pinion yoke.

6   Carefully pry out the transfer case pinion seal with a seal removal tool or a large screwdriver. Be careful not to damage or scratch the seal bore.

7   Using a seal installer or a large deep socket as a drift, install the new oil seal. Drive it into the bore squarely and make sure it's completely seated.

8   Installation is the reverse of removal, noting the following points:

a) Tighten all fasteners to the torque listed in this Chapter's Specifications.

b) Check the transfer case fluid level and fill as necessary (see Chapter 1).

3 Driveaxle oil seal (right side) - replacement

1 Loosen the right front wheel driveaxle/hub nut and wheel lug nuts. Raise the vehicle and support it securely on jackstands. Remove the wheel.

2 Remove the right side driveaxle and intermediate shaft (if equipped) (see Chapter 8).

3 Carefully pry out the driveaxle oil seal with a seal removal tool or a large screwdriver. Be careful not to damage or scratch the seal bore.

4 Using a seal installer or a large deep socket as a drift, install the new oil seal. Drive it into the bore squarely and make sure it's completely seated.

5 Lubricate the lip of the new seal with multi-purpose grease.

6 Install the right side driveaxle and intermediate shaft (if equipped).

7 Check the transfer case fluid level and add some if necessary, to bring it to the appropriate level (see Chapter 1).

4 Input shaft oil seal - replacement

1 Loosen the right front wheel driveaxle/hub nut and the front wheel lug nuts. Raise the vehicle and support it securely on jackstands. Remove the wheels.

2 Remove the transfer case (see Section 5).

3 Drill holes in the input shaft seal and use screws and a slide hammer to remove the seal. Use care not to get metal shavings inside the transfer case.

4 Lubricate the lip of the new seal with multi-purpose grease.

5 Using a seal installer or a large deep socket as a drift, install the new oil seal. Drive it into the bore squarely and make sure it's completely seated.

6 Installation is the reverse of removal, noting the following points:

a) Tighten all fasteners to the torque listed in this Chapter's Specifications.

b Check the transfer case and transaxle fluid levels and fill as necessary (see Chapter 1).

5 Transfer case - removal and installation

Removal

1 Loosen the right front wheel driveaxle/hub nut and the front wheel lug nuts. Raise the vehicle and support it securely on jackstands. Remove the wheels.

2 Remove the engine undercover, if equipped.

3 Drain the transfer case lubricant (see Chapter 1).

4 Drain the transaxle fluid (see Chapter 1).

5 Remove the 3 rear transaxle bracket mounting bolts.

6 Disconnect the lower balljoints from the lower control arms (see Chapter 10).

7 Support the engine from above with a hoist or engine support fixture, or place a jack and a block of wood under the transaxle or engine oil pan.

8 Remove the 6 bolts attaching the under-engine subframe to the body and the engine mount subframe.

9 Remove the engine mount through-bolt, and remove the front engine mount subframe from the vehicle (see Chapter 10).

10 Remove the rear engine mount (see Chapter 2A, 2B, 2C or 2D).

11 Remove the screws attaching the power steering hose to the engine mount subframe.

12 Remove the heat shield from the steering gear.

13 Remove the 2 steering gear mounting bolts and secure the steering gear to the vehicle so it is out of the way when the engine mount subframe is removed (see Chapter 10).

14 Remove the bolts attaching the stabilizer bar bracket to the engine mount subframe.

Note: Before removing the suspension subframe bolts and lowering the engine mount subframe, mark the location of the engine mount subframe to the vehicle body to aid installation and minimize the need for an alignment after installation.

15 Support the engine mount subframe using a floor jack. Remove the 4 bolts attaching the engine mount subframe to the vehicle.

16 Remove the engine mount subframe support bracket mounting screws (one bracket on each side), and remove the brackets.

17 Carefully lower the engine mount subframe and remove it from under the vehicle (see Chapter 10).

18 Remove the rear exhaust manifold, converter and exhaust crossover pipe (see Chapter 4). Support the rear exhaust pipe with wire.

19 Use chalk or a scribe to index the relationship of the driveshaft to the pinion yoke at the transfer case.

Caution: DO NOT allow the driveshaft to hang from the front, rear or center support bearing - always support the driveshaft. Damage to the joints, boots and or center support bearing may occur, resulting in vibration.

20 Remove the front Constant Velocity (CV) joint bolts and washers from the transfer case. Support the front of the driveshaft.

21 Remove the right side driveaxle and intermediate shaft (if equipped) (see Chapter 8).

22 Support the transfer case with a jack, jackstand or equivelent. Secure the transfer case so it does not fall and get damaged.

23 Disconnect any electrical connectors or vent hoses.

24 Remove the 2 upper transfer case mounting bolts and 2 heat shield retainers.

25 Remove the lower heat shield attaching bolt and remove the heat shield from the transfer case.

26 Remove the 2 lower transfer case mounting bolts.

27 Carfully rotate the engine and transmission forward and remove the transfer case.

Installation

28 Installation is the reverse of removal, noting the following points:

a) Install a new O-ring seal between the transfer case and transaxle.

b) Tighten the exhaust system fasteners to the torque listed in the Chapter 4 Specifications.

c) Tighten the driveshaft fasteners to the torque listed in the Chapter 8 Specifications.

d) Tighten the transfer case mounting bolts to the torque listed in this Chapter's Specifications.

e) Refill the transfer case and transaxle with the proper type and amount of lubricant (see Chapter 1).

f) Tighten the driveaxle/hub nut to the torque listed in the Chapter 8 Specifications.

g) Tighten the wheel lug nuts to the torque listed in the Chapter 1 Specifications.

Chapter 8   Driveline

Contents

	Section			Section
Driveaxle - removal and installation	2		Rear differential assembly (AWD models) - removal and installation	9
Driveaxle boot replacement	3		Rear differential driveaxle oil seals (AWD models) - replacement	7
Driveaxles - general information and inspection	1			
Driveshaft (AWD models) - removal and installation	4		Rear differential electronic clutch - removal and installation	10
Driveshaft center support bearing (AWD models) - replacement	6		Rear differential pinion seal (AWD models) - replacement	8
Driveshaft universal and constant velocity joints (AWD models) - general information and check	5			

Specifications

Torque specifications

Ft-lbs (unless otherwise indicated)

Note: *One foot-pound (ft-lb) of torque is equivalent to 12 inch-pounds (in-lbs) of torque. Torque values below approximately 15 ft-lbs are expressed in inch-pounds, because most foot-pound torque wrenches are not accurate at these smaller values.*

Driveaxles

Driveaxle/hub nut*	
2014 and earlier models	118
2015 and later models	
Front	148
Rear	129
Intermediate shaft bearing bracket fasteners (2014 and earlier models)	
8mm bolts	17
10mm bolts	28
Intermediate shaft bearing bracket fasteners (2015 and later models)	
Bracket-to-engine bolts	48
Bearing-to-bracket bolts	17
Wheel lug nuts	See Chapter 1

Driveshaft (AWD models)

2014 and earlier models	
Center bearing mounting nuts	30
Driveshaft-to-transfer case bolts	22
Driveshaft-to-rear differential bolts	43
2015 and later models	
Driveshaft flange bolts (both ends)	15
Center bearing bracket bolts	17

Rear differential assembly (AWD models)

2014 and earlier models	
Differential drain/fill plugs	15
Differential-to-crossmember front bolts	75
Differential-to-crossmember rear bolt	75
Electronic clutch bolts	44
Pinion flange nut*	100
2015 and later models	
Differential fill plug	26
Hydraulic fluid cover bolts	53 in-lbs
Differential bracket-to-differential bolts	44
Differential front through-bolt/nut	66
Differential rear through-bolt	89

* Nut must be replaced

1 Driveaxles - general information and inspection

Front-wheel drive models

1   Power is transmitted from the transaxle to the wheels through a pair of driveaxles. The driveaxles are either two equal length halfshafts, consisting of short halfshafts on both sides connected to a mid-shaft on the right side, or two unequal length halfshafts with a short halfshaft on the left side and long halfshaft on the right. The inner end of each driveaxle is splined to the differential side gears or the mid-shaft. The driveaxles can be pulled out to replace the oil seals (see Chapter 7A). The outer ends of the driveaxles are splined to the front hubs and locked in place by a large nut.

All-wheel drive models

2   All-wheel drive models also transmit power from the transaxle to the rear wheels via a transfer case, driveshaft, rear differential, and a pair of driveaxles similar to the ones up front.

All models

3   Each driveaxle assembly consists of an inner and outer constant velocity (CV) joint connected together by a driveaxle shaft. The inner ends of the driveaxles are equipped with a tri-pot type joint. The design is capable of both angular and axial motion. In other words, the inner CV joints are free to slide in-and-out as the driveaxle moves up-and-down with the wheel.

4   The outer CV joints use a ball-and-cage design, or "Rzeppa" joint, capable of angular but not axial movement.

5   The boots should be inspected periodically for damage and leaking lubricant. Torn CV joint boots must be replaced immediately or the joints can be damaged. Boot replacement involves removal of the driveaxle (see Section 2).

6   The most common symptom of worn or damaged CV joints, besides lubricant leaks, is

2.2 Loosen the driveaxle/hub nut before you raise the vehicle and remove the wheel

a clicking noise in turns, a clunk when accelerating after coasting and vibration at highway speeds. To check for wear in the CV joints and driveaxle shafts, grasp each axle (one at a time) and rotate it in both directions while holding the CV joint housings, feeling for play indicating worn splines or sloppy CV joints. Also check the axleshafts for cracks, dents and distortion.

2 Driveaxle - removal and installation

Front

Removal

1   Set the parking brake. Remove the front wheel cover or hubcap.

2   Loosen, but do not remove, the driveaxle/hub nut **(see illustration)**.

Note: *If the opening in the wheel is too small to accommodate your socket, wait until after the wheel is removed to loosen the nut. You can prevent the brake disc from turning by inserting a long punch into a disc cooling vane and letting it come to rest against the caliper mounting bracket, or have an assistant apply the brake.*

3   Loosen, but do not remove, the wheel lug nuts.

4   Raise the vehicle and support it securely on jackstands, then remove the lug nuts and the wheel.

5   Remove the brake caliper and disc (see Chapter 9). Support the brake caliper so as not to damage the brake hose.

6   Remove the two steering knuckle-to-strut bracket bolts (see Chapter 10) and separate the strut from the steering knuckle.

Caution: *The steering knuckle-to-strut bolts are serrated and must not be turned during removal.*

Note: *If the strut assembly is attached to the steering knuckle using a cam bolt in the lower slotted hole, mark the relationship of the cam bolt to the strut to preserve the wheel alignment setting on reassembly.*

7   Remove the driveaxle/hub nut and pull the steering knuckle out and away from the outer CV joint of the driveaxle **(see illustration)**. Strike the end of the stub shaft with a soft-faced hammer to separate the splines from the hub and bearing assembly, if necessary. Retrieve the washer from the stub shaft **(see illustration)**.

Caution: *Discard the driveaxle/hub nut and obtain a new one for reassembly.*

2.7a Pull the steering knuckle away from the outer CV joint

2.7b Remove the washer from the end of the driveaxle

2.8 Using a large pry bar, pry the inner CV joint out sharply to disengage the snap-ring from the differential gears inside the transaxle

3.3 Cut off the boot clamps and discard them

8   Support the outer end of the driveaxle and insert a prybar between the inner CV joint and the transaxle case **(see illustration)**. Pry out sharply to disengage the inner CV joint from the transaxle. Make sure you have a drain pan under the transaxle, as some oil will leak out.

9   On models equipped with a passenger's side intermediate shaft, insert a prybar between the inner CV joint and the mid-shaft bearing housing and pry out sharply to disengage the inner CV joint from the intermediate shaft bearing.

10   Carefully withdraw the inner CV joint from the transaxle. Do not let the spline or the snap-ring drag across the sealing lip of the driveaxle oil seal.

11   On models equipped with an intermediate shaft bearing on the passenger's side, remove the heat shield fasteners (if equipped), then remove the bearing mounting bolts and pull the mid-shaft from the transaxle.

Installation

12   Installation is the reverse of removal, noting the following additional points:

  a) *Thoroughly clean the splines and bearing shield on the outer CV joint. This is very important, as the bearing shield protects the wheel bearings from water and contamination. Also clean the wheel bearing area of the steering knuckle. Son't forget to install the washer on the stub shaft.*

  b) *Thoroughly clean the splines and oil seal sealing surface on the inner CV joint. Apply an even bead of multi-purpose grease around the oil seal sealing surface of the inner CV joint.*

  c) *When installing the driveaxle, push it in sharply to seat the snap-ring on the inner CV joint stub shaft into its groove in the differential gears inside the transaxle. Pull out on the inner CV joint housing to ensure it's seated.*

  d) *On models equipped with a passenger's side intermediate shaft, insert the*

splined shaft assembly into the splined differential gears inside the transaxle, then install the bearing mounting bolts and tighten the bolts to the torque listed in this Chapter's Specifications. Install the bearing heat shield and tighten the mounting bolts securely.

  e) *Tighten the steering knuckle-to-strut bolts to the torque listed in the Chapter 10 Specifications.*

  f) *Tighten the NEW driveaxle/hub nut to the torque listed in this Chapter's Specifications.*

  g) *Tighten the lug nuts to the torque listed in the Chapter 1 Specifications.*

Rear

Removal

13   Set the parking brake. Remove the rear wheel cover or hubcap.

14   Loosen, but do not remove, the driveaxle/hub nut **(see illustration 2.2)**.

15   Raise the vehicle and support it securely on jackstands. Place the shift selector lever in Neutral.

16   Remove the rear differential assembly from the vehicle (see Section 9).

17   Remove the driveaxle/hub nut and pull the the driveaxle from the hub and bearing assembly towards the center of the vehicle. Strike the end of the stub shaft with a soft-faced hammer to separate the splines from the hub and bearing assembly, if necessary.

Installation

18   Installation is the reverse of removal, noting the following additional points:

  a) *Thoroughly clean the splines and bearing shield on the outer CV joint. This is very important, as the bearing shield protects the wheel bearings from water and contamination. Also clean the wheel bearing area of the steering knuckle.*

  b) *Thoroughly clean the splines and oil seal sealing surface on the inner CV joint.*

Apply an even bead of multi-purpose grease around the oil seal sealing surface of the inner CV joint.

  c) *When installing the driveaxle, push it in sharply to seat the snap-ring on the inner CV joint stub shaft into its groove in the rear differential gears. Pull out on the inner CV joint housing to ensure it's seated.*

  d) *Tighten the driveaxle/hub nut to the torque listed in this Chapter's Specifications.*

  e) *Tighten the lug nuts to the torque listed in the Chapter 1 Specifications.*

---

3   Driveaxle boot replacement

Note: *If the CV joints or boots must be replaced, explore all options before beginning the job. Complete, rebuilt driveaxles are available on an exchange basis, eliminating much time and work. Whichever route you choose to take, check on the cost and availability of parts before disassembling the vehicle.*

Front driveaxle

1   Remove the driveaxle (see Section 2).

2   Mount the driveaxle in a vise with wood-lined jaws, to prevent damage to the axleshaft. Check the CV joints for excessive play in the radial direction, which indicates worn parts. Check for smooth operation throughout the full range of motion for each CV joint. If a boot is torn, the recommended procedure is to disassemble the joint, clean the components and inspect for damage due to loss of lubrication and possible contamination by foreign matter. If the CV joint is in good condition, lubricate it with CV joint grease and install a new boot.

Outer CV joint

Disassembly

3   Cut the boot clamps with side-cutters, then remove and discard them **(see illustration)**.

3.5 Strike the edge of the CV joint housing sharply with a soft-faced hammer to dislodge the CV joint from the shaft

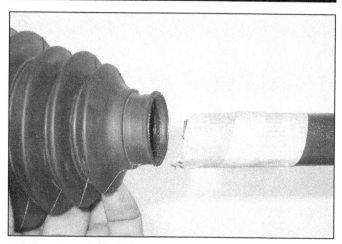

3.8 Wrap the axleshaft splines with tape to prevent damaging the boot as it's slid onto the shaft

3.9a Pack the outer CV joint assembly with grease . . .

3.9b . . . then apply grease to the inside of the boot

4   Using a screwdriver, pry up on the edge of the boot, pull it off the CV joint housing and slide it down the axleshaft.

5   Strike the edge of the CV joint housing sharply with a soft-face hammer to dislodge the outer CV joint from the axleshaft **(see**

3.12 Equalize the pressure inside the boot by inserting a screwdriver between the boot and the CV joint housing

illustration). Remove and discard the bearing retainer clip from the axleshaft.

6   Remove the large stop-ring from the axleshaft, then slide the boot off the shaft.

Inspection

7   Clean the components with solvent to remove all traces of grease. Inspect the cage and races for pitting, score marks, cracks and other signs of wear and damage. Shiny, polished spots are normal and won't adversely affect CV joint operation.

Reassembly

8   Wrap the splines on the inner end of the axleshaft with tape to protect the boots from the sharp edges of the splines, and slide the clamps, then the boot onto the axleshaft **(see illustration)**. Remove the tape and install a **NEW** stop-ring and circlip on the axleshaft.

9   Place half the grease provided in the sealing boot kit into the outer CV joint assembly housing **(see illustration)**. Put the remaining grease into the sealing boot **(see illustration)**.

10   Align the splines on the axleshaft with the splines on the outer CV joint assembly and gently drive the CV joint onto the axleshaft using a soft-faced hammer until the CV

joint is seated to the axleshaft.

11   Slide the boot into place, making sure the raised bead on the inside of the seal boot is positioned in the groove on the interconnecting shaft. If the driveaxle has multiple locating grooves on the shaft, position the boot so only one of the grooves (the thinnest) is exposed. Position the sealing boot into the groove on the outer CV joint housing.

12   Using a thin screwdriver, equalize the pressure in the boot **(see illustration)**.

13   Make sure each end of the boot is seated properly, and the boot is not distorted.

14   Install the boot clamps. There are three types of clamps you're likely to encounter: the band type, which requires a special tightening tool, the crimp type (which also requires a special tool), or the fold-over type **(see illustrations)**.

15   Install the driveaxle (see Section 2).

Inner CV joint

16   At the time of writing, replacement inner CV joint boots were not available, nor did the manufacturer provide a replacement procedure. If the inner boot becomes torn, the entire driveaxle may have to be replaced.

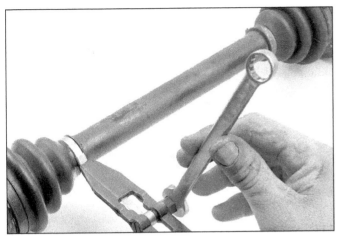

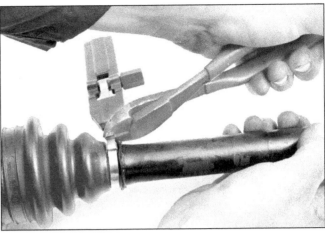

3.14a You'll need a special tightening tool to install "band" type boot clamps: Install the band with its end pointing in the direction of axle rotation and tighten it securely . . .

3.14b . . . then bend down the end of the clamp back and cut off the excess

3.14c If you're installing crimp-type boot clamps, you'll need a pair of special crimping pliers (available at most auto parts stores)

3.14d To install fold-over type boot clamps, bend the tang down . . .

3.14e . . . then tap the tabs over to hold it in place

Check with your local auto parts store and dealer parts department to see if replacement boots have become available.

Rear driveaxle (AWD models)

17   At the time of writing, replacement CV joint boots for the rear driveaxles were not available, nor did the manufacturer provide replacement procedures. If a boot on a rear driveaxle becomes torn, the entire drive-axle may have to be replaced. Check with your local auto parts store and dealer parts department to see if replacement boots have become available.

---

4   Driveshaft (AWD models) - removal and installation

Caution: *DO NOT allow the driveshaft to hang from the front, rear or center support bearing - always suppport the driveshaft. Damage to the joints, boots and/or center support bearing may occur, resulting in vibration.*

Note: *The manufacturer recommends replacing driveshaft fasteners with new ones when installing the driveshaft.*

1   Raise the vehicle and support it securely on jackstands. Place the shift selector lever in Neutral.
2   Use chalk or a scribe to index the relationship of the driveshaft to the differential pinion yoke and the flange at the transfer case.
3   Remove the three fasteners and disconnect the rear driveshaft rubber coupler from the rear axle flange. Support the rear of the driveshaft.
4   Remove the center support heat shield and remove the bolts attaching the driveshaft center support to the vehicle. Support the center of the driveshaft.
5   Remove the front Constant Velocity (CV) joint bolts and washers from the transfer case. Support the front of the driveshaft.
Note: *If the front CV joint will not break free from the transfer case flange, use a brass drift and a hammer to carefully drive the CV joint out from the flange.*
6   Remove the driveshaft from the vehicle.

7   Installation is the reverse of removal, noting the following points:
a) *Make sure the marks you made previously are aligned.*
b) *Tighten the fasteners to the torques listed in this Chapter's Specifications.*

---

5   Driveshaft universal and constant velocity joints (AWD models) - general information and check

Universal joints

1   Universal joints are mechanical couplings which connect two rotating components that meet each other at different angles.
2   These joints are composed of a yoke on each side connected by a crosspiece called a trunnion. Cups at each end of the trunnion contain needle bearings which provide smooth transfer of the torque load. Snap-rings, either inside or outside of the bearing cups, hold the assembly together.

3   Wear in the needle roller bearings is characterized by vibration in the driveline, noise during acceleration, and in extreme cases of lack of lubrication, metallic squeaking and ultimately grating and shrieking sounds as the bearings disintegrate.

4   It is easy to check if the needle bearings are worn with the driveshaft in position, by trying to turn the shaft with one hand, the other hand holding the rear differential pinion flange when the rear universal joint is being checked, and the front half coupling when the front universal joint is being checked. Any movement between the driveshaft and the front half couplings, and around the rear half couplings, is indicative of considerable wear. Another method of checking for universal joint wear is to use a prybar inserted into the gap between the universal joint and the driveshaft or flange. Leave the vehicle in gear and try to pry the joint both radially and axially. Any looseness should be apparent with this method. A final test for wear is to attempt to lift the shaft and note any movement between the yokes of the joints.

5   If any of the above conditions exist, replace the driveshaft.

Constant velocity joint

6   The constant velocity joint on the forward end of the driveshaft is not serviceable. If the joint wears out or the boot becomes damaged, the front portion of the driveshaft must be replaced.

6   Driveshaft center support bearing (AWD models) - replacement

The center support bearing is not replaceable separately; if it is in need of replacement, the entire driveshaft must be replaced.

7   Rear differential driveaxle oil seals (AWD models) - replacement

1   Raise the vehicle and support it securely on jackstands. Place the transaxle in Neutral with the parking brake off.

2   Remove the rear differential (see Section 9).

3   Carefully pry out the driveaxle oil seal with a seal removal tool or a large screwdriver. Be careful not to damage or scratch the seal bore.

4   Using a seal installer or a large deep socket as a drift, install the new oil seal. Drive it into the bore squarely and make sure it's completely seated.

5   Lubricate the lip of the new seals with multi-purpose grease.

6   Install the rear differential (see Section 9).

7   Check the differential lubricant level and add some, if necessary, to bring it to the appropriate level.

8   Rear differential pinion seal (AWD models) - replacement

1   Raise the vehicle and support it securely on jackstands.

2   Use chalk or a scribe to index the relationship of the driveshaft to the differential pinion yoke.

Caution: *DO NOT allow the driveshaft to hang from the front, rear or center support bearing - always support the driveshaft. Damage to the joints, boots and or center support bearing may occur, resulting in vibration.*

3   Remove the three fasteners and disconnect the rear driveshaft rubber coupler from the differential pinion yoke. Support the rear of the driveshaft.

4   Remove the differential pinion yoke nut and washer.

5   Using a puller, remove the differential pinion yoke.

6   Use a screwdriver, seal removal tool or a slide hammer to remove the differential pinion yoke case seal from the differential.

7   Remove the differential pinion yoke shaft seal from the shaft or pinion flange.

8   Using a seal installer or a large deep socket as a drift, install the new differential pinion yoke case seal. Drive it into the bore squarely and make sure it's completely seated.

9   Lubricate the lip of the new seal with multi-purpose grease.

10 Install the new differential pinion yoke shaft seal half-way down the threads on the pinion shaft.

11 Installation is the reverse of removal, noting the following points:

a) *Tighten all fasteners to the torque listed in this Chapter's Specifications.*

b) *Check the differential lubricant level and add some, if necessary, to bring it to the appropriate level.*

9   Rear differential assembly (AWD models) - removal and installation

1   Raise the vehicle and support it securely on jackstands.

2   Remove the center hanger for the exhaust system near the driveshaft center support bearing. Disconnect the rear exhaust hangers.

3   Lower and support the exhaust system using jackstands or appropriate wire, with a minimum of 10 inches of clearance below the vehicle to allow for removal of the rear differential.

4   Drain the differential fluid to prevent spilling when removing.

5   Support the rear differential using a floor jack or similar.

6   Use chalk or a scribe to index the relationship of the driveshaft to the differential pinion yoke and the flange at the transfer case.

Caution: *DO NOT allow the driveshaft to hang from the front, rear or center support bearing - always suppport the driveshaft. Damage to the joints, boots and or center support bearing may occur, resulting in vibration.*

7   Remove the three fasteners and disconnect the rear driveshaft rubber coupler from the rear axle flange. Support the rear of the driveshaft.

8   Disconnect the electronic clutch control module electrical connector.

9   Use a prybar or similar to dislodge the driveaxles part way from the rear differential housing.

10 Remove the three rear differential-to-crossmember mounting bolts.

11 Lower the rear differential partially and remove the driveaxles from the rear differential housing.

12 Carefully lower and remove the differential from the vehicle.

13 Installation is the reverse of removal, noting the following points:

a) *Don't tighten any of the mounting fasteners until all of them have been installed.*

b) *Tighten all fasteners to the torque listed in this Chapter's Specifications.*

10   Rear differential electronic clutch - removal and installation

1   Raise the vehicle and support it securely on jackstands.

2   Use chalk or a scribe to index the relationship of the driveshaft to the differential pinion yoke.

Caution: *DO NOT allow the driveshaft to hang from the front, rear or center support bearing - always suppport the driveshaft. Damage to the joints, boots and or center support bearing may occur, resulting in vibration.*

3   Remove the three fasteners and disconnect the rear driveshaft rubber coupler from the rear axle flange. Support the rear of the driveshaft.

4   Disconnect the electronic clutch control module electrical connector.

5   Remove the mounting bolts attaching the electronic clutch assembly to the rear differential.

6   Remove the electronic clutch assembly from the vehicle.

7   Installation is the reverse of removal, noting the following points:

a) *Ensure the electronic clutch and rear differential splines engage properly.*

b) *Tighten all fasteners to the torque listed in this Chapter's Specifications.*

Chapter 9 Brakes

Contents

	Section
Anti-lock Brake System (ABS) - general information	3
Brake disc - inspection, removal and installation	6
Brake hoses and lines - inspection and replacement	10
Brake light switch - check, replacement and adjustment	15
Brake system - bleeding	11
Brake vacuum pump - removal and installation	16
Disc brake caliper - removal and installation	5
Disc brake pads - replacement	4
Drum brake shoes - replacement	7

	Section
General information	1
Master cylinder - removal and installation	9
Parking brake - adjustment	14
Parking brake shoes (models with rear disc brakes) - replacement	13
Power brake booster - check, removal and installation	12
Troubleshooting	2
Wheel cylinder - removal and installation	8

Specifications

General

Brake fluid type	See Chapter 1

Disc brakes

Brake pad minimum thickness	See Chapter 1
Disc lateral runout limit	
Front	0.002 inch
Rear	
14-inch disc	0.0024 inch
16-inch disc	0.0016 inch
Disc minimum thickness	Cast into disc
Thickness variation (parallelism)	
Front	0.0002 inch
Rear	0.0006 inch

Drum brakes

Minimum shoe lining thickness	See Chapter 1
Maximum radial runout	0.0024 inch
Maximum drum diameter	Cast into drum

Torque specifications

Ft-lbs (unless otherwise indicated)

Note: *One foot-pound (ft-lb) of torque is equivalent to 12 inch-pounds (in-lbs) of torque. Torque values below approximately 15 ft-lbs are expressed in inch-pounds, because most foot-pound torque wrenches are not accurate at these smaller values.*

Brake booster mounting nuts...	17
Brake hose banjo bolt-to-front caliper	
2009 and earlier models..	18
2010 to 2014 models...	26
2015 and later models...	156 in-lbs
Brake hose bolt-to-rear caliper	
2014 and earlier models..	132 in-lbs
2015 and later models...	156 in-lbs
Caliper mounting (guide pin) bolts	
2014 and earlier	32
2015 and later	26
Caliper mounting bracket bolts	
Front	
2014 and earlier..	80
2015 and later...	129
Rear	
2010 and earlier models	52
2011 to 2014 models	44
2015 and later models...	66
Master cylinder-to-brake booster mounting nuts	18
Wheel speed sensor mounting screw..	108 in-lbs
Wheel cylinder mounting bolts..	115 in-lbs
Wheel lug nuts...	See Chapter 1
Vacuum pump mounting fasteners	
Electric pump	
Pump-to-bracket	71 in-lbs
Bracket-to-engine	17
Mechanical pump (driven by camshaft)	
Pump fitting-to-pump	71 in-lbs
Vacuum pump-to-cylinder head	17

---

1   General information

The vehicles covered by this manual are equipped with hydraulically operated front and rear brake systems. The front and rear brakes are disc type. Both the front and rear brakes are self adjusting. The disc brakes automatically compensate for pad wear.

Hydraulic system

The hydraulic system consists of two separate circuits. The master cylinder has separate reservoirs for the two circuits, and, in the event of a leak or failure in one hydraulic circuit, the other circuit will remain operative.

Power brake booster

The power brake booster, utilizing engine manifold vacuum and atmospheric pressure to provide assistance to the hydraulically operated brakes, is mounted on the firewall in the engine compartment.

Parking brake

The parking brake operates the rear brakes only, through cable actuation. It's activated by a lever mounted in the center console.

Service

After completing any operation involving disassembly of any part of the brake system, always test drive the vehicle to check for proper braking performance before resuming normal driving. When testing the brakes, perform the tests on a clean, dry, flat surface. Conditions other than these can lead to inaccurate test results.

Test the brakes at various speeds with both light and heavy pedal pressure. The vehicle should stop evenly without pulling to one side or the other. Avoid locking the brakes, because this slides the tires and diminishes braking efficiency and control of the vehicle.

Tires, vehicle load and wheel alignment are factors which also affect braking performance.

Precautions

There are some general cautions and warnings involving the brake system on this vehicle:

a) *Use only brake fluid conforming to DOT 3 specifications.*

b) *The brake pads contain fibers which are hazardous to your health if inhaled. Whenever you work on brake system components, clean all parts with brake system cleaner. Do not allow the fine dust to become airborne. Also, wear an approved filtering mask.*

c) *Safety should be paramount whenever any servicing of the brake components is performed. Do not use parts or fasteners which are not in perfect condition, and be sure that all clearances and torque specifications are adhered to. If you are at all unsure about a certain procedure, seek professional advice. Upon completion of any brake system work, test the brakes carefully in a controlled area before putting the vehicle into normal service. If a problem is suspected in the brake system, don't drive the vehicle until it's fixed.*

2   Troubleshooting

PROBABLE CAUSE	CORRECTIVE ACTION

No brakes - pedal travels to floor

1 Low fluid level	1 and 2 Low fluid level and air in the system are symptoms of another problem
2 Air in system a leak somewhere in the hydraulic system.	2 Locate and repair the leak
3 Defective seals in master cylinder	3 Replace master cylinder
4 Fluid overheated and vaporized due to heavy braking	4 Bleed hydraulic system (temporary fix). Replace brake fluid (proper fix)

Brake pedal slowly travels to floor under braking or at a stop

1 Defective seals in master cylinder	1 Replace master cylinder
2 Leak in a hose, line, caliper or wheel cylinder	2 Locate and repair leak
3 Air in hydraulic system	3 Bleed the system, inspect system for a leak

Brake pedal feels spongy when depressed

1 Air in hydraulic system	1 Bleed the system, inspect system for a leak
2 Master cylinder or power booster loose	2 Tighten fasteners
3 Brake fluid overheated (beginning to boil)	3 Bleed the system (temporary fix). Replace the brake fluid (proper fix)
4 Deteriorated brake hoses (ballooning under pressure)	4 Inspect hoses, replace as necessary (it's a good idea to replace all of them if one hose shows signs of deterioration)

Brake pedal feels hard when depressed and/or excessive effort required to stop vehicle

1 Power booster faulty	1 Replace booster
2 Engine not producing sufficient vacuum, or hose to booster clogged, collapsed or cracked	2 Check vacuum to booster with a vacuum gauge. Replace hose if cracked or clogged, repair engine if vacuum is extremely low
3 Brake linings contaminated by grease or brake fluid	3 Locate and repair source of contamination, replace brake pads or shoes
4 Brake linings glazed	4 Replace brake pads or shoes, check discs and drums for glazing, service as necessary
5 Caliper piston(s) or wheel cylinder(s) binding or frozen	5 Replace calipers or wheel cylinders
6 Brakes wet	6 Apply pedal to boil-off water (this should only be a momentary problem)
7 Kinked, clogged or internally split brake hose or line	7 Inspect lines and hoses, replace as necessary

Excessive brake pedal travel (but will pump up)

1 Drum brakes out of adjustment	1 Adjust brakes
2 Air in hydraulic system	2 Bleed system, inspect system for a leak

Excessive brake pedal travel (but will not pump up)

1 Master cylinder pushrod misadjusted	1 Adjust pushrod
2 Master cylinder seals defective	2 Replace master cylinder
3 Brake linings worn out	3 Inspect brakes, replace pads and/or shoes
4 Hydraulic system leak	4 Locate and repair leak

Troubleshooting (continued)

PROBABLE CAUSE	CORRECTIVE ACTION

Brake pedal doesn't return

1 Brake pedal binding	1 Inspect pivot bushing and pushrod, repair or lubricate
2 Defective master cylinder	2 Replace master cylinder

Brake pedal pulsates during brake application

1 Brake drums out-of-round	1 Have drums machined by an automotive machine shop
2 Excessive brake disc runout or disc surfaces out-of-parallel	2 Have discs machined by an automotive machine shop
3 Loose or worn wheel bearings	3 Adjust or replace wheel bearings
4 Loose lug nuts	4 Tighten lug nuts

Brakes slow to release

1 Malfunctioning power booster	1 Replace booster
2 Pedal linkage binding	2 Inspect pedal pivot bushing and pushrod, repair/lubricate
3 Malfunctioning proportioning valve	3 Replace proportioning valve
4 Sticking caliper or wheel cylinder	4 Repair or replace calipers or wheel cylinders
5 Kinked or internally split brake hose	5 Locate and replace faulty brake hose

Brakes grab (one or more wheels)

1 Grease or brake fluid on brake lining	1 Locate and repair cause of contamination, replace lining
2 Brake lining glazed	2 Replace lining, deglaze disc or drum

Vehicle pulls to one side during braking

1 Grease or brake fluid on brake lining	1 Locate and repair cause of contamination, replace lining
2 Brake lining glazed	2 Deglaze or replace lining, deglaze disc or drum
3 Restricted brake line or hose	3 Repair line or replace hose
4 Tire pressures incorrect	4 Adjust tire pressures
5 Caliper or wheel cylinder sticking	5 Repair or replace calipers or wheel cylinders
6 Wheels out of alignment	6 Have wheels aligned
7 Weak suspension spring	7 Replace springs
8 Weak or broken shock absorber	8 Replace shock absorbers

Brakes drag (indicated by sluggish engine performance or wheels being very hot after driving)

1 Brake pedal pushrod incorrectly adjusted	1 Adjust pushrod
2 Master cylinder pushrod (between booster and master cylinder) incorrectly adjusted	2 Adjust pushrod
3 Obstructed compensating port in master cylinder	3 Replace master cylinder
4 Master cylinder piston seized in bore	4 Replace master cylinder
5 Contaminated fluid causing swollen seals throughout system	5 Flush system, replace all hydraulic components
6 Clogged brake lines or internally split brake hose(s)	6 Flush hydraulic system, replace defective hose(s)
7 Sticking caliper(s) or wheel cylinder(s)	7 Replace calipers or wheel cylinders
8 Parking brake not releasing	8 Inspect parking brake linkage and parking brake mechanism, repair as required
9 Improper shoe-to-drum clearance	9 Adjust brake shoes
10 Faulty proportioning valve	10 Replace proportioning valve

PROBABLE CAUSE   **CORRECTIVE ACTION**

Brakes fade (due to excessive heat)

1 Brake linings excessively worn or glazed	1 Deglaze or replace brake pads and/or shoes
2 Excessive use of brakes	2 Downshift into a lower gear, maintain a constant slower speed (going down hills)
3 Vehicle overloaded	3 Reduce load
4 Brake drums or discs worn too thin	4 Measure drum diameter and disc thickness, replace drums or discs as required
5 Contaminated brake fluid	5 Flush system, replace fluid
6 Brakes drag	6 Repair cause of dragging brakes
7 Driver resting left foot on brake pedal	7 Don't ride the brakes

Brakes noisy (high-pitched squeal)

1 Glazed lining	1 Deglaze or replace lining
2 Contaminated lining (brake fluid, grease, etc.)	2 Repair source of contamination, replace linings
3 Weak or broken brake shoe hold-down or return spring	3 Replace springs
4 Rivets securing lining to shoe or backing plate loose	4 Replace shoes or pads
5 Excessive dust buildup on brake linings	5 Wash brakes off with brake system cleaner
6 Brake drums worn too thin	6 Measure diameter of drums, replace if necessary
7 Wear indicator on disc brake pads contacting disc	7 Replace brake pads
8 Anti-squeal shims missing or installed improperly	8 Install shims correctly

Brakes noisy (scraping sound)

1 Brake pads or shoes worn out; rivets, backing plate or brake shoe metal contacting disc or drum	1 Replace linings, have discs and/or drums machined (or replace)

Brakes chatter

1 Worn brake lining	1 Inspect brakes, replace shoes or pads as necessary
2 Glazed or scored discs or drums	2 Deglaze discs or drums with sandpaper (if glazing is severe, machining will be required)
3 Drums or discs heat checked	3 Check discs and/or drums for hard spots, heat checking, etc. Have discs/drums machined or replace them
4 Disc runout or drum out-of-round excessive	4 Measure disc runout and/or drum out-of-round, have discs or drums machined or replace them
5 Loose or worn wheel bearings	5 Adjust or replace wheel bearings
6 Loose or bent brake backing plate (drum brakes)	6 Tighten or replace backing plate
7 Grooves worn in discs or drums	7 Have discs or drums machined, if within limits (if not, replace them)
8 Brake linings contaminated (brake fluid, grease, etc.)	8 Locate and repair source of contamination, replace pads or shoes
9 Excessive dust buildup on linings	9 Wash brakes with brake system cleaner
10 Surface finish on discs or drums too rough after machining (especially on vehicles with sliding calipers)	10 Have discs or drums properly machined
11 Brake pads or shoes glazed	11 Deglaze or replace brake pads or shoes

Brake pads or shoes click

1 Shoe support pads on brake backing plate grooved or excessively worn	1 Replace brake backing plate
2 Brake pads loose in caliper	2 Loose pad retainers or anti-rattle clips
3 Also see items listed under Brakes chatter	

Troubleshooting (continued)

PROBABLE CAUSE   CORRECTIVE ACTION

Brakes make groaning noise at end of stop

1 Brake pads and/or shoes worn out	1 Replace pads and/or shoes
2 Brake linings contaminated (brake fluid, grease, etc.)	2 Locate and repair cause of contamination, replace brake pads or shoes
3 Brake linings glazed	3 Deglaze or replace brake pads or shoes
4 Excessive dust buildup on linings	4 Wash brakes with brake system cleaner
5 Scored or heat-checked discs or drums	5 Inspect discs/drums, have machined if within limits (if not, replace discs or drums)
6 Broken or missing brake shoe attaching hardware	6 Inspect drum brakes, replace missing hardware

Rear brakes lock up under light brake application

1 Tire pressures too high	1 Adjust tire pressures
2 Tires excessively worn	2 Replace tires
3 Defective proportioning valve	3 Replace proportioning valve

Brake warning light on instrument panel comes on (or stays on)

1 Low fluid level in master cylinder reservoir (reservoirs with fluid level sensor)	1 Add fluid, inspect system for leak, check the thickness of the brake pads and shoes
2 Failure in one half of the hydraulic system	2 Inspect hydraulic system for a leak
3 Piston in pressure differential warning valve not centered	3 Center piston by bleeding one circuit or the other (close bleeder valve as soon as the light goes out)
4 Defective pressure differential valve or warning switch	4 Replace valve or switch
5 Air in the hydraulic system	5 Bleed the system, check for leaks
6 Brake pads worn out (vehicles with electric wear sensors - small probes that fit into the brake pads and ground out on the disc when the pads get thin)	6 Replace brake pads (and sensors)

Brakes do not self adjust

Disc brakes

1 Defective caliper piston seals	1 Replace calipers. Also, possible contaminated fluid causing soft or swollen seals (flush system and fill with new fluid if in doubt)
2 Corroded caliper piston(s)	2 Same as above

Rapid brake lining wear

1 Driver resting left foot on brake pedal	1 Don't ride the brakes
2 Surface finish on discs or drums too rough	2 Have discs or drums properly machined
3 Also see Brakes drag	

3   Anti-lock Brake System (ABS) - general information

General information

1   The Anti-lock Brake System is designed to maintain vehicle steerability, directional stability and optimum deceleration under severe braking conditions on most road surfaces. It does so by monitoring the rotational speed of each wheel and controlling the brake line pressure to each wheel during braking. This prevents the wheels from locking up.

2   The ABS system has three main components: the wheel speed sensors, an electronic control unit and a hydraulic unit. Four wheel speed sensors - one at each wheel - send a variable voltage signal to the control unit, which monitors these signals, compares them to its program and determines whether a wheel is about to lock up. When a wheel is about to lock up, the control unit signals the hydraulic unit to reduce hydraulic pressure (or not increase it further) at that wheel's brake caliper. Pressure modulation is handled by electrically-operated solenoid valves within the hydraulic control unit **(see illustration)**.

3   If a problem develops within the system, an ABS warning light will glow on the dashboard. Sometimes, a visual inspection of the ABS system can help you locate the problem. Carefully inspect the ABS wiring harness. Pay particularly close attention to the harness and connections near each wheel. Look for signs of chafing and other damage caused by incorrectly routed wires. If a wheel sensor harness is damaged, it must be replaced along with the sensor (if they are assembled together).

Warning: *Do NOT try to repair an ABS wiring harness. The ABS system is sensitive to even the smallest changes in resistance. Repairing the harness could alter resistance values and cause the system to malfunction. If the ABS wiring harness is damaged in any way, it must be replaced.*

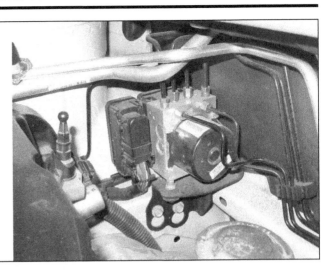

3.2 The integrated electronic and hydraulic control unit is located in the right rear corner of the engine compartment

Caution: *Make sure the ignition is turned off before unplugging or reattaching any electrical connections.*

Diagnosis and repair

4   If the ABS warning light comes on and stays on while the vehicle is in operation, the ABS system requires attention. Although special electronic ABS diagnostic testing tools are necessary to properly diagnose the system, you can perform a few preliminary checks before taking the vehicle to a dealer service department.

a) *Check the brake fluid level in the reservoir.*
b) *Verify that the computer electrical connectors are securely connected.*
c) *Check the electrical connectors at the hydraulic control unit.*
d) *Check the fuses.*
e) *Follow the wiring harness to each wheel and verify that all connections are secure and that the wiring is undamaged.*

5   If the above preliminary checks do not solve the problem, the vehicle should be diagnosed by a dealer service department or other qualified repair shop. Due to the complex nature of the ABS system, all actual repair work must be done by a qualified automotive technician.

Wheel speed sensor - removal and installation

6   Loosen the wheel lug nuts, raise the vehicle and support it securely on jackstands. Remove the wheel.

7   Make sure the ignition key is turned to the Off position.

8   Trace the wiring back from the sensor, detaching all brackets and clips while noting its correct routing, then disconnect the electrical connector.

9   Remove the mounting fastener and carefully detach the sensor from the knuckle (front) or hub and bearing assembly (rear) **(see illustrations)**.

Note: *The rear sensor on AWD models is held in place by a spring-loaded retainer.*

10   Installation is the reverse of the removal procedure. Tighten the bolt to the torque listed in this Chapter's Specifications.

11   Install the wheel and lug nuts, lower the vehicle and tighten the lug nuts to the Chapter 1 Specifications.

3.9a The front wheel speed sensor is mounted to the steering knuckle

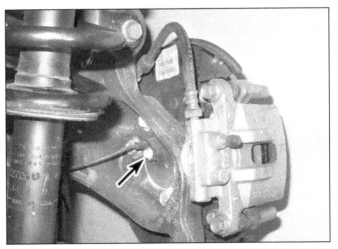

3.9b The rear wheel speed sensor is mounted to the rear hub and bearing assembly (FWD model shown)

4.4 Spray the disc and brake pads with brake cleaner to remove brake dust; DO NOT blow brake dust off with compressed air - collect the contaminated fluid in a suitable container and dispose of it properly!

4.5 Use a C-clamp to press the caliper piston into its bore

4.6a Remove the caliper lower mounting bolt (guide pin) . . .

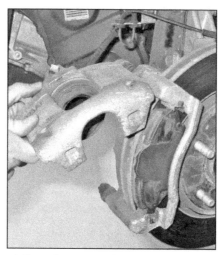

4.6b . . . then rotate the caliper up on the mounting bracket

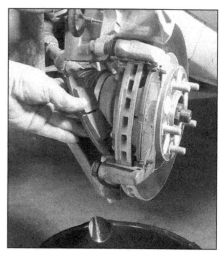

4.6c Remove the inner brake pad

4   Disc brake pads - replacement

Warning: *Disc brake pads must be replaced on both front or both rear wheels at the same time; never replace the pads on only one side. Also, the dust created by the brake system is harmful to your health. Never blow it out with compressed air and don't inhale any of it. An approved filtering mask should be worn when working on the brakes. Do not, under any circumstances, use petroleum-based solvents to clean brake parts. Use brake system cleaner only!*
Caution: *Don't depress the brake pedal with the caliper removed.*

1   Using a syringe or equivalent, remove approximately two-thirds of the fluid from the master cylinder reservoir and discard it.
Caution: *Brake fluid will damage paint. If any fluid is spilled, wash it off immediately with plenty of clean, cold water.*

2   Loosen the wheel lug nuts, raise the end of the vehicle you will be working on and support it securely on jackstands. Block the wheels that remain on the ground.
3   Remove the wheels. Work on one brake assembly at a time, using the assembled brake for reference if necessary.

Front

4   Position a drain pan under the brake assembly and clean the caliper and surrounding area with brake system cleaner **(see illustration)**.
5   Push the piston back into its bore using a C-clamp **(see illustration)**. As the piston is depressed to the bottom of the caliper bore, the fluid in the master cylinder will rise as the brake fluid is displaced. Make sure it doesn't overflow. If necessary, remove more of the fluid.

6   To replace the brake pads, follow the accompanying photos, beginning with **illustration 4.6a**. Stay in order and read the caption under each illustration.
7   While the pads are removed, inspect the caliper for brake fluid leaks and ruptures of the piston dust boot. Replace the caliper if necessary (see Section 5). Also inspect the brake disc carefully (see Section 6). If machining is necessary, follow the information in that Section to remove the disc. Inspect the brake hoses for damage and replace if necessary (see Section 10).
8   Before installing the caliper, clean and inspect the guide pin bolts for corrosion and damage. If they're significantly corroded or damaged, replace them. Also check the guide pin rubber bushings for wear. When installing the caliper, Tighten the guide pin bolts to the torque listed in this Chapter's Specifications.

4.6d Remove the outer brake pad

4.6e Remove the brake pad support plates from the caliper bracket

4.6f Slide the guide pin out and remove the caliper from the mounting bracket

9   Repeat the procedure on the opposite wheel, then install the wheels and lug nuts, lower the vehicle and tighten the lug nuts to the torque listed in the Chapter 1 Specifications.

10   Add the specified type of brake fluid to the reservoir until it's full (see Chapter 1).

11   Pump the brake pedal a few times to bring the pads into contact with the disc. Check the level of the brake fluid, adding some if necessary.

12   Check the operation of the brakes carefully before placing the vehicle into normal service. Try to avoid heavy brake application until the brakes have been applied lightly several times to seat the pads.

Rear

13   Position a drain pan under the brake assembly and clean the caliper and surrounding area with brake system cleaner (**see illustration 4.4**).

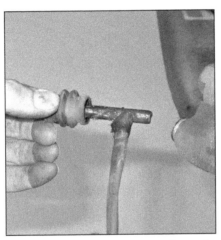

4.6g Pull out the caliper guide pins and clean them, then apply a coat of high-temperature grease to the pins and reinstall the pins in the caliper bracket

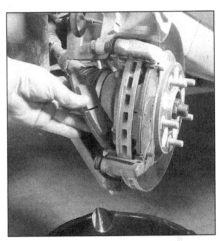

4.6h Clean the support plates, lubricate the wear points with high-temp brake grease and reinstall on the caliper mounting bracket then place the pads in the caliper mounting bracket

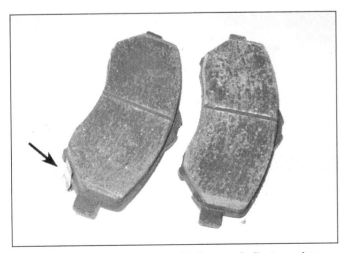

4.6i The inner pad is the one with the wear indicator - when installed, the wear indicator must be positioned at the top

4.6j Install the caliper and tighten the caliper mounting bolts to the torque listed in this Chapter's Specifications

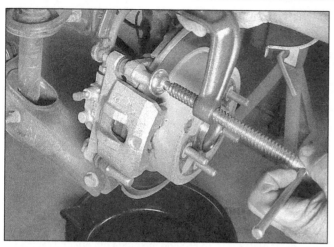

4.14a Use a C-clamp, positioned alternately on each caliper mounting bolt (guide pin) and the caliper bracket, and evenly compress the piston into its bore (but only a little bit at this time)

4.14b Remove the caliper lower mounting (guide pin) bolt

4.14c Rotate the caliper up and hold it in place with wire or a Bungee cord

4.14d Remove the inner pad from the caliper mounting bracket

4.14e Remove the outer pad from the caliper

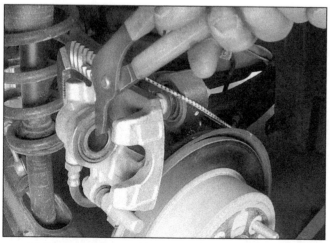

4.14f Compress the piston into its bore to make room for the new pads

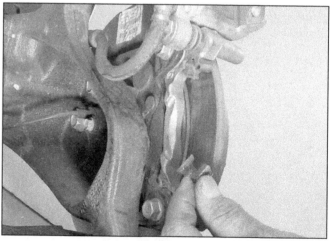

4.14g Remove the pad support plates from the caliper mounting bracket and inspect them. If they are cracked or fit loosely, replace them

4.14h Lubricate the shanks of the caliper mounting bolts (guide pins) with high-temperature brake grease

4.14i Install the shim to the new outer pad . . .

4.14j . . . then install the outer pad and shim to the caliper, making sure the projections on the pad backing plate engage properly with the holes in the caliper frame

4.14k Install the shim to the new inner pad . . .

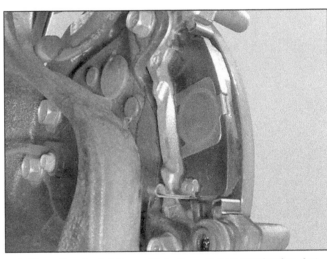

4.14l . . . then install the pad into the caliper mounting bracket

14   To replace the brake pads, follow the accompanying photos, beginning with **illustration 4.14a**. Stay in order and read the caption under each illustration.

15   While the pads are removed, inspect the caliper for brake fluid leaks and ruptures of the piston dust boot. Replace the caliper if necessary (see Section 5). Also inspect the brake disc carefully (see Section 6). If machining is necessary, follow the information in that Section to remove the disc. Inspect the brake hoses for damage and replace if necessary (see Section 10).

16   Before installing the caliper mounting bolts (guide pins), clean and inspect them for corrosion and damage. If they're significantly corroded or damaged, replace them. Also check the guide pin rubber bushings for wear.

Tighten them to the torque listed in this Chapter's Specifications.

17   Repeat the procedure on the opposite wheel, then install the wheels and lug nuts, lower the vehicle and tighten the lug nuts to the torque listed in the Chapter 1 Specifications.

18   Add the specified type of brake fluid to the reservoir until it's up to the appropriate level (see Chapter 1).

19   Pump the brake pedal a few times to bring the pads into contact with the disc. Check the level of the brake fluid, adding some if necessary.

20   Check the operation of the brakes carefully before placing the vehicle into normal service. Try to avoid heavy brake application until the brakes have been applied lightly several times to seat the pads.

4.14m Rotate the caliper back down into place, install the lower mounting bolt (guide pin) and tighten it to the torque listed in this Chapter's Specifications

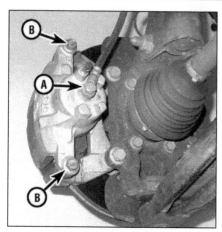

5.2 Remove the banjo bolt (A), then remove the caliper mounting (guide pin) bolts (B) (front caliper)

6.3 The brake pads on this vehicle were obviously neglected - they wore down completely and cut deep grooves into the disc (wear this severe means the disc must be replaced)

6.4a Use a dial indicator to measure disc runout - if the reading exceeds the maximum allowable runout limit, the disc will have to be machined or replaced

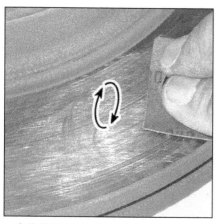

6.4b Using a swirling motion, remove the glaze from the disc surface with sandpaper or emery cloth

5   Disc brake caliper - removal and installation

Warning: *Dust created by the brake system is harmful to your health. Never blow it out with compressed air and don't inhale any of it. An approved filtering mask should be worn when working on the brakes. Do not, under any circumstances, use petroleum-based solvents to clean brake parts. Use brake system cleaner only.*

Note: *If replacement is indicated (usually because of fluid leakage), it is recommended that the calipers be replaced, not overhauled. New and factory rebuilt units are available on an exchange basis. Always replace the calipers in pairs - never replace just one of them.*

Removal

1   Loosen the wheel lug nuts, raise the vehicle and support it securely on jackstands. Remove the wheel.

2   If you're removing a front caliper, remove the banjo bolt and disconnect the brake hose from the caliper **(see illustration)**. Plug the brake hose to keep contaminants out of the brake system and to prevent losing any more brake fluid than is necessary. Discard the sealing washers - new ones should be used during installation. If you're removing a rear caliper, remove the brake hose (see Section 10)
Note: *If the caliper is being removed for access to other components, don't disconnect the hose.*

3   Remove the caliper mounting bolts (guide pins).

Installation

4   Install the caliper by reversing the removal procedure. On front calipers, remember to replace the sealing washers at the brake hose-to-caliper connection. Tighten the caliper mounting bolts (guide pins) to the torque listed in this Chapter's Specifications.

5   Bleed the brake circuit (see Section 11) (only if the brake hose was disconnected). Make sure there are no leaks from the hose connections. If you didn't disconnect the hose, pump the brake pedal several times to bring the pads into contact with the disc.

6   Test the brakes carefully before returning the vehicle to normal service.

6   Brake disc - inspection, removal and installation

Warning: *Dust created by the brake system is harmful to your health. Never blow it out with compressed air and don't inhale any of it. An approved filtering mask should be worn when working on the brakes. Do not, under any circumstances, use petroleum-based solvents to clean brake parts. Use brake system cleaner only.*

Note: *This procedure applies to both front and rear brake discs.*

Inspection

1   Loosen the wheel lug nuts, raise the vehicle and support it securely on jackstands. Remove the wheel and reinstall the lug nuts to hold the disc in place (washers may be required). If the rear brake disc is being worked on, release the parking brake.

2   Remove the brake caliper and pads (see Sections 4 and 5). Don't disconnect the brake hose from the caliper, or you'll have to bleed the brakes when everything is reassembled. After removing the caliper, suspend it out of the way with a piece of wire.

3   Visually inspect the disc surface for score marks and other damage. Light scratches and shallow grooves are normal after use and may not always be detrimental to brake operation, but deep scoring requires disc removal and refinishing by an automotive machine shop. Check both sides of the disc **(see illustration)**. If pulsating has been noticed during application of the brakes, suspect excessive disc runout.

4   To check disc runout, place a dial indicator at a point about 1/2-inch from the outer edge of the disc **(see illustration)**. Set the indicator to zero and turn the disc. The indicator reading should not exceed the specified allowable runout limit. If it does, the disc should be refinished by an automotive machine shop.

Note: *The discs should be resurfaced regardless of the dial indicator reading, as this will impart a smooth finish and ensure a perfectly flat surface, eliminating any brake pedal pulsation or other undesirable symptoms related to questionable discs. At the very least, if you elect not to have the discs resurfaced, remove the glaze from the surface with emery cloth using a swirling motion **(see illustration)**.*

5   It's absolutely critical that the disc not be machined to a thickness under the specified minimum allowable thickness. The minimum thickness is cast into the inside of the disc

6.5a The minimum wear dimension is cast into the back side of the disc (typical)

6.5b Use a micrometer to measure disc thickness

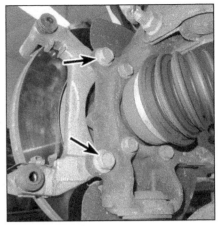

6.6 Remove the caliper mounting bracket bolts and remove the bracket (front caliper shown, rear similar)

(see illustration). The disc thickness can be checked with a micrometer **(see illustration)**. **Note:** *The rear disc has the minimum thickness specification cast into it as well, although the location may vary.*

Removal

6   Remove the caliper mounting bracket bolts **(see illustration)**. Remove the lug nuts which were put on to hold the disc in place and remove the disc from the hub. **Note:** *Remove and discard any retaining clips on the wheel studs that hold the disc to the hub. These clips are not necessary for reinstallation of the brake disc.*

Installation

7   Place the disc in position over the wheel studs. Install the mounting bracket, tightening the bolts to the torque listed in this Chapter's Specifications
8   Install the brake pads and caliper (see

Section 4). Tighten the caliper mounting bolts (guide pins) to the torque listed in this Chapter's Specifications.
9   Install the wheel, lower the vehicle and tighten the lug nuts to the torque listed in the Chapter 1 Specifications.
10   Pump the brake pedal a few times to bring the brake pads into contact with the disc. Bleeding won't be necessary unless the brake hose was disconnected from the caliper. Check the operation of the brakes carefully before driving the vehicle.

7   Drum brake shoes - replacement

Warning: *Drum brake shoes must be replaced on both wheels at the same time - never replace the shoes on only one wheel. Also, the dust created by the brake system is harmful to your health. Never blow it out with compressed air and don't inhale any of it. An approved filtering mask should be worn when working on the*

brakes. Do not, under any circumstances, use petroleum-based solvents to clean brake parts. Use brake system cleaner only!
Caution: *Whenever the brake shoes are replaced, the return and hold-down springs should also be replaced. Due to the continuous heating/cooling cycle the springs are subjected to, they lose tension over a period of time and may allow the shoes to drag on the drum and wear at a much faster rate than normal.*
1   Loosen the wheel lug nuts, raise the rear of the vehicle and support it securely on jackstands. Block the front wheels to keep the vehicle from rolling. Remove the wheels.
2   Release the parking brake and remove the brake drums. If the brake drum is difficult to remove, remove the access plug from the backing plate, insert a small screwdriver through the hole, lift the adjuster lever off the adjusting wheel and turn the wheel with another screwdriver to back off the brake shoes **(see illustrations)**. The drum should now come off.

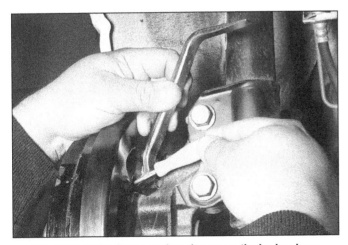

7.2a If the brake drums are hanging up on the brake shoes because of excessive wear, remove the access plug from the backing plate, insert a pair of small screwdrivers (or a screwdriver and a brake adjuster tool, as shown) and retract the shoes

7.2b Lift the adjuster lever off the adjuster wheel with the screwdriver and rotate the adjuster wheel until the shoes are retracted sufficiently to allow drum removal (brake drum removed for clarity)

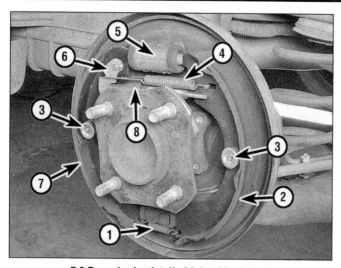

7.3 Drum brake details (right side shown)

1	Lower return spring	5	Wheel cylinder
2	Leading shoe	6	Adjuster lever
3	Hold-down spring	7	Trailing shoe
4	Adjuster spring	8	Adjuster

7.4a Before disassembling the brake, wash it thoroughly with brake system cleaner and allow it to dry - position a drain pan under the brake to catch the residue - DO NOT use compressed air to blow the brake dust off!

3   All four rear brake shoes must be replaced at the same time, but to avoid mixing up parts, work on only one brake assembly at a time **(see illustration)**.

4   Before disassembling anything, wash off the brake assembly with brake system cleaner **(see illustration)**. Follow the accompanying illustrations for the brake shoe replacement procedures **(see illustrations)**. Be sure to stay in order and read the caption under each illustration.

5   Before reinstalling the drum, it should be checked for cracks, score marks, deep scratches and hard spots, which will appear as small discolored areas. If the hard spots cannot be removed with fine emery cloth or if any of the other conditions listed above exist, the drum must be taken to an automotive

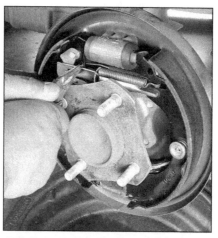

7.4b Unhook the adjuster spring from the adjuster lever

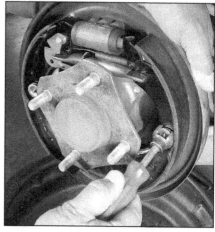

7.4c Remove the hold-down spring from the leading shoe

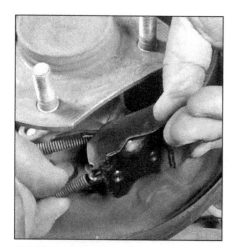

7.4d Unhook the lower return spring from the leading shoe

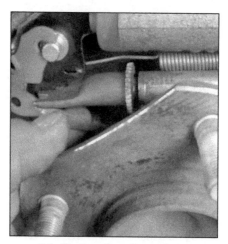

7.4e Remove the adjuster lever

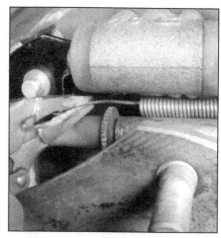

7.4f Unhook the upper return spring from the trailing shoe

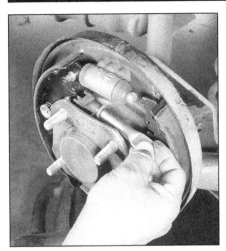

7.4g Remove the leading shoe and adjuster

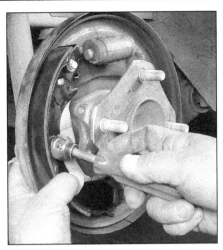

7.4h Remove the hold-down spring from the trailing shoe

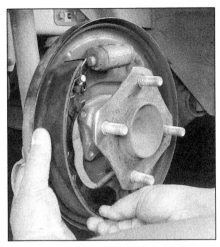

7.4i Remove the trailing shoe from the backing plate . . .

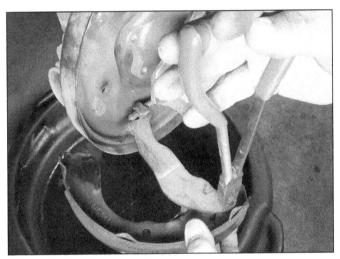

7.4j . . . then remove the retaining clip and detach the parking brake lever from the shoe

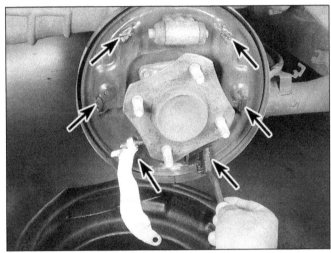

7.4k Clean the backing plate, then apply a thin film of high-temperature brake grease to the brake shoe contact areas

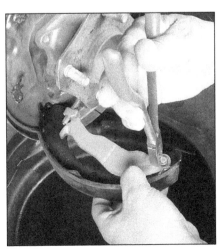

7.4l Connect the parking brake lever to the new trailing shoe and secure it with a new retaining clip

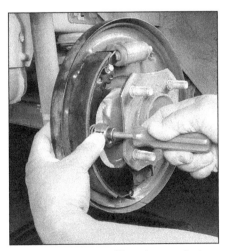

7.4m Place the trailing shoe against the backing plate and install the hold-down spring

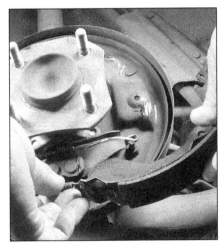

7.4n Attach the lower return spring to the bottom of each shoe

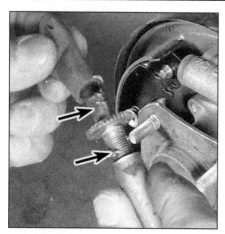

7.4o Clean and lubricate the moving parts of the adjuster . . .

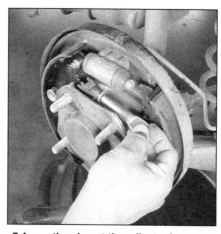

7.4p . . . then insert the adjuster between the two shoes, making sure it properly engages the notches

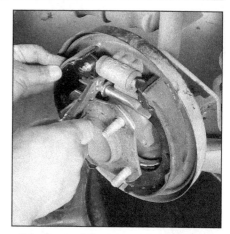

7.4q Connect the upper return spring to both shoes

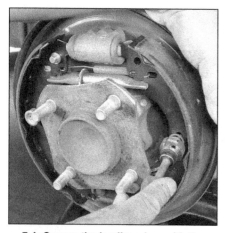

7.4r Secure the leading shoe with the hold-down spring

7.4s Install the adjuster lever - make sure it's properly engaged with the pivot pin and the adjuster

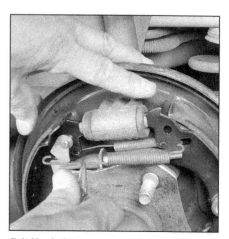

7.4t Hook the rear end of the return spring to the adjuster lever and the front end into the hole on the leading shoe

machine shop to have it resurfaced.
Caution: *Professionals recommend resurfacing the drums each time a brake job is done. Resurfacing will eliminate the possibility of*

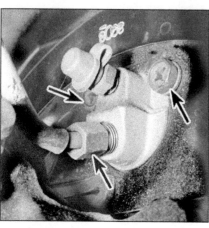

8.2 To detach the wheel cylinder from the brake backing plate, disconnect the brake line fitting, then remove the wheel cylinder bolts

out-of-round drums. If the drums are worn so much that they can't be resurfaced without exceeding the maximum allowable diameter, which is stamped into the drum, then new ones will be required. At the very least, if you elect not to have the drums resurfaced, remove the glaze from the surface with emery cloth using a swirling motion.

6   Install the brake drum.
7   To make a preliminary adjustment of the brake, turn the adjuster star wheel until the brakes just begin to drag on the drum as it is turned, then back-off the star wheel until no dragging can be heard when you rotate the drum. Depress the brake pedal firmly several times, then rotate the drum to ensure that the brakes are not dragging. If they are, back off the star wheel a little more.
8   Mount the wheel, install the lug nuts, then lower the vehicle. Tighten the wheel lug nuts to the torque listed in the Chapter 1 Specifications.
9   Make a number of forward and reverse stops and operate the parking brake to adjust the brakes until satisfactory pedal action is obtained.

10   Check the operation of the brakes carefully before driving the vehicle.

8   Wheel cylinder - removal and installation

Note: *If replacement is warranted (usually because of fluid leakage or sticky operation) explore all options before beginning the job. New wheel cylinders are available, which makes this job quite easy. Never replace only one wheel cylinder. Always replace both of them at the same time.*

Removal

1   Remove the rear brake shoes (see Section 7).
2   Using a flare-nut wrench (if available), disconnect the brake line fitting from the wheel cylinder **(see illustration)**. Plug the end of the brake line to prevent fluid loss and contamination.
3   Remove the two bolts securing the wheel cylinder to the backing plate and remove the wheel cylinder.

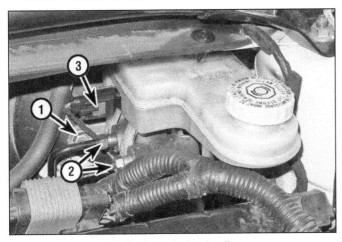

9.6 Master cylinder details:

1   *Mounting nut (other nut not visible)*
2   *Brake line fittings*
3   *Brake fluid level sensor connector*

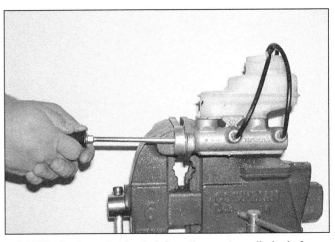

9.10 The best way to bleed air from the master cylinder before installing it on the vehicle is with a pair of bleeder tubes that direct brake fluid into the reservoir during bleeding

Installation

4   Installation is the reverse of removal. Tighten the wheel cylinder mounting bolts to the torque listed in this Chapter's Specifications. Tighten the line fitting securely.
5   Install the brake shoes and brake drum (see Section 7).
6   Bleed the brakes (see Section 11). Carefully test brake operation before resuming normal operation.

9   Master cylinder - removal and installation

Caution: *Brake fluid will quickly damage paint. Cover all body parts and be careful not to spill fluid during any of the following procedures. Wipe up any spilled fluid immediately and then flush the area thoroughly with water.*

Removal

Note: *The master cylinder is located in the engine compartment, mounted to the power brake booster.*
1   Disconnect the negative battery cable from the remote ground terminal or battery (see Chapter 5).
2   With the engine off, pump the brake pedal several times to relieve the vacuum reserve inside the power brake booster.
Note: *This step will prevent contaminants from being sucked into the power brake booster when the cylinder is removed.*
3   Remove the Powertrain Control Module (PCM) and its mounting bracket (see Chapter 6).
4   Using a syringe or equivalent, siphon the brake fluid from the master cylinder reservoir and dispose of it properly.
5   Disconnect the brake fluid level switch connector and move the harness out of the way.

6   Place rags under the fluid fittings and prepare caps or plastic bags to cover the ends of the lines once they are disconnected. Loosen the fittings at the ends of the brake lines where they enter the master cylinder **(see illustration)**. Pull the brake lines slightly away from the master cylinder and quickly plug the ends to prevent contamination.
Note: *To prevent rounding off the corners on these nuts, the use of a flare-nut wrench, which wraps around the nut, is preferred.*
7   Thoroughly clean the area where the master cylinder mounts to the power booster. Remove the nuts attaching the master cylinder to the power booster **(see illustration 9.6)**. Pull the master cylinder off the studs and out of the engine compartment. Again, be careful not to spill any fluid as this is done.
8   If necessary, disengage the retaining tabs holding the reservoir and transfer the reservoir to the new master cylinder.
Note: *Install new seals between the master cylinder and reservoir.*

Installation

9   Bench bleed the new master cylinder before installing it. Mount the master cylinder in a vise, with the jaws of the vise clamping on the mounting flange.
10   Attach a pair of master cylinder bleeder tubes to the outlet ports of the master cylinder **(see illustration)**.
11   Fill the reservoir with brake fluid of the recommended type (see Chapter 1).
12   Slowly push the pistons into the master cylinder (a large Phillips screwdriver can be used for this) - air will be expelled from the pressure chambers and into the reservoir. Because the tubes are submerged in fluid, air can't be drawn back into the master cylinder when you release the pistons.
13   Repeat the procedure until no more air bubbles are present.
14   Remove the bleed tubes, one at a time,

and install plugs in the open ports to prevent fluid leakage and air from entering. Install the reservoir cap.
15   Install a new vacuum seal onto the master cylinder where it mates with the power booster.
Warning: *Do not skip this step or a vacuum leak could occur and render the power booster ineffective; this results in greatly increased pedal effort and longer stopping distances.*
16   Install the master cylinder over the studs on the power brake booster and tighten the attaching nuts only finger tight at this time.
17   Carefully thread the brake line fittings into the master cylinder. Since the master cylinder is still a bit loose, it can be moved slightly in order for the fittings to thread in easily. Do not strip the threads as the fittings are tightened.
18   Fully tighten the mounting nuts, and then the brake line fittings. Tighten the nuts to the torque listed in this Chapter's Specifications.
19   Fill the master cylinder reservoir with fluid, then bleed the master cylinder and the brake system (see Section 11). To bleed the cylinder on the vehicle, have an assistant depress the brake pedal and hold the pedal to the floor. Loosen the fitting to allow air and fluid to escape. Repeat this procedure on both fittings until the fluid is clear of air bubbles.
Caution: *Have plenty of rags on hand to catch the fluid - brake fluid will ruin painted surfaces. After the bleeding procedure is completed, rinse the area under the master cylinder thoroughly with clean water.*
20   Test the operation of the brake system carefully before placing the vehicle into normal service.
Warning: *Do not operate the vehicle if you are in doubt about the effectiveness of the brake system. It is possible for air to become trapped in the anti-lock brake system hydraulic control unit; if the pedal continues to feel spongy after*

10.3 Front brake hose/line details

1 Brake line fitting
2 Brake hose retaining clip
3 Brake hose-to-strut bracket
4 Brake line banjo fitting at caliper

10.12 Rear brake hose/line details

1 Brake line fitting
2 Brake hose retaining clip
3 Brake hose threaded fitting

repeated bleedings or the BRAKE or ANTI-LOCK light stays on, have the vehicle towed to a dealer service department or other qualified shop to be bled with the aid of a scan tool.

---

10 Brake hoses and lines - inspection and replacement

Brake hose inspection

1   Whenever the vehicle is raised and supported securely on jackstands, the rubber hoses which connect the steel brake lines with the front and rear brake assemblies should be inspected for cracks, chafing of the outer cover, leaks, blisters and other damage. These are important and vulnerable parts of the brake system and inspection should be thorough. A light and mirror will be helpful for a complete check. If a hose exhibits any of the above conditions, replace it immediately.

Flexible hose replacement

Front

2   Clean all dirt away from the hose and line fittings.
3   Using a flare-nut wrench, disconnect the metal brake line from the hose fitting and immediately plug the metal line to prevent excessive leakage and contamination **(see illustration)**. Be careful not to bend the metal line. If the threaded fitting is corroded, spray it with a penetrating oil and allow it to soak in for about 10 minutes, then try again. If you try to break loose a brake tube nut that's stuck, you will kink the metal line, which will then have to be replaced.
4   Remove the clip retaining the brake hose to the bracket.

5   Remove the bolt retaining the brake hose bracket to the strut.
6   Unscrew the banjo bolt at the caliper and remove the hose, discarding the sealing washers on either side of the fitting.
7   Attach the new brake hose to the caliper.
Note: *When replacing the brake hoses, always use new sealing washers.*
8   Tighten the banjo bolt to the torque listed this Chapter's Specifications.
9   Attach the brake hose bracket to the strut, making sure the hose isn't kinked or twisted. Then connect the hose to the bracket on the chassis and install the retaining clip.
10   Connect the metal line to the hose fitting, tightening the fitting securely.

Rear

11   Clean all dirt away from the hose and line fittings.
12   Using a flare-nut wrench, disconnect the metal brake line from the hose fitting and immediately plug the metal line to prevent excessive leakage and contamination **(see illustration)**. Be careful not to bend the metal line. If the threaded fitting is corroded, spray it with a penetrating oil and allow it to soak in for about 10 minutes, then try again. If you try to break loose a brake tube nut that's stuck, you will kink the metal line, which will then have to be replaced.
13   Remove the clip retaining the brake hose to the bracket.
14   Unscrew the brake hose from the caliper.
15   Thread the new hose into the caliper and tighten it securely.
16   Connect the hose to the bracket, making sure it isn't twisted, then install the retaining clip.

17   Connect the brake line fitting to the hose and tighten it securely.

Front or rear

18   Carefully check to make sure the suspension or steering components don't make contact with the hose. Have an assistant push down on the vehicle while you watch to see whether the hose interferes with suspension operation. If you're replacing a front hose, have your assistant turn the steering wheel lock-to-lock while you make sure the hose doesn't interfere with the steering linkage or the steering knuckle.
19   After installation, check the master cylinder fluid level and add fluid as necessary. Bleed the brakes (see Section 11). Carefully test brake operation before resuming normal operation.

Metal brake line replacement

20   When replacing brake lines, be sure to use the correct parts. Do not use copper tubing for any brake system components. Purchase steel brake lines from a dealer parts department or auto parts store.
21   Prefabricated brake lines, with the tube ends already flared and fittings installed, are available at auto parts stores and dealer parts departments. These lines can be bent to the proper shapes using a tubing bender.
22   When installing the new line make sure it's well supported in the brackets and has plenty of clearance between moving or hot components. Make sure you tighten the fittings securely.
23   After installation, check the master cylinder fluid level and add fluid as necessary. Bleed the brakes (see Section 11). Carefully test brake operation before resuming normal operation.

11   Brake system - bleeding

Warning: *The following procedure is a manual bleeding procedure. This is the only bleeding procedure which can be performed at home without special tools. However, if air has found its way into the hydraulic control unit, the entire system must be bled manually, then with a DRB scan tool (or equivalent), then manually a second time. If the brake pedal feels spongy even after bleeding the brakes, or the ABS light on the instrument panel does not go off, or if you have any doubts whatsoever about the effectiveness of the brake system, have the vehicle towed to a dealer service department or other repair shop equipped with the necessary tools for bleeding the system.*

Warning: *Wear eye protection when bleeding the brake system. If the fluid comes in contact with your eyes, immediately rinse them with water and seek medical attention.*

Note: *Bleeding the hydraulic system is necessary to remove any air that manages to find its way into the system when it's been opened during removal and installation of a hydraulic component.*

1   It will be necessary to bleed the complete system if air has entered the system due to low fluid level, or if the brake lines have been disconnected at the master cylinder.

2   If a brake line was disconnected only at a wheel, then only that caliper or wheel cylinder must be bled.

3   If a brake line is disconnected at a fitting located between the master cylinder and any of the brakes, that part of the system served by the disconnected line must be bled. The following procedure describes bleeding the entire system, however.

4   Remove any residual vacuum from the brake power booster by applying the brake several times with the engine off.

5   Remove the cap from the master cylinder reservoir and fill the reservoir with brake fluid. Reinstall the cap.

Note: *Check the fluid level often during the bleeding operation and add fluid as necessary to prevent the fluid level from falling low enough to allow air bubbles into the master cylinder.*

6   Have an assistant on hand, as well as a supply of new brake fluid, a clear container partially filled with clean brake fluid, a length of clear tubing to fit over the bleeder valve and a wrench to open and close the bleeder valve.

7   Begin the bleeding process by bleeding the first wheel in the bleeding sequence, loosen the bleeder valve slightly, then tighten it to a point where it is snug but can still be loosened quickly and easily. The bleeding sequence is as follows:

   Left rear
   Right front
   Right rear
   Left front

8   Place one end of the hose over the bleeder valve and submerge the other end in brake fluid in the container **(see illustration)**.

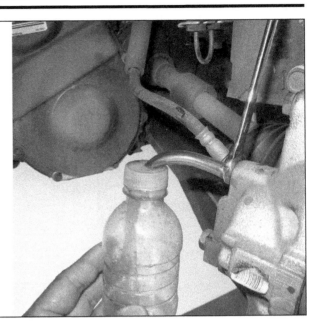

11.8 When bleeding the brakes, a hose is connected to the bleed screw at the caliper or wheel cylinder and then submerged in clean brake fluid - air will be seen as bubbles exiting the tube (all air must be expelled before moving to the next wheel)

9   Have the assistant push the brake pedal slowly to the floor, then hold the pedal firmly depressed.

10   While the pedal is held depressed, open the bleeder valve just enough to allow a flow of fluid to leave the valve. Watch for air bubbles to exit the submerged end of the tube. When the fluid flow slows after a couple of seconds, close the valve and have your assistant release the pedal.

11   Repeat Steps 9 and 10 until no more air is seen leaving the tube, then tighten the bleeder valve and proceed to bleed the other calipers/wheel cylinders, in the proper sequence, using the same procedure. Check the fluid in the master cylinder reservoir frequently.

Note: *Be careful not to over-tighten the bleeder valve.*

12   Never use old brake fluid. It contains moisture which can boil, rendering the brakes inoperative.

13   Refill the master cylinder with fluid at the end of the operation.

14   Check the operation of the brakes. The pedal should feel solid when depressed, with no sponginess. If necessary, repeat the entire process.

Warning: *If, after bleeding the system, you do not have a firm brake pedal, if the ABS light on the instrument panel does not go off, or if you have any doubts whatsoever about the effectiveness of the brake system, have it towed to a dealer service department or other repair shop to have the system bled.*

12   Power brake booster - check, removal and installation

Operating check

1   Depress the pedal and start the engine. If the pedal goes down slightly, operation is normal.

2   Depress the brake pedal several times with the engine running and make sure that there is no change in the pedal reserve distance.

Airtightness check

3   Start the engine and turn it off after one or two minutes. Depress the brake pedal several times slowly. If the pedal goes down farther the first time but gradually rises after the second or third depression, the booster is airtight.

4   Depress the brake pedal while the engine is running, then stop the engine with the pedal depressed. If there is no change in the pedal reserve travel after holding the pedal for 30 seconds, the booster is airtight.

Removal

5   The power brake booster unit requires no special maintenance apart from periodic inspection of the vacuum hoses and the case. The booster should never be disassembled. If a problem develops, it must be replaced with a new one.

6   Remove any vacuum from the booster by pumping the pedal several times with the engine off, until the pedal feels hard to push.

7   Disconnect the cable from the remote ground terminal or battery (see Chapter 5).

8   On 2014 and earlier models, unbolt the Powertrain Control Module (PCM and set it aside (see Chapter 9).

9   On 2015 and later models, unbolt the Power Distribution Center (PDC, or fuse box) and its bracket. Also remove the Anti-lock Brake System Integrated Control Unit (ICU)

10   Clean the area where the master cylinder attaches to the power brake booster.

11   Remove the master cylinder (see Section 9). Also disconnect the brake lines from the junction block below the power brake booster. Plug all open lines.

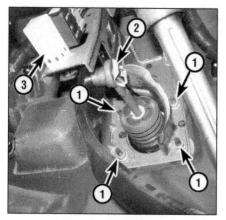

12.14 Power brake booster mounting details:

1   Mounting nuts
2   Pushrod-to-brake pedal retaining clip (DO NOT reuse)
3   Brake light switch

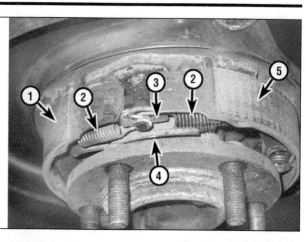

13.4a Parking brake shoe details (right side, viewed from above)

1   Trailing shoe
2   Upper return spring
3   Anchor plate
4   Strut
5   Leading shoe

13.4b Parking brake shoe details (right side, viewed from below)

1   Hold-down spring and retainer
2   Trailing shoe
3   Lower return spring
4   Adjuster screw assembly
5   Leading shoe

12   Disconnect the vacuum hose from the check valve that's located on the outside of the brake booster.
Warning: *Do not remove the check valve from the booster.*
13   Working under the dash, disconnect and remove the brake light switch (see Section 15).
14   Disconnect the brake pedal pushrod from the top of the brake pedal by inserting a small screwdriver into the center of the retaining clip and carefully prying the tang back **(see illustration)**. For safety reasons, discard the old pushrod retaining clip and buy a new clip for reassembly.
15   Remove the nuts attaching the booster to the firewall.
16   Working inside the engine compartment, carefully withdraw the brake booster unit from the firewall and out of the engine compartment.

Installation

17   To install the booster, place it into position on the firewall, then tighten the retaining nuts to the torque listed in this Chapter's Specifications. Connect the brake pedal to the brake booster pushrod using a new retaining clip.
Warning: *DO NOT reuse the old booster pushrod retaining clip.*
18   The remainder of installation is the reverse of removal. Install a new vacuum seal on the master cylinder before reinstalling it to the power brake booster.
19   Bleed the brake system (see Section 11).
20   Carefully test the operation of the brakes before placing the vehicle into normal operation.

13   Parking brake shoes (models with rear disc brakes) - replacement

Warning: *Parking brake shoes must be replaced on both wheels at the same time - never*

replace the shoes on only one wheel. Also, the dust created by the brake system is harmful to your health. Never blow it out with compressed air and don't inhale any of it. An approved filtering mask should be worn when working on the brakes. Do not, under any circumstances, use petroleum-based solvents to clean brake parts. Use brake system cleaner only!

1   Loosen the rear wheel lug nuts, raise the rear end of the vehicle and support it securely on jackstands. Block the front wheels to keep the vehicle from rolling. Release the parking brake and remove the rear wheels.
2   Remove the brake caliper and brake disc (see Sections 5 and 6).
3   Once the disc is removed, clean the parking brake assembly with brake system cleaner.
4   Using locking pliers, unhook and remove the springs **(see illustrations)**.
5   Remove the adjuster assembly (noting which end is facing forward) and strut from between the brake shoes.
6   Grasp one of the shoe hold-down cups with pliers and push it toward the brake backing plate to compress the hold-down spring. Twist the cup 1/4-turn to align the slot in the hold-down pin with the cup, then release the spring pressure (the pin will pass through the cup slot) and take off the cup and spring. Repeat this with the cup and spring on the other parking brake shoe.
7   Take the shoes off the backing plate. Disengage the parking brake lever from the cable.

8   Check all parts for wear and damage, paying special attention to metal-to-metal contact points. Replace worn or damaged parts. The parking brake lever is integral with the shoe it's attached to.
Note: *If the vehicle has high mileage, it's a good idea to replace all of the springs as well as any parts that have visible problems.*
9   Check the parking brake drum surface inside the brake disc for score marks, cracks, deep scratches and hard spots, which will appear as small discolored areas. If the hard spots cannot be removed with emery cloth or if any of the other conditions are seen, the drum must be resurfaced by an automotive machine shop.
Note: *If you don't have the drums resurfaced, remove the glazing from the surface with emery cloth or sandpaper using a swirling motion.*
10   Apply a small amount of high-temperature brake grease to the friction points of the backing plate and adjuster screw assembly.
11   Reverse the removal steps to install the brake shoes. The shoe-to-anchor spring with the paint mark is installed with the paint mark toward the rear of the vehicle (for both left and right sides). Expand the shoes, using the automatic adjuster, until the drum will just fit over them.
12   Install the brake disc and caliper. Remove the rubber plug in the hub portion of the disc, then turn the adjuster screw until the disc will not turn. Now turn the adjuster in the opposite direction five notches.

14.4 Parking brake adjustment nut

15.1 Brake light switch location

13   Adjust the parking brake (see Section 14).
14   Install the wheel and lug nuts. Lower the vehicle and tighten the lug nuts to the torque listed in the Chapter 1 Specifications.
15   Check the operation of the parking brake.

14   Parking brake - adjustment

1   Loosen the wheel lug nuts. Raise the rear of the vehicle and support it securely on jackstands, then remove the wheels.
2   Adjust the parking brake shoes (models with rear disc brakes; see Section 13) or the brake shoes (models with rear drum brakes; see Section 7).
3   Remove the center console (see Chapter 11).
4   Turn the parking brake cable adjustment nut until the parking brake lever travels 5 to 7 clicks when applied with moderate force **(see illustration)**.
5   Release the parking brake lever and verify that the parking brake shoes (models with rear disc brakes) or brake shoes (models with rear drum brakes) don't drag as the rear discs or drums are turned.
6   Reinstall the console and the wheels. Tighten the wheel lug nuts to the torque listed in the Chapter 1 Specifications.

15   Brake light switch - check, replacement and adjustment

1   The brake light switch is located along the arm of the brake pedal and is attached to a bracket near the power brake booster mount **(see illustration)**. When the brake pedal is applied, the pedal arm moves away from the switch and a spring-loaded plunger closes the circuit to the brake lights.
2   Models equipped with cruise control use a dual-purpose brake light switch that also deactivates the cruise control system when the brake pedal is depressed.

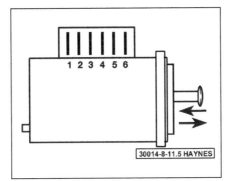

15.5a With the switch closed (plunger extended fully), there should be continuity between terminals 1 and 2. When the pedal is at rest (plunger depressed) there should be continuity between terminals 3 and 4, and 5 and 6 (2007 and 2008 models)

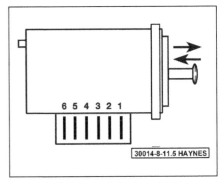

15.5b With the switch closed (plunger extended fully), there should be continuity between terminals 1 and 2. When the pedal is at rest (plunger depressed) there should be continuity between terminals 3 and 4, and 5 and 6 (2009 and later models)

Check

Caution: *This switch can only be adjusted once and this occurs when the switch is installed. Do not move the small lever on the switch or remove the switch unless you intend on replacing it. Once the switch is removed, it cannot be reused.*
3   Check the brake light fuse (see Chapter 12). If the fuse has blown, replace it. If it blows again, look for a short in the brake light circuit.
4   If the fuse is okay, use a test light or voltmeter to verify that there's voltage to the switch. If there's no voltage to the switch, look for an open or short in the power wire to the switch. Repair as necessary.
5   If the brake lights still don't come on when the brake pedal is applied, unplug the electrical connector from the brake light switch and, using an ohmmeter, verify that there is continuity between the switch terminals when the brake pedal is applied (the switch is closed and the plunger is extended). If continuity is not detected, replace the switch **(see illustrations)**.

6   If there is continuity between the switch terminals when the brake is applied (it closes the circuit), but the brake lights don't come on when the brake pedal is applied, check for power to the brake light bulb sockets when the pedal is depressed. If voltage is present, replace the bulbs (it isn't very likely that all of them would fail simultaneously, but it is possible that they could be burned out). If voltage is not available, check the wiring between the switch and the brake lights for an open circuit and repair as necessary.

Replacement and adjustment

7   Disconnect the negative battery cable from the remote ground terminal or battery (see Chapter 5).
8   Depress and hold the brake pedal, then rotate the brake light switch about 30-degrees in a counterclockwise direction and remove it from the mounting bracket.
9   Unplug the electrical connector from the switch, remove the switch from the vehicle and discard it.

10   Pull the plunger out on the new switch, depress the brake pedal and install the switch into the bracket by aligning the slots, inserting the switch and rotating it about 30-degrees clockwise.

Caution: *Do not move the small lever on the switch before it is fully installed.*

11   Release the brake pedal and gently pull it back to make certain that it is seated against the pedal striker. The switch plunger will ratchet backward to the correct position.

12   Plug the electrical connector into the switch.

13   Reconnect the battery and test the brake lights for proper operation.

---

16   Brake vacuum pump - removal and installation

Note: *2015 and later models are equipped with either a mechanical or electric brake vacuum pump. The mechanical pump is driven by the rear end of the camshaft and mounted on the left end of the cylinder head on 2.4L engines, and the electric pump on 3.6L engines is located at the left side of the engine and secured to a bracket by two mounting bolts.*

1   Disconnect the cable from the negative battery terminal (see Chapter 5).

Mechanical vacuum pump (camshaft driven) (2.4L engines)

Note: *Be careful not to damage the plastic vacuum pump hose connector fitting. Also, oil may leak from the pump so be sure to have plenty of rags ready to soak up spills.*

2   Remove the engine cover, then disconnect the quick-connect hose fitting from the pump.

3   Remove the vacuum pump mounting bolts, then carefully separate the pump from the cylinder head. Discard the old O-ring seals.

4   Remove the retaining screw and separate the plastic connector piece from the pump. Check and replace the O-ring on the fitting, if necessary.

Chapter 10
Suspension and steering systems

Contents

	Section			Section
Balljoints - replacement	6		Shock absorber/coil spring assembly (rear) - removal and installation .	9
Control arm - removal, inspection, and installation	5		Stabilizer bar and bushings (front) - removal and installation	2
Front load beam (2015 and later models) - removal and installation	27		Steering column - removal and installation	20
			Steering gear - removal and installation	21
General information	1		Steering knuckle - removal and installation	8
Hub and bearing assembly (front) - removal and installation	7		Steering wheel - removal and installation	18
Hub and bearing assembly (rear) - removal and installation	14		Strut assembly (front) - removal, inspection and installation	3
Lower control arm (rear) - removal and installation	17		Strut/coil spring - replacement	4
Power steering pump (2014 and earlier models) - removal and installation	22		Subframe - removal and installation	24
			Tie-rod ends - removal and installation	19
Power steering system - bleeding (2014 and earlier models)	23		Toe link - removal and installation	12
Rear stabilizer bar - removal and installation	13		Trailing link - removal and installation	11
Rear suspension knuckle - removal and installation	15		Upper control arm (rear) - removal and installation	16
Shock absorber/coil spring assembly (rear) - component replacement	10		Wheel alignment - general information	26
			Wheels and tires - general information	25

Specifications

Torque specifications

Ft-lbs (unless otherwise indicated)

Note: *One foot-pound (ft-lb) of torque is equivalent to 12 inch-pounds (in-lbs) of torque. Torque values below approximately 15 ft-lbs are expressed in inch-pounds, because most foot-pound torque wrenches are not accurate at these smaller values.*

Front suspension

2014 and earlier

Lower control arm front pivot bolt	129
Lower control arm rear pivot bolt nut	107
Hub and bearing mounting bolts	60
Stabilizer bar link nuts	35
Stabilizer bar bracket bolts	44
Strut-to-steering knuckle bolt/nuts	103
Strut damper shaft nut	44
Strut upper mounting nuts	41
Tie-rod end jam nut	55
Tie-rod end-to-steering knuckle nut	63
Balljoint stud nut	70
Driveaxle/hub nut	See Chapter 8
Subframe reinforcement bracket bolts	37
Subframe mounting bolts	100
Longitudinal crossmember bolts	41
Wheel lug nuts	See Chapter 1

2015 and later

Lower control arm horizontal pivot bolt/nut*	
Step 1	74
Step 2	Tighten an additional 135-degrees
Lower control arm vertical bolt/nut*	
Step 1	55
Step 2	Tighten an additional 80-degrees
Balljoint pinch bolt/nut*	
Step 1	44
Step 2	Tighten an additional 48-degrees
Hub and bearing mounting bolts*	70
Stabilizer bar bracket bolts	48
Stabilizer bar link nuts	36
Strut-to-steering knuckle bolt/nut*	81
Strut rod-to-strut mount nut	52
Subframe-to-body upper bolts	96
Subframe-to-body lower bolts	89
Load beam bolts	33
Load beam bracket-to-frame bolts	70

* DO NOT reuse these fasteners.

Torque specifications (continued)

Ft-lbs (unless otherwise indicated)

Note: *One foot-pound (ft-lb) of torque is equivalent to 12 inch-pounds (in-lbs) of torque. Torque values below approximately 15 ft-lbs are expressed in inch-pounds, because most foot-pound torque wrenches are not accurate at these smaller values.*

Rear suspension

2014 and earlier

Hub and bearing mounting bolts	77
Lower control arm-to-knuckle bolt/nut	77
Lower control arm-to-crossmember bolt/nut	77
Shock absorber lower mounting bolt/nut	73
Shock absorber upper mounting nuts	41
Shock absorber damper shaft nut	30
Stabilizer bar bracket bolts	18
Stabilizer link mounting nuts	35
Toe link-to-knuckle mounting bolt/nut	77
Toe link cam bolt/nut	74
Trailing link-to-body mounting bolts	81
Trailing link-to-knuckle bolts	
2009 and earlier models	44
2010 to 2014 models	55
Upper control arm-to-knuckle bolt/nut	77
Upper control arm-to-crossmember bolt nut	77
Wheel lug nuts	See Chapter 1

2015 and later

Hub and bearing mounting bolts*	70
Upper control arm bolts/nuts (either end)*	
Step 1	44
Step 2	Tighten an additional 90-degrees
Shock absorber-to-knuckle bolt	133
Shock absorber-to-body bolts	52
Shock absorber center rod nut	22
Lower control arm-to-knuckle bolt/nut*	
Step 1	59
Step 2	Tighten an additional 90-degrees
Lower control arm-to-frame bolt/nut*	96
Stabilizer bar bracket bolts	37
Stabilizer link-to-lower control arm nut	41
Stabilizer link-to-stabilizer bar nut	33
Toe link-to-knuckle bolt/nut*	
Step 1	66
Step 2	Tighten an additional 90-degrees
Toe link-to-frame bolt/nut*	111
Trailing link-to-body mounting bolts*	77
Trailing link-to-knuckle mounting bolts*	81
Wheel lug nuts	See Chapter 1

Steering system

Steering wheel bolt	37
Airbag module bolts (if equipped)	120 in-lbs
Steering column mounting fasteners	
2014 and earlier	21
2015 and later	18
Steering gear mounting bolts	
2010 and earlier models	74
2011 and later models	85
Fluid hose tube nut-to-steering gear (pressure and return)	24
Fluid hose tube nut-to-power steering pump	24
Tie-rod end-to-knuckle nut	
2010 and earlier models	63
2011 to 2014 models	48
2015 and later models*	
Step 1	22
Step 2	Tighten an additional 90-degrees
Tie-rod end jam nut	55
Power steering pump mounting bolts (2014 and earlier models)	
2.4L engine	
2010 and earlier models	19
2011 to 2014 models	
25 mm bolts	15
90 mm bolts	19
2.7L and 3.5L engines	22
3.6L engine	15
Intermediate shaft coupling bolt	
2009 and earlier models	35
2010 to 2014 models	31
2015 and later models	22

* *DO NOT reuse these fasteners.*

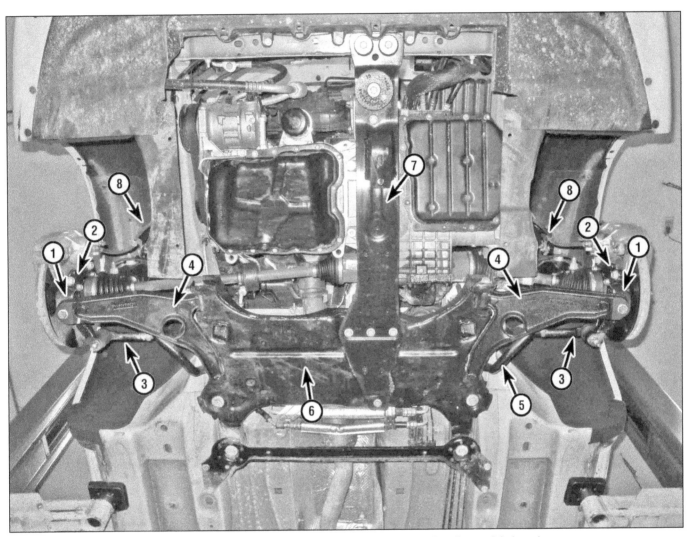

1.1 Front suspension and steering components (2014 and earlier model shown)

1	Balljoint	4	Control arm	7	Longitudinal crossmember
2	Steering knuckle	5	Stabilizer bar	8	Strut/coil spring assembly
3	Tie-rod end	6	Front suspension crossmember		

1   General information

1   The front suspension **(see illustration)** on these vehicles is a MacPherson strut design. The upper end of each strut is attached to the vehicle's body strut support. The lower end of the strut is connected to the upper end of the steering knuckle. The lower end of the steering knuckle is attached by a balljoint mounted to the outer end of the suspension control arm. The control arm is connected to the front suspension crossmember. A stabilizer bar, mounted to the crossmember and connected to the strut, reduces body roll while cornering.

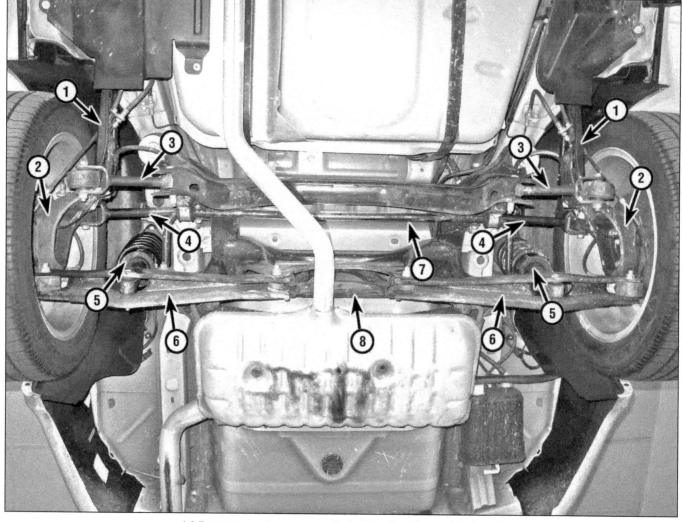

1.2 Rear suspension components (2014 and earlier model shown)

1	Trailing link	4	Upper control arm	7	Stabilizer bar
2	Knuckle	5	Shock absorber/coil spring assembly	8	Crossmember
3	Toe link	6	Lower control arm		

2   The rear suspension **(see illustration)** is of a multi-link design, using upper and lower control arms, toe-links, trailing links, coil-over shock absorber assemblies, and a stabilizer bar.

3   The power-assisted rack-and-pinion steering gear is attached to the front suspension main crossmember. The steering gear actuates the tie-rods, which are attached to the steering knuckles. The steering column is designed to collapse in the event of an accident.

4   Frequently, when working on the suspension or steering system components, you may come across fasteners that seem impossible to loosen. These fasteners on the underside of the vehicle are continually subjected to water, road grime, mud, etc., and can become rusted or frozen, making them extremely difficult to remove. In order to unscrew these stubborn fasteners without damaging them (or

other components), Use lots of penetrating oil and allow it to soak in for a while. Using a wire brush to clean exposed threads will also ease removal of the nut or bolt and prevent damage to the threads. Sometimes a sharp blow with a hammer and punch will break the bond between a nut and bolt threads, but care must be taken to prevent the punch from slipping off the fastener and ruining the threads. Heating the stuck fastener and surrounding area with a torch sometimes helps too, but isn't recommended because of the obvious dangers associated with fire. Long breaker bars and extension, or cheater, pipes will increase leverage, but never use an extension pipe on a ratchet - the ratcheting mechanism could be damaged. Sometimes tightening the nut or bolt first will help to break it loose. Fasteners that require drastic measures to remove should always be replaced with new ones.

5   Since most of the procedures dealt with in this Chapter involve jacking up the vehicle and working underneath it, a good pair of jackstands will be needed. A hydraulic floor jack is the preferred type of jack to lift the vehicle, and it can also be used to support certain components during various operations.
Warning: *Never, under any circumstances, rely on a jack to support the vehicle while working on it.*

6   Whenever any of the suspension or steering fasteners are loosened or removed they must be inspected and, if necessary, replaced with new ones of the same part number or of original equipment quality and design. Torque specifications must be followed for proper reassembly and component retention. Never attempt to heat or straighten any suspension or steering components. Instead, replace any bent or damaged part with a new one.

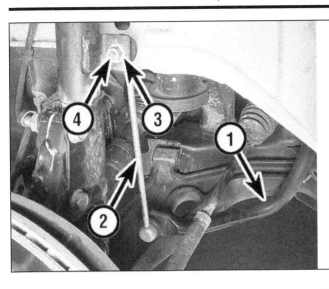

2.3 Stabilizer bar link details

1   Stabilizer bar
2   Stabilizer bar link
3   Link nut
4   Link stud

2   Stabilizer bar and bushings (front) - removal and installation

Removal

1   Disconnect the negative cable from the battery or the remote battery terminal (see Chapter 5).

2   Loosen the front wheel lug nuts, raise the front of the vehicle and support it securely on jackstands. Apply the parking brake and block the rear wheels to keep the vehicle from rolling off the stands. Remove the front wheels.

Note: *There is no need to remove the sub-frame if you are just replacing the stabilizer bar links.*

3   Remove the stabilizer bar link nuts to detach the link from the stabilizer bar **(see illustration).**

Note: *Hold the stud on the link with a wrench so that it does not rotate when removing the nut. The link can be detached from the sta-bilizer bar and the strut assembly entirely if replacement is necessary.*

4   Remove the under-vehicle splash shield.

2014 and earlier models

5   Remove the rear engine mount (see Chapter 2A, 2B, 2C or 2D).

6   Remove the front engine mount through-bolt.

7   Remove the two screws attaching the power steering hose clamps to the rear of the crossmember.

8   Remove the heat shield.

9   Remove the two bolts that secure the steering gear to the crossmember. Support the steering gear with a rope, bungee cord or wire to hold the it in place as the crossmember is lowered.

All models

10   Remove the subframe (see Section 24). Remove the stabilizer bar bracket bolts and remove the brackets and bushings **(see illus-tration).** Remove the stabilizer bar.

Installation

11   Install the stabilizer bar, bushings and brackets onto the crossmember, but do not tighten them yet.

12   Center the stabilizer bar on the crossmember. Tighten the bracket bolts to the torque listed in this Chapter's Specifications.

13   Align the crossmember with the marks you made earlier and install the mounting bolts and screws. Tighten to the torque listed in this Chapter's Specifications.

14   If removed, install the engine mount and power steering line brackets/fasteners.

15   Install the stabilizer links and tighten the nuts to the torque listed in this Chapter's Specifications.

16   Install the wheels and lug nuts. Lower the vehicle and tighten the wheel lug nuts to the torque listed in the Chapter 1 Specifications.

3   Strut assembly (front) - removal, inspection and installation

Removal

Warning: *Always replace the struts and/or coil springs in pairs - never replace just one strut or one coil spring; this could cause dangerous handling peculiarities.*

Note: *If both strut assemblies are going to be removed, mark the assemblies Right and Left so they will be reinstalled on the cor-rect side.*

1   Loosen the wheel lug nuts, raise the vehicle and support it securely on jackstands. Remove the wheels and detach the flex hose mounting bracket from the strut.

2   Disconnect the stabilizer bar link from the strut assembly (see Section 2).

2014 and earlier models

3   Mark the position of the strut to the steer-ing knuckle (this is only necessary if special camber adjusting bolts have been installed in place of the regular strut-to-knuckle bolts).

4   Remove the nuts and bolts, then sepa-rate the strut from the steering knuckle **(see illustration).**

Caution: *The bolts are serrated and must not be turned. Hold the bolts with a wrench, then remove the strut-to-knuckle nuts. Knock the bolts out with a brass hammer, noting which way the bolt heads face.*

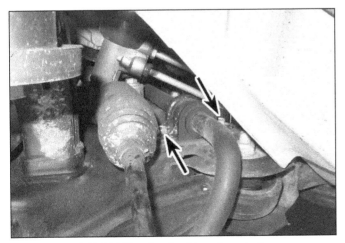

2.10 Stabilizer bar bushing clamp bolts

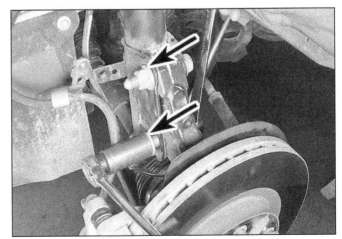

3.4 Hold the bolts with a wrench while turning the nuts (2014 and earlier model shown)

3.5 Strut assembly upper mounting nuts (2014 and earlier models)

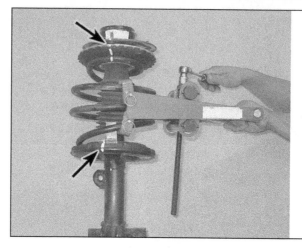

4.3 The strut assembly is carefully clamped in a vise with necessary components marked for disassembly and with a spring compressor installed

5   Secure the steering knuckle safely aside. Have an assistant support the strut and spring assembly, then remove the three strut-to-body nuts **(see illustration)**. Remove the assembly from the fenderwell.

2015 and later models

Warning: *2015 and later models are not equipped with strut upper mounting bolts - instead, there is a special clip that holds the upper strut mount in place to the vehicle body. DO NOT reuse this clip after removing it, and obtain a new one for installation.*

6   Remove the steering knuckle (see Section 8).

7   With the help of an assistant to hold up the strut and prevent it from falling downward, use two flathead screwdrivers to pry apart the two sections of the strut upper mount retaining clip until it breaks. Discard the clip.

8   Remove the strut assembly from the fenderwell.

Inspection

9   Check the strut body for leaking fluid, dents, cracks and other obvious damage which would warrant repair or replacement.

10   Check the coil spring for chips or cracks in the spring coating (this will cause premature spring failure due to corrosion). Inspect the spring seat for cuts, hardness and general deterioration.

11   If any undesirable conditions exist, proceed to the strut disassembly procedure (see Section 4).

Installation

2014 and earlier models

12   Guide the strut assembly up into the fenderwell and insert the upper mounting studs through the holes in the body. Once the studs protrude, install the nuts so the strut won't fall back through. This is most easily accomplished with the help of an assistant, as the strut is quite heavy and awkward.

13   Tighten the upper mounting nuts to the torque listed in this Chapter's Specifications.

14   Slide the steering knuckle into the strut flange and insert the two bolts. Install the nuts, align the previously made match marks (if applicable) and tighten them to the torque listed in this Chapter's Specifications.
Note: *Make certain that the bolts are installed in their original direction; the direction is different for each side of the vehicle.*

2015 and later models

15   Guide the strut assembly up into the fenderwell and align the hole/tab in the strut mount to the hole on the vehicle body.

16   Once strut is aligned and held up against the body with the help of an assistant, install the NEW strut retaining clip, pushing each half into place. Note that the grooves on each half of the clip must fully engage.

17   Install the steering knuckle (see Section 8). Tighten all steering knuckle and related fasteners to the torque listed in this Chapter's Specifications.

All models

18   Install the flex hose mounting bracket.

19   Connect the stabilizer bar link to the strut. Tighten the nut to the torque listed in this Chapter's Specifications.

20   Install the wheel and lug nuts, then lower the vehicle and tighten the lug nuts to the torque listed in the Chapter 1 Specifications.

21   Have the front end alignment checked, and if necessary, adjusted.

4   Strut/coil spring - replacement

Warning: *Struts and/or coil springs must be replaced in pairs - never replace just one of them.*
Note: *You'll need a spring compressor for this procedure. Spring compressors are available on a daily rental basis at most auto parts stores or equipment rental yards.*

1   If the struts or coil springs exhibit the tell-tale signs of wear (leaking fluid, loss of damping capability, chipped, sagging or cracked coil springs) explore all options before beginning any work. The strut body is not serviceable and must be replaced if a problem develops. However, complete strut assemblies (with springs) may be available on an exchange basis, which eliminates much time and work. Whichever route you choose to take, check on the cost and availability of parts before disassembling your vehicle.
Warning: *Disassembling a strut assembly is potentially dangerous and utmost attention must be directed to the job, or serious injury may result. Use only a high-quality spring compressor and carefully follow the manufacturer's instructions furnished with the tool. After removing the coil spring from the strut, set it aside in a safe, isolated area.*

Disassembly

2   Remove the strut and spring assembly (see Section 3).

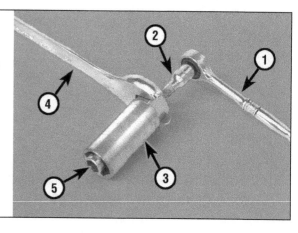

4.5 Here's the setup that can be used to unscrew the damper shaft nut

1   1/4-inch drive ratchet
2   Extension
3   18 mm deep socket
4   Wrench to turn socket
5   8 mm socket (to hold damper shaft)

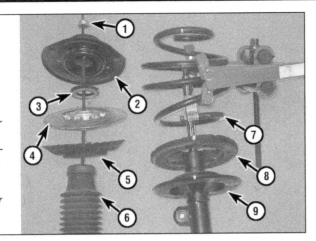

4.6 Front strut/coil assembly details

1   Nut
2   Upper mount
3   Pivot bearing
4   Upper spring seat
5   Upper spring isolator
6   Dust boot (rubber bumper underneath - hidden by dust boot)
7   Coil spring (with spring compressor)
8   Lower spring isolator
9   Damper unit

15   Install the nut on the damper shaft and tighten it to the torque listed in this Chapter's Specifications.
16   Loosen the coil spring compressor until the top coil is properly seated against the upper spring seat and upper mount. Relieve all tension from the spring compressor and remove the tool from the coil spring.
17   Install the strut/spring assembly (see Section 3).

5   Control arm - removal, inspection, and installation

Removal

1   Loosen the lug nuts on both front wheels, raise the front of the vehicle and support it securely on jackstands. Remove the wheels.

2014 and earlier models

2   Remove the nut that attaches the balljoint to the control arm (see illustration).
3   Release the balljoint from the control arm by using the proper balljoint separation tool (see illustration). Do not separate the knuckle from the control arm yet.
Note: *Once the balljoint is separated, do not pull outward on the knuckle; this can lead to separation and damage to the driveaxle inner CV joint.*
4   Detach the stabilizer links from each end of the stabilizer bar (see Section 2). Rotate the stabilizer bar upward and out of the way.

2015 and later models

Note: *NEW control arm mounting fasteners MUST be used when installing on these models.*
5   Remove the lower engine cover.
6   Remove the lower control arm-to-steering knuckle balljoint pinch bolt and nut, noting the direction in which they were installed.
7   Place a prybar into the opening in the control arm, then pry downward to separate the balljoint from the steering knuckle.
8   Remove the front vehicle load beam (see Section 27).

3   Mount the strut clevis bracket portion of the strut assembly in a vise and mark the components for reassembly **(see illustration)**.
Caution: *Do not clamp any other portion of the strut assembly in the vise; it will be damaged. Line the vise jaws with wood or rags to prevent damage to the unit and don't tighten the vise excessively.*
4   Following the tool manufacturer's instructions, install the spring compressor on the spring and compress it sufficiently to relieve all pressure from the upper mount **(see illustration 4.3)**. This can be verified by wiggling the spring.
5   While holding the damper shaft from turning, loosen the shaft nut with a socket. A special tool is available to do this, but a substitute can be made from a 18 mm socket (with a hex surface at the top), a ratchet, a 1/4-inch drive extension inserted through the hole in the spark plug socket, and an 8 mm socket attached to the extension **(see illustration)**.
6   Remove the nut and upper mount **(see illustration)**. Inspect the pivot bearing for smooth operation. If it doesn't turn smoothly, replace it. Remove the upper spring seat and check the upper spring isolator for cracking and general deterioration. Replace any parts that are damaged or worn.

7   Carefully lift the compressed spring from the assembly.
Warning: *When removing the compressed spring, lift it off very carefully and set it in a safe place. Keep the ends of the spring away from your body.*
8   Remove the dust boot from the damper shaft.
9   Slide the rubber bumper off the damper shaft. Check the lower spring isolator for cracking and hardness; replace it if necessary **(see illustration 4.6)**.

Reassembly

10   Extend the damper rod to its full length and install the rubber bumper.
11   Install the dust boot onto the damper.
12   Carefully place the coil spring onto the damper. Align the coil spring on the damper using the reference marks made during disassembly. If a new spring or strut damper unit is being installed, use the marks on the old component to help you orient the spring properly.
13   Install the upper spring isolator and seat onto the damper shaft, noting reference marks.
14   Install the upper mount to the damper shaft, noting its alignment.

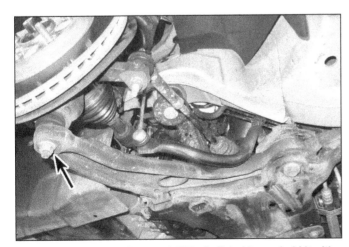

5.2 Balljoint to control arm nut. If the ballstud turns, hold it with an Allen wrench

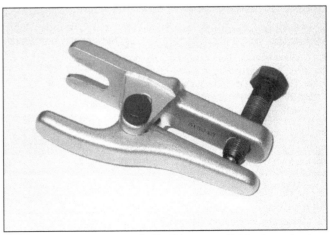

5.3 Balljoint separator tools like this are available at most auto parts stores and, when used properly, won't damage the balljoint boot

5.9 Control arm-to-subframe bolts

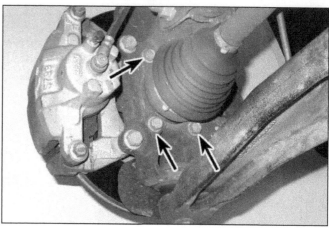

7.4 Remove the hub/bearing mounting bolts
(one of the bolts is not visible in this photo)

All models

9   Remove the control arm-to-subframe bolts **(see illustration)** and remove the control arm.

Inspection

10   Make sure the control arm is straight. If it is bent, replace it. Do not attempt to straighten a bent control arm.
11   Inspect all bushings for cracks, distortion, and tears. If a bushing is torn or worn, replace the control arm.

Installation

12   Position the control arm in the subframe and install the mounting bolts and nuts, but do not tighten them yet. Note that it is easiest to install the vertical-facing bolt and nut first, then the horizontal-facing bolt/nut after.
Note: *Make certain that the rear bushing is positioned correctly in the crossmember.*

2014 and earlier models

13   Reconnect the balljoint to the control arm and tighten the NEW nut to the torque listed in this Chapter's Specifications.
14   Tighten the front and rear control arm mounting bolts to the torque listed in this Chapter's Specifications.

2015 and later models

15   Join the lower control arm balljoint stud into the steering knuckle, then raise the lower control arm with a jack until it's at normal ride height. When the control arm is in this raised position, tighten the control arm-to-subframe bolts to this Chapter's Specifications. Remove the floor jack.
16   Install the front load beam (see Section 27).
17   Making sure the control arm ballstud is inserted fully into the steering knuckle, install a new balljoint pinch bolt/nut and tighten it to the torque listed in this Chapter's Specifications.

All models

18   Install the wheel and lug nuts, lower the vehicle and tighten the lug nuts to the torque listed in the Chapter 1 Specifications.
19   Have the front wheel alignment checked and, if necessary, adjusted.

6   Balljoints - replacement

Note: *On 2015 and later models, the balljoint is an integral part of the control arm and is not serviced separately. If the balljoint is faulty, the entire control arm must be replaced.*

1   Loosen the wheel lug nuts and driveaxle/hub nut (see Chapter 8), raise the vehicle and support it securely on jackstands. Remove the wheel.
2   Remove the brake caliper and disc (see Chapter 9).
3   Remove the clip that holds the wheel speed sensor onto the knuckle. Remove the speed sensor bolt (see Chapter 9) and pull the sensor up and out of the knuckle.
4   Remove the balljoint-to-control arm nut **(see illustration 5.2)**, then separate the balljoint from the control arm using a balljoint separator tool **(see illustration 5.3)**.
5   Carefully pull the knuckle outwards and off of the driveaxle, being careful not to overextend the inner CV joint. If the splines stick in the hub, press the driveaxle out with a puller. Support the driveaxle with a length of wire or rope.
6   There is a small hole in the knuckle at the balljoint cap. Use a drift punch to knock the snap-ring off the balljoint.
7   Using a balljoint press tool and the appropriate adapters (available for rent at many auto parts stores or equipment rental yards), place the tools over the balljoint and proceed to tighten the press tool until the balljoint comes out of the knuckle.
8   Installation is the reverse of removal. Use a new balljoint stud nut and tighten all fasteners to the torque listed in this Chapter's Specifications.

7   Hub and bearing assembly (front) - removal and installation

Removal

Note: *If the hub/bearing assembly cannot be removed easily and appears frozen in the steering knuckle, it will have to be pressed out*

of the steering knuckle. If this is the case, remove the steering knuckle (see Section 8) and take it to an automotive machine shop or other repair facility for service.
1   Loosen the driveaxle/hub nut (see Chapter 8).
2   Loosen the wheel lug nuts, raise the vehicle, support it securely on jackstands and remove the wheel. Remove the hub nut.
3   Remove the brake caliper, the caliper mounting bracket, brake disc and the backing plate from the hub (see Chapter 9). Also disconnect the ABS wheel speed sensor from the knuckle (see Chapter 9).
Note: *Be sure to support the brake caliper as described in Chapter 9.*
4   Remove the hub/bearing assembly mounting bolts from the rear of the steering knuckle **(see illustration)**.
5   Remove the hub/bearing assembly from the steering knuckle.
Note: *If the driveaxle splines stick in the hub, push the driveaxle out of the hub with a two-jaw puller. Be careful not to overextend the driveaxle inner CV joint.*

Installation

6   Make sure that the mounting surfaces inside the steering knuckle and on the driveaxle splines are smooth and free of burrs and nicks prior to installing the hub/bearing assembly.
7   Lubricate the driveaxle splines with multi-purpose grease. Install the hub/bearing assembly onto the driveaxle and into the steering knuckle until it is seated on the steering knuckle.
8   Install the hub/bearing assembly-to-steering knuckle bolts. Tighten the bolts equally in a criss-cross pattern until the hub/bearing assembly is seated securely against the steering knuckle. Tighten the bolts to the torque listed in this Chapter's Specifications.
9   Install the driveaxle/hub nut and tighten it to the torque listed in the Chapter 8 Specifications.
10   Install the brake disc, the caliper mounting bracket and the caliper; tighten the fasteners to the torque values listed in the Chapter 9 Specifications.

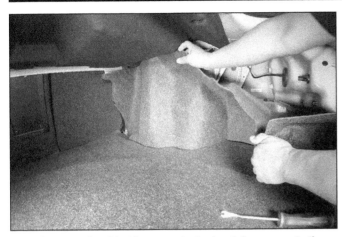

9.1 Remove the side trim panes in the trunk for access to the shock absorber upper mounting fasteners (2014 and earlier)

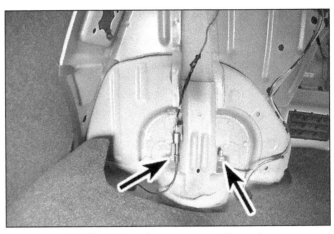

9.3 Shock absorber upper mounting nuts (2014 and earlier)

11   Install the wheel and lug nuts, remove the jackstands, and lower the vehicle.
12   Tighten the lug nuts to the torque listed in the Chapter 1 Specifications.

8   Steering knuckle - removal and installation

Note: *The steering knuckle is not a repairable component. It must be replaced if it is damaged in any way.*

Removal

1   Loosen the wheel lug nuts and the driveaxle/hub nut (see Chapter 8). Raise the vehicle and support it securely on jackstands.
2   Remove the wheel.
3   Remove the driveaxle/hub nut.
4   Remove the brake disc and the ABS front wheel speed sensor (see Chapter 9).
5   Detach the tie-rod end from the steering knuckle (see Section 19).
6   Separate the balljoint from the control arm (see Section 5).
7   Remove the steering knuckle-to-strut bolt(s), noting their direction (see Section 3).
Caution: *On 2014 and earlier models, the steering knuckle-to-strut assembly bolts are serrated and must not be turned during removal - turn the nuts only.*
Note: *On 2014 and earlier models, if the strut assembly is attached to the steering knuckle using a cam bolt in the lower slotted hole, mark the relationship of the cam bolt to the strut to preserve the wheel alignment setting on reassembly (see Section 3).*
8   Remove the steering knuckle from the strut, separate the driveaxle from the hub splines and suspend it safely aside.
Caution: *Do not separate the inner CV joint during this operation. Do not allow the driveshaft to hang by the inner CV joint. If the driveaxle splines stick in the hub, push the driveaxle out with a two-jaw puller.*

Installation

9   Installation is the reverse of removal. Tighten all suspension fasteners to the torque

listed in this Chapter's Specifications.
10   Install the wheel, lower the vehicle and tighten the lug nuts to the torque listed in the Chapter 1 Specifications. Tighten the driveaxle/hub nut to the torque listed in the Chapter 8 Specifications.
11   Have the front end alignment checked and, if necessary, adjusted.

9   Shock absorber/coil spring assembly (rear) - removal and installation

Warning: *Always replace the shock absorbers and/or coil springs in pairs - never replace just one shock absorber or one coil spring; this could cause dangerous handling peculiarities.*

2014 and earlier models

1   Remove trim panels from the sides of the trunk **(see illustration)**.
2   Loosen the rear wheel lug nuts. Raise the vehicle and support it securely on jackstands. Remove the rear wheels.
3   Remove the two nuts in the trunk that hold the shock absorber to the vehicle body **(see illustration)**.
4   Remove the trim panels covering the lower control arm, if equipped.
5   Remove the shock absorber lower mounting nut and bolt and remove the shock unit **(see illustration)**.

2015 and later models

Note: *2015 and later models are equipped with shock absorbers alone, without the coil spring attached. For the replacement of the coil spring on these models, see Section 10.*
6   Raise the rear of the vehicle and support it securely on jackstands. Remove the wheels. Remove the rear fenderwell splash shield (see Chapter 11).
7   Using a floor jack, support the knuckle and raise it slightly to take the load off the shock mounting bolts. The jack must remain under the knuckle until the shock is installed.

8   Remove the shock absorber lower (shock-to-knuckle) mounting bolt.
Note: *If necessary, loosen the fuel filler tube bolt and position aside the filler tube to gain better access to the shock upper mounting bolts.*
9   Remove the shock absorber upper (shock-to-body) mounting bolts, then remove the shock absorber assembly from the vehicle.
10   If necessary, remove the nut and transfer over the shock upper mount to the new shock. Tighten the center nut to the torque listed in this Chapter's Specifications, once installed.

All models

11   Installation is the reverse of removal. It is a good idea to grease the lower mount with white lithium grease to facilitate installation and prolong the life of the lower shock bushing. If you plan on removing the springs and overhauling the shock unit, follow the procedures in Section 10.
Note: *When installed, the side of the upper mount that protrudes more must be facing towards the outside of the vehicle on 2014 and earlier models.*

9.5 Shock absorber lower mounting bolt/nut (2014 and earlier)

12   On 2015 and later models, tighten the shock absorber upper bolts to the torque listed in this Chapter's Specifications, at this stage.

13   Raise the rear suspension with a floor jack positioned under the knuckle to simulate normal ride height, then tighten the lower mounting bolt to the torque listed in this Chapter's Specifications.

14   Install the wheel and lug nuts. Lower the vehicle and tighten the lug nuts to the torque listed in the Chapter 1 Specifications.

15   On 2014 and earlier models, tighten the shock absorber upper mounting nuts to the torque listed in this Chapter's Specifications. Reinstall the trim panels in the trunk.

---

10   Shock absorber/coil spring assembly (rear) - component replacement

Warning: *Always replace the shock absorbers and/or coil springs in pairs - never replace just one shock absorber or one coil spring; this could cause dangerous handling peculiarities.*

Removal

1   Loosen the rear wheel lug nuts. Raise the vehicle and support it securely on jackstands. Remove the rear wheels.

2   Remove the rear shock absorber/coil spring assembly (see Section 9).

Note: *If you plan on servicing both shocks at the same time, mark the parts right and left to ensure proper reassembly.*

3   Compress the coil spring with a spring compressor; read and follow the tool manufacturer's instructions on placement of the hooks and proper usage.

4   Compress the spring until the tension is removed from the upper bracket.

5   After the spring has been compressed, hold the damper shaft with a wrench while you remove the damper shaft nut.

Warning: *Do not use air or power tools on the nut or damper shaft. Damage may occur or even severe injury.*

6   While keeping the spring compressed, remove the shock through the bottom of the spring and remove all of the parts from the shock body, inspecting each one for cracks, wear, or other signs of damage. If damage is found, replace that part.

Installation

7   Install the parts on the new shock as follows.

> Lower spring isolator
> Spring
> Upper spring isolator
> Dust shield
> Jounce bumper
> Washer
> Bushing
> Sleeve
> Upper mounting bracket
> Bushing
> Washer
> Nut

Note: *The upper mount must be oriented so that the side that protrudes more faces outward when installed in the vehicle.*

8   Install the washer and damper shaft nut. Hold the damper shaft with a wrench and tighten the nut to the torque listed in this Chapter's Specifications.

9   Once the nut is tight and tools removed, slowly release the tension from the spring by backing off on the spring compressor. Ensure that your spring has settled in the locator holes and ensure the brackets/mounts are not twisted or off center.

10   Install the shock absorber/coil spring assembly (see Section 9).

Coil spring (2015 and later models)

Warning: *Always replace the coil springs in pairs - never replace just one coil spring; this could cause dangerous handling peculiarities.*

Note: *2015 and later models are not equipped with rear "coil-over" shock absorber assemblies; they have stand-alone coil springs.*

11   Raise the rear of the vehicle and support it securely on jackstands. Remove the wheels.

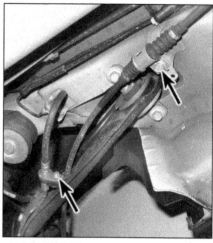

11.2 Brake hose and parking brake cable bracket bolts

12   Disconnect the stabilizer bar link from the lower control arm (see Section 13).

13   Using a floor jack, support the knuckle and raise it slightly to take the load off the lower control arm bolt/nut.

14   Remove the lower control arm-to-knuckle bolt and nut (see Section 17).

15   Lower the control arm slowly until the spring can safely be removed.

16   Installation is the reverse of removal.

---

11   Trailing link - removal and installation

1   Loosen the wheel lug nuts. Raise the vehicle and support it securely on jackstands. Remove the wheels.

2   Remove the bolts that hold the brake hose to the trailing link and the bolt that holds the parking brake cable near the forward end of the trailing arm **(see illustration)**. Remove the brake line from the clip on the trailing link. Also remove the wheel speed sensor wire from the clips on the trailing arm.

11.7 Trailing link-to-rear knuckle bolts
(2014 and earlier model shown, 2015 and later model similar)

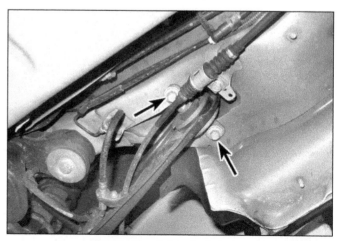

11.8 Trailing link-to-body bolts
(2014 and earlier model shown, 2015 and later model similar)

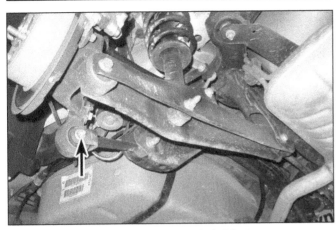

**12.2 Toe link-to-knuckle bolt/nut
(2014 and earlier model shown)**

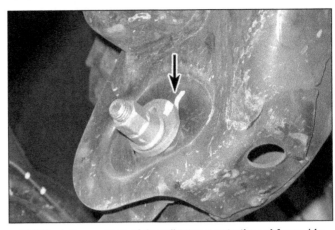

**12.3 Mark the position of the adjuster cam to the subframe (do
this on the other side of the subframe, too)**

3   If your vehicle is equipped with disc brakes in the rear, remove the caliper and the adapter as a full assembly. Hang it out of the way using wire, zip-ties, or equivalent. Do not allow the caliper to hang or over-extend the brake line; this can lead to brake failure.

4   Remove the parking brake cable from the parking brake lever on the shoe by pushing in on the cable spring and lifting the cable up and out of the lever.

5   Locate the hairpin just behind the support plate; pull this pin to remove the parking brake cable from the support plate.

6   The support plate with the parking brake shoes still attached should slide off the vehicle, and the cable can be fully removed from the trailing link.

7   Remove the three bolts that secure the knuckle to the trailing link **(see illustration)**.

8   Remove the two bolts that attach the trailing link to the vehicle body **(see illustration)**. Remove the trailing link.

9   Install the body side of the trailing link first and tighten the bolts to the torque listed in this Chapter's Specifications.

10   Installation is otherwise the reverse of removal. Tighten the trailing arm-to-knuckle

bolts to the torque listed in this Chapter's Specifications.

12   Toe link - removal and installation

Removal

1   Loosen the wheel lug nuts. Raise the vehicle and support it securely on jackstands. Remove the wheels.

2   Remove the nut and bolt that secure the link to the knuckle **(see illustration)**.

3   Mark the position of the inner pivot bolt adjuster cam to the subframe using a marker or crayon **(see illustration)**. Do not scratch the finish of the crossmember to make your mark. Hold the bolt in place and remove the nut.

4   Remove the bolt and detach the toe link from the subframe.

Installation

5   Install the inner pivot and adjuster cams, lining up the marks you made earlier.

6   Raise the rear suspension with a floor jack placed under the knuckle to simulate normal

ride height, then tighten the fasteners to the torque listed in this Chapter's Specifications.

7   Install the wheels and lug nuts, lower the vehicle and tighten the lug nuts to the torque listed in the Chapter 1 Specifications.

13   Rear stabilizer bar - removal and installation

Removal

1   Loosen the rear wheel lug nuts. Raise the rear of vehicle and support it securely with jackstands. Remove the rear wheels.

2   Remove the upper stud nut from the link on each side of the stabilizer bar **(see illustration)**. If the ballstud turns, hold it with an Allen wrench. On 2015 and later models, also remove the rear crossmember brackets from each side - they are located near the stabilizer bar mounting brackets.

3   Remove the stabilizer bar bushing bracket bolts **(see illustration)**. On 2015 and later models, also remove the rear crossmember brackets from each side - they are located near the stabilizer bar mounting brackets.

13.2 Detach the stabilizer bar links from the bar

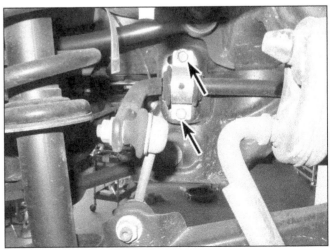

13.3 Stabilizer bar bushing bracket bolts

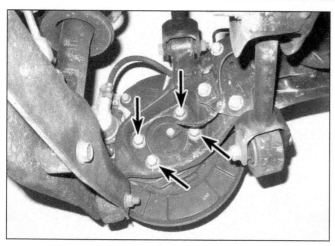

**14.4 Rear hub and bearing assembly mounting bolts
(2014 and earlier model shown, 2015 and later model similar)**

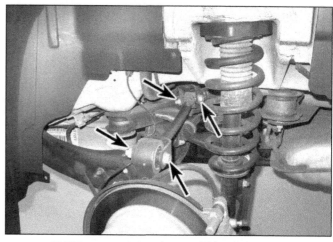

**16.2 Upper control arm mounting nuts and bolts
(2014 and earlier model shown, 2015 and later model similar)**

4   Remove the bushing brackets and pull out the stabilizer bar.

Installation

5   Place the stabilizer bar on the crossmember and reinstall the bolts.
6   Install the stabilizer links and tighten the bolts to the torque listed in this Chapter's Specifications.
7   Install the wheel and lug nuts. Lower the vehicle and tighten the lug nuts to the torque listed in the Chapter 1 Specifications.

14   Hub and bearing assembly (rear) - removal and installation

1   Loosen the rear wheel lug nuts, raise the rear of the vehicle, support it securely on jackstands and remove the wheels.
2   Remove the ABS wheel speed sensor (see Chapter 9). On AWD models, loosen the hub nut and free the driveaxle splines from the hub (see Chapter 8).
3   Remove the brake disc (see Chapter 9) and parking brake cable bracket fasteners.
4   Remove the bolts retaining the hub/bearing assembly to the rear knuckle (see illustration).
5   Remove the hub/bearing assembly.
6   Installation is the reverse of removal. Tighten the hub/bearing mounting bolts to the torque listed in this Chapter's Specifications. Tighten the caliper mounting bracket bolts to the torque listed in the Chapter 9 Specifications.
7   Install the wheel and lug nuts. Lower the vehicle and tighten the lug nuts to the torque listed in the Chapter 1 Specifications.

15   Rear suspension knuckle - removal and installation

Note: *On models equipped with an electric parking brake actuator, release the parking*

brake and disconnect the cable from the negative battery terminal (see Chapter 5).
1   Loosen the rear wheel lug nuts, raise the rear of the vehicle, support it securely on jackstands and remove the wheels.
2   If equipped, disconnect the electrical connector and remove the electric parking brake actuator from the rear brake assembly. Detach the wiring harness from any retaining clips on the knuckle.
3   Remove the ABS wheel speed sensor wiring harness from the knuckle and free the wiring harness from any retaining clips.
4   On AWD models, remove the driveaxle/hub nut and free the driveaxle outer joint splines from the hub (see Chapter 8).
5   Remove the brake caliper and mounting bracket (see Chapter 9). Support the caliper with a length of wire - don't let it hang by the brake hose. Remove the brake disc and backing plate (see Chapter 9).
6   Disconnect the stabilizer bar link from the lower control arm.
7   Remove the knuckle-to-trailing link bolts (see Section 11).
8   On 2015 and later models, support the outer end of the lower control arm with a floor jack.
9   Remove the nuts and bolts that secure the toe link, lower control arm, and the upper control arm to the knuckle.
10   On 2015 and later models, slowly lower the lower control arm/floor jack until spring tension is relieved, then remove the shock absorber-to-knuckle bolt (see Section 9).
11   Remove the knuckle.
12   Installation is the reverse of removal. tighten the suspension fasteners to the torque listed in this Chapter's Specifications. Tighten the caliper mounting bracket bolts to the torque listed in the Chapter 9 Specifications. Tighten the hub nut (AWD models) to the torque listed in Chapter 8 Specifications. Tighten the wheel lug nuts to the torque listed in the Chapter 1 Specifications.

16   Upper control arm (rear) - removal and installation

1   Loosen the rear wheel lug nuts. Raise the rear of the vehicle and support it securely with jackstands. Remove the wheel.
2   Remove the upper control arm-to-knuckle fasteners and upper control arm-to-crossmember fasteners **(see illustration)**. Remove the upper control arm.
3   Installation is the reverse of removal.
4   Raise the rear suspension with a floor jack placed under the knuckle to simulate normal ride height, then tighten the mounting fasteners to the torque listed in this Chapter's Specifications.
5   Install the wheel and lug nuts. Lower the vehicle and tighten the lug nuts to the torque listed in the Chapter 1 Specifications.

17   Lower control arm (rear) - removal and installation

1   Raise the rear of the vehicle and support it securely on jackstands.

2014 and earlier models
2   Remove the cover from the lower control arm, if equipped.
3   Remove the lower shock mount bolt and the stabilizer bar link-to-lower control arm nut **(see illustration)**. Separate the link from the arm.
4   Remove the lower control arm-to-knuckle bolt.

2015 and later models
5   Disconnect the stabilizer bar link from the lower control arm (see Section 13).
6   Support the outer end of the lower control arm with a floor jack.
7   Remove the lower control arm-to-knuckle bolt and nut.
8   Slowly lower the lower control arm/floor jack until spring tension is relieved.

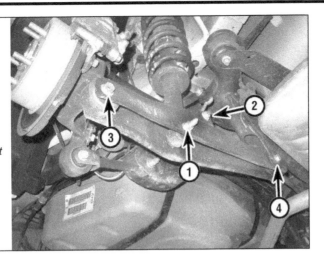

17.3 Lower control arm details (2014 and earlier model shown)

1　*Shock absorber lower mounting nut/ bolt*
2　*Stabilizer bar link nut*
3　*Lower control arm-to-knuckle bolt*
4　*Lower control arm-to-subframe bolt*

18.2 Pry up the trim caps, then remove the airbag module retaining bolts (2014 and earlier models)

All models

9　Remove the lower control arm-to-subframe fastener and remove the lower control arm.

Note: *On 2015 and later models, mark the position of the lower control arm-to-crossmember inner cam bolt with paint (see illustration 12.3), then remove the cam bolt and nut. FWD models may require a heat shield to be removed before the bolt can be withdrawn. AWD models require the old bolt to be cut, and the new bolt/nut to be installed in the opposite direction from which it was originally installed.*

10　Installation is the reverse of removal. Before tightening any of the fasteners, raise the suspension with a floor jack placed under the knuckle to simulate normal ride height, then tighten the fasteners to the torque listed in this Chapter's Specifications.

18　Steering wheel - removal and installation

Warning: *These models have airbags. Always disarm the airbag system before working in*

the vicinity of the impact sensors, steering column, or instrument panel to avoid accidental deployment of the airbag, which could cause personal injury (see Chapter 12).

Warning: *Do not use a memory saving device to preserve the PCM's memory when working on or near airbag system components.*

Removal

1　Park the vehicle with the wheels pointing straight ahead and the steering wheel centered. Disconnect the negative battery cable from the battery or remote ground terminal (see Chapter 5).

Warning: *Wait at least two minutes before proceeding with the following steps.*

2　Remove the airbag module from the steering wheel by removing the two bolts **(see illustration)**. On 2015 and later models, insert an appropriate size Allen wrench into the airbag release spring hole in the steering wheel, then push inward to depress the spring and release one side of the airbag module. Repeat this for the opposite side.

3　Disconnect the electrical connectors from the airbag module and clockspring **(see illustration)**.

4　Set the module aside in a safe, isolated area, with the airbag side of the module facing UP.

Warning: *When carrying the airbag module, keep the driver's (trim) side facing away from you.*

5　While firmly holding the steering wheel in the straight ahead position, loosen the steering wheel retaining bolt a few turns.

6　Break loose the steering wheel from the shaft using a puller with hooked legs **(see illustration)**. The puller screw must contact the steering wheel bolt. Once the wheel has broken loose, remove the puller and bolt, then mark the position of the steering wheel to the shaft.

Caution: *Do not hammer on the steering shaft or the steering wheel in an attempt to free the steering wheel from the shaft. Also, do not use a slide hammer puller to remove the steering wheel.*

Caution: *While the steering wheel is removed, DO NOT turn the steering shaft. If you do so, the airbag clockspring could be damaged when the vehicle is put back in service.*

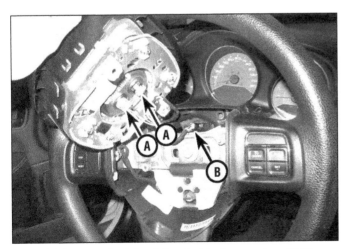

18.3 Squeeze the tabs of the electrical connectors (A) to release them from the airbag. Also disconnect the clockspring electrical connector (B)

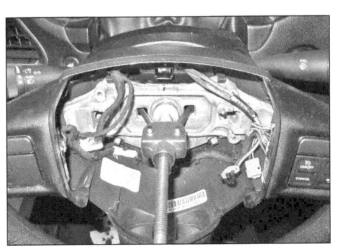

18.6 Use a puller with hooked legs to remove the steering wheel from the steering shaft; ensure that you hold the wheel in the straight ahead position while doing so

18.7 Apply a piece of tape over the clockspring so it can't turn and become uncentered

19.2 Hold the tie-rod end while breaking loose the jam nut

7   After the steering wheel has been removed, tape the clockspring so it can't turn **(see illustration)**.

Installation

8   Make sure the clockspring is still centered, the wheels are straight ahead, and the turn signal stalk is in the middle (neutral) position.

9   If the clockspring has become uncentered, turn the clockspring rotor clockwise until it stops (don't apply too much force, though), then turn it counterclockwise approximately 2-1/2. The dowel pin should be at the 6 o'clock position and the wiring should be at the top.

10   Install the wheel on the steering shaft, aligning the wide spline on the shaft with the gap in the steering wheel splines. Push the wheel onto the shaft and ensure the clockspring lines up on the back of the steering wheel.

Note: *Make sure the clockspring wires are routed correctly through the steering wheel.*

11   Install a **NEW** steering wheel retaining bolt and tighten it to the torque listed in this Chapter's Specifications.

12   Reconnect the clockspring connector.

13   Reconnect the electrical connectors to the airbag module, then install the module to the steering wheel. Tighten the fasteners to the torque listed in this Chapter's Specifications.

14   Connect the cable to the negative battery terminal (see Chapter 5).

15   Turn the ignition key On and verify that the airbag system is operating properly by watching the airbag warning light in the instrument cluster (see Chapter 12).

Warning: *If the airbag system is not operating properly (as indicated by the airbag warning light), DO NOT drive the vehicle. Have the airbag system repaired at a dealership service department or other qualified repair shop.*

19   Tie-rod ends - removal and installation

Removal

1   Loosen the wheel lug nuts, raise the front of the vehicle and support it securely on jackstands. Apply the parking brake and block the rear wheels to keep the vehicle from roll-

ing off the jackstands. Remove the wheel.

2   Loosen the tie-rod end jam nut **(see illustration)**.

3   Mark the relationship of the tie-rod end to the threaded portion of the tie-rod. This will ensure the toe-in setting is restored when reassembled **(see illustration)**.

4   Loosen the nut on the tie-rod end ballstud a few turns **(see illustration)**. Break loose the tie-rod end ballstud from the steering knuckle arm with a puller **(see illustration)**.

5   Remove the nut and separate the tie-rod end from the steering knuckle, then unscrew the tie-rod end from the tie-rod.

Installation

6   Thread the tie-rod end onto the tie-rod to the marked position and connect the tie-rod end to the steering knuckle arm. Install the nut on the ballstud and tighten it to the torque listed in this Chapter's Specifications.

7   Tighten the jam nut securely and install the wheel. Lower the vehicle and tighten the lug nuts to the torque listed in the Chapter 1 Specifications.

8   Have the front end alignment checked and, if necessary, adjusted.

19.3 Back-off the jam nut and mark the exposed threads

19.4a If the tie-rod end stud spins when loosening the nut, hold the stud with a wrench or socket

19.4b Disconnect the tie-rod end from the steering knuckle with a puller

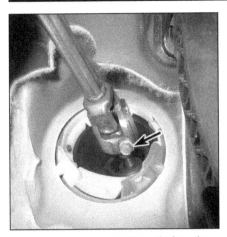

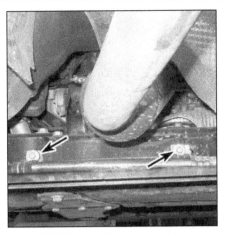

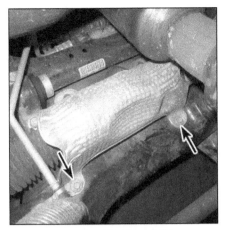

20.8 Mark the intermediate shaft to the steering gear input shaft, then remove the coupler pinch bolt

21.8 Remove the hose bracket mounting bolts (2014 and earlier model)

21.11 Arrows show the two visible screws securing the heat shield. There are two more on the back side that are not visible in this photo

20 Steering column - removal and installation

Warning: *These models have airbags. Always disarm the airbag system before working in the vicinity of the impact sensors, steering column, or instrument panel to avoid accidental deployment of the airbag, which could cause personal injury (see Chapter 12).*
Warning: *Do not use a memory saving device to preserve the PCM's memory when working on or near airbag system components.*

Removal

1 Park the vehicle with the wheels pointing straight ahead. Disconnect the cable from the negative terminal of the battery (see Chapter 5).
2 Position the tilt steering column in the full up position, and remove the steering column cover and shroud (see Chapter 11)
3 Remove the column reinforcement plate and check for a shaft locking module.
4 If the vehicle is equipped with a shaft locking module, remove the unit by inserting the key into the lock and rotating it to the unlocked position. Use a drift or drill bit of the proper size (4 mm max) insert it into the holes located at the top of the module, push in until the punch or drill bit bottom out.
5 Slide the module toward the base of the column. The module should clear the bracket at this point.
6 Remove the module, leave it plugged in and keep the unit supported and away from the column.
7 Move the carpeting to allow access to the base of the column and the coupler bolt.
8 Ensure the wheels are in the straight ahead position, then turn the steering wheel to the right and remove the intermediate shaft pinch bolt **(see illustration)**.
9 Turn the steering column back to the straight ahead position and remove the airbag.
10 Remove the steering wheel (see Section 18).

Warning: *Do not move the steering shaft after the steering wheel has been removed or damage to the clockspring could occur when the vehicle is put back into service. To ensure this doesn't happen, make sure the steering column is locked.*
11 Using the tilt lever, move the steering column down to its lowest position.
12 Remove the screws that hold the steering column cover and lower shroud, and remove the lower shroud (see Chapter 11).
13 Position the tilt steering column in the up position, and lock it in place.
14 Disconnect all electrical connectors coming from the large harness on the side of the steering column and any ground wires that may be attached to the steering column from the other side.
15 Remove the steering column mounting fasteners, carefully lower the column, making sure nothing is still connected, and remove it.

Installation

16 Guide the steering column into position, then install the steering column mounting fasteners and tighten them to the torque listed in this Chapter's Specifications.
Note: *Begin the tightening sequence by starting with the left upper nut and working in a clockwise order.*
17 Connect the intermediate shaft and install the pinch bolt, tightening it to the torque listed in this Chapter's Specifications.
18 The remainder of installation is the reverse of removal. Reconnect the negative battery cable (see Chapter 5).

21 Steering gear - removal and installation

Warning: *Lock the steering wheel to keep it from moving while the steering shaft is disconnected, or damage to the airbag system could occur when the vehicle is placed back in service. With the ignition key in the LOCK position, turn the steering wheel just enough to lock it or secure it with the seatbelt.*

Removal

1 Park the vehicle with the wheels pointing straight ahead. Disconnect the negative battery cable from the battery or remote ground terminal (see Chapter 5).
2 Remove as much fluid as possible from the power steering pump reservoir.
3 Loosen the wheel lug nuts, raise the vehicle, support it securely on jackstands, and remove the wheels.
4 Turn the steering wheel to the right until the intermediate shaft pinch bolt is accessible, then remove the bolt (see Section 20). Return the steering wheel to the straight ahead position, securing the steering wheel in place, then detach the intermediate shaft from the steering gear input shaft.
5 Disconnect both tie-rod ends from the steering knuckles (see Section 19).
6 Remove the through-bolt from the rear engine mount.
7 Remove the longitudinal crossmember (see Section 24).
8 Remove the fasteners that hold the power steering hose brackets to the crossmember **(see illustration)**.
9 Unscrew the return line from the steering gear by loosening the tube nut.
10 Remove the pressure line from the steering gear by loosening the tube nut.
11 Remove the steering gear heat shield mounting screws **(see illustration)** and remove the heat shield.
12 Remove the subframe (see Section 24).
13 Remove any firewall seals that may be present, remove the steering gear-to-sub-frame bolts, and remove the steering gear.

Installation

14 Installation is the reverse of removal, noting the following points:
  a) *If a new steering gear is being installed, center the gear by turning the input shaft clockwise until it stops. Turn the input shaft counterclockwise and count the number of rotations until it stops. Divide*

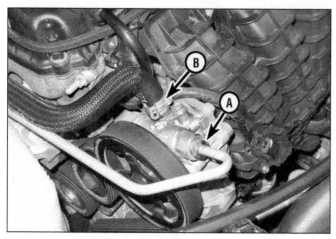

22.5 Power steering pressure line fitting (A) and supply hose (B) (2.4L engine)

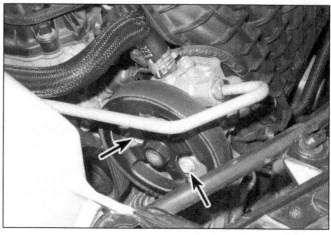

22.7 Power steering pump mounting bolts (2.4L engine)

that number by two and turn the input shaft clockwise that amount.

b) *Install the mounting bolts and tighten them to the torque listed in this Chapter's Specifications.*
c) *Align the index marks on the steering gear shaft and intermediate shaft coupler before installing the bolt in the coupler.*
d) *Install the fluid lines to the steering gear in the correct position and tighten them securely.*
e) *Tighten the engine mount bolt, cross-member bolts, and reinforcement brackets to the torque listed in this Chapter's Specifications.*
f) *Install the wheels and lug nuts. Lower the vehicle, then tighten the lug nuts to the torque listed in the Chapter 1 Specifications.*
g) *Fill the power steering pump reservoir with the recommended fluid (see Chapter 1), then bleed the power steering system (see Section 23).*
h) *Have the front wheel alignment checked and, if necessary, adjusted.*

22   Power steering pump (2014 and earlier models) - removal and installation

Removal

1   Disconnect the cable from the negative terminal of the battery (see Chapter 5).

2.4L engine
2   Remove the fluid from the reservoir using a suction pump or equivalent.
Caution: *Be careful not to tear any mesh filter that may be present just under the surface of the fluid.*
3   Raise the vehicle and support it securely on jackstands.
4   Remove the engine beauty cover and disconnect the pressure hose bracket.
5   Disconnect the pressure line and supply hose from the pump **(see illustration)**.
6   Remove the drivebelt (see Chapter 1).
7   Remove the pump mounting bolts through the pulley openings **(see illustration)**.
8   Remove the power steering pump.

2.7L engine
9   Remove the fluid from the reservoir using a suction pump or equivalent, then disconnect the fluid supply hose at the pump.
10   Loosen the right front wheel lug nuts. Raise the vehicle and support it securely on jackstands. Remove the right front wheel.
11   Remove the drivebelt (see Chapter 1).
12   Remove the driveaxle/hub nut while an assistant depresses the brake pedal.
13   Detach the tie-rod end from the steering knuckle (see Section 19).
14   Remove the balljoint nut and separate the balljoint from the knuckle using tool 9360 or a suitable equivalent.
15   Lift the knuckle up and out of the control arm and pull the driveaxle out of the knuckle.
Note: *Do not let the driveaxle hang free or strike any part of the vehicle. This will damage the inner CV joint.*
16   Remove the driveaxle (see Chapter 8).
17   Disconnect the power steering pressure line from the pump and remove the pump mounting bracket bolt.
18   Remove the other two pump bracket bolts and remove the pump and bracket from the engine by rotating the pump so the pulley faces the engine. The pump will come out through the gap made when the half shaft was removed.
19   Remove the pump bolts through the gaps in the pump pulley. Separate the pump from the bracket.

3.5L engine
20   Remove the fluid from the reservoir using a suction pump or equivalent.
21   Remove the engine cover and unscrew the tube nut on the pressure line at the pump side. Remove the pressure line bracket from the cylinder head.

22   Detach the supply hose from the power steering pump fitting.
23   Loosen the right front wheel lug nuts. Raise the vehicle and support it securely on jackstands. Remove the right front wheel.
24   Remove the right engine mount bracket nuts. Place a floor jack with a block of wood under the engine just below the oil pan. Raise the jack until the wood block just comes in contact with the oil pan.
25   Slowly raise the jack until all three power steering mounting bolts are visible, remove the bolts, then remove the pump.

3.6L engine
26   Remove the fluid from the reservoir using a suction pump or equivalent.
27   Remove the drivebelt (see Chapter 1).
28   Raise the vehicle and support it securely on jackstands.
29   Remove the exhaust pipe (see Chapter 4) and the heat shield mounting bolts.
30   Remove the supply hose and the pressure hose from the pump body.
31   Remove the 3 pump bolts through the pump pulley. Rotate the pulley until the holes line up to facilitate this.
32   Remove the pump from the underside of the engine.

Installation

33   Installation is the reverse of removal.
34   Tighten all power steering mounting fasteners to the torque listed in this Chapter's Specifications.
35   Tighten the pressure line-to-pump fitting securely.
36   Make sure the hoses are properly routed and all hose clamps are tightened securely.
37   Fill the power steering fluid reservoir with the recommended fluid (see Chapter 1).
38   Connect the negative battery cable to the remote ground terminal (see Chapter 5).
39   Bleed the power steering system (see Section 23). Stop the engine, check the fluid level, and inspect the system for leaks.

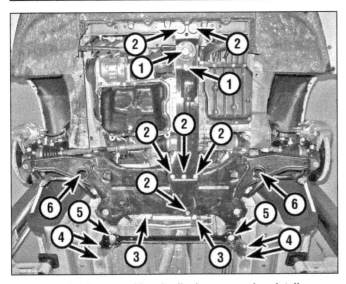

24.6 Subframe and longitudinal crossmember details (2014 and earlier models)

1 Front engine mount bolts
2 Longitudinal crossmember bolts
3 Power steering hose brackets
4 Subframe reinforcement bracket bolts
5 Rear subframe mounting bolts
6 Front subframe mounting bolts (accessed through control arms)

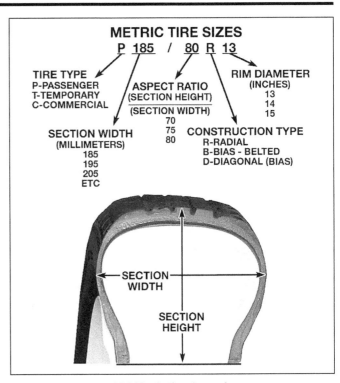

25.1 Metric tire size code

23 Power steering system - bleeding (2014 and earlier models)

1   Following any operation in which the power steering fluid lines have been disconnected, the power steering system must be bled to remove all air and obtain proper steering performance.
2   With the front wheels in the straight ahead position, check the power steering fluid level (see Chapter 1).
3   Start the engine and allow it to run at fast idle. Recheck the fluid level and add fluid if necessary.
4   Bleed the system by turning the wheels from side to side, without hitting the stops. This will work the air out of the system. Maintain the proper fluid level as this is done.
5   When the air is worked out of the system, return the wheels to the straight ahead position and leave the vehicle running for several more minutes before shutting it off.
6   Road test the vehicle to be sure the steering system is functioning normally and noise free.
7   Recheck the fluid level to be sure it is correct; add fluid if necessary

24 Subframe - removal and installation

Note: *Disconnect the cable from the negative terminal of the battery before beginning (see Chapter 5).*

1   Loosen the front wheel lug nuts. Apply the parking brake, then raise the front of the vehi-

cle and support it securely on jackstands. Block the rear wheels. Remove the front wheels.
2   Mark all four corners of the subframe to the vehicle (this will preserve alignment when assembling).

2014 and earlier models

3   Remove the power steering hose fasteners from the subframe.
4   Disconnect the control arms from the steering knuckles (see Section 5).
5   Disconnect the stabilizer bar links from the stabilizer bar (see Section 2).
6   Remove the longitudinal (engine mount) crossmember **(see illustration)**.
7   Remove the steering gear mounting bolts and separate the gear from the subframe. Once the gear is separated, tie it up out of the way.
8   Support the subframe with a floor jack and remove the rear reinforcement bolts and the mounting bolts.

2015 and later models

Note: *On AWD models, it will be necessary to remove the transfer case (see Chapter 7) before the subframe can be lowered.*

9   Keep the steering wheel in the straight-ahead position by securing it with the seatbelt. This will prevent the clockspring from rotating during the procedure.
10   Remove the intermediate shaft coupling bolt, then separate the intermediate shaft from the pinion shaft (see Section 20).
11   Remove the engine under cover and bumper cover lower panel.
12   Separate the stabilizer bar link from the front strut (see Section 2).

13   Separate the tie-rod end from the steering knuckle (see Section 19).
14   Separate the lower control arm from the steering knuckle (see Section 5), remove the front load beams (see Section 27), then remove the front portion of the exhaust pipe.
15   Disconnect the steering gear electrical connectors and detach the wiring harness retainers.
16   Disconnect the manual park release cable (shift cable) from the steering gear bracket.
17   Remove the rear engine mount bracket through-bolt.
18   Support the subframe with a floor jack, then remove the rear reinforcement bolts and remaining subframe mounting bolts.
19   Slowly lower the subframe, making sure nothing is still attached.
20   If the subframe is being replaced, separate the control arms once it is on the ground (see Section 5).
21   Installation is the reverse of removal. Tighten all fasteners to the torque listed in this Chapter's Specifications.

25 Wheels and tires - general information

1   All vehicles covered by this manual are equipped with metric-sized fiberglass or steel belted radial tires **(see illustration)**. Use of other size or type of tires may affect the ride and handling of the vehicle. Don't mix different types of tires, such as radials and bias belted, on the same vehicle as handling may be seriously affected. It's recommended that tires be

replaced in pairs on the same axle, but if only one tire is being replaced, be sure it's the same size, structure and tread design as the other.

2   Because tire pressure has a substantial effect on handling and wear, the pressure on all tires should be checked at least once a month or before any extended trips (see Chapter 1).

3   Wheels must be replaced if they are bent, dented, leak air, have elongated bolt holes, are heavily rusted, out of vertical symmetry or if the lug nuts won't stay tight. Wheel repairs that use welding or peening are not recommended.

4   Tire and wheel balance is important in the overall handling, braking and performance of the vehicle. Unbalanced wheels can adversely affect handling and ride characteristics as well as tire life. Whenever a tire is installed on a wheel, the tire and wheel should be balanced by a shop with the proper equipment.

26   Wheel alignment - general information

1   A wheel alignment refers to the adjustments made to the wheels so they are in proper angular relationship to the suspension and the ground. Wheels that are out of proper alignment not only affect vehicle control, but also increase tire wear. The front end angles normally measured are camber, caster and toe-in **(see illustration)**. Toe-in is the only routine adjustment made; Both camber and caster are set at the factory and are not considered to be adjustable. However, camber can be slightly adjusted by using special adjustment parts from the manufacturer. Caster is set at the factory and is the result of sub frame and engine crossmemember location.

2   Getting the proper wheel alignment is a very exacting process, one in which complicated and expensive machines are necessary to perform the job properly. Because of this, you should have a technician with the proper equipment perform these tasks. We will, however, use this space to give you a basic idea of what is involved with a wheel alignment so you can better understand the process and deal intelligently with the shop that does the work.

3   Toe-in is the turning in of the wheels. The purpose of a toe specification is to ensure parallel rolling of the wheels. In a vehicle with zero toe-in, the distance between the front edges of the wheels will be the same as the distance between the rear edges of the wheels. The actual amount of toe-in is normally only a fraction of an inch. Toe-in is controlled by the tie-rod end position on the tie-rod. Incorrect toe-in will cause the tires to wear improperly by making them scrub against the road surface.

4   Camber is the tilting of the wheels from vertical when viewed from one end

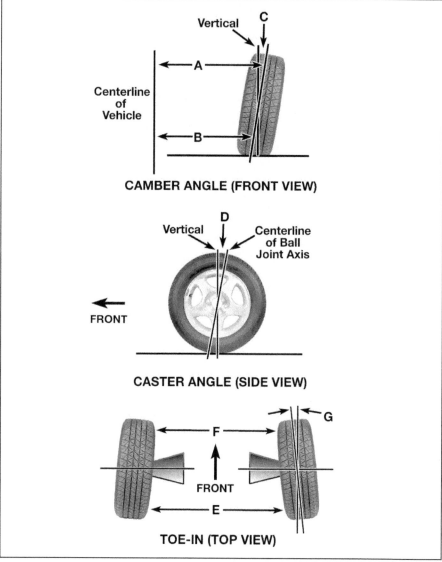

CAMBER ANGLE (FRONT VIEW)

CASTER ANGLE (SIDE VIEW)

TOE-IN (TOP VIEW)

26.1 Wheel alignment details

A minus B = C (degrees camber)
D = caster (expressed in degrees)

E minus F = toe-in (measured in inches)
G = toe-in (expressed in degrees)

of the vehicle. When the wheels tilt out at the top, the camber is said to be positive (+). When the wheels tilt in at the top the camber is negative (-). The amount of tilt is measured in degrees from vertical and this measurement is called the camber angle. This angle affects the amount of tire tread which contacts the road and compensates for changes in the suspension geometry when the vehicle is cornering or traveling over an undulating surface. On the front end it is adjusted using special camber adjusting bolts and a lower strut simulation device, which alter the relationship between the strut and the steering knuckle.

5   Caster is the tilting of the front steering axis from the vertical. A tilt toward the rear is positive caster and a tilt toward the front is negative caster. This is not considered to be adjustable on these vehicles.

27   Front load beam (2015 and later models) - removal and installation

Note: *The load beam fasteners are accessed by working through the fenderwell.*

1   Raise the vehicle and support it securely on jackstands. Remove the wheels. Remove the engine under cover and bumper cover lower panel.

2   Remove the fenderwell splash shield-to-load beam fastener.

3   Remove the load beam bracket bolts and remove the bracket.

4   Remove the load beam-to-load beam extension mounting bolts. Remove the load beam.

5   Installation is the reverse of removal. Tighten the bolts to the torque listed in this Chapter's Specifications.

Chapter 11   Body

Contents

	Section
Body repair - major damage	4
Body repair - minor damage	3
Bumper covers - removal and installation	10
Center floor console - removal and installation	24
Cowl cover - removal and installation	12
Dashboard trim panels - removal and installation	21
Door - removal and installation	14
Door module, latch, lock and handle - removal and installation	15
Door trim panels - removal and installation	13
Door window glass - removal and installation	16
Fastener and trim removal	6
Front fender - removal and installation	11

	Section
General information	1
Hood - removal, installation and adjustment	7
Hood latch and cable - removal and installation	8
Instrument panel - removal and installation	22
Mirrors - removal and installation	18
Radiator grille - removal and installation	9
Repairing minor paint scratches	2
Seats - removal and installation	23
Steering column covers - removal and installation	20
Trunk lid, latch and support struts - removal and installation	19
Upholstery, carpets and vinyl trim - maintenance	5
Window glass regulators - removal and installation	17

Specifications

Torque specifications

Seat belt mounting bolts .... 29 ft-lbs

---

1   General information

Warning: *The models covered by this manual are equipped with Supplemental Restraint Systems (SRS), more commonly known as airbags. Always disable the airbag system before working in the vicinity of any airbag system components to avoid the possibility of accidental deployment of the airbags, which could cause personal injury (see Chapter 12).*

1   Certain body components are particularly vulnerable to accident damage and can be unbolted and repaired or replaced. Among these parts are the hood, doors, tailgate, liftgate, bumpers and front fenders.

2   Only general body maintenance practices and body panel repair procedures within the scope of the do-it-yourselfer are included in this Chapter.

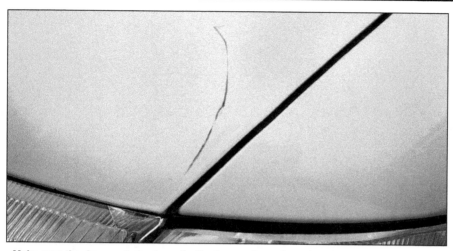

Make sure the damaged area is perfectly clean and rust free. If the touch-up kit has a wire brush, use it to clean the scratch or chip. Or use fine steel wool wrapped around the end of a pencil. Clean the scratched or chipped surface only, not the good paint surrounding it. Rinse the area with water and allow it to dry thoroughly

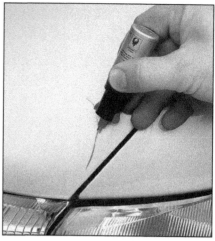

Thoroughly mix the paint, then apply a small amount with the touch-up kit brush or a very fine artist's brush. Brush in one direction as you fill the scratch area. Do not build up the paint higher than the surrounding paint

2   Repairing minor paint scratches

No matter how hard you try to keep your vehicle looking like new, it will inevitably be scratched, chipped or dented at some point. If the metal is actually dented, seek the advice of a professional. But you can fix minor scratches and chips yourself. Buy a touch-up paint kit from a dealer service department or an auto parts store. To ensure that you get the right color, you'll need to have the specific make, model and year of your vehicle and, ideally, the paint code, which is located on a special metal plate under the hood or in the door jamb.

3   Body repair - minor damage

Plastic body panels

The following repair procedures are for minor scratches and gouges. Repair of more serious damage should be left to a dealer service department or qualified auto body shop. Below is a list of the equipment and materials necessary to perform the following repair procedures on plastic body panels.

Wax, grease and silicone removing solvent
Cloth-backed body tape
Sanding discs
Drill motor with three-inch disc holder
Hand sanding block
Rubber squeegees
Sandpaper
Non-porous mixing palette
Wood paddle or putty knife
Curved-tooth body file
Flexible parts repair material

Flexible panels (bumper trim)

1   Remove the damaged panel, if necessary or desirable. In most cases, repairs can be car-

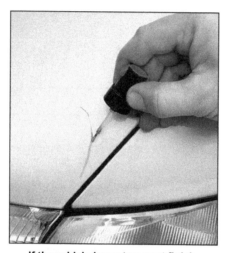

If the vehicle has a two-coat finish, apply the clear coat after the color coat has dried

ried out with the panel installed.
2   Clean the area(s) to be repaired with a wax, grease and silicone removing solvent applied with a water-dampened cloth.
3   If the damage is structural, that is, if it extends through the panel, clean the backside of the panel area to be repaired as well. Wipe dry.
4   Sand the rear surface about 1-1/2 inches beyond the break.
5   Cut two pieces of fiberglass cloth large enough to overlap the break by about 1-1/2 inches. Cut only to the required length.
6   Mix the adhesive from the repair kit according to the instructions included with the kit, and apply a layer of the mixture approximately 1/8-inch thick on the backside of the panel. Overlap the break by at least 1-1/2 inches.
7   Apply one piece of fiberglass cloth to the adhesive and cover the cloth with additional adhesive. Apply a second piece of fiberglass

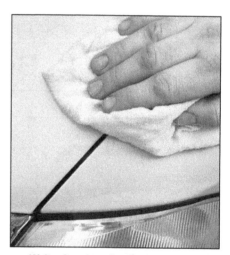

Wait a few days for the paint to dry thoroughly, then rub out the repainted area with a polishing compound to blend the new paint with the surrounding area. When you're happy with your work, wash and polish the area

cloth to the adhesive and immediately cover the cloth with additional adhesive in sufficient quantity to fill the weave.
8   Allow the repair to cure for 20 to 30 minutes at 60-degrees to 80-degrees F.
9   If necessary, trim the excess repair material at the edge.
10   Remove all of the paint film over and around the area(s) to be repaired. The repair material should not overlap the painted surface.
11   With a drill motor and a sanding disc (or a rotary file), cut a "V" along the break line approximately 1/2-inch wide. Remove all dust and loose particles from the repair area.
12   Mix and apply the repair material. Apply a light coat first over the damaged area; then

continue applying material until it reaches a level slightly higher than the surrounding finish.

13   Cure the mixture for 20 to 30 minutes at 60-degrees to 80-degrees F.

14   Roughly establish the contour of the area being repaired with a body file. If low areas or pits remain, mix and apply additional adhesive.

15   Block sand the damaged area with sandpaper to establish the actual contour of the surrounding surface.

16   If desired, the repaired area can be temporarily protected with several light coats of primer. Because of the special paints and techniques required for flexible body panels, it is recommended that the vehicle be taken to a paint shop for completion of the body repair.

Steel body panels

See photo sequence

Repair of dents

17   When repairing dents, the first job is to pull the dent out until the affected area is as close as possible to its original shape. There is no point in trying to restore the original shape completely as the metal in the damaged area will have stretched on impact and cannot be restored to its original contours. It is better to bring the level of the dent up to a point that is about 1/8-inch below the level of the surrounding metal. In cases where the dent is very shallow, it is not worth trying to pull it out at all.

18   If the backside of the dent is accessible, it can be hammered out gently from behind using a soft-face hammer. While doing this, hold a block of wood firmly against the opposite side of the metal to absorb the hammer blows and prevent the metal from being stretched.

19   If the dent is in a section of the body which has double layers, or some other factor makes it inaccessible from behind, a different technique is required. Drill several small holes through the metal inside the damaged area, particularly in the deeper sections. Screw long, self-tapping screws into the holes just enough for them to get a good grip in the metal. Now pulling on the protruding heads of the screws with locking pliers can pull out the dent.

20   The next stage of repair is the removal of paint from the damaged area and from an inch or so of the surrounding metal. This is easily done with a wire brush or sanding disk in a drill motor, although it can be done just as effectively by hand with sandpaper. To complete the preparation for filling, score the surface of the bare metal with a screwdriver or the tang of a file or drill small holes in the affected area. This will provide a good grip for the filler material. To complete the repair, see the Section on filling and painting.

Repair of rust holes or gashes

21   Remove all paint from the affected area and from an inch or so of the surrounding metal using a sanding disk or wire brush mounted in a drill motor. If these are not available, a few sheets of sandpaper will do the job just as effectively.

22   With the paint removed, you will be able to determine the severity of the corrosion and decide whether to replace the whole panel, if possible, or repair the affected area. New body panels are not as expensive as most people think and it is often quicker to install a new panel than to repair large areas of rust.

23   Remove all trim pieces from the affected area except those which will act as a guide to the original shape of the damaged body, such as headlight shells, etc. Using metal snips or a hacksaw blade, remove all loose metal and any other metal that is badly affected by rust. Hammer the edges of the hole in to create a slight depression for the filler material.

24   Wire-brush the affected area to remove the powdery rust from the surface of the metal. If the back of the rusted area is accessible, treat it with rust inhibiting paint.

25   Before filling is done, block the hole in some way. This can be done with sheet metal riveted or screwed into place, or by stuffing the hole with wire mesh.

26   Once the hole is blocked off, the affected area can be filled and painted. See the following subsection on filling and painting.

Filling and painting

27   Many types of body fillers are available, but generally speaking, body repair kits which contain filler paste and a tube of resin hardener are best for this type of repair work. A wide, flexible plastic or nylon applicator will be necessary for imparting a smooth and contoured finish to the surface of the filler material. Mix up a small amount of filler on a clean piece of wood or cardboard (use the hardener sparingly). Follow the manufacturer's instructions on the package, otherwise the filler will set incorrectly.

28   Using the applicator, apply the filler paste to the prepared area. Draw the applicator across the surface of the filler to achieve the desired contour and to level the filler surface. As soon as a contour that approximates the original one is achieved, stop working the paste. If you continue, the paste will begin to stick to the applicator. Continue to add thin layers of paste at 20-minute intervals until the level of the filler is just above the surrounding metal.

29   Once the filler has hardened, the excess can be removed with a body file. From then on, progressively finer grades of sandpaper should be used, starting with a 180-grit paper and finishing with 600-grit wet-or-dry paper. Always wrap the sandpaper around a flat rubber or wooden block, otherwise the surface of the filler will not be completely flat. During the sanding of the filler surface, the wet-or-dry paper should be periodically rinsed in water. This will ensure that a very smooth finish is produced in the final stage.

30   At this point, the repair area should be surrounded by a ring of bare metal, which in turn should be encircled by the finely feathered edge of good paint. Rinse the repair area with clean water until all of the dust produced by the sanding operation is gone.

31   Spray the entire area with a light coat of primer. This will reveal any imperfections in the surface of the filler. Repair the imperfections with

fresh filler paste or glaze filler and once more smooth the surface with sandpaper. Repeat this spray-and-repair procedure until you are satisfied that the surface of the filler and the feathered edge of the paint are perfect. Rinse the area with clean water and allow it to dry completely.

32   The repair area is now ready for painting. Spray painting must be carried out in a warm, dry, windless and dust free atmosphere. These conditions can be created if you have access to a large indoor work area, but if you are forced to work in the open, you will have to pick the day very carefully. If you are working indoors, dousing the floor in the work area with water will help settle the dust that would otherwise be in the air. If the repair area is confined to one body panel, mask off the surrounding panels. This will help minimize the effects of a slight mismatch in paint color. Trim pieces such as chrome strips, door handles, etc., will also need to be masked off or removed. Use masking tape and several thickness of newspaper for the masking operations.

33   Before spraying, shake the paint can thoroughly, then spray a test area until the spray painting technique is mastered. Cover the repair area with a thick coat of primer. The thickness should be built up using several thin layers of primer rather than one thick one. Using 600-grit wet-or-dry sandpaper, rub down the surface of the primer until it is very smooth. While doing this, the work area should be thoroughly rinsed with water and the wet-or-dry sandpaper periodically rinsed as well. Allow the primer to dry before spraying additional coats.

34   Spray on the top coat, again building up the thickness by using several thin layers of paint. Begin spraying in the center of the repair area and then, using a circular motion, work out until the whole repair area and about two inches of the surrounding original paint is covered. Remove all masking material 10 to 15 minutes after spraying on the final coat of paint. Allow the new paint at least two weeks to harden, then use a very fine rubbing compound to blend the edges of the new paint into the existing paint. Finally, apply a coat of wax

4   Body repair - major damage

1   Major damage must be repaired by an auto body shop specifically equipped to perform body and frame repairs. These shops have the specialized equipment required to do the job properly.

2   If the damage is extensive, the frame must be checked for proper alignment or the vehicle's handling characteristics may be adversely affected and other components may wear at an accelerated rate.

3   Due to the fact that all of the major body components (hood, fenders, etc.) are separate and replaceable units, any seriously damaged components should be replaced rather than repaired. Sometimes the components can be found in a wrecking yard that specializes in used vehicle components, often at considerable savings over the cost of new parts.

These photos illustrate a method of repairing simple dents. They are intended to supplement *Body repair - minor damage* in this Chapter and should not be used as the sole instructions for body repair on these vehicles.

1 If you can't access the backside of the body panel to hammer out the dent, pull it out with a slide-hammer-type dent puller. Tap with a hammer near the edge of the dent to help 'pop' the metal back to its original shape, about 1/8-inch below the surface of the surrounding metal

2 Using coarse-grit sandpaper, remove the paint down to the bare metal. Clean the repair area with wax/silicone remover.

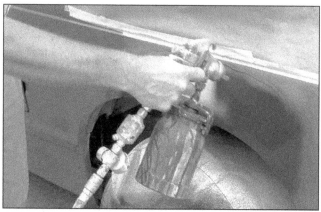

3 Following label instructions, mix up a batch of plastic filler and hardener, then quickly press it into the metal with a plastic applicator. Work the filler until it matches the original contour and is slightly above the surrounding metal

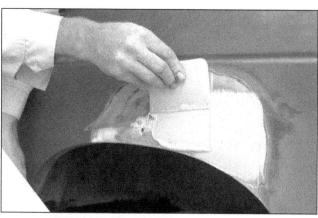

4 Let the filler harden until you can just dent it with your fingernail. File, then sand the filler down until it's smooth and even. Work down to finer grits of sandpaper - always using a board or block - ending up with 360 or 400 grit

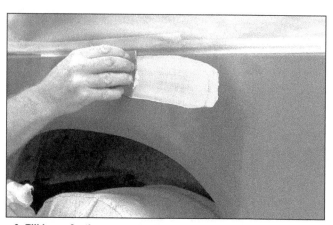

5 When the area is smooth to the touch, clean the area and mask around it. Apply several layers of primer to the area. A professional-type spray gun is being used here, but aerosol spray primer works fine

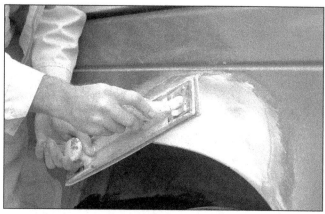

6 Fill imperfections or scratches with glazing compound. Sand with 360 or 400-grit and re-spray. Finish sand the primer with 600 grit, clean thoroughly, then apply the finish coat. Don't attempt to rub out or wax the repair area until the paint has dried completely (at least two weeks)

5 Upholstery, carpets and vinyl trim - maintenance

Upholstery and carpets

1 Every three months remove the floor-mats and clean the interior of the vehicle (more frequently if necessary). Use a stiff whiskbroom to brush the carpeting and loosen dirt and dust, then vacuum the upholstery and carpets thoroughly, especially along seams and crevices.

2 Dirt and stains can be removed from carpeting with basic household or automotive carpet shampoos available in spray cans. Follow the directions and vacuum again, then use a stiff brush to bring back the nap of the carpet.

3 Most interiors have cloth or vinyl upholstery, either of which can be cleaned and maintained with a number of material-specific cleaners or shampoos available in auto supply stores. Follow the directions on the product for usage, and always spot-test any upholstery cleaner on an inconspicuous area (bottom edge of a backseat cushion) to ensure that it doesn't cause a color shift in the material.

4 After cleaning, vinyl upholstery should be treated with a protectant.

Note: *Make sure the protectant container indicates the product can be used on seats - some products may make a seat too slippery.*

Caution: *Do not use protectant on vinyl-covered steering wheels.*

5 Leather upholstery requires special care. It should be cleaned regularly with saddle-soap or leather cleaner. Never use alcohol, gasoline, nail polish remover or thinner to clean leather upholstery.

6 After cleaning, regularly treat leather upholstery with a leather conditioner, rubbed in with a soft cotton cloth. Never use car wax on leather upholstery.

7 In areas where the interior of the vehicle is subject to bright sunlight, cover leather seating areas of the seats with a sheet if the vehicle is to be left out for any length of time.

Vinyl trim

8 Don't clean vinyl trim with detergents, caustic soap or petroleum-based cleaners. Plain soap and water works just fine, with a soft brush to clean dirt that may be ingrained. Wash the vinyl as frequently as the rest of the vehicle.

9 After cleaning, application of a high-quality rubber and vinyl protectant will help prevent oxidation and cracks. The protectant can also be applied to weather-stripping, vacuum lines and rubber hoses, which often fail as a result of chemical degradation, and to the tires.

6 Fastener and trim removal

1 There is a variety of plastic fasteners used to hold trim panels, splash shields and other parts in place in addition to typical screws, nuts and bolts. Once you are familiar with them, they can usually be removed without too much difficulty.

2 The proper tools and approach can prevent added time and expense to a project by minimizing the number of broken fasteners and/or parts.

3 The following illustrations show various types of fasteners that are typically used on most vehicles and how to remove and install them **(see illustrations)**. Replacement fasteners are commonly found at most auto parts stores, if necessary.

Fasteners

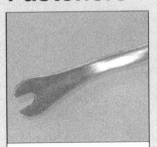

This tool is designed to remove special fasteners. A small pry tool used for removing nails will also work well in place of this tool

A Phillips head screwdriver can be used to release the center portion, but light pressure must be used because the plastic is easily damaged. Once the center is up, the fastener can easily be pried from its hole

Here is a view with the center portion fully released. Install the fastener as shown, then press the center in to set it

This fastener is used for exterior panels and shields. The center portion must be pried up to release the fastener. Install the fastener with the center up, then press the center in to set it

This type of fastener is used commonly for interior panels. Use a small blunt tool to press the small pin at the center in to release it . . .

. . . the pin will stay with the fastener in the released position

Reset the fastener for installation by moving the pin out. Install the fastener, then press the pin flush with the fastener to set it

This fastener is used for exterior and interior panels. It has no moving parts. Simply pry the fastener from its hole like the claw of a hammer removes a nail. Without a tool that can get under the top of the fastener, it can be very difficult to remove

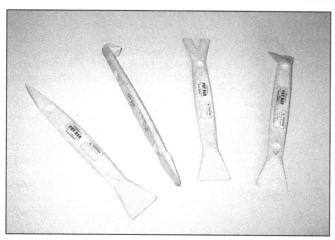

6.4 These small plastic pry tools are ideal for prying off trim panels

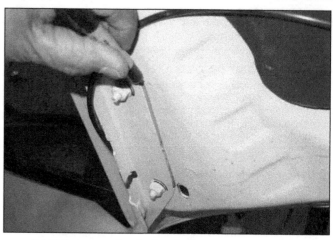

7.2 Before removing the hood, draw a mark around the hinge plates

7.4 Support the hood with your shoulder while removing the hood bolts

7.11 Adjust the hood height by screwing the hood bumpers in or out

3   Remove the top bolts holding the hood to the hinge and loosen the bottom bolts until they can be removed by hand.

4   Have an assistant on the opposite side of the vehicle support the weight of the hood. Simultaneously remove the bottom bolts holding the hood to the hinge and lift off the hood **(see illustration)**.

5   Remove the under hood lamp wire connector to the engine compartment wire harness, if so equipped.

6   Detach the windshield washer hose to the hood.

7   Installation is the reverse of removal.

Adjustment

8   Front-and-back and side-to-side adjustment of the hood is done by moving the hood in relation to the hinge plate after loosening the bolts.

9   Scribe or trace a line around the entire hinge plate so you can judge the amount of movement.

10   Loosen the bolts or nuts and move the hood into correct alignment. Move it only a little at a time. Tighten the hinge bolts or nuts and carefully lower the hood to check the alignment.

11   Adjust the hood bumpers on the radiator support so the hood is flush with the fenders when closed **(see illustration)**.

4   Trim panels are typically made of plastic and their flexibility can help during removal. The key to their removal is to use a tool to pry the panel near its retainers to release it without damaging surrounding areas or breaking-off any retainers. The retainers will usually snap out of their designated slot or hole after force is applied to them. Stiff plastic tools designed for prying on trim panels are available at most auto parts stores **(see illustration)**. Tools that are tapered and wrapped in protective tape, such as a screwdriver or small pry tool, are also very effective when used with care.

7   Hood - removal, installation and adjustment

Note: *The hood is heavy and somewhat awkward to remove and install - at least two people should perform this procedure.*

Removal and installation

1   Use blankets or pads to cover the cowl area of the body and the fenders. This will protect the body and paint as the hood is lifted off.

2   Scribe alignment marks around the bolt heads and hinge attachment locations to insure proper alignment during installation - a permanent-type felt-tip marker also will work for this **(see illustration)**.

Note: *On models with a support strut instead of a prop rod: While supporting the hood, release the upper support strut mounting clip with a screwdriver, then pull the strut off the balljoint on the hood* **(see illustration 19.7)**.

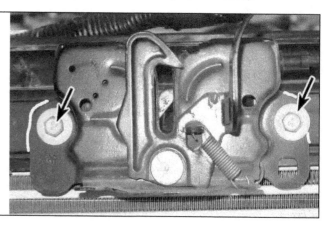

7.12 Mark the position of the hood latch, loosen the bolts and move the latch to adjust the hood in the closed position

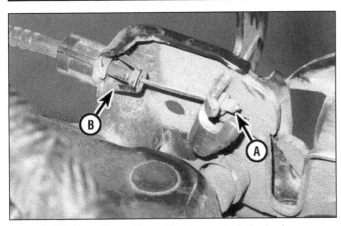

8.5 Release the cable end (A), then pinch the barb on the cable case

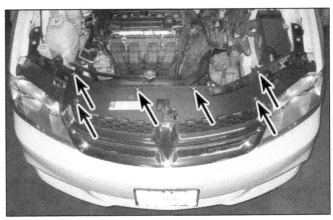

10.1 Remove the front bumper cover upper mounting fasteners

12   Mark the hood latch as a guide for adjustment (or removal and replacement). The hood latch assembly can also be adjusted up-and-down and side-to-side after loosening the bolts (see illustration).

13   The hood latch assembly, as well as the hinges, should be periodically lubricated with white lithium-base grease to prevent sticking and wear.

8   Hood latch and cable - removal and installation

Latch

Note: *Some models may have an additional hood latch release cable - it is disconnected the same way as in Step 5.*

1   If equipped, disconnect the hood ajar sensor electrical connector.

2   Remove the bolts holding the hood latch to the radiator support and detach the latch assembly (see illustration 7.12).

3   Detach the hood release cable (see Step 5), then remove the latch from the radiator support.

4   Installation is the reverse of removal.

Cable

5   Release the cable end, then slide the cable case end sideways in the keyhole slot of the hood latch while pinching the barb on the cable case closed (see illustration).

6   Remove the cable from the latch.

7   In the passenger compartment, remove the screws and detach the hood release cable and handle assembly. Detach the cable end from the hood release handle.

8   Under the dash, remove the rubber cable insulator from the hole in the dash panel.

9   Connect a string or piece of wire to the engine compartment end of the cable, then detach the cable and pull it through the firewall into the passenger compartment.

10   Connect the string or wire to the new cable and pull it through the firewall into the engine compartment.

11   The remainder of installation is the reverse of removal.

9   Radiator grille - removal and installation

1   Remove the push-pins that are holding the grille.

2   Remove the fascia fasteners from the grille.

3   Carefully lift the grille away from the bumper cover.

4   Installation is the reverse of removal.

10   Bumper covers - removal and installation

Note: *Disconnect the negative battery cable from the remote ground terminal or battery before beginning (see Chapter 5).*

Note: *On 2015 and later models, there are upper and lower plastic covers that need to be removed before all of the mounting bolts can be accessed.*

Front bumper cover

1   Release the hood latch and open the hood. Remove the fasteners attaching the grille to the radiator support (see illustration).

2   On 2014 and earlier models, remove the bolts holding the bumper cover to the headlight mounting panel at each side of the grille.

3   Loosen the front wheel lug nuts, then raise the vehicle and support it securely on jackstands.

4   Remove the front wheels.

5   Remove the front inner fender splash shield fasteners as necessary to gain access to the front cover-to-fender bolts (see Section 11).

6   Remove the fasteners securing the bottom of the bumper cover (see illustrations).

7   Disconnect the fog light electrical connectors, if necessary.

10.6a Remove the front bumper cover lower mounting fasteners

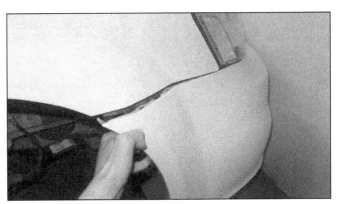

10.6b Pull out on the side of bumper cover to disengage it

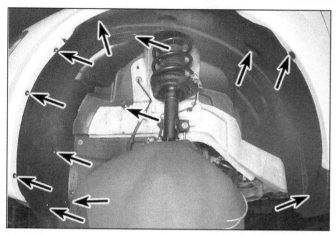

11.3 Remove the fender inner splash shield mounting fasteners

11.4 Open the door and remove this fastener

8   Disconnect the side marker light electrical connectors.
9   Carefully remove the bumper cover from the vehicle.
10   Installation is the reverse of removal.

Rear bumper cover

11   Open the trunk.
12   Raise the vehicle and support it securely on jackstands.
13   Remove the upper mounting fasteners.
14   Remove the fasteners securing the bottom of the bumper cover.
15   Release the hooks on the sides of the bumper cover from the tabs in the rear cover brackets.
16   Carefully remove the rear bumper cover from the vehicle.
17   Installation is the reverse of removal.

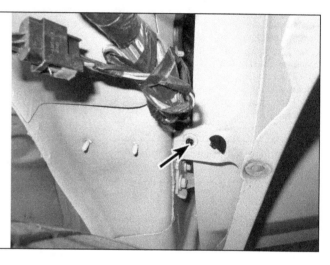

11.5 Remove the fender-to-door pillar lower mounting fastener

11   Front fender - removal and installation

1   Remove the front bumper cover (see Section 10), and fog lights if so equipped.

2   Remove the headlight housing (see Chapter 12).
3   Remove the fasteners retaining the fender inner splash shield **(see illustration)**.
4   Remove the fender-to-door pillar upper mounting fastener **(see illustration)**.
5   Working inside the fenderwell, remove the fender-to-door pillar lower mounting fastener **(see illustration)**.
6   Remove the fender brace **(see illustration)**.
7   Remove the fasteners securing the fender to the lower rocker panel **(see illustration)**.

11.6 Remove the fasteners securing the fender brace

11.7 Remove the fender-to-rocker panel cover, then remove the fasteners that secure the fender

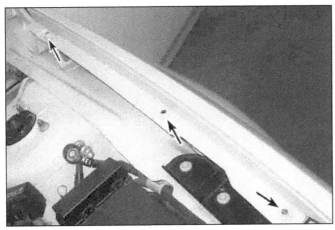

11.8 Remove the fender upper mounting bolts and lift off the fender

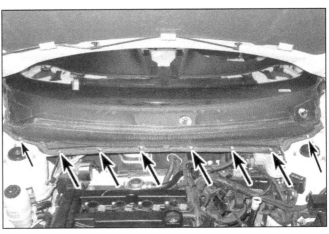

12.3 Cowl cover mounting fasteners

8   Remove the upper fender mounting bolts **(see illustration)**.
9   Detach the fender. It's a good idea to have an assistant support the fender while it's being moved away from the vehicle to prevent damage to the surrounding body panels.
10   Installation is the reverse of removal.

12   Cowl cover - removal and installation

1   Open the hood.
2   Remove the wiper arms (see Chapter 12). On 2015 and later models, release the clips and pull the side trim pieces forward from each end of the cowl cover.
3   Remove the clips that attach the cowl cover to the cowl **(see illustration)**.
4   Lift the cowl cover off the vehicle.
5   Installation is the reverse of removal.

13   Door trim panels - removal and installation

Note: *This procedure applies to front and rear door panels.*

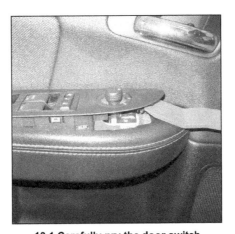

13.1 Carefully pry the door switch panel from the door and disconnect the electrical connectors

Removal

1   Use a small flat-bladed pry tool to remove the door switch panel and the control switch assembly **(see illustration)**.
2   Remove all door trim panel retaining fasteners located near the door pull handle and inside door handle **(see illustrations)**.

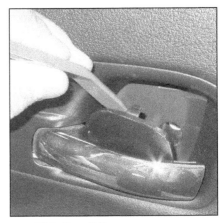

13.2a Pry off the screw cover . . .

3   Using a small flat-bladed pry tool, pry upward on the door trim panel and disengage the panel from the door run. Start from the bottom of the trim panel and work around the perimeter until the panel has been released from the door **(see illustration)**.

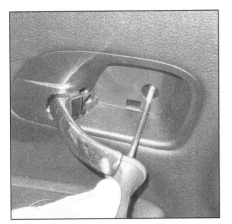

13.2b . . .then remove the fastener securing the inside door handle

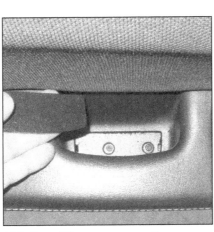

13.2c Pry off the screw cover, then remove the fasteners

13.3 Carefully pry the clips free so the door trim panel can be removed

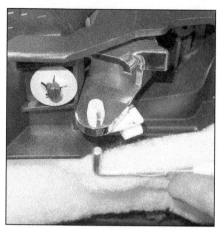

13.4 Disconnect the inner door handle lock cable

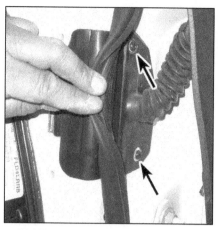

14.1 Remove these two fasteners, then disconnect the door harness wire connector

14.3 Remove the fasteners holding the lower hinge to the door

4   Disconnect the inner door handle lock cable from the handle **(see illustration)**. Disconnect any electrical connectors remaining and remove the door panel.

Installation

5   Position the wire harness connectors for the control switch and the power window switch (if so equipped) on the back of the panel, then place the panel in position in the door. Reconnect the inner door handle lock cable, then press the door panel into place until it is seated.

6   The remainder of installation is the reverse of removal.

---

14   Door - removal and installation

Caution: *If the hinge pin must be removed from the hinge, do not reuse the original pin. If you plan to remove the pin, be sure you have a new one before starting the job. The retaining clips used on the door hinge pins should also be replaced with new ones.*

1   Disconnect the door harness connector from the body harness **(see illustration)**.

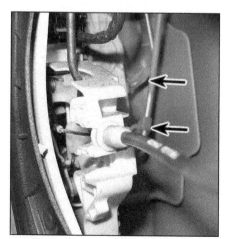

15.5 Release the clips, the disconnect the linkages from the door latch

2   Scribe around both the hinges with a marking pen, then place a jack under the door or have an assistant on hand to support it when the hinge fasteners are removed.
Note: *If a jack is used, place a towel between it and the door to protect the door's painted surfaces.*

3   Remove the fasteners holding the lower hinge to the door end frame **(see illustration)**. Keeping the door steady, remove the fasteners holding the upper hinge to the door end frame and carefully lift off the door.

4   Installation is the reverse of removal; align the hinge with the marks made during removal before tightening the fasteners.

5   Following installation of the door, check the alignment and adjust it if necessary as follows:

a)   *Up-and-down and in-and-out adjustments are made by loosening the hinge-to-door fasteners and moving the door as necessary.*

b)   *Forward-and-backward adjustments are made by loosening the hinge-to-body fasteners and moving the door as necessary.*

c)   *The door lock striker can also be adjusted both up-and-down and sideways to provide positive engagement with the lock mechanism. This is done by loosening the mounting screws and moving the striker as necessary.*

---

15   Door module, latch, lock and handle - removal and installation

---

Warning: *Models covered by this manual are equipped with a Supplemental Restraint System (SRS), more commonly known as airbags. Always disable the airbag system before working in the vicinity of any airbag system component to avoid the possibility of accidental deployment of the airbag, which could cause personal injury (see Chapter 12).*

Front door module

1   Disconnect the cable from the remote ground terminal or battery (see Chapter 5).

2   Remove the door trim panel (see Section 13).

3   Disconnect the door harness connector from the body harness (see Section 14).

4   Disconnect the window glass from the window regulator and remove the window glass.

5   Release the clips securing the linkage to the door latch, and disconnect the linkages from the door latch **(see illustration)**.

6   Remove the fasteners securing the door module **(see illustration)**. Remove the door module.

7   Installation is the reverse of removal.

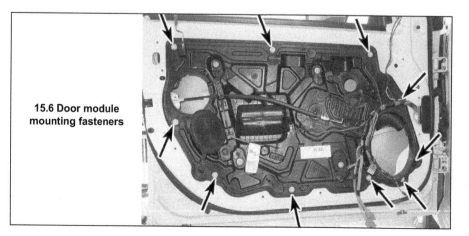

15.6 Door module mounting fasteners

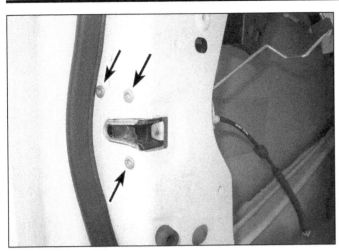

15.11 Remove the latch assembly fasteners

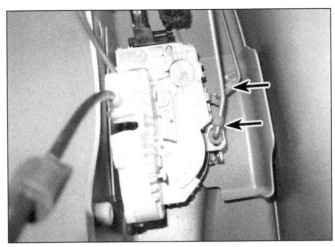

15.13 Pry the end of the lock and handle rod retaining clips up and pull the rods out of the latch

Front door latch

8   Raise the window to the full-up position.
9   Remove the door trim panel.
10 Remove the rubber door latch access cover which is located on the door module.
Note: *A two inch opening must be cut into the door module latch access opening to permit the removal of the door latch.*
11 Remove the latch retaining screws **(see illustration).**
12 Disconnect the exterior door handle and lock cylinder linkage from the latch.
13 If you're working on the driver's door, disconnect the lock cylinder linkage at the latch **(see illustration).**
14 Push upward on the latch locator bracket to release the bracket from the door module (if equipped).
15 Remove the interior door handle release cable from the door latch.
16 Disconnect the electrical connector to the door latch.
17 Remove the latch from the vehicle.
18 Installation is the reverse of removal.

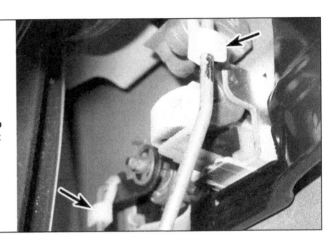

15.21 Pry the end of the lock and handle rod retaining clips up and pull the rods out of the handle

Front door outside handle

19 Raise the window to the full up position.
20 Remove the door trim panel (see Section 13).
21 Disconnect the latch rods from the door

handle and lock cylinder **(see illustration).**
22 Remove the fasteners securing the door handle to the outer door panel and remove the handle from the vehicle **(see illustrations).**
23 Installation is the reverse of removal.

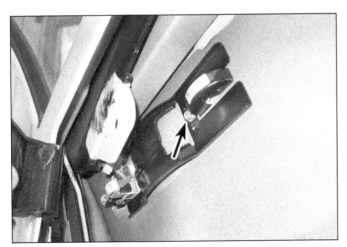

15.22a Through the access hole remove the handle front mounting fastener . . .

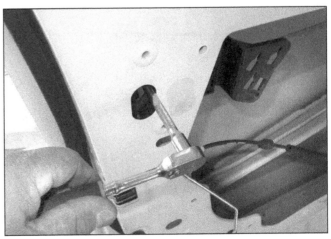

15.22b . . . then peal the tape back and working through the acces hole, remove the handle front mounting fastener

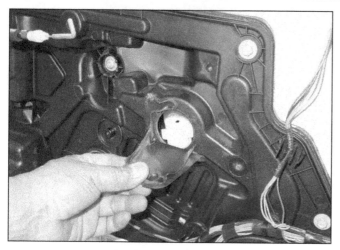

16.2a Pull out the front . . .

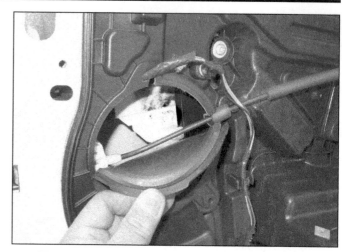

16.2b . . . and rear close-out plugs from the door module

Front door lock cylinder

24   Remove the front door outside handle (see Steps 19 through 22) and pull the lock cylinder from the door handle.

25   Installation is the reverse of removal.

16   Door window glass - removal and installation

Note: *This procedure applies to front and rear doors.*

1   Remove the door trim panel (see Section 13).

2   Pull out the close-out plugs from the door module **(see illustrations)**.

3   Lower the window glass about halfway down, lining up the glass with holes in the door module.

4   On 2014 and earlier sedan models, use a punch or screwdriver to press the tabs releasing the glass from the window regulator **(see illustrations)**.

5   On convertible models, loosen the front retaining nut and remove the rear mounting bolt. The window on 2015 and later models is retained by bolts.

6   Raise the glass upward from the channel. Separate the glass from the channel, then lift the glass upward and out of the exterior opening at the top of the door.

7   Installation is the reverse of removal.

17   Window glass regulators - removal and installation

1   Remove the door trim panel (see Section 13).

Note: *On sedan models, the window glass does not have to be removed. You can lift it up to the weather strip and use tape to hold the window in place.*

2   Remove the door module (see Section 15).

3   Remove the power window motor **(see illustration)**.

4   Remove the regulator-to-door module fasteners **(see illustration)**.

5   Remove the regulator from the door module.

6   Installation is the reverse of removal.

18   Mirrors - removal and installation

Outside mirrors

1   Use a pry tool to remove the side view mirror trim panel **(see illustration)**.

2   On vehicles with power mirrors, disconnect the power mirror electrical connector **(see illustration)**.

3   One vehicles with manual mirror adjuster, disconnect the adjuster from the side mirror cover.

4   Remove the outside mirror mounting fasteners.

5   Remove the outside mirror from the vehicle.

6   Installation is the reverse of removal.

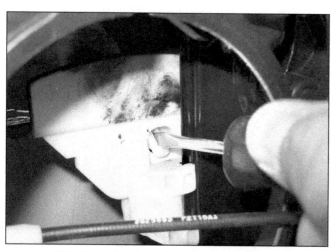

16.4a Press the plastic locking tabs inwards . . .

16.4b . . . past the glass, then slide the regulator down until the locking tab is clear

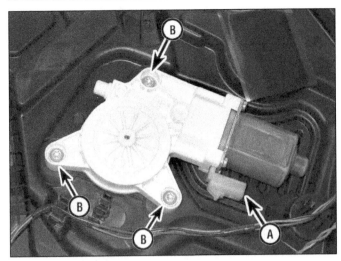

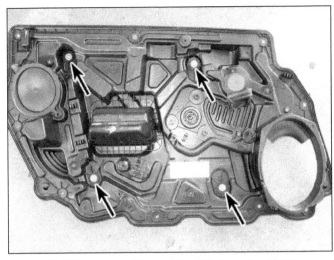

17.3 Unplug the electrical connector (A) and remove the window motor fasteners (B)

17.4 Window regulator guide rail fastener locations

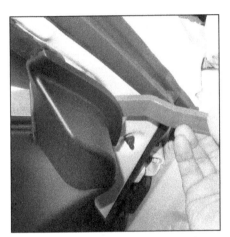

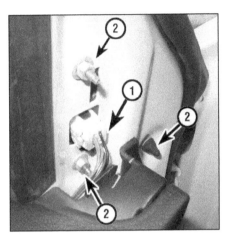

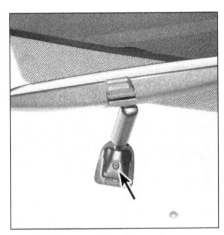

18.1 Carefully pry off the trim panel from the corner of the door

18.2 Power mirror details

1   Power mirror electrical connector
2   Mirror mounting fasteners

18.8 Remove the mirror set screw

Inside mirror

7   If the vehicle is equipped with either the electrochromic or telematic mirror options, first disconnect the battery (see Chapter 5).

Then disconnect the electrical connector at the back of the mirror.

8   Use a Philips head screwdriver to remove the set screw on the support/bracket

used on the windshield **(see illustration)**.

9   Slide the mirror off the mount plate and remove the mirror.

Note: *If the mount plate itself has come off the windshield, adhesive kits are available at auto parts stores to re-secure it. Follow the instructions included with the kit.*

10   Installation is the reverse of removal.

---

19   Trunk lid, latch and support struts - removal and installation

Trunk lid

Note: *The trunk lid is heavy and awkward to remove and install, so have an assistant available to help you.*

1   Remove the trunk lid trim panel fasteners **(see illustration).**

2   Disconnect any cables and electrical connectors attached to the trunk lid that would interfere with removal.

3   Use a marking pen to make alignment

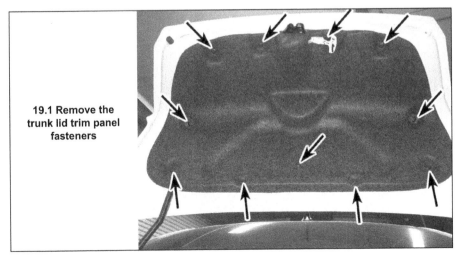

19.1 Remove the trunk lid trim panel fasteners

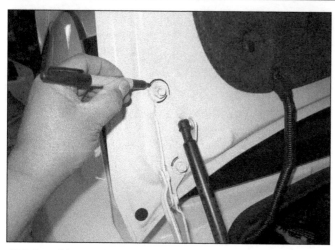

19.3 A marking pen is useful for drawing alignment marks around the hinge posts

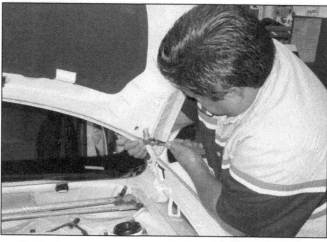

19.4 Remove the hinge bolts from both sides of the trunk lid

marks around the hinge bolts **(see illustration)**.
4   With an assistant supporting the weight of the trunk lid, remove the hinge bolts from both sides **(see illustration)** and lift off the trunk lid.
5   Installation is the reverse of removal.

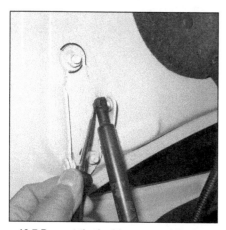

19.7 Pry out the locking caps at the top and bottom of the support struts

Trunk lid support struts
Note: *The lid supports are attached to the upper and lower brackets of the trunk lid. They do not need to be removed to remove the trunk lid.*
6   Open the trunk lid and support it with a prop.
7   Pry out the locking caps at the top and bottom of the support struts and pull the struts from their mounting studs **(see illustration)**.
Note: *Always replace support struts as a pair, but remove and install them from one side of the vehicle at a time. Always have one strut in place.*
8   Installation is the reverse of removal.

Trunk lid latch
9   Remove the trunk lid trim panel.
10   Disconnect the electrical connector from the trunk lid latch.
11   Remove the latch mounting fasteners on the trunk lid and remove the latch **(see illustration)**.
12   Installation is the reverse of removal.

20 Steering column covers - removal and installation

Warning: *Models covered by this manual are equipped with a Supplemental Restraint System (SRS), more commonly known as airbags. Always disable the airbag system before working in the vicinity of any airbag system component to avoid the possibility of accidental deployment of the airbag, which could cause personal injury (see Chapter 12).*
1   Place the steering column tilt in the full upward position and remove the steering column opening cover.
2   Detach the steering column cover mounting screws and remove the support.
3   Place the column tilt at the full downward position as you leave the tilt lever in the released position.
4   Remove the fasteners and release the mounting clips that hold each shroud, and remove the lower shroud **(see illustration)**.

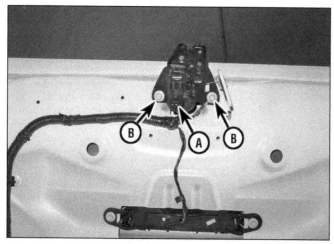

19.11 Disconnect the electrical connector (A), then remove the latch mounting fasteners (B)

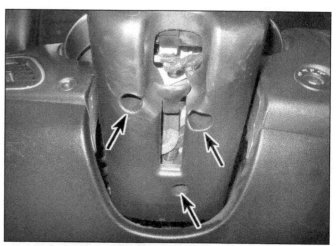

20.4 Remove the fasteners securing the lower steering column cover (2014 and earlier model shown, later models similar)

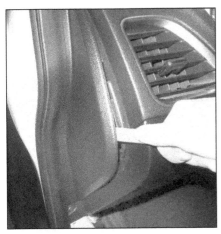

21.2 Pry the instrument panel end cap out
to disconnect the four mounting clips -
left side shown, right side identical

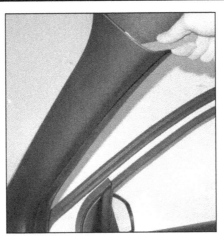

21.4a Pry off the A-pillar trim panels . . .

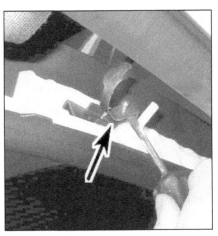

21.4b . . . then disengage the tether clip
using a small screwdriver

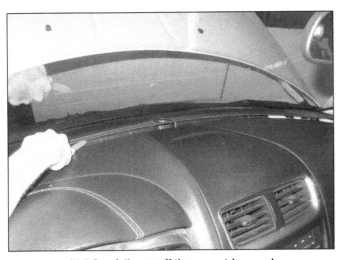

21.5 Carefully pry off the upper trim panel

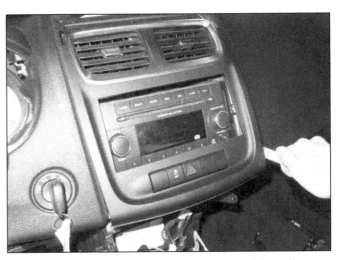

21.7 Pry the radio trim bezel out from around the radio

Note: *Be careful as the gap hider on the upper shroud is attached to the instrument cluster by four retaining tabs which can be released by hand.*

5   Place the steering column in the maximum extended position.

6   Reach up behind the shroud gap hider and release the tabs that retain the shroud gap hider and instrument cluster.

7   Installation is the reverse of removal.

---

21 Dashboard trim panels - removal and installation

Warning: *Models covered by this manual are equipped with a Supplemental Restraint System (SRS), more commonly known as airbags. Always disable the airbag system before working in the vicinity of any airbag system component to avoid the possibility of accidental deployment of the airbag, which could cause personal injury (see Chapter 12).*

1   Disconnect the cable from the remote ground terminal or battery (see Chapter 5).

Instrument panel end caps

2   Use a trim panel removal tool to release the clips holding the end cap cover on the left or right side of the instrument panel **(see illustration)**.

3   Installation is the reverse of removal.

Instrument panel top cover

4   Carefully pry off the A-pillar trim panels **(see illustrations)**.

5   Using a trim tool, start from the bottom of the trim panel and work around the perimeter until all the fasteners have been released, then remove the cover **(see illustration)**.

6   Installation is the reverse of removal.

Radio trim panel bezel

7   Using a trim tool, carefully pry around the bezel to release the mounting clips **(see illustration)** and remove the bezel from the instrument panel.

8   Installation is the reverse of removal.

Instrument panel center trim bezel

9   Using a trim tool, carefully pry around the bezel to release the mounting clips **(see illustration)** and remove the bezel from the instrument panel.

10   Installation is the reverse of removal.

21.9 Carefully pry the center trim bezel out

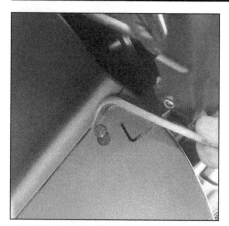

21.11 Pull down the glove box dampener to release it from the side of the glovebox

21.17 Remove the fastener from the side of the instrument panel

21.18 Pull the cover straight out

Glove box

11   Disconnect the glove box dampener from the glove box **(see illustration)**.

12   Open the glove box, push in on the sides of the glove box, then lower the door.

13   Pivot the glove box in a downward position, disengage the hinge hooks from the instrument panel and remove the glove box.

14   Installation is the reverse of removal.

Knee bolster

Note: *On 2015 and later models - once the knee bolster is removed, there is a knee bolster airbag behind it. Remove the four retaining nuts, then carefully rotate the knee bolster to access the airbag connector (make sure the battery has been disconnected for over two minutes). Disconnect the connector - it has a similar connector type to the steering wheel airbag connector as pictured in Chapter 10, Section 18.*

15   Remove the steering column cover (see Section 20).

16   Remove the driver's side instrument panel end cap **(see illustration 21.2)**.

17   Remove the knee bolster cover-to-instrument panel mounting fastener **(see illustration)**.

18   Pull the cover straight out to disengage the panel from the instrument panel **(see illustration)**.

19   Installation is the reverse of removal.

22   Instrument panel - removal and installation

Warning: *Models covered by this manual are equipped with a Supplemental Restraint System (SRS), more commonly known as airbags. Always disable the airbag system before working in the vicinity of any airbag system component to avoid the possibility of accidental deployment of the airbag, which could cause personal injury (see Chapter 12).*

Note: *This is a difficult procedure for the home mechanic. There are many hidden fasteners, difficult angles to work in and many electrical connectors to tag and disconnect/connect. We recommend that this procedure be done only by an experienced do-it-yourselfer.*

Note: *During removal of the instrument panel, make careful notes of how each piece comes off, where it fits in relation to other pieces and what holds it in place. If you note how each part is installed before removing it, getting the instrument panel back together again will be much easier.*

1   Disconnect the cable from the remote ground terminal or battery (see Chapter 5).

2   Remove the steering wheel and column as a unit (see Chapter 10).

3   Remove all dashboard trim panels and the glove box (see Section 21).

4   Remove the air conditioning and heater control assembly (see Chapter 3).

5   Remove the instrument cluster, radio and instrument panel speakers (see Chapter 12).

6   Remove the front seats (see Section 23) (though not absolutely necessary, removing both front seats allows more room to work and eliminates the possibility of the occurrence of damage to the seats during this procedure).

7   Remove the center floor console (see Section 24).

8   Remove the shifter knob and disconnect the shift cable (see Chapter 7A).

9   Disconnect the dashboard wiring harnesses at each A-pillar.

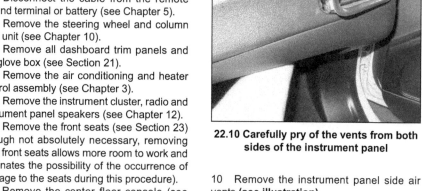

22.10 Carefully pry of the vents from both sides of the instrument panel

10   Remove the instrument panel side air vents **(see illustration)**.

11   Remover the radio support bracket **(see illustration)**.

12   Remove the passenger's side lower trim panel fasteners and remove the panel **(see illustration)**.

22.11 Remove the fasteners securing the support bracket, then remove the bracket

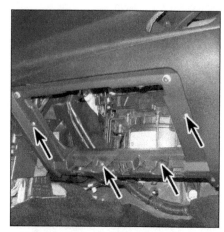

22.12 Passenger's side lower trim panel mounting fasteners

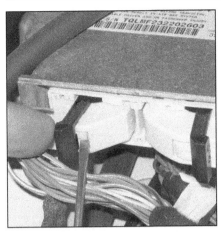

22.13 Push down the tab, then slide the lever to release the connectors

22.14a Disconnect the electrical connectors from the right . . .

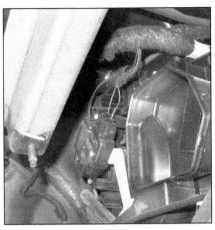

22.14b . . . and left sides of the air conditioning/heater housing

22.15 Carefully pry the harness off the housing

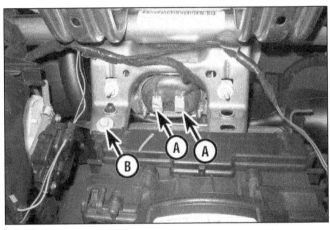

22.16 Disconnect the electrical connectors for the passenger airbag (A), then remove the fastener securing the bracket to the A/C and heating housing (B)

13   Disconnect the electrical connectors to the occupant restraint controller **(see illustration)**.
14   Disconnect the A/C and heating wiring harness electrical connectors **(see illustrations)**.

15   Disconnect the A/C and heating wiring harness from the A/C and heating housing **(see illustration)**.
16   Disconnect the passenger airbag, the remove the instrument panel-to-air conditioning/heater housing fastener **(see illustration)**.

17   Remove the fastener securing the top of the instrument panel **(see illustration)**.
18   Remove the fasteners that secure the A/C and heating housing to the bottom of the instrument panel support **(see illustration)**.
19   Disconnect the antenna harness and

22.17 Remove the fastener securing the top of the instrument panel

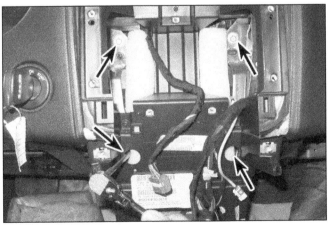

22.18 Remove these four fasteners

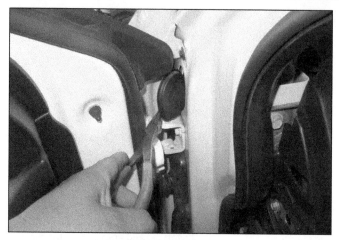

22.21a Remove the plug . . .

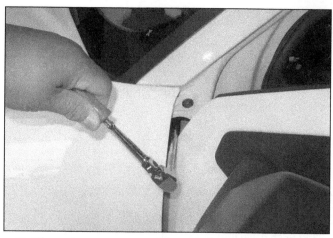

22.21b . . . then remove the bolt (left side shown, right similar)

22.22 Remove the fastener securing the
support (left side shown,
right side similar)

22.23 Remove the fasteners holding the
instrument panel to the vehicle; repeat the
process on the other side

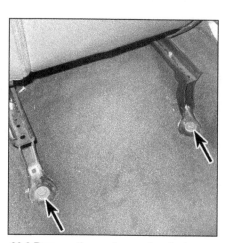

23.2 Remove the seat mounting fasteners

main vehicle harnesses. Remove the ground
wire and separate the data link connector
from the instrument panel.
20   Remove the hood release cable handle.
21   Located between the door and the fender's edge, remove the bolt holding the instrument panel at the side **(see illustrations)**.
Repeat on the other side.
22   Remove the instrument panel support

mounting fasteners **(see illustration)**.
23   Remove the instrument panel mounting
bolts from each end of the instrument panel
(see illustration).
Caution: *To avoid damage to the instrument
panel when removing these last bolts, have an
assistant support the instrument panel. You'll
also need an assistant's help when installing
the instrument panel and these bolts.*

24   With an assistant helping you, remove
the instrument panel from the vehicle.
25   Installation is the reverse of removal.

23   Seats - removal and installation

Warning: *Models covered by this manual
are equipped with a Supplemental Restraint
System (SRS), more commonly known as
airbags. Always disable the airbag system before working in the vicinity of any airbag system component to avoid the possibility of accidental deployment of the airbag, which could
cause personal injury (see Chapter 12).*
1   Disconnect the cable from the remote
ground terminal or battery (see Chapter 5).

Front seats
2   Slide the front seat forward and remove
the mounting fasteners **(see illustration)**.
3   Slide the seat backwards and remove the
front mounting fasteners **(see illustration)**.
4   If you're working on a sedan model, use
a flat tool and carefully pry off the seat belt
anchor trim. Remove the seat belt anchor bolt.

**23.3 Remove the front
mounting fasteners**

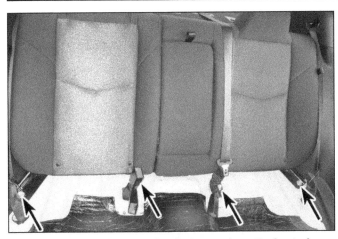

23.11 Remove the mounting fasteners along the front of the seat back

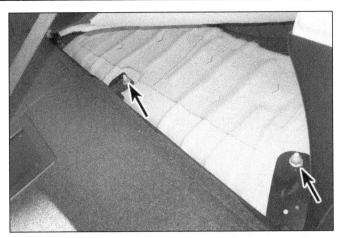

23.12 With the seat backs folded down, remove the mounting fasteners (left side shown, right side similar)

5   Tilt the seat forward and disconnect all electrical connectors attached to the seat.
Note: *Do not handle the seat by the adjuster release bar, as the bar is spring loaded.*
6   Remove the front seat from the vehicle.
7   Installation is the reversal of removal. Tighten the seat belt mounting bolts to the torque listed in this Chapter's Specification.

Rear seats

Sedan models

8   Pull upward at the forward edge of the cushion to disengage the retainer loops from the cups set in the floor.
9   As the seat belt is guided through the loops, the rear seat cushion can be taken from the vehicle.
10   Remove the rear seat belt buckles and anchors.
11   Remove the seat back front mounting fasteners **(see illustration).**
12   Release the folding sear seat back latches and fold the seats down completely; the seat back fasteners that secured the seat back to the floor can now be removed **(see illustration).**

13   Remove the seat back from the vehicle.
Warning: *Discard and replace the seat back bolts with new bolts when servicing. Installing used bolts may result in the loosening of the seat back bolts after repair. Failure to follow these instructions may result in serious or fatal injury.*
14   Installation is the reverse of removal. Tighten the seat belt mounting bolts to the torque listed in this Chapter's Specification.

Convertible models

15   Remove the rear shelf trim panel.
16   Pull upward at each retainer loop of the rear seat cushion in order to disengage the retainer loops in the floor.
17   Guide the seat belts through the loops.
18   Remove the rear seat cushion from the vehicle.
19   Remove the top seat-back fasteners.
20   On the center stud, remove the seat belt fastener.
21   Fold down the seat-back and remove the side bottom fastener.
22   Remove the seat-back from the vehicle.
Warning: *Discard and replace the seat-back bolts with new bolts when servicing. Installing used bolts may result in the loosening of the*

bolts after repair. Failure to follow these instructions may result in serious or fatal injury.
23   Installation is the reverse of removal.

24   Center floor console - removal and installation

Warning: *Models covered by this manual are equipped with a Supplemental Restraint System (SRS), more commonly known as airbags. Always disable the airbag system before working in the vicinity of any airbag system component to avoid the possibility of accidental deployment of the airbag, which could cause personal injury (see Chapter 12).*
1   Disconnect the cable from the remote ground terminal or battery (see Chapter 5).
2   Remove the heater/air conditioner control assembly (see Chapter 3).

2014 and earlier models

3   Remove the console storage tray **(see illustration).**
4   Remove the shifter bezel **(see illustration).**

24.3 Carefully pry out the storage tray and disconnect the electrical connector (2014 and earlier models)

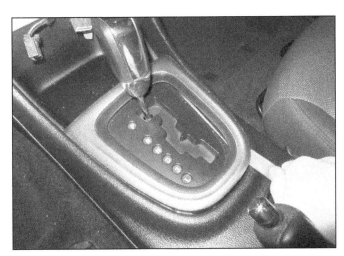

24.4 Carefully pry off the shifter bezel (2014 and earlier models)

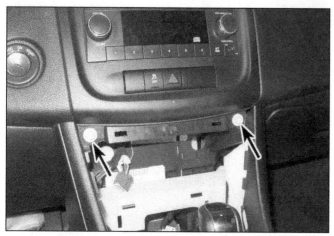

**24.5 Remove the fasteners at the front of the console
(2014 and earlier model shown, 2015 and later similar)**

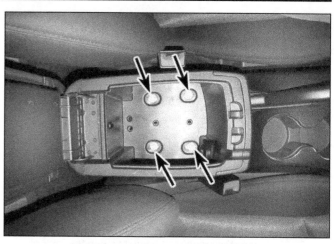

**24.6 Remove these mounting fasteners
(2014 and earlier model)**

5   At the front of the console, remove the two fasteners securing the front of the center console **(see illustration).**

6   Inside the console compartment, remove the fasteners securing the rear of the center console **(see illustration).**

2015 and later models

7   Reconnect the negative battery cable, then move the seat all the way forward, then remove the trim covers and bolts from the rear ends of the console.

8   Move the seat all the way back, then disconnect the negative battery cable.

9   Release the clips and remove the center console side panels using a trim tool.

10   Pry up the trim panel and remove the center console front mounting screws **(see illustration 24.5).**

11   Remove the driver's knee bolster (see Section 21).

12   Pull the console rearward slightly, enough to gain access to the electrical connectors.

All models

13   Disconnect any electrical connectors for the center console.

14   Engage the parking brake.

15   Lift the console up and remove the center console from the vehicle.

16   Installation is the reverse of removal.

Chapter 12
Chassis electrical system

Contents

	Section
Airbag system - general information and precautions	25
Antenna - removal and installation	12
Bulb replacement	17
Circuit breakers - general information	4
Cruise control system – description and check	20
Daytime Running Lights (DRL) - general information	23
Electric side view mirrors - general information	19
Electrical connectors - general information	6
Electrical troubleshooting - general information	2
Fuse box/power distribution center (2015 and later models) - removal and installation	26
Fuses and fusible links - general information	3
General information	1
Headlight bulb - replacement	15
Headlight housing - replacement	14

	Section
Headlights - adjustment	16
Horn - replacement	18
Ignition switch - replacement	8
Instrument cluster - removal and installation	9
Multi-function switch - replacement	7
Power door lock and keyless entry system - description and check	22
Power seat - general information and procedures	24
Power window system - description and check	21
Radio and speakers - removal and installation	11
Rear window defogger - check and repair	13
Relays - general information	5
Wiper motor - replacement	10
Wiring diagrams - general information	27

1 General information

1 The electrical system is a 12-volt, negative ground type. Power for the lights and all electrical accessories is supplied by a lead/acid-type battery that is charged by the alternator.

2 This Chapter covers repair and service procedures for the various electrical components not associated with the engine. Information on the battery, alternator, ignition system and starter motor can be found in Chapter 5.

3 It should be noted that when portions of the electrical system are serviced, the negative battery cable should be disconnected from the remote ground terminal to prevent electrical shorts and/or fires.

2 Electrical troubleshooting - general information

1 A typical electrical circuit consists of an electrical component, any switches, relays, motors, fuses, fusible links or circuit breakers related to that component and the wiring and connectors that link the component to both the battery and the chassis. To help you pinpoint an electrical circuit problem, wiring diagrams are included at the end of this Chapter.

2 Before tackling any troublesome electrical circuit, first study the appropriate wiring diagrams to get a complete understanding of what makes up that individual circuit. Trouble spots, for instance, can often be narrowed down by noting if other components related to the circuit are operating properly. If several components or circuits fail at one time, chances are the problem is in a fuse or ground connection, because several circuits are often routed through the same fuse and ground connections.

3 Electrical problems usually stem from simple causes, such as loose or corroded connections, a blown fuse, a melted fusible link or a failed relay. Visually inspect the condition of all fuses, wires and connections in a problem circuit before troubleshooting the circuit.

4 If test equipment and instruments are going to be utilized, use the diagrams to plan ahead of time where you will make the necessary connections in order to accurately pinpoint the trouble spot.

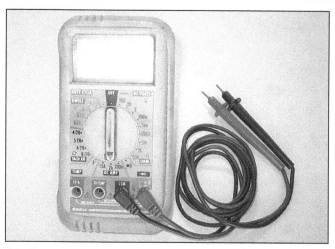

2.5a The most useful tool for electrical troubleshooting is a digital multimeter that can check volts, amps, and test continuity

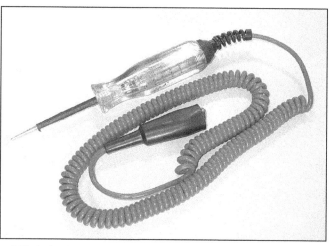

2.5b A simple test light is a very handy tool for testing voltage

Note: *Along with using the wire diagrams, tools, and scanners to evaluate a problem. It is always a good idea to read through the diagnostic procedures fully before starting any service work. Many times the diagnostic process may include something that you won't be able to accomplish without the aide of a certian tool, scanner, or software, and this may not be all that clear until you're well into trying to solve the problem. Read all information available throughly before attempting any repair.*

5   The basic tools needed for electrical troubleshooting include a circuit tester or voltmeter (a 12-volt bulb with a set of test leads can also be used), a continuity tester, which includes a bulb, battery and set of test leads, and a jumper wire, preferably with a circuit breaker incorporated, which can be used to bypass electrical components **(see illustrations)**. Before attempting to locate a problem with test instruments, use the wiring diagram(s) to decide where to make the connections.

Voltage checks

6   Voltage checks should be performed if a circuit is not functioning properly. Connect one lead of a circuit tester to either the negative battery terminal or a known good ground. Connect the other lead to a connector in the circuit being tested, preferably nearest to the battery or fuse **(see illustration)**. If the bulb of the tester lights, voltage is present, which means that the part of the circuit between the connector and the battery is problem free. Continue checking the rest of the circuit in the same fashion. When you reach a point at which no voltage is present, the problem lies between that point and the last test point with voltage. Most of the time the problem can be traced to a loose connection.
Note: *Keep in mind that some circuits receive voltage only when the ignition key is in the Accessory or Run position.*

Finding a short

7   One method of finding shorts in a circuit is to remove the fuse and connect a test

light or voltmeter in place of the fuse terminals. There should be no voltage present in the circuit. Move the wiring harness from side-to-side while watching the test light. If the bulb goes on, there is a short to ground somewhere in that area, probably where the insulation has rubbed through. The same test can be performed on each component in the circuit, even a switch.

Ground check

8   Perform a ground test to check whether a component is properly grounded. Disconnect the battery and connect one lead of a continuity tester or multimeter (set to the ohms scale), to a known good ground. Connect the other lead to the wire or ground connection being tested. If the resistance is low (less than 5 ohms), the ground is good. If the bulb on a self-powered test light does not go on, the ground is not good.
Note: *A lot of problems with electrical circuits*

2.6 In use, a basic test light's lead is clipped to a known good ground, then the pointed probe can test connectors, wires or electrical sockets - if the bulb lights, the circuit being tested has battery voltage

can be attributed to weak or faulty grounds. Pay close attention to ground leads and their connections. A weak connection can result in a voltage drop across a given circuit. This voltage drop can cause some systems to become slow or not work at all. An example would be headlights that appear dim or have a low intensity. In this case, connections - or the ground itself - can be the cause.

Continuity check

9   A continuity check is done to determine if there are any breaks in a circuit - if it is passing electricity properly. With the circuit off (no power in the circuit), a self-powered continuity tester or multimeter can be used to check the circuit. Connect the test leads to both ends of the circuit (or to the power end and a good ground), and if the test light comes on, the circuit is passing current properly **(see illustration)**. If the resistance is low (less than 5 ohms), there is continuity; if the reading is

2.9 With a multimeter set to the ohm scale, resistance can be checked across two terminals - when checking for continuity, a low reading indicates continuity, a high reading or infinity indicates lack of continuity

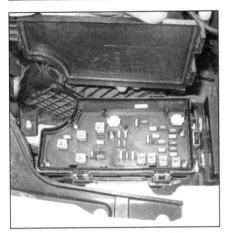

3.1 The engine compartment fuse and relay box (TIPM)

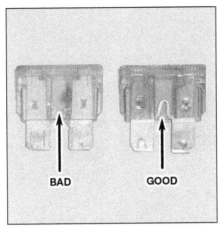

3.3 When a fuse blows, the element between the terminals melts

Note: *Not all fuses are energized with the key in the Off position. Turn the key to the On position, then check those fuses for voltage.*

4   Be sure to replace blown fuses with the correct type. Fuses (of the same physical size) of different ratings may be physically interchangeable, but only fuses of the proper rating should be used. Replacing a fuse with one of a higher or lower value than specified is not recommended. Each electrical circuit needs a specific amount of protection. The amperage value of each fuse is molded into the top of the fuse body.

5   If the replacement fuse immediately fails, don't replace it again or use a larger amperage fuse. More damage can occur until the cause of the problem is isolated and corrected. In most cases, this will be a short circuit in the wiring caused by a positive feed wire lead reaching ground potential before it has reached its module or component. Internal component failure or computer failure can also be a cause of a fuse to blow. Look for obvious signs of wire harnesses against metal components or near hot exahust. Abrasions to wiring harnesses is the most common electrical failures.

Fusible links

6   Some circuits are protected by fusible links. The links are used in circuits which are not ordinarily fused, or which carry high current, such as the circuit between the alternator and the battery. Fusible links, which are usually several wire gauges smaller in size than the circuit that they protect, are designed to melt if the circuit is subjected to more current than it was designed to carry. If you have to replace a blown fusible link, make sure that you replace it with one of the same specification. If the replacement fusible link blows in the same circuit, make sure that you troubleshoot the circuit in which the fusible link melted BEFORE installing another fusible link.

4   Circuit breakers - general information

1   Circuit breakers protect certain circuits, such as the power windows, power seats, windshield wipers, and seat heaters. Usually any high amperage and voltage system will generally have a breaker vs. a fuse as their protection device. Depending on the vehicle's accessories, there may be one or two circuit breakers, located in the fuse/relay box in the engine compartment.

2   Because the circuit breakers reset automatically, an electrical overload in a circuit breaker-protected system will cause the circuit to fail momentarily, then come back on.

Caution: *Most breakers are silver in color and much larger than a standard fuse. If you suspect the breaker is the problem, be very careful touching the actual breaker. An electrical short generates a great deal of heat, so the breaker itself can be very hot to the touch.*

10,000 ohms or higher, there is a break somewhere in the circuit. The same procedure can be used to test a switch, by connecting the continuity tester to the switch terminals. With the switch turned On, the test light should come on (or low resistance should be indicated on a meter).

Parasitic battery drain

10   Parasitic drain (sometimes referred to as a parasitic load) is the tendency of the electrical system to drain down it's usable energy over a certain amount of time while the ignition is turned off. The small amount of current draw is described in milliamps (mA). On these vehicles, this draw should not be more than 30 milliamps (0.030 amps). An amp meter is placed between the post and the battery clamp to obtain the readings. Each model, and each model type (depending on accessories) will have a different acceptable parasitic drain that is allowable.

11   Parasitic load is different from an open or a short in the system. This parasitic load is supposed to be there to provide necessary voltage to certain modules, computers, and other devices to maintain internal requirements. Some modules will power down after a few minutes, some can take up to 45 minutes or more to completely power down (sleep mode). Checking for a battery drain on these models takes a highly trained technician with the proper equipment. It is not advisable to attempt this without proper training.

Finding an open circuit

12   When diagnosing for possible open circuits, it is often difficult to locate them by sight because the connectors hide oxidation or terminal misalignment. Merely wiggling a connector on a sensor or in the wiring harness may correct the open circuit condition. Remember this when an open circuit is indicated when troubleshooting a circuit. Intermittent problems may also be caused by oxidized or loose connections.

13   Electrical troubleshooting is simple if you keep in mind that all electrical circuits

are basically electricity running from the battery, through the wires, switches, relays, fuses and fusible links to each electrical component (light bulb, motor, etc.) and to ground, from which it is passed back to the battery. Any electrical problem is an interruption in the flow of electricity to and from the battery.

Note: *When using a test light or any other sharp instrument to probe a wiring circuit, avoid stabbing directly into the wire. Use the back side of the connector to gain access when at all possible. If you do have to expose a section of wiring, be sure to insulate and protect the exposed section from corrosion after your testing is completed. This will avoid future problems from oxidation, wire rotting, and short circuits.*

3   Fuses and fusible links - general information

Fuses

1   The electrical circuits of the vehicle are protected by a combination of fuses, circuit breakers and fusible links. The main fuse/relay panel is in the engine compartment **(see illustration)**. On these vehicles, the fuse/relay box is referred to as the Totally Integrated Power Module (TIPM). Each of the fuses is designed to protect a specific circuit, and the various circuits are identified on the fuse panel itself.

2   Several sizes of fuses are employed in the fuse blocks. There are small, medium and large sizes of the same design, all with the same blade terminal design. The medium and large fuses can be removed with your fingers, but the small fuses require the use of pliers or the small plastic fuse-puller tool found in most fuse boxes.

3   If an electrical component fails, always check the fuse first. The best way to check the fuses is with a test light. Check for power at the exposed terminal tips of each fuse. If power is present at one side of the fuse but not the other, the fuse is blown. A blown fuse can also be identified by visually inspecting it **(see illustration)**.

Electrical connectors

Most electrical connectors
have a single release
tab that you depress to
release the connector

Some electrical
connectors have a
retaining tab which must
be pried up to free
the connector

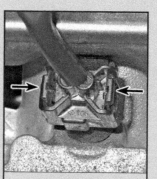

Some connectors have
two release tabs that you
must squeeze to release
the connector

Some connectors use
wire retainers that you
squeeze to release
the connector

Critical connectors often
employ a sliding lock (1)
that you must pull out
before you can depress
the release tab (2)

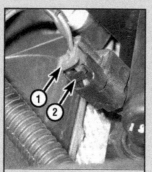

Here's another sliding-lock
style connector, with the
lock (1) and the release
tab (2) on the side of the
connector

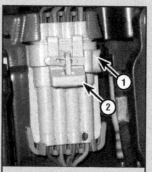

On some connectors the
lock (1) must be pulled out
to the side and removed
before you can lift the
release tab (2)

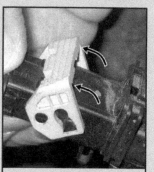

Some critical connectors,
like the multi-pin
connectors at the
Powertrain Control Module
employ pivoting locks that
must be flipped open

Proceed with caution when removing a hot breaker. If the breaker circuit is too hot, disconnect the battery and allow the breaker to cool down before removing it.

3   For a basic check, pull the circuit breaker up out of its socket on the fuse panel, but just far enough to probe with a voltmeter. The breaker should still contact the sockets. With the voltmeter negative lead on a good chassis ground, touch each end prong of the circuit breaker with the positive meter probe. There should be battery voltage at each end. If there is battery voltage only at one end, the circuit breaker must be replaced.

4   Some circuit breakers must be reset manually.

5   Relays - general information

Several electrical accessories in the vehicle, such as the fuel injection system, horns, starter, and fog lamps use relays to transmit the electrical signal to the component. Relays use a low-current circuit (the control circuit) to open and close a high-current circuit (the power circuit). If the relay is defective, that component will not operate properly. Relays are mounted in the engine compartment fuse/relay box (see illustration 3.1). Relays can be found in other locations; check your wiring diagram and component locator for exact locations.

Note: *Some relays are energized by a computer using a negative or positive feed. Do*

NOT bypass a relay from these connections unless you are sure of the source of the signal. Refer to the wiring diagram for the routing and exact location of these input voltages.

6   Electrical connectors - general information

1   Most electrical connections on these vehicles are made with multiwire plastic connectors. The mating halves of many connectors are secured with locking clips molded into the plastic connector shells. The mating halves of some large connectors, such as some of those under the instrument panel,

7.3 Multi-function switch mounting screw (steering wheel removed for clarity)

8.6 Carefully pry out the radio bezel with a flat prying tool

8.7 Pry the ignition switch bezel off with a flat pry tool

are held together by a bolt through the center of the connector.

2   To separate a connector with locking clips, use a small screwdriver to pry the clips apart carefully, then separate the connector halves. Pull only on the shell, never pull on the wiring harness, as you may damage the individual wires and terminals inside the connectors. Look at the connector closely before trying to separate the halves. Often the locking clips are engaged in a way that is not immediately clear. Additionally, many connectors have more than one set of clips.

3   Each pair of connector terminals has a male half and a female half. When you look at the end view of a connector in a diagram, be sure to understand whether the view shows the harness side or the component side of the connector. Connector halves are mirror images of each other, and a terminal shown on the right side end-view of one half will be on the left side end-view of the other half.

4   It is often necessary to take circuit voltage measurements with a connector connected. Whenever possible, carefully insert a small straight pin (not your meter probe) into the rear of the connector shell to contact the terminal inside, then clip your meter lead to the pin. This kind of connection is called "backprobing." When inserting a test probe into a terminal, be careful not to distort the terminal opening. Doing so can lead to a poor connection and corrosion at that terminal later. Using the small straight pin instead of a meter probe results in less chance of deforming the terminal connector. "T" pins are a good choice as temporary meter connections. They allow for a larger surface area to attach the meter leads too.

7   Multi-function switch - replacement

Warning: *The models covered by this manual are equipped with Supplemental Restraint Systems (SRS), more commonly known as airbags. Always disable the airbag system before working in the vicinity of any airbag*

system components to avoid the possibility of accidental deployment of the airbags, which could cause personal injury (see Section 25).
Note: *The multi-function switch is on the left side of the steering wheel. This switch controls the interior lights as well as the exterior lighting systems. The portion of the switch that is visible (without removing the steering column trim) is NOT the actual switch but merely the operating knob for the systems. The actual switch is inside the steering column. There are no serviceable parts inside the knob or the switch assembly. Replace as a unit.*
Note: *The multi-function switch controls headlights (both High and Low beam), fog lights (if applicable), park lamps, turn signals, interior lamp defeat, interior lights, and panel lamp dimming control. Diagnosis of the multi-function switch requires special diagnostic tools. Refer to your local dealership or qualified independent repair shop.*
Note: *The multi-function switch works in conjunction with serial data signals. Do NOT attempt to bypass the switch with voltage or ground signals with a jumper wire. Testing of the leads is best performed with the proper scanning equipment.*

1   Disconnect the negative battery cable from the remote ground terminal or battery (see Chapter 5).
Warning: *Wait at least two minutes for the reserve voltage to discharge before servicing or working near any airbag or steering column components.*
2   Remove the knee bolster, and the upper and lower steering column covers (see Chapter 11).
3   Remove the screw securing the multi-function switch **(see illustration)**. Pull the switch straight out from the steering column and disconnect the electrical connector.
4   Installation is the reverse of removal.

8   Ignition switch - replacement

Warning: *The models covered by this manual are equipped with Supplemental*

Restraint Systems (SRS), more commonly known as airbags. Always disable the airbag system before working in the vicinity of any airbag system components to avoid the possibility of accidental deployment of the airbags, which could cause personal injury (see Section 25).
Note: *On some models, the key lock cylinder and ignition switch have been replaced by a Wireless Ignition Node (WIN). The WIN is an integrated electronic receiver that is the center of communication for the key transmitter, remote starting system, security system and even the tire pressure monitoring system. If the WIN is damaged it cannot be repaired and must be replaced.*

Wireless Ignition Node (WIN)
Note: *If a new WIN is installed, it must be programmed to the vehicle and the ignition keys. This can only done with a factory scan tool. The vehicle will not operate unless the WIN is programmed to the vehicle.*
1   Disconnect the negative battery cable from the remote ground terminal or battery (see Chapter 5).
2   Remove the knee bolster and the lower driver's side trim panel (see Chapter 11).
3   Disconnect the coaxial cable connector and the electrical connector to the WIN.
4   Remove the fascia trim around the key opening in the dash to expose the WIN fasteners. The fascia trim is held in place with snap clips. Pull it away from the dash while applying firm pressure between the dash and the fascia with a smooth plastic blade. Remove the mounting fasteners and remove the WIN from the instrument panel.
5   Installation is the reverse of removal.

Ignition switch
6   Remove the radio trim bezel **(see illustration)**.
7   Remove the ignition switch bezel **(see illustration)**.
8   Remove the instrument cluster (see Section 9).

8.9a Gently pry the trim panel from around the ignition switch

8.9b Pull instrument cluster trim out by leaning it towards the steering wheel

8.11 Remove the ignition switch mounting fasteners

8.12 Pull down on the lock tabs to remove the ignition switch

8.13 Push on the tab to release the interlock cable

9   Remove the instrument cluster trim panel **(see illustrations)**.
10   Remove the driver's side knee bolster (see Chapter 11).

11   Remove the switch mounting fasteners **(see illustration)**.
12   Using a thin flat blade screwdriver, gently pry the clips from the ignition switch and push it out of the dash **(see illustration)**.

13   Disconnect the shift interlock cable and remove the switch assembly **(see illustration)**.
14   Installation is the reverse of removal.

9   Instrument cluster - removal and installation

Warning: *The models covered by this manual are equipped with Supplemental Restraint Systems (SRS), more commonly known as airbags. Always disable the airbag system before working in the vicinity of any airbag system components to avoid the possibility of accidental deployment of the airbags, which could cause personal injury (see Section 25).*

1   Disconnect the negative battery cable from the remote ground terminal or battery (see Chapter 5).
2   Remove the instrument cluster bezel **(see illustrations)**.
3   Remove the instrument cluster upper cover **(see illustrations)**.

9.2a Carefully pry around the bezel...

9.2b . . .then pull it out to remove the bezel

9.3a Carefully pry up the cover . . .

9.3b . . .then lift it off

9.4 Instrument cluster mounting fasteners

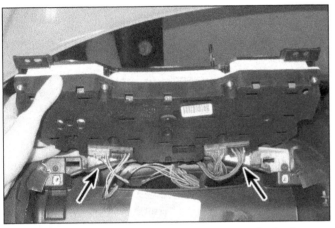

9.5 Disconnect the electrical connectors on the back
of the instrument cluster

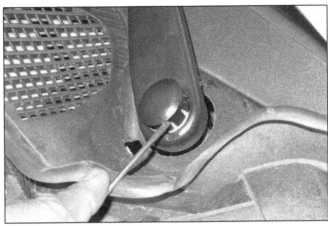

10.1 To access each front wiper arm retaining nut, carefully pry
off the protective cap that covers up the nut

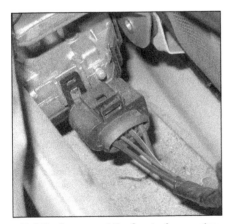

10.5 To disconnect the front wiper motor
electrical connector, push the lock up,
then depress the release tab and pull
off the connector

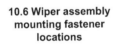

10.6 Wiper assembly
mounting fastener
locations

4   Remove the instrument cluster mounting
fasteners **(see illustration)**. Pull out the instrument cluster far enough to access the electrical
connector on the backside of the cluster.
5   Disconnect the electrical connector and
remove the cluster **(see illustration)**.
6   Installation is the reverse of removal.

10   Wiper motor - replacement

1   Place the wipers in the park position.
Mark the location of the wiper arms before
removing them. Pry off the protective caps covering the wiper arm nuts **(see illustration)**.
2   Remove the nuts that attach the wiper
arms to their spindle shafts.
3   Mark the position of each wiper arm in

relation to its shaft, then remove the wiper
arms. If the arm is difficult to remove from the
shaft, use a small two-jaw puller.
4   Remove the cowl cover (see Chapter 11).
5   Disconnect the wiper motor electrical
connector **(see illustration)**.
6   Remove the wiper assembly-to-firewall
bracket fasteners **(see illustration)**.
7   Lift the wiper assembly from the mounting brackets.

10.8 Pry off the link arm from the wiper motor crank arm (do NOT remove the crank arm from the motor)

10.9 Wiper motor mounting fasteners

11.2 Use a trim tool to pry the bezel from the instrument panel

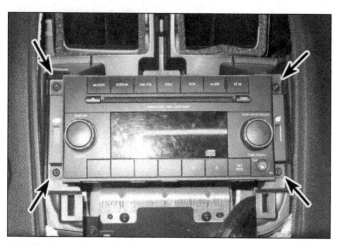

11.3 To detach the radio from the instrument panel, remove these four mounting screws

11.4 After pulling the radio out of the instrument panel, depress the release tabs and disconnect the electrical connectors and the antenna lead from the backside of the radio

8   Detach the wiper link from the wiper motor crank arm **(see illustration)**. Do NOT remove the crank arm from the motor.

9   Remove the windshield wiper motor-to-

11.8 To detach a front door speaker from the door, remove these four screws

linkage module fasteners **(see illustration)**, and remove the motor.

10   Installation is the reverse of removal.

11   Radio and speakers - removal and installation

Warning: *The models covered by this manual are equipped with Supplemental Restraint Systems (SRS), more commonly known as airbags. Always disable the airbag system before working in the vicinity of any airbag system components to avoid the possibility of accidental deployment of the airbags, which could cause personal injury (see Section 25).*

1   Disconnect the negative battery cable from the remote ground terminal or battery (see Chapter 5).

Radio

Warning: *Do not disassemble the radio. The radios covered by this manual are equipped with invisible laser radiation when the unit is*

opened and interlock failed or defeated. Avoid direct exposure to the beam, which could cause personal injury.

2   Using a trim tool, remove the radio trim bezel **(see illustration)**.

3   Remove the radio retaining screws **(see illustration)**, then pull the radio out of the instrument panel.

4   Disconnect the electrical connectors and the antenna lead from the backside of the radio **(see illustration)** and remove the radio.

5   Installation is the reverse of removal.

Speakers

6   These models are equipped with five speakers: one in the upper center of the instrument panel, one in each door, and two mounted in each rear quarter panel (Convertibles) or the rear shelf (Sedans).

Front door speakers

7   Remove the front door trim panel (see Chapter 11).

8   Remove the speaker mounting screws **(see illustration)**.

9   Pull out the speaker, disconnect the electrical connector, and remove the speaker.
10   Installation is the reverse of removal.

Instrument panel speakers

11   Remove the instrument panel top cover (see Chapter 11).
12   Remove the speaker mounting screws **(see illustration)**.
13   Pull out the speaker, disconnect the electrical connector and remove the speaker.
14   Installation is the reverse of removal.

12 Antenna - removal and installation

Antenna mast

1   The antenna mast simply unscrews from the base. In most cases it can be unscrewed by hand, but if necessary, use a pair of pliers to break it loose.

Antenna base

2   Remove the rear quarter trim panels (see Chapter 11) and the rear garnish molding at the rear of the headliner.
3   Pull the rear of the headliner down for access to the antenna base.
4   Unscrew the antenna base nut, disconnect the cable and detach the antenna base from the roof.
5   Installation is the reverse of removal.

13 Rear window defogger - check and repair

1   The rear window defogger consists of a number of horizontal elements baked onto the glass surface.
2   Small breaks in the element can be repaired without removing the rear window.

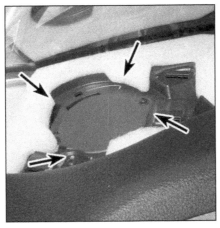

11.12 Remove the speaker mounting screws

Check

3   Turn the ignition switch and defogger system switches to the ON position. Using a voltmeter, place the positive probe against the defogger grid positive terminal and the negative probe against the ground terminal. If battery voltage is not indicated, check the fuse, defogger switch and related wiring. If voltage is indicated, but all or part of the defogger doesn't heat, proceed with the following tests.
4   When measuring voltage during these tests, wrap a piece of aluminum foil around the tip of the voltmeter positive probe and press the foil against the heating element with your finger **(see illustration)**. Place the negative probe on the defogger grid ground terminal.
5   Check the voltage at the center of each heating element **(see illustration)**. If the voltage is 5 or 6-volts, the element is okay (there is no break). If there is not voltage, the element is broken between the center of the element and the positive end. If the voltage is 10

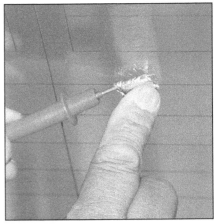

13.4 When measuring the voltage at the rear window defogger grid, wrap a piece of aluminum foil around the positive probe of the voltmeter and press the foil against the wire with your finger

to 12 volts, the element is broken between the center of the element and ground. Check each heating element.
6   Connect the negative lead to a good body ground. The reading should stay the same. If it doesn't, the ground connection is bad.
7   To find the break, place the voltmeter negative probe against the defogger ground terminal. Place the voltmeter positive probe with the foil strip against the heating element at the positive terminal end and slide it toward the negative terminal end. The point at which the voltmeter deflects from several volts to zero is the point at which the heating element is broken **(see illustration)**.

Repair

8   Repair the break in the element using a repair kit specifically recommended for this

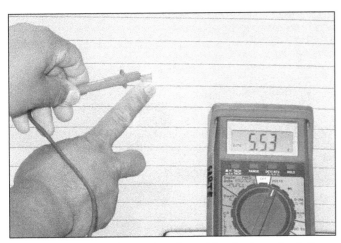

13.5 To determine if a heating element has broken, check the voltage at the center of each element. If the voltage is 5 or 6-volts, the element is unbroken; if the voltage is 10 or 12-volts, the element is broken between the center and the ground side; if there is no voltage, the element is broken between the center and the positive side

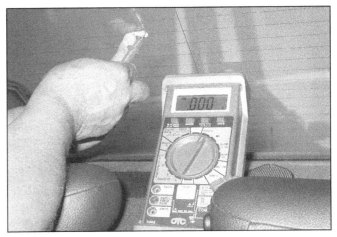

13.7 To find the break, touch the voltmeter negative lead to the defogger ground terminal, place the voltmeter positive lead with the foil strip against the heating element at the positive terminal end and slide it toward the negative terminal end. The point at which the voltmeter reading changes abruptly is the point at which the element is broken

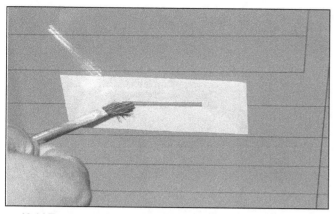

13.14 To use a defogger repair kit, apply masking tape to the inside of the window at the damaged area, then brush on the special conductive coating

14.3a Remove the two fasteners at the top of the housing...

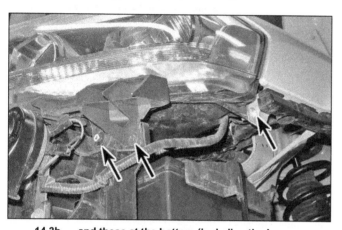

14.3b . . .and these at the bottom (including the bumper cover bracket) (2014 and earlier models)

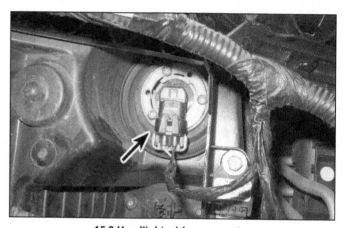

15.2 Headlight wiring connector

purpose, available at most auto parts stores. Included in this kit is plastic conductive epoxy.

9   Be sure to have adequate ventilation and that the glass be at a moderate temperature. Do NOT try this repair on an extremely hot or cold glass. Glass should be at or near 72 degrees F (room temperature).

10   Prior to repairing a break, turn off the system and allow it to cool off for a few minutes.

11   Lightly buff the element area with fine steel wool, then clean it thoroughly with rubbing alcohol.

12   Use masking tape to mask off the area being repaired.

13   Thoroughly mix the epoxy, following the instructions provided with the repair kit.

14   Apply the epoxy material to the slit in the masking tape, overlapping the undamaged area about 3/4-inch on either end **(see illustration)**.

15   Allow the repair to cure for 24 hours before removing the tape and using the system.

---

14   Headlight housing - replacement

Warning: *These vehicles are equipped with halogen gas-filled headlight bulbs, which are*

under pressure and may shatter if the surface is damaged or the bulb is dropped. Wear eye protection and handle the bulbs carefully, grasping only the base whenever possible. Do not touch the surface of the bulb with your fingers because the oil from your skin could cause it to overheat and fail prematurely. If you do touch the bulb surface, clean it with rubbing alcohol.

1   Disconnect the negative battery cable from the remote ground terminal or battery (see Chapter 5).

2   Remove the front bumper cover (see Chapter 11).

2014 and earlier models

3   Remove the headlight housing mounting fasteners **(see illustrations)**.

2015 and later models

4   On the bracket behind the headlight housing, remove the screws on the upper face of the bracket, then loosen the screw on the side of the bracket and remove the bracket.

5   Remove the headlight housing-to-fender retaining screws.

All models

6   Pull the headlight housing forward enough to disconnect the electrical connectors

from the housing, then remove the housing.

7   Installation is the reverse of removal. Adjust the headlights (see Section 16).

---

15   Headlight bulb - replacement

Warning: *Halogen bulbs are gas-filled and under pressure and might shatter if the surface is scratched or the bulb is dropped. Wear eye protection and handle the bulbs carefully, grasping only the base whenever possible. Don't touch the surface of the bulb with your fingers because the oil from your skin could cause it to overheat and fail prematurely. If you do touch the bulb surface, clean it with rubbing alcohol.*

Halogen headlights

Note: *On Dodge models, the headlights use a dual headlamp bulb; on Chrysler models, a "quad system" is used, with one bulb for high beam and one bulb for low beam.*

1   Disconnect the negative battery cable from the remote ground terminal or battery (see Chapter 5).

2   Open the hood and, working from inside the engine compartment, disconnect the electrical connector(s) from the back side of the housing **(see illustration)**.

15.3 Rotate the headlight bulb counterclockwise and remove it

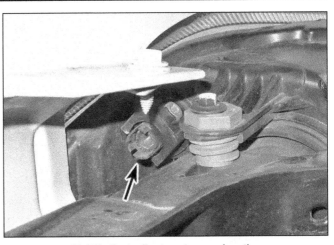

**16.1 Vertical adjustment screw location
(right headlight shown, left headlight identical)**

3   Rotate the headlight bulb counterclockwise and pull the bulb from the housing (**see illustration**).

4   When installing the new bulb, align the three slots on the circumference of the bulb socket's mounting flange with the three bosses on the inside edge of the mounting hole in the headlight assembly. Install the bulb into the housing, then rotate the bulb clockwise to lock it in place.

Caution: *Don't touch the surface of the bulb with your fingers because the oil from your skin could cause it to overheat and fail prematurely. If you accidentally touch the bulb surface, clean it with rubbing alcohol.*

5   Installation is otherwise the reverse of removal.

Xenon (HID) headlights

Warning: *Some models use High Intensity Discharge (HID) bulbs instead of conventional halogen bulbs. According to the manufacturer, the high voltages produced by this system can be fatal in the event of shock. Also, the voltage can remain in circuit even after the headlight switch has been turned to OFF and the ignition key has been removed. Therefore, for your safety, we don't recommend that you try to replace one of these bulbs yourself Instead, have this service performed by a dealer service department or other qualified repair shop.*

16   Headlights - adjustment

Warning: *The headlights must be aimedcorrectly. If adjusted incorrectly, they could temporarily blind the driver of an oncoming vehicle and cause an accident or seriously reduce your ability to see the road. The headlights should be checked for proper aim every 12 months and any time a new headlight is installed or front-end bodywork is performed. The following procedure is only an interim step to provide temporary adjustment until the headlights can be adjusted by a properly equipped shop.*

1   The vertical headlight adjustment screw, located on the inside corner of the housing (**see**

illustration), controls up-and-down movement. The adjustment screw on the outside corner controls the left-and-right movement.

2   There are several methods of adjusting the headlights. The simplest method requires a blank wall 25 feet in front of the vehicle and a level floor (**see illustration**).

3   Position masking tape on the wall in reference to the vehicle centerline and the centerlines of both headlights.

4   Measure the height of the headlight reference marks (in the centers of the headlight lenses) from the ground. Position a horizontal

tape line on the wall at the same height as the headlight reference marks.

Note: *It may be easier to position the tape on the wall with the vehicle parked only a few inches away.*

5   Adjustment should be made with the vehicle sitting level, the gas tank half-full and no unusually heavy load in the vehicle.

6   Turn on the low beams. Turn the adjusting screw to position the high intensity zone so it is two inches below the horizontal line.

7   Have the headlights adjusted by a dealer service department at the earliest opportunity.

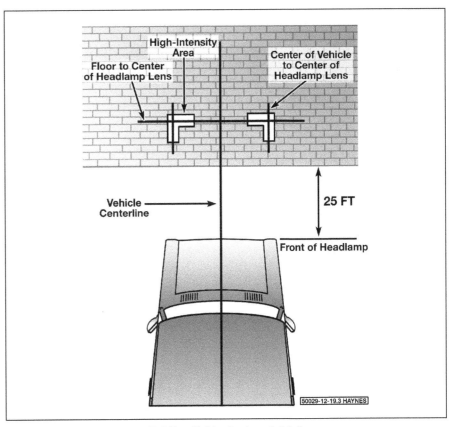

16.2 Headlight adjustment details

Bulb removal

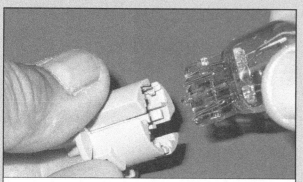

To remove many modern exterior bulbs from their holders, simply pull them out

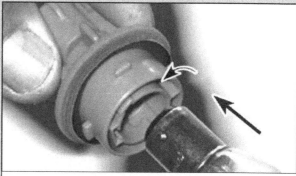

On bulbs with a cylindrical base ("bayonet" bulbs), the socket is spring-loaded; a pair of small posts on the side of the base hold the bulb in place against spring pressure. To remove this type of bulb, push it into the holder, rotate it 1/4-turn counterclockwise, then pull it out

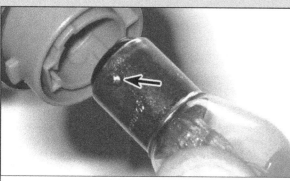

If a bayonet bulb has dual filaments, the posts are staggered, so the bulb can only be installed one way

To remove most overhead interior light bulbs, simply unclip them

---

17   Bulb replacement

Exterior light bulbs

Front park and turn signal bulbs

1   Reach behind the headlight housing, rotate the bulb holder counterclockwise and pull it out of the headlight housing.
2   Remove the bulb from the socket.
3   Install the new bulb in the socket.
4   The remainder of installation is the reverse of removal.

Fog light bulbs

Warning: *Halogen bulbs are gas-filled and under pressure and might shatter if the surface is scratched or the bulb is dropped. Wear eye protection and handle the bulbs carefully, grasping only the base whenever possible. Don't touch the surface of the bulb with your fingers because the oil from your skin could cause it to overheat and fail prematurely. If you do touch the bulb surface, clean it with rubbing alcohol.*

5   Disconnect the negative battery cable

from the remote ground terminal or battery (see Chapter 5).
6   Raise the front of the vehicle and place it securely on jackstands.
7   Remove the filler panel behind the bumper cover to gain access to the fog lamps. Disconnect the bulb electrical connector.
8   Rotate the bulb socket counterclockwise and pull it out of the fog light housing.

9   Replace the bulb and socket as a unit.
10   Installation is the reverse of removal.

Center high-mounted brake light

Note: *The high-mounted brake light LEDs are an integral part of the housing and are not available separately.*

11   Remove the trunk lid trim panel **(see illustration).** On 2015 and later models,

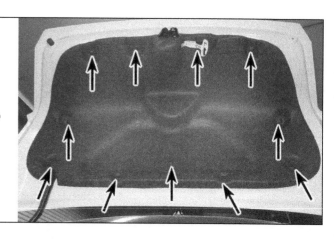

17.11 Pry out the push-pin fasteners to remove the panel

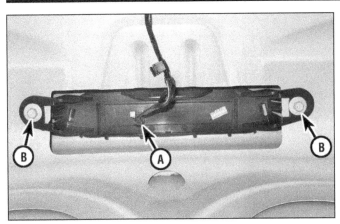

17.12 Disconnect the electrical connector (A), then remove the mounting fasteners (B) (2014 and earlier model shown)

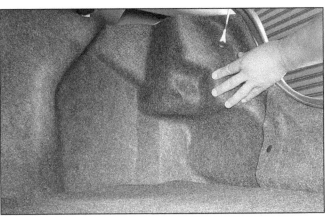

17.15 Remove the push-pin fastener, then pull back the trunk inner liner

remove the rear shelf paneling and the rear seat cushion (see Chapter 11).

12   On 2014 and earlier models, disconnect the electrical connector, then remove the mounting fasteners **(see illustration)**. On 2015 and later models, disconnect the connector, then disengage the tabs from the underside of the rear shelf panel.

13   Remove the center high-mounted brake light.

14   Installation is the reverse of removal.

Taillight bulbs

Brake light bulbs

Note: *The brake light LEDs are an integral part of the lamp unit and are not separately serviceable. If the light is not working, the lamp unit must be replaced.*

Back-up light bulb

Note: *On 2015 and later models, the taillight housing assembly is retained by two bolts, located underneath a cover between the trunk rubber lining and taillight housing. Pry up this plastic cover with a trim tool to access the bolts.*

15   Open the trunk and pull back the inner liner **(see illustration)**.

16   Unscrew the plastic fasteners attaching the taillight assembly to the quarter panel **(see illustration)**.

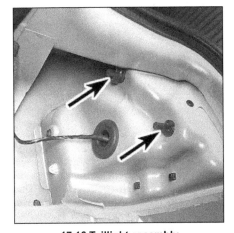

17.16 Taillight assembly mounting fasteners

17   Pull the taillight assembly rearward, away from the vehicle **(see illustration)**.

18   Disconnect the electrical connectors from the taillight sockets and remove the housing **(see illustration)**.

19   Remove the bulb socket by turning it counterclockwise and pulling it out from the taillight housing.

17.17 Pull straight away from the vehicle to remove the lens assembly

20   Remove the bulb from the socket **(see illustration)**.

21   Installation is the reverse of removal.

License plate light bulbs

22   Remove the two screws attaching the license plate light lens to the bumper cover.

23   Pull the bulb from the socket.

24   Installation is the reverse of removal.

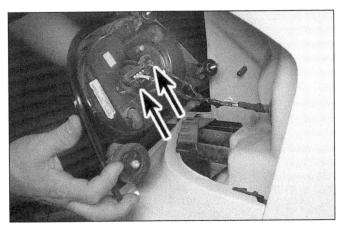

17.18 Disconnect the electrical connectors

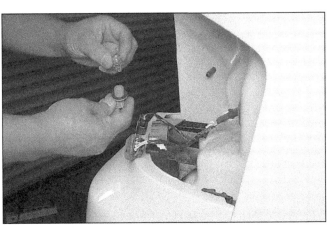

17.20 Remove the taillight marker bulb from the socket

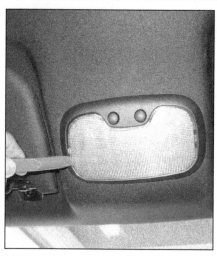

17.26 Carefully pry off the lens cover to access the bulb

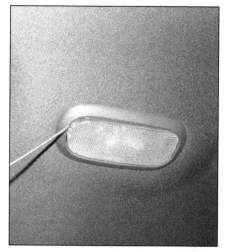

17.29 Carefully pry off the lens cover to access the bulb

18.3 Remove the fastener securing the horn (right side shown, left side similar)

Interior lights

Instrument cluster illumination bulbs

25   The instrument cluster LEDs are an integral part of the cluster, and not separately serviceable. If the LEDs stop working, the cluster must be replaced (see Section 9).

Reading lights

26   Using a trim panel removal tool or a screwdriver, pry off the lens cover **(see illustration)**.
27   Remove the bulb.
28   Installation is the reverse of removal.

Dome light

29   Using a trim panel removal tool or a screwdriver, pry off the dome light lens cover **(see illustration)**.
30   Remove the bulb.
31   Installation is the reverse of removal.

18   Horn - replacement

Warning: *The models covered by this manual are equipped with Supplemental Restraint Systems (SRS), more commonly known as airbags. Always disable the airbag system before working in the vicinity of any airbag system components to avoid the possibility of accidental deployment of the airbags, which could cause personal injury (see Section 25).*
Note: *There are two horns located in the front area of the vehicle, just outside of the main frame rail, behind the front bumper, on the right and left sides of the vehicle.*
1   Remove the front bumper cover (see Chapter 11).
2   Disconnect the electrical connector from the horn.
3   Remove the horn mounting fastener and the horn assembly **(see illustration)**.
4   Installation is the reverse of removal.

19   Electric side view mirrors - general information

Note: *On these models, the outside mirrors are controlled by the Front Door Power Switch Assembly (FDPSA), which houses the mirror control switches and the Front Door Control Module (FDCM). The FDPSA interprets the mirror control switch movements, while the FDCM supplies the battery voltage and ground to the motors to move the mirrors.*
1   The electric rear view mirrors use two motors to move the glass; one for up and down adjustments and one for left-right adjustments.
2   The control switch has a selector portion which sends voltage to the left or right side mirror. With the ignition in the ACC position and the engine OFF, roll down the windows and operate the mirror control switch through all functions (left-right and up-down) for both the left and right side mirrors.
3   Listen carefully for the sound of the electric motors running in the mirrors.
4   If the motors can be heard but the mirror glass doesn't move, there's probably a problem with the drive mechanism inside the mirror. A loud but distinct clicking can be heard with most failures associated with the drive motors failing to move the actual mirrors. Power mirrors have no user-serviceable parts inside - a defective mirror must be replaced as a unit (see Chapter 11).
5   If the mirrors don't operate and no sound comes from the mirrors, check the fuses (see Section 3).
6   If the fuses are OK, remove the mirror control switch. Have the switch continuity checked by a dealer service department or other qualified shop.
7   Check the ground connections.
8   If the mirror still doesn't work, remove the mirror and check the wires at the mirror for voltage.
9   If there's not voltage in each switch position, check the mirror and control switch for a

disconnected (open) lead. It should be noted that a weak or loose connection is often overlooked as the problem. Be sure to check all electrical connections as well.
10   If there is voltage and grounds at the leads on their respective positions and the mirror still does not respond, replace the mirror assembly.

20   Cruise control system – description and check

1   The cruise control system maintains vehicle speed with the Antilock Brake Module (ABM), Powertrain Control Module (PCM), instrument cluster, throttle actuator control motor, Brake Pedal Position (BPP) sensor, wheel speed sensors, control switches and associated wiring. There is no mechanical connection, such as a vacuum servo or cable. Some features of the system require special testers and diagnostic procedures that are beyond the scope of the home mechanic. Listed below are some general procedures that may be used to locate common problems.
2   Check the fuses (see Section 3).
3   The BPP switch (or brake light switch) deactivates the cruise control system. Have an assistant press the brake pedal while you check the brake light operation.
4   If the brake lights do not operate properly, correct the problem and retest the cruise control.
5   Check the wiring between the PCM and throttle actuator motor for any obvious opens or shorts and repair as necessary.
6   The cruise control system uses information from the PCM, including the wheel speed sensors, located on the knuckle of each wheel. Refer to Chapter 9 for more information on the wheel speed sensors.
7   Test drive the vehicle to determine if the cruise control is now working. If it isn't, take it to a dealer service department or other qualified repair shop for further diagnosis.

22.11 Using the emergency key or small screwdriver, carefully pry the halves apart

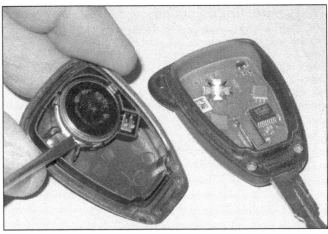

22.12 Remove the battery, noting which direction the battery faces

21 Power window system - description and check

Note: *These models are equipped with door control modules in each door. These modules allow for more features to be added to the window operation than traditional positive/negative arrangement. Each window switch is connected to the door module by way of a serial data line. Convertible models use the TIPM for power distributions as well. Proper scanning and diagnostic equipment is needed to properly test and evaluate window operation issues.*

1   The power window system operates electric motors, mounted in the doors, which lower and raise the windows. The system consists of the control switches, the motors, regulators, glass mechanisms, the Door Module and associated wiring.

2   The power windows can be lowered and raised from the master control switch by the driver or by remote switches located at the individual windows. Each window has a separate motor that is reversible. The position of the control switch determines the polarity and therefore the direction of operation.

3   The circuit is protected by a fuse and a circuit breaker. Each motor is also equipped with an internal circuit breaker; this prevents one stuck window from disabling the whole system. However, it should be noted that each door module will reverse the window direction if the current flow becomes greater than its preset values.

4   The power window system will only operate when the ignition switch is ON, and for a period of time after the ignition key has been turned Off (unless one of the doors is opened). In addition, many models have a window lockout switch at the master control switch which, when activated, disables the switches at the rear windows and, sometimes, the switch at the passenger's window also. Always check these items before troubleshooting a window problem.

5   These procedures are general in nature, so if you can't find the problem using them,

take the vehicle to a dealer service department or other properly equipped repair facility.

6   If the power windows won't operate, always check the fuse and circuit breaker first.

7   If only the driver's window is functioning, check to see if the master switch window lockout button - located on the driver's window switch panel - has been depressed.

8   If only one window is inoperative from the master control switch, try the other control switch at the window.

9   If one window is inoperative from both switches see a professional repair shop and have it properly tested.

22 Power door lock and keyless entry system - description and check

Note: *These models use input and output signals from various electronic components. These systems are linked to each other in several ways to the CCN (Cab Compartment Node), which allows simple and accurate troubleshooting, but only with a professional-grade scan tool. The door lock systems involve the ignition switch, wireless control module, instrument cluster, TIPM, door latches, door control switches and the respective door modules. Proper diagnostics requires special training and equipment. Have the vehicle diagnosed by a dealership service department or other qualified automotive repair facility.*

1   The power door lock system operates the door lock actuators mounted in each door. Diagnosis can usually be limited to simple checks of the wiring connections and actuators for minor faults that can be easily repaired.

2   Power door lock systems are operated by bi-directional solenoids located in the doors. The lock switches have two operating positions: Lock and Unlock.

3   If you are unable to locate the trouble using the following general steps, consult your dealer service department or qualified independent repair facility.

4   Always check the circuit protection first.

Some vehicles use a combination of circuit breakers and fuses. Refer to the wiring diagrams at the end of this Chapter.

5   If all but one lock solenoids operates, remove the trim panel from the affected door (see Chapter 11). Using the wiring diagram, locate the positive voltage wire/connection that leads to the lock actuator and check for voltage while operating the lock. The lock actuator is an integral part of the door latch.

6   If the inoperative solenoid is receiving voltage, replace the solenoid or latch assembly.

7   If the inoperative solenoid isn't receiving voltage, check for an open or short in the wire between the lock solenoid and the instrument cluster.

8   If the above tests do not pinpoint the problem, take the vehicle to a qualified dealership repair facility or qualified independent shop with the correct scanning equipment for proper testing and repair.

Keyless entry system

9   The keyless entry system consists of a remote control transmitter that sends a coded infrared signal to a receiver, which then operates the door lock system.

10   Replace the battery when the transmitter doesn't operate the locks at a distance of ten feet. Normal range should be about 30 feet.

Key remote control battery replacement

Note: *The key remote control replacement battery is a CR2032 battery.*

11   Use the tip of the emergency key, a plastic trim tool or coin to carefully separate the case halves **(see illustration)**.

Caution: *Do not touch the battery terminals that are on the back of the transmitter housing or the printed circuit board. The transmitter contains "perchlorate material" that may require special handling for disposal. See www.dtsc.ca.gov/hazardouswaste/perchlorate for more information.*

12   Replace the battery **(see illustration)**.

13   Snap the case halves together.

Transmitter programming

14 Programming replacement transmitters requires the use of a specialized scan tool. Take the vehicle and the transmitter(s) to a dealer service department or other qualified repair shop equipped with the necessary tool to have the transmitter(s) programmed to the vehicle.

23 Daytime Running Lights (DRL) - general information

The Daytime Running Lights (DRL) system, which is required on new Canadian models and an option on models made for the United States, illuminates the headlights when the engine is running. The DRL system supplies reduced power to the headlights so they won't be too bright for daytime use, which also prolongs headlight life.

24 Power seat - general information and procedures

1 The power seat recieves its voltage from a fuse in the Totally Integrated Power Module (TIPM) located in the engine compartment. The power seats can be operated in eight different directions, not including the recliner (the recliner is on its own switch, incorporated with the seat switch).

Power seat system check

2 If the power seat fails to function properly, follow these test procedures:
3 Check under the seat for any obstructions or anything that might have pulled the electrical connections apart. Secure all electrical connections. If the connections are good and the seat still does not move, go to the next Step.
4 With engine off (to reduce noise level), operate seat controls in all directions and listen for any sounds coming from the seat motors.
5 A grinding sound when the motor is on indicates either a broken gear (usually plastic gears) or the cable between the motor and the seat transmission (if applicable) has stripped. If you do not hear any noise or motor movement, this could mean there is no voltage/ground to the motor assembly, or the motor/seat transmission assembly has frozen. A quick check is to leave a door open and watch the interior lights as you operate the switch. If the lights dim as you move the lever, but the seat fails to move, chances are it is stuck in that position for some reason.
6 Check the power seat fuse in the TIPM. If the fuse is good, proceed to the next Step.
7 Remove the power seat switch, and check for positive and negative signals on their respective terminals (following the appropriate wiring diagram). If switch is OK, check the leads between the switch and motor assembly. If the wires are OK and no signal is

leaving the switch when moved to the appropriate position, replace it with a known good switch and recheck. If voltage is leaving the switch, wire connectors and harness is good the track assembly will need replaced.

Track assembly replacement

8 Remove the driver's seat from the vehicle (see Chapter 11).
9 Remove the seat side trim and the fasteners securing the cushion to the track assembly. Remove all wiring harness connections and leads from the track assembly. Remove the seat back and the track assembly. Transfer any components that are going to be reused to the new seat.
10 Installation is the reverse of removal.

Power seat switch replacement

11 Disconnect the negative battery cable from the remote ground terminal (see Chapter 5), and isolate it from touching anything.
12 Remove the side cushion side shields. Disconnect the electrical connections. Using a small flat screwdriver, gently pry the four mounting tabs that secure the seat switch to the side shield.
13 Snap the switch onto the side shield. The remainder of installation is the reverse of removal.

Heated seat system - general information and procedures

14 The seat heater circuit receives its power from a fuse in the TIPM mounted in the engine compartment. The system is designed to operate only with the key on. The seat heater system is controlled by the seat heater module mounted below the driver's seat. When the seat heater switch is depressed, a data message is sent via the LIN (Local Interface Network) to the instrument cluster (CCN - Cab Compartment Node). This message is then transferred to the seat heater modules. The module sends positive voltage from an internal solid state relay to the seat heater elements.
15 Both the seat heater module and the carbon fiber seat heater elements can be serviced by removing the seat. The seat heater module is bolted to the bottom of the lower cushion. To access the seat heater elements, the seat must be removed and the seat cushions must be disassembled. The elements can be peeled from the seat cushion and replaced separately, rather than having to replace the entire seat and/or cushions.

25 Airbag system - general information and precautions

General information

1 All models are equipped with a frontal-impact airbag system, which is referred to as the Supplemental Restraint System (SRS).

The SRS is designed to protect the driver and the front seat passenger from serious injury in the event of a head-on or frontal collision. The SRS is controlled by the Occupant Restraint Controller (ORC), also referred to as the Airbag Control Module (ACM), which is mounted in the center of the vehicle, on the floor transmission tunnel, below the center of the instrument panel. The SRS uses an array of airbags to protect the front-seat occupants (and on models equipped with side curtain airbags, the rear seat passengers, too): the driver's airbag in the steering wheel; the passenger airbag, which is located in the right end of the instrument panel, beneath the instrument panel top pad and above the glove box; and, on models so equipped, the side curtain airbags, which are located above the side windows, in the outer edges of the headliner, between the A- and C-pillars. The SRS is activated by a pair of front impact sensors located on the front of the frame, just behind the bumper attachments. Other important components in the SRS include the clockspring (a wind-up coil that delivers battery voltage to the steering wheel airbag), and the AIRBAG readiness light on the instrument cluster.
2 In addition to the airbags, seat belt pretensioners are incorporated into the front seat belt retractor mechanisms. These are pyrotechnic (explosive) devices which retract the seat belts up to four inches when the airbag system is activated.

Driver's airbag

3 The airbag inflator module contains a housing incorporating the cushion (airbag) and inflator unit, mounted in the center of the steering wheel. The inflator assembly is mounted on the back of the housing over a hole through which gas is expelled, inflating the bag almost instantaneously when an electrical signal is sent from the system. The clockspring assembly on the steering column under the steering wheel carries this signal to the module. The clockspring assembly can transmit an electrical signal regardless of steering wheel position. The igniter in the airbag converts the electrical signal to heat and ignites the powder, which inflates the bag.

Passenger's airbag

4 The airbag is mounted in the right end of the instrument panel, beneath the instrument panel top pad and above the glove box. It consists of an inflator containing an igniter, a bag assembly, a reaction housing and a trim cover. The passenger airbag is considerably larger than the steering wheel-mounted unit and is supported by the steel reaction housing. The trim cover is textured and painted to match the instrument panel and has a molded seam that splits when the bag inflates.

Side-impact window airbags

5 Optional side-impact window airbags protect vehicle occupants in the event of a side impact. The side-impact window airbags are located above the windows, between the A- and C-pillars. If you're not sure whether

your vehicle is equipped with side-impact window airbags, look for the words "SRS AIR-BAG" imprinted on a small identification trim button located above the B-and C-pillars.

6   Vehicles equipped with side-impact window airbags use six side-impact sensors: three on the left side of the vehicle and three on the right side. The front row side-impact sensors are located inside the B-pillars, above the front seatbelt retractors. The second-row side-impact sensors are located in the sliding door track openings, just ahead of the C-pillars. The third-row sensors are located behind the quarter-trim panels, between the C- and D-pillars, above the rear wheelwells.

Seat airbags

7   These airbags are optional equipment in sedans and standard equipment in convertibles. They are located in the outsides of the front seat backs.

Seat belt pre-tensioners

8   Some models are equipped with pyrotechnic (explosive) units in the front seat belt retracting mechanisms. During an impact that would trigger the airbag system, the airbag control unit also triggers the seat belt retractors. When the pyrotechnic charges go off, they accelerate the retractors to instantly take up any slack in the seat belt system to more fully prepare the driver and front seat passenger for impact.

9   The airbag system should be disabled any time work is done to or around the seats. **Warning:** *Never strike the pillars or floorpan with a hammer or use an impact-driver tool in these areas unless the system is disabled.*

Occupant Restraint Controller (ORC) or Airbag Control Module (ACM)

10   The ORC or ACM supplies current to the SRS in the event of a collision, even if battery power is cut off. The ORC/ACM checks the SRS every time the vehicle is started, and indicates that it is doing so by turning on the AIRBAG readiness light. If the SRS is operating properly, the ORC/ACM turns off the AIRBAG readiness light. If it detects a fault in the system, the AIRBAG

readiness light will remain on. If this condition occurs, take the vehicle to your dealer immediately for service.

Disarming the system and other precautions

Warning: *Failure to follow these precautions could result in accidental deployment of the airbag and personal injury.*

11   Whenever you are working in the vicinity of the driver's airbag in the steering wheel or any of the other airbags on your vehicle, DISARM THE SYSTEM. To disarm the system:

- *Point the wheels straight ahead and turn the ignition key to the LOCK position.*
- *Disconnect the negative battery cable from the remote ground terminal (see Chapter 5). Isolate the cable terminal so it won't accidentally contact the battery post.*
- *Wait at least two minutes for the back-up power supply to be depleted. Back-up power is supplied by a capacitor that takes about two minutes to fully discharge. During this two-minute interval, the SRS is still capable of deploying.*

12   Whenever handling an airbag, always keep the airbag opening (the trim side) pointed away from your body. Never place the airbag on a bench or other surface with the airbag opening facing the surface. Always place the airbag module in a safe location with the airbag opening facing up.

13   Never measure the resistance of any SRS component. An ohmmeter has a built-in battery supply that could accidentally deploy the airbag.

14   Never dispose of a live airbag. Return it to a dealer service department or other qualified repair shop for safe deployment and disposal.

Component removal and installation

Driver airbag and clockspring

15   Refer to Chapter 10, Section 18 for the driver's side airbag module and clockspring removal and installation procedures.

Other airbag modules

16   We don't recommend removing any of the other airbag modules. These jobs are best left to a professional.

---

26   Fuse box/power distribution center (2015 and later models) - removal and installation

---

1   Disconnect the cable from the negative battery terminal (see Chapter 5).

2   Remove the fuse box cover, then unscrew the nut(s) and disconnect the power cable connections at the end of the upper fuse box. Be sure to label the cables for correct installation.

3   Fully loosen the three fuse box-to-lower fuse box retaining screws.

4   Release the plastic side clips, then pull up the upper fuse box, separating it from the connectors below.

5   Installation is the reverse of removal.

---

27   Wiring diagrams - general information

---

1   Since it isn't possible to include all wiring diagrams for every year and model covered by this manual, the following diagrams are those that are typical and most commonly needed.

2   Prior to troubleshooting any circuits, check the fuse and circuit breakers (if equipped) to make sure they're in good condition. Make sure the battery is properly charged and check the cable connections (see Chapter 5).

3   When checking a circuit, make sure that all connectors are clean, no signs of burnt or melted connections, with no broken or loose terminals. When disconnecting a connector, do not pull on the wires. Pull only on the connector housings themselves.

Notes

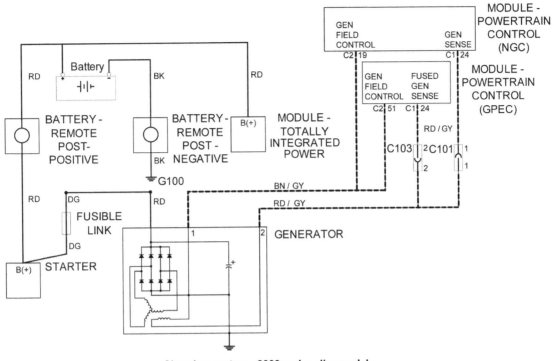

Charging system - 2009 and earlier models

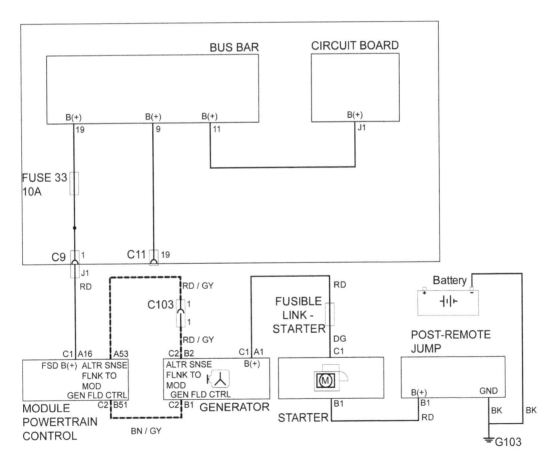

Charging system - 2010 and later models

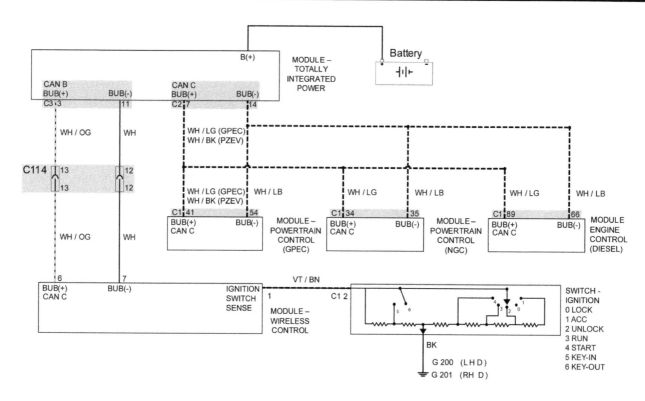

Starting system - 2009 and earlier models (1 of 3)

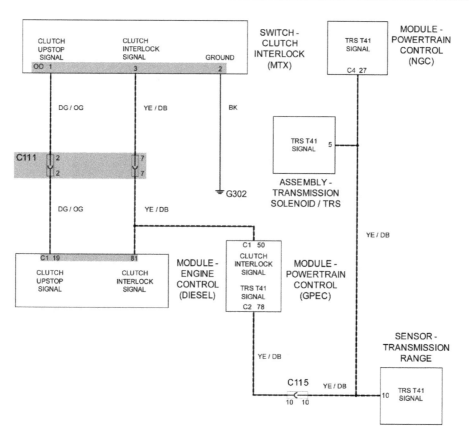

Starting system - 2009 and earlier models (2 of 3)

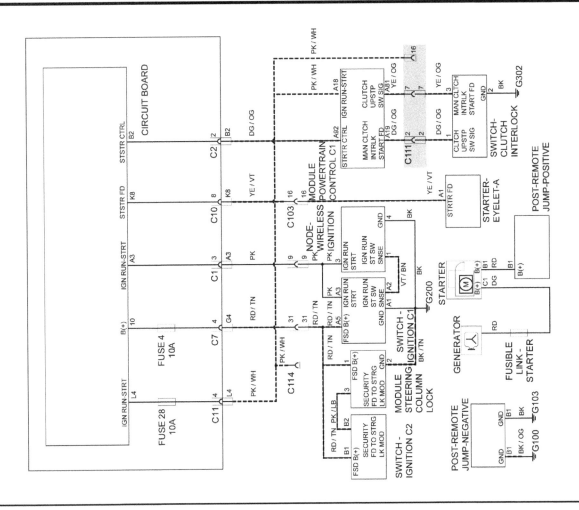

Starting system - 2010 and later models

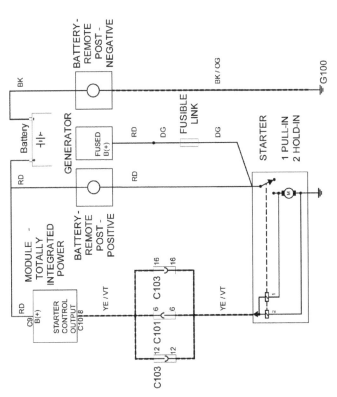

Starting system - 2009 and earlier models (3 of 3)

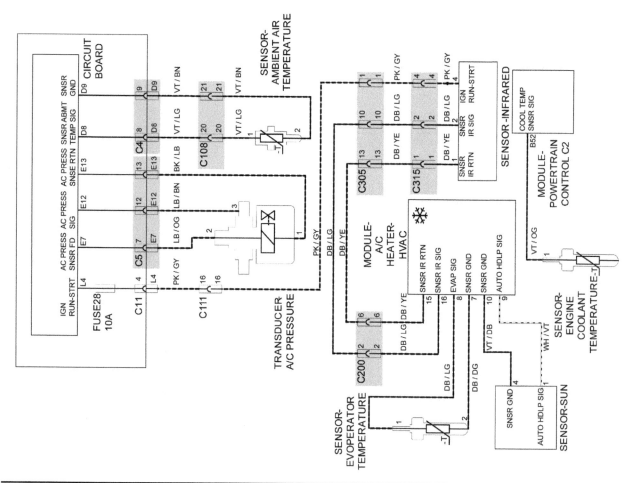

Heating, air conditioning and engine cooling fan systems - 2010 and later models (2 of 7)

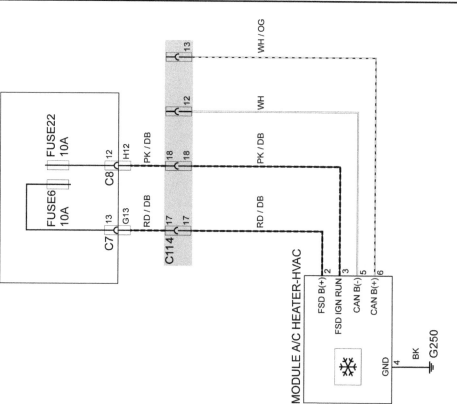

Heating, air conditioning and engine cooling fan systems - 2010 and later models (1 of 7)

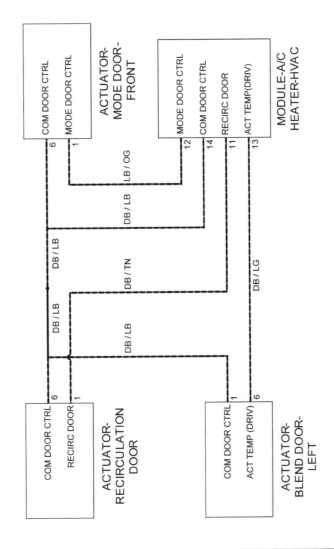

Heating, air conditioning and engine cooling fan systems - 2010 and later models (4 of 7)

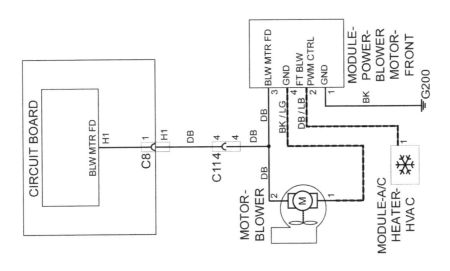

Heating, air conditioning and engine cooling fan systems - 2010 and later models (3 of 7)

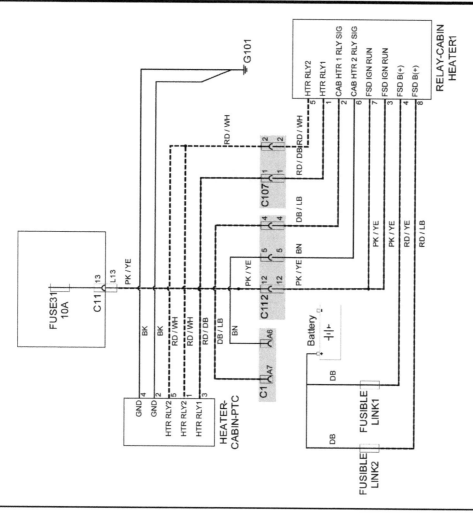

Heating, air conditioning and engine cooling fan systems - 2010 and later models (6 of 7)

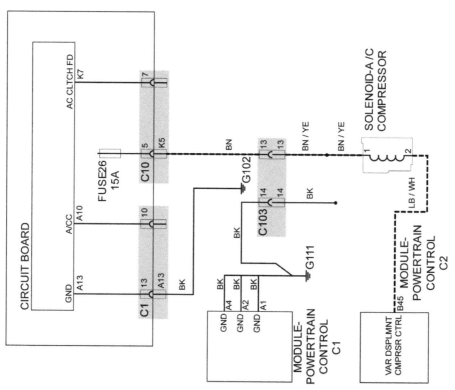

Heating, air conditioning and engine cooling fan systems - 2010 and later models (5 of 7)

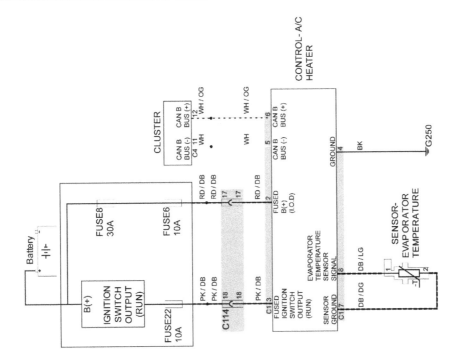

Heating, air conditioning and engine cooling fan systems - 2009 and earlier models (1 of 4)

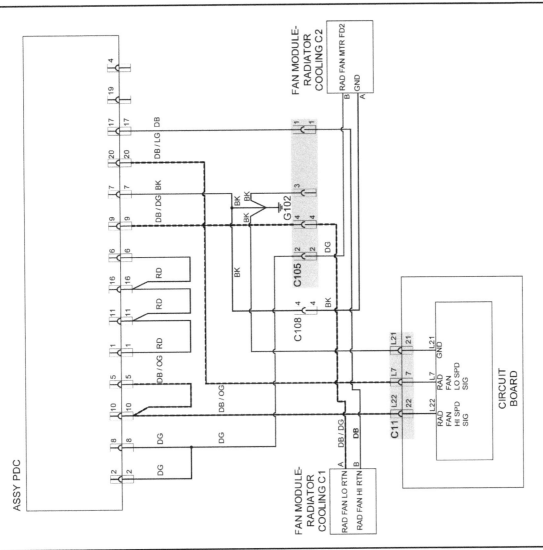

Heating, air conditioning and engine cooling fan systems - 2010 and later models (7 of 7)

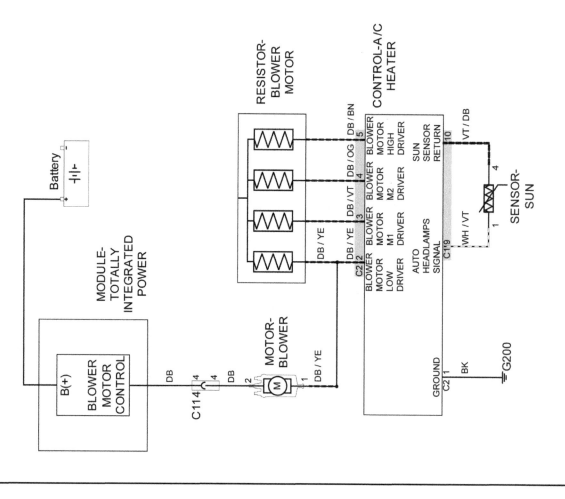

Heating, air conditioning and engine cooling fan systems - 2009 and earlier models (3 of 4)

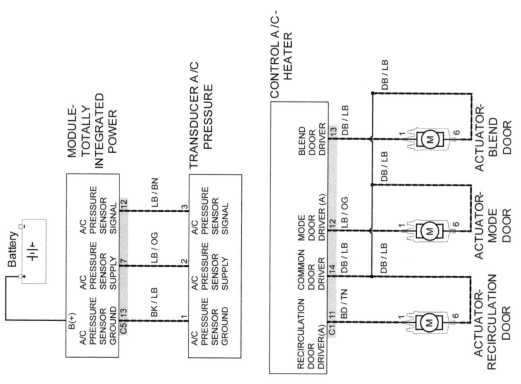

Heating, air conditioning and engine cooling fan systems - 2009 and earlier models (2 of 4)

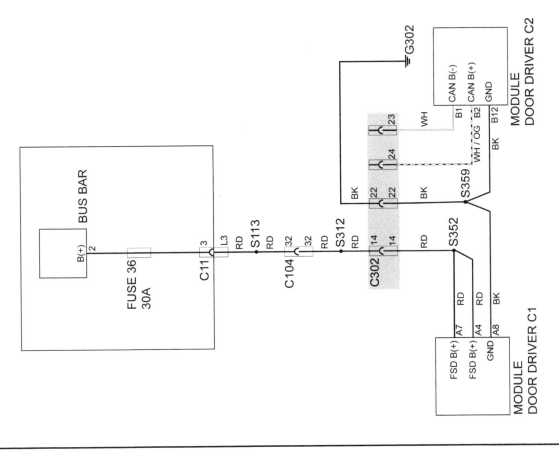

Power window system - 2010 and later models (1 of 6)

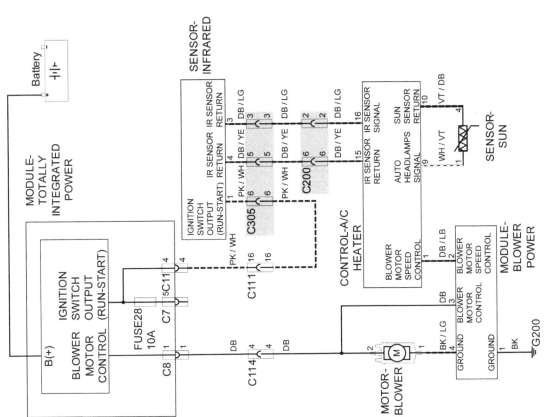

Heating, air conditioning and engine cooling fan systems - 2009 and earlier models (4 of 4)

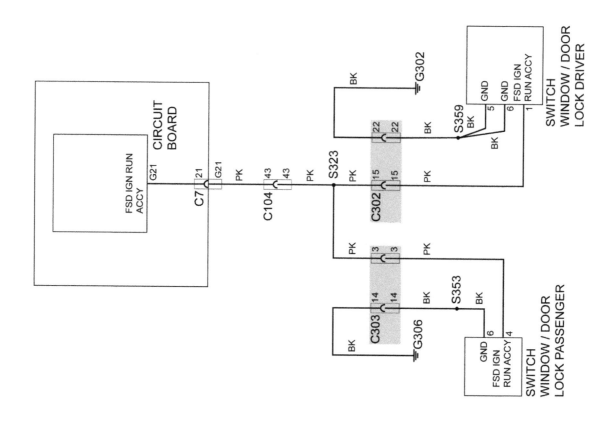

Power window system - 2010 and later models (3 of 6)

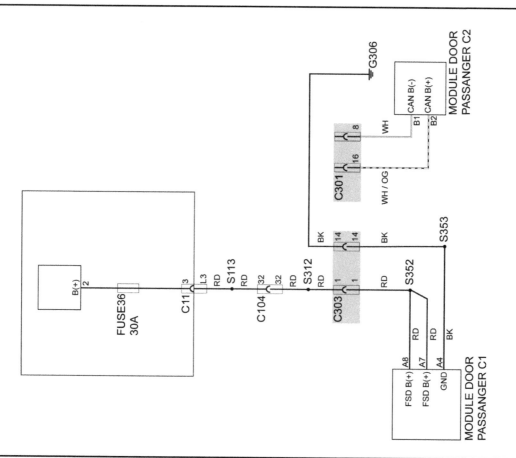

Power window system - 2010 and later models (2 of 6)

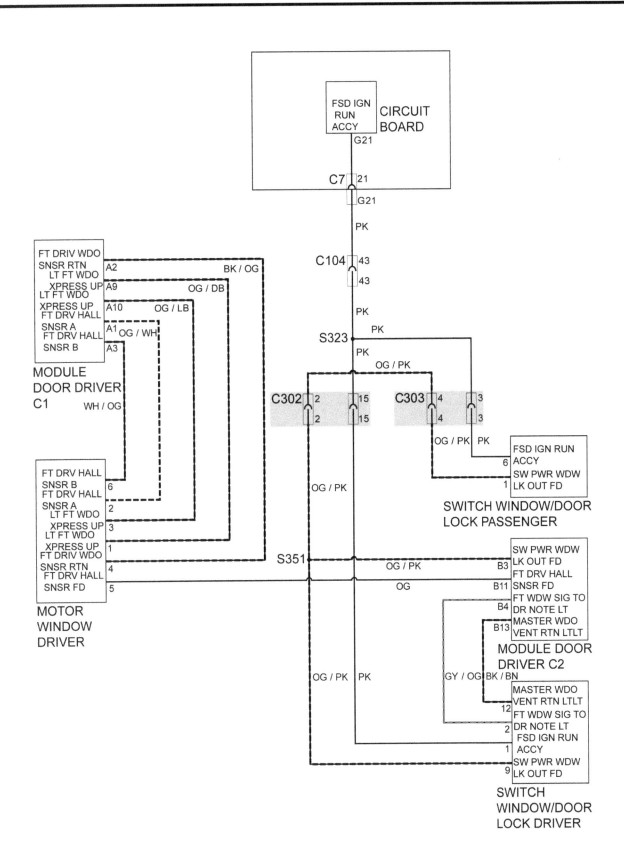

Power window system - 2010 and later models (4 of 6)

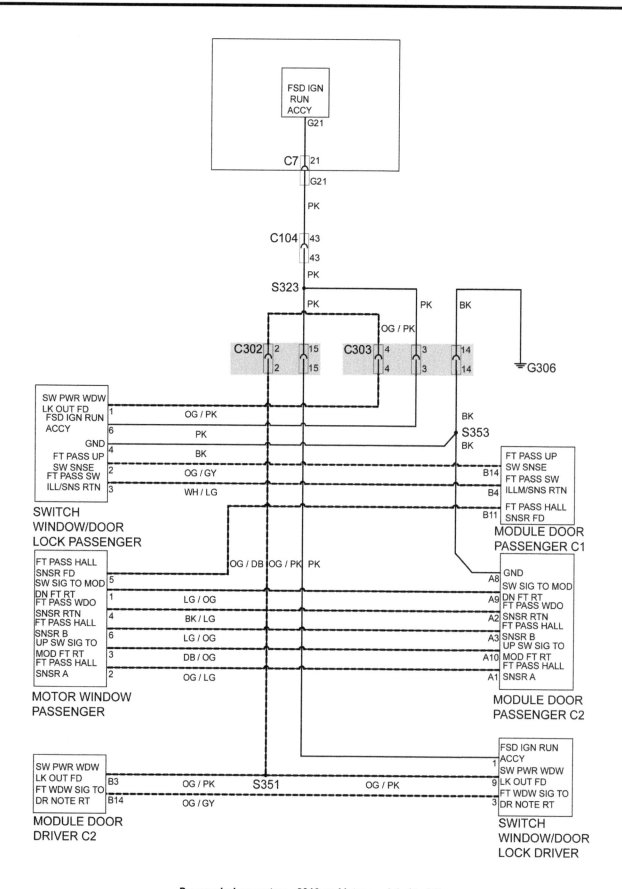

Power window system - 2010 and later models (5 of 6)

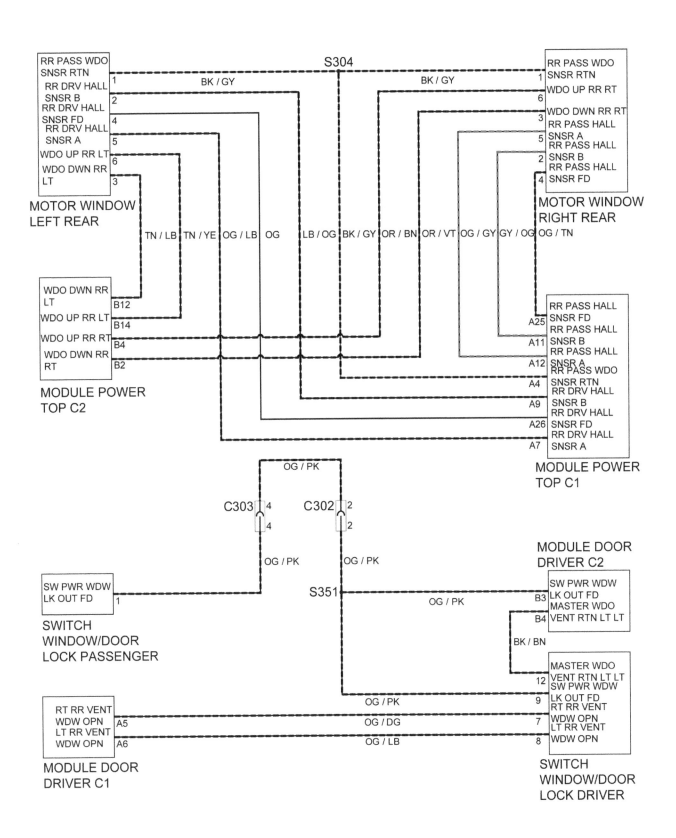

Power window system - 2010 and later models (6 of 6)

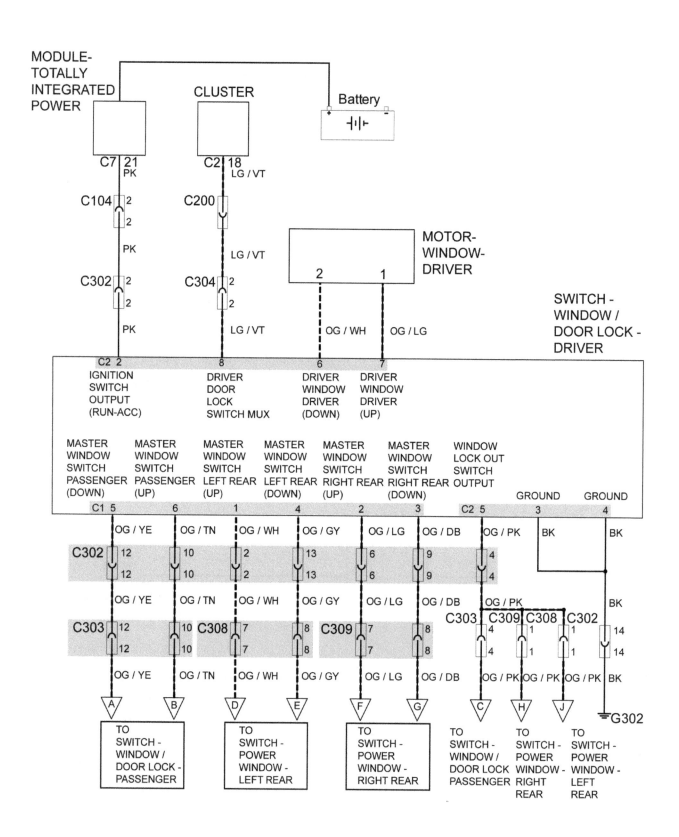

Power window system (w/o express windows) - 2009 and earlier models (1 of 2)

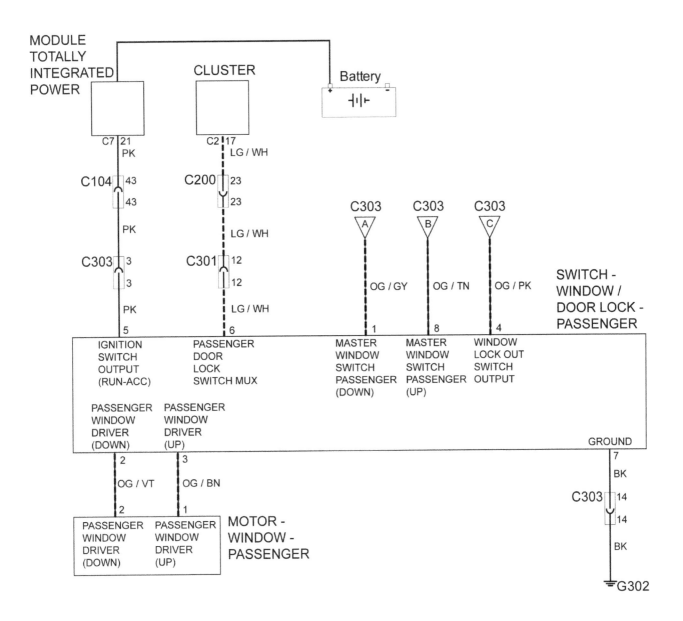

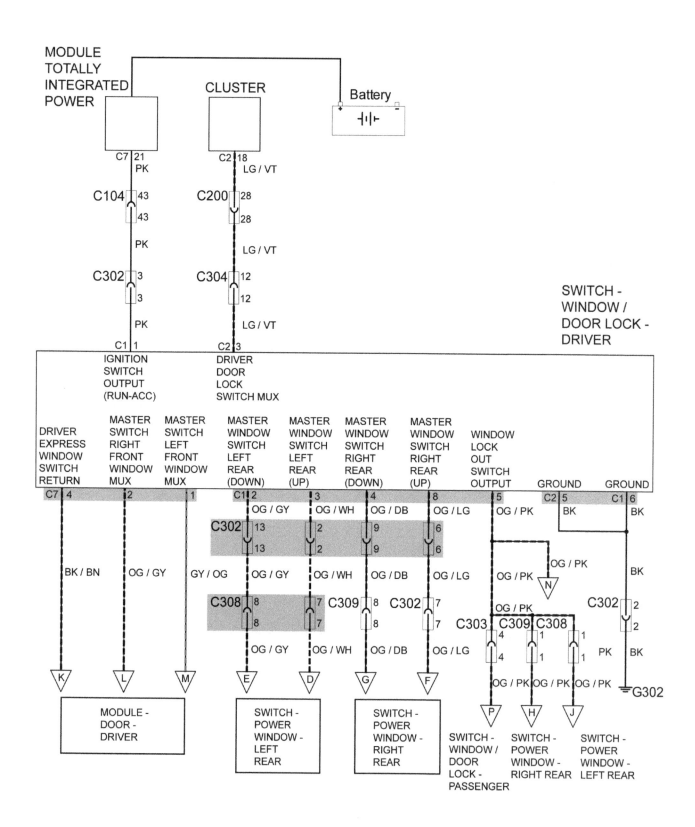

Power window system (express windows) - 2009 and earlier models (1 of 2)

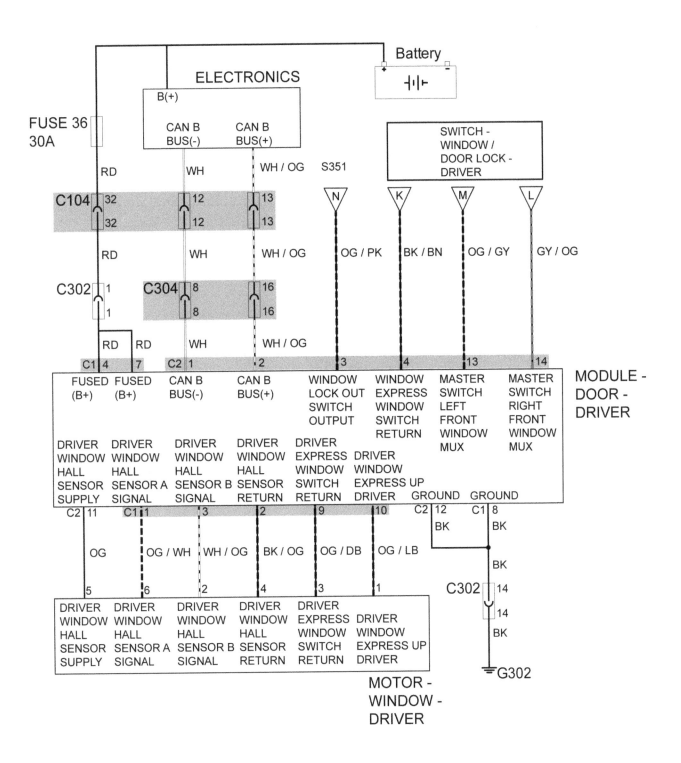

Power window system (express windows) - 2009 and earlier models (2 of 2)

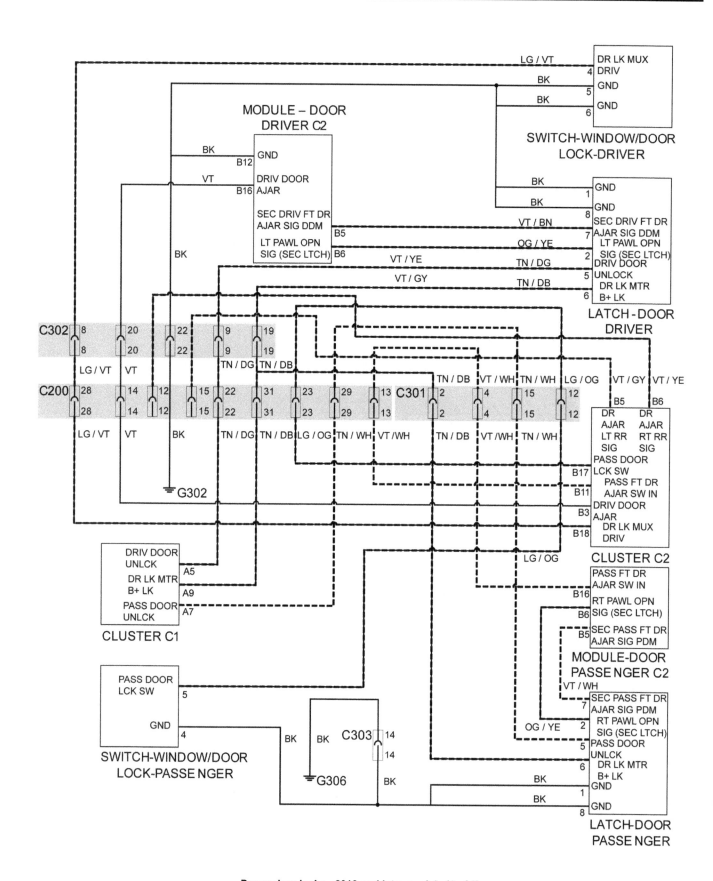

Power door locks - 2010 and later models (1 of 2)

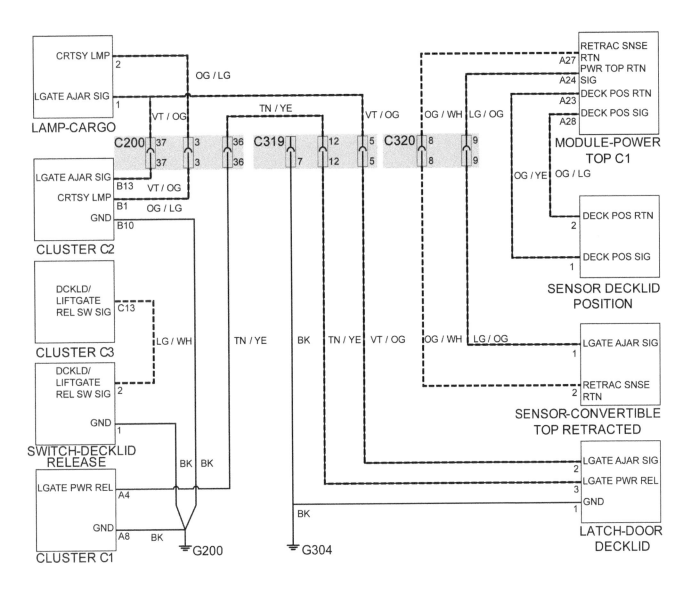

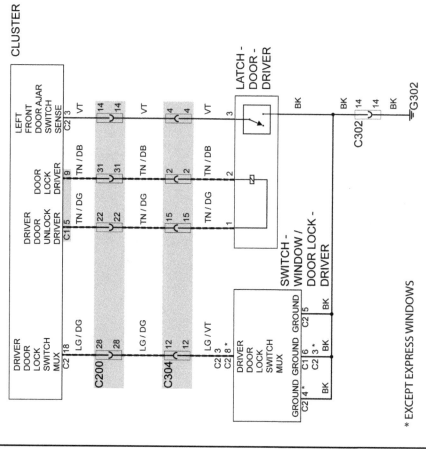

Power door locks - 2009 and earlier models (2 of 5)

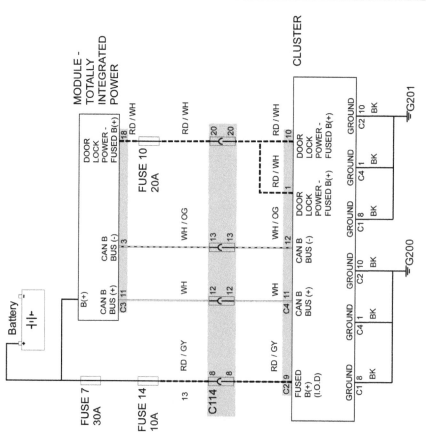

Power door locks - 2009 and earlier models (1 of 5)

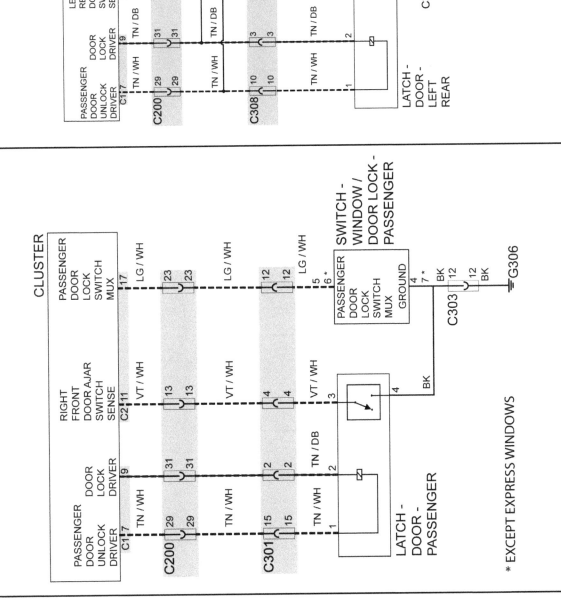

Power door locks - 2009 and earlier models (4 of 5)

Power door locks - 2009 and earlier models (3 of 5)

* EXCEPT EXPRESS WINDOWS

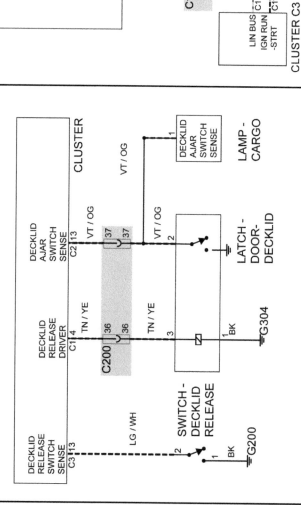

Wiper/washer system - 2010 and later models (1 of 4)

Power door locks - 2009 and earlier models (5 of 5)

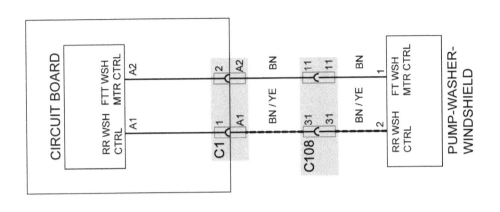

Wiper/washer system - 2010 and later models (3 of 4)

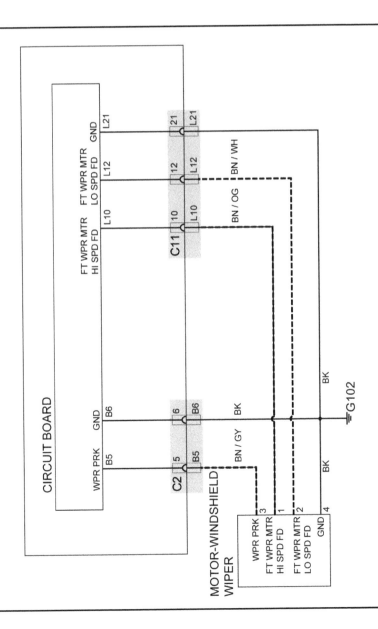

Wiper/washer system - 2010 and later models (2 of 4)

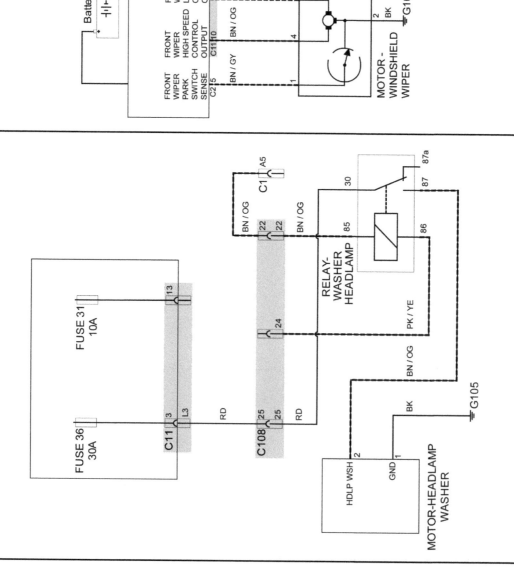

Wiper/washer system - 2009 and earlier models (1 of 3)

Wiper/washer system - 2010 and later models (4 of 4)

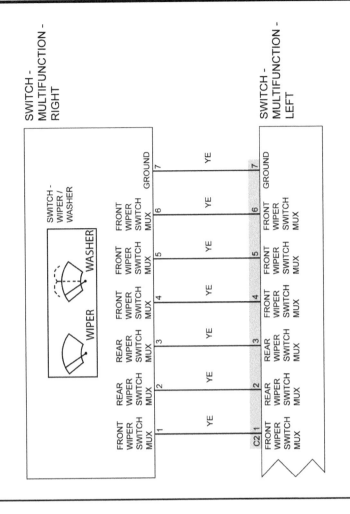

Wiper/washer system - 2009 and earlier models (3 of 3)

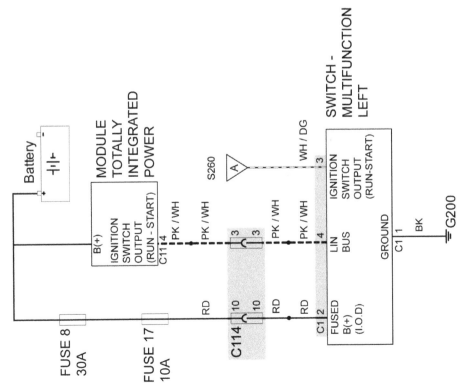

Wiper/washer system - 2009 and earlier models (2 of 3)

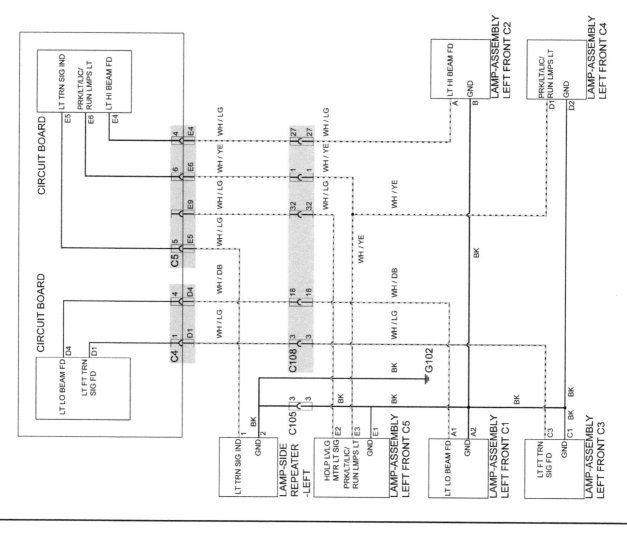

Exterior lighting system - 2010 and later models (1 of 6)

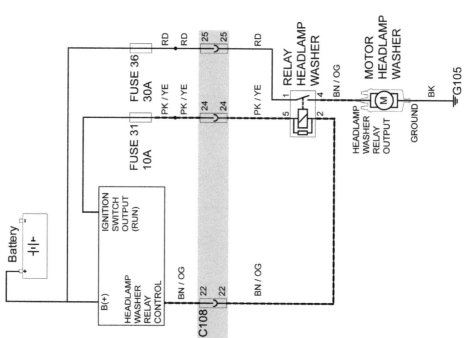

Wiper/washer system - 2009 and earlier export models

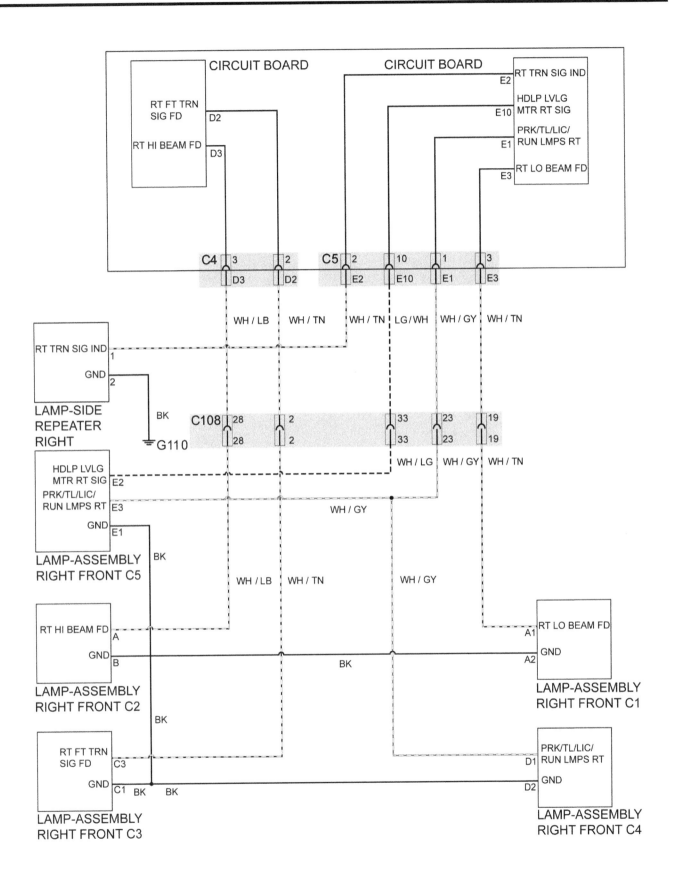

Exterior lighting system - 2010 and later models (2 of 6)

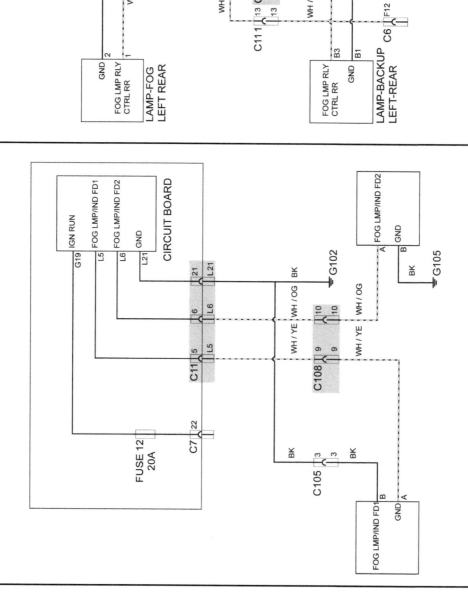

Exterior lighting system - 2010 and later models (4 of 6)

Exterior lighting system - 2010 and later models (3 of 6)

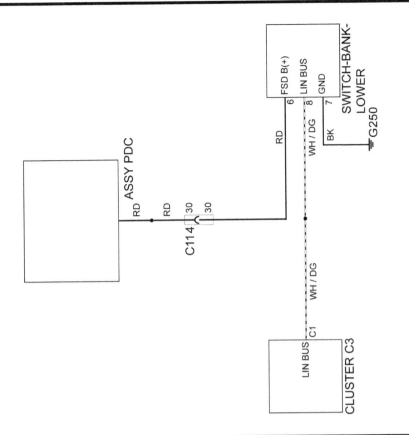

Exterior lighting system - 2010 and later models (6 of 6)

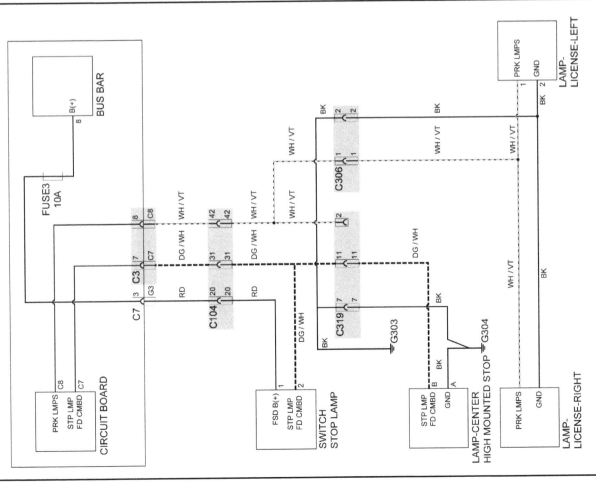

Exterior lighting system - 2010 and later models (5 of 6)

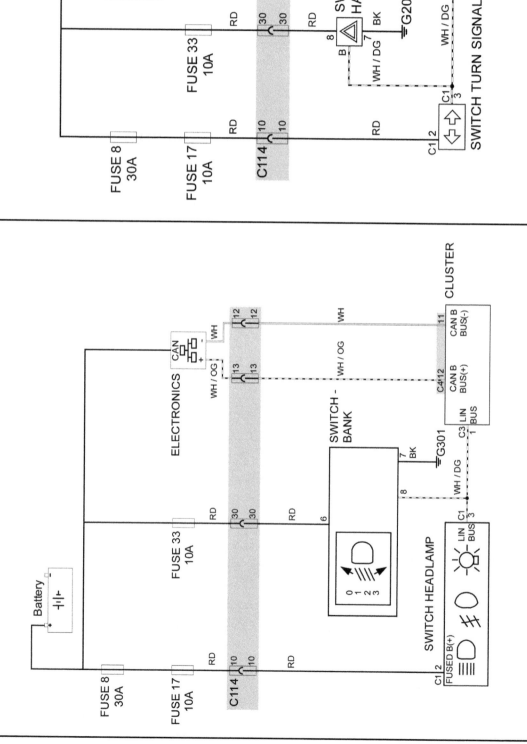

Exterior lighting system - 2009 and earlier models (2 of 3)

Exterior lighting system - 2009 and earlier models (1 of 3)

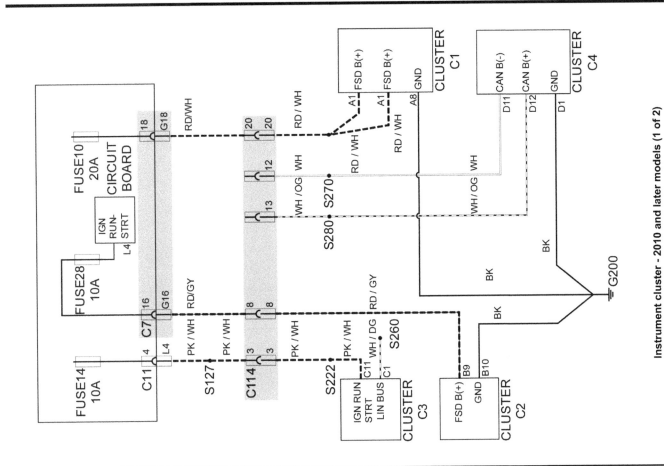

Instrument cluster - 2010 and later models (1 of 2)

Exterior lighting system - 2009 and earlier models (3 of 3)

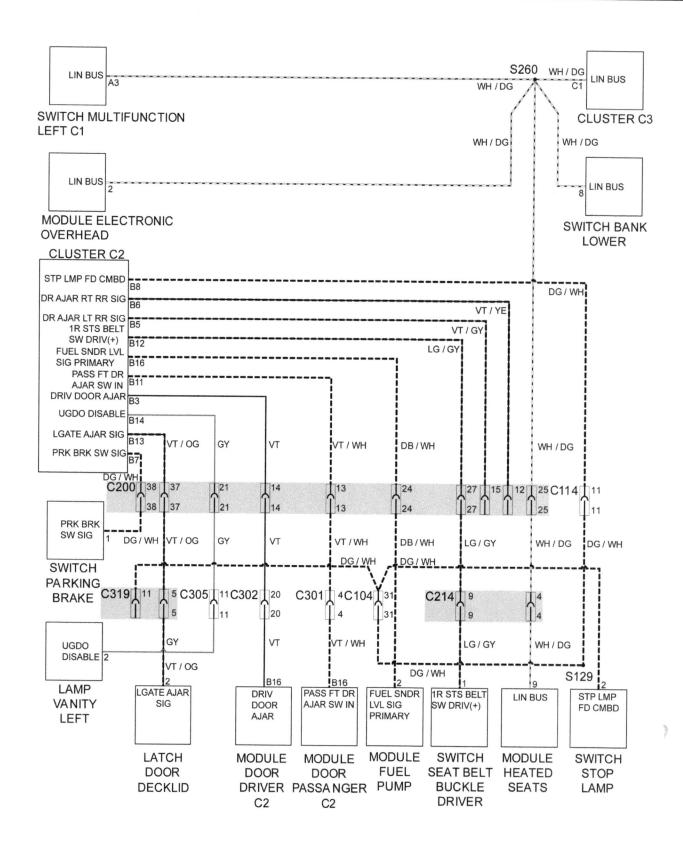

Instrument cluster - 2010 and later models (2 of 2)

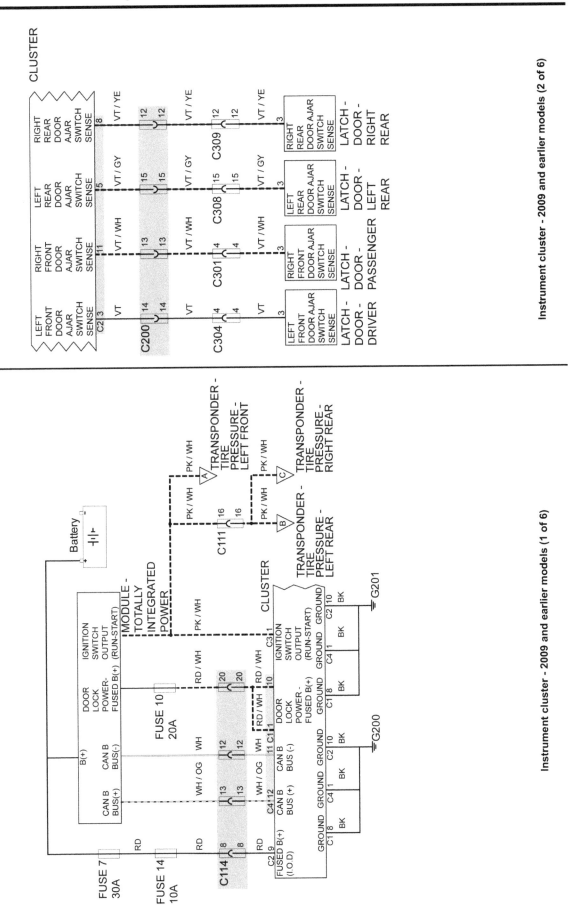

Instrument cluster - 2009 and earlier models (2 of 6)

Instrument cluster - 2009 and earlier models (1 of 6)

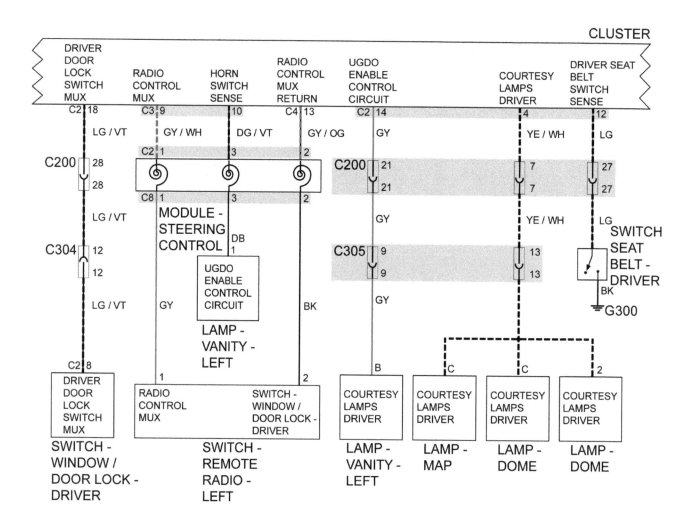

Instrument cluster - 2009 and earlier models (3 of 6)

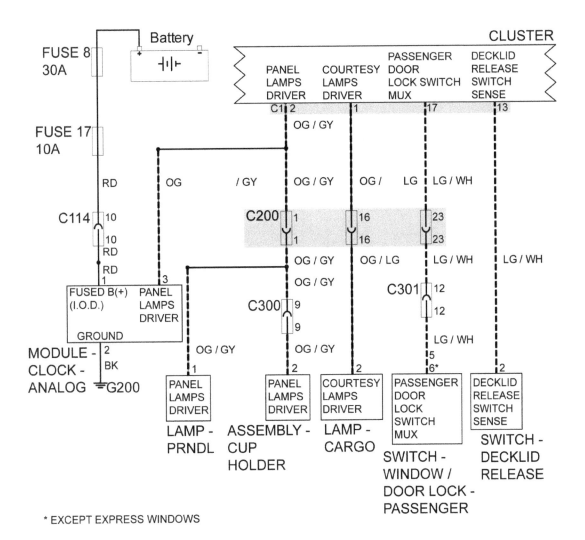

* EXCEPT EXPRESS WINDOWS

Instrument cluster - 2009 and earlier models (4 of 6)

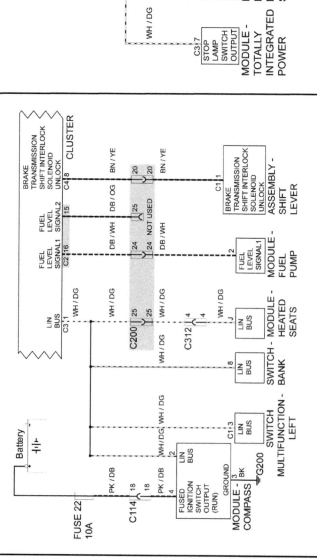

Instrument cluster - 2009 and earlier models (6 of 6)

Instrument cluster - 2009 and earlier models (5 of 6)

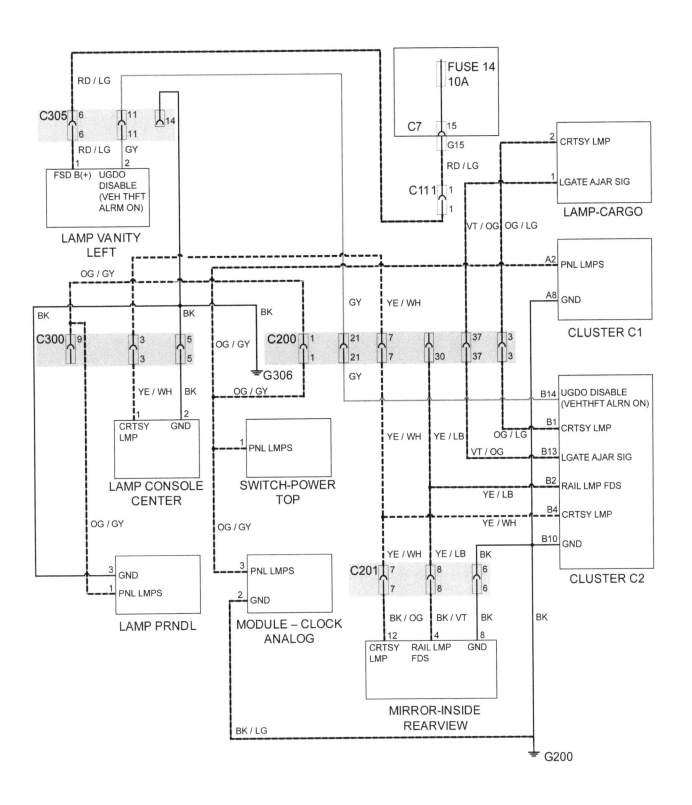

Interior lighting system - 2010 and later models

Interior lighting system - 2009 and earlier models (2 of 5)

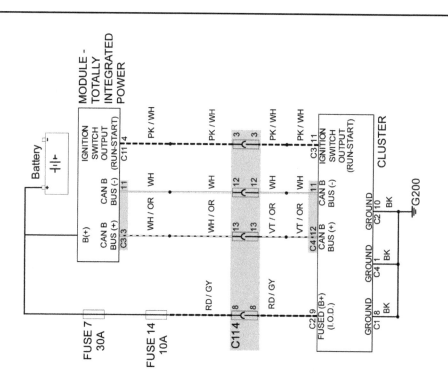

Interior lighting system - 2009 and earlier models (1 of 5)

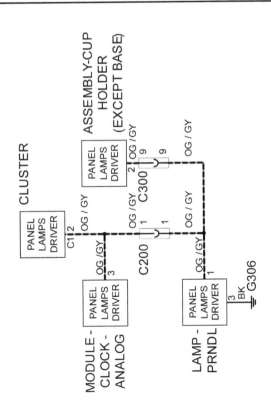

Interior lighting system - 2009 and earlier models (4 of 5)

Interior lighting system - 2009 and earlier models (3 of 5)

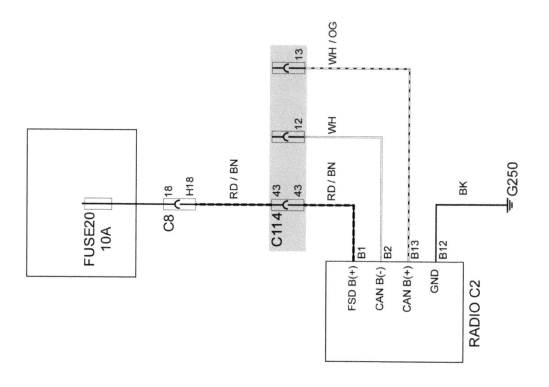

Audio system - 2010 and later models (1 of 7)

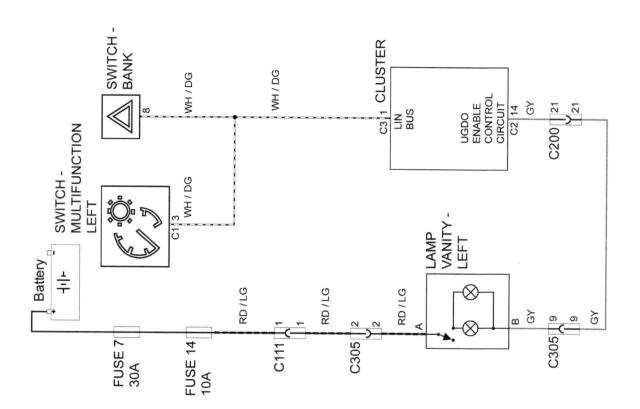

Interior lighting system - 2009 and earlier models (5 of 5)

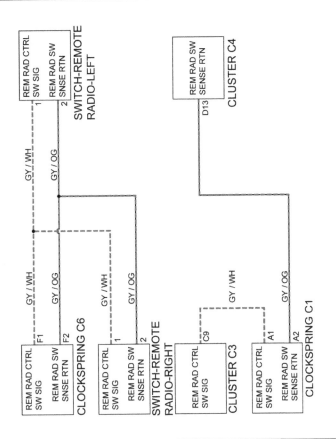

Audio system - 2010 and later models (3 of 7)

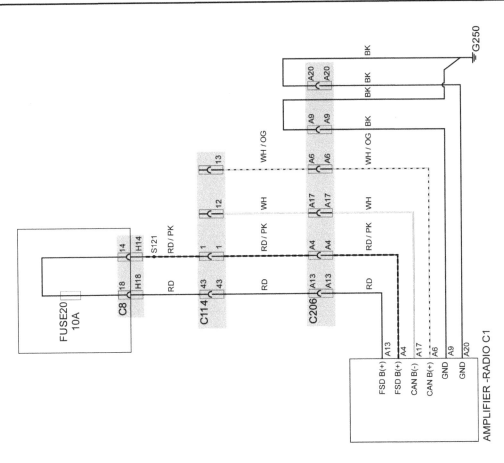

Audio system - 2010 and later models (2 of 7)

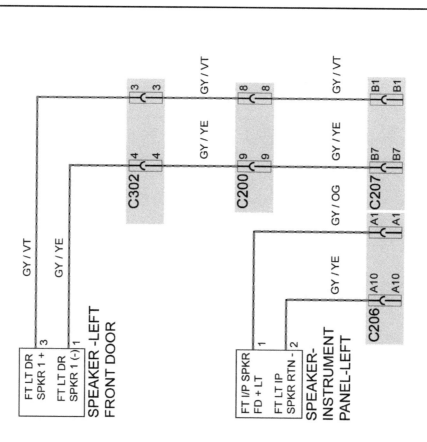

Audio system - 2010 and later models (5 of 7)

Audio system - 2010 and later models (4 of 7)

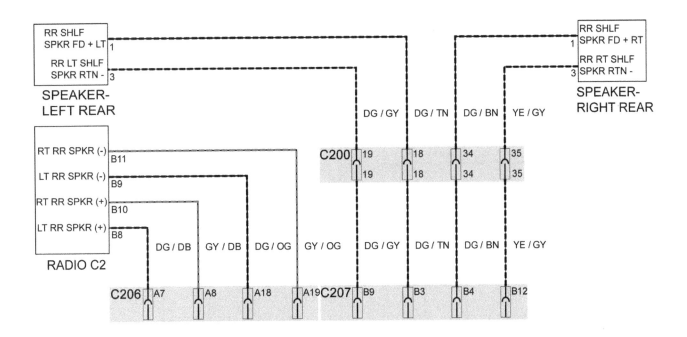

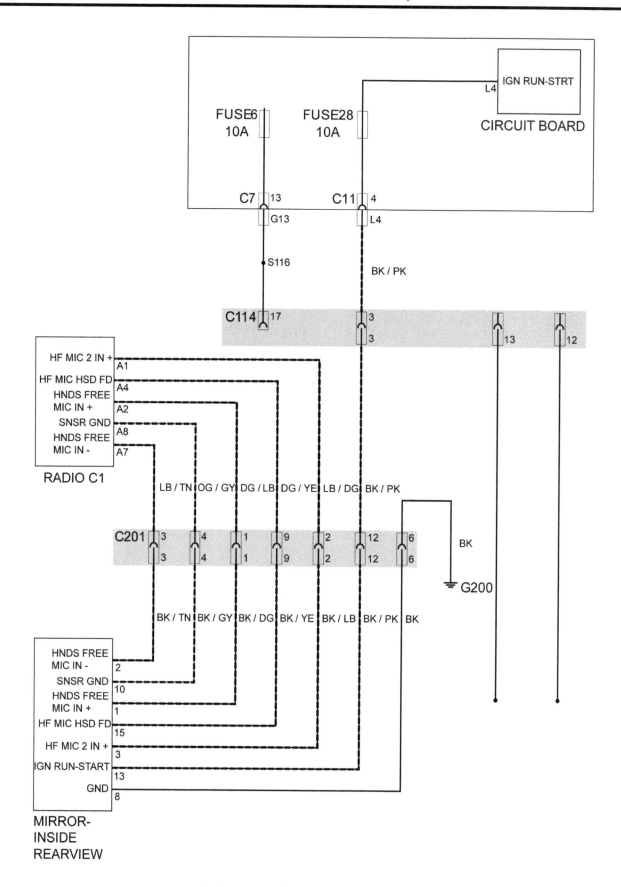

Audio system - 2010 and later models (7 of 7)

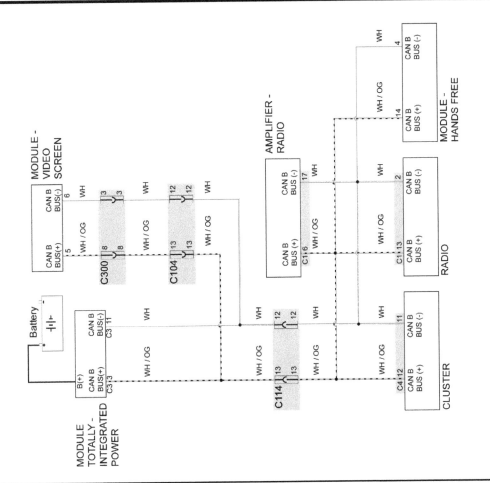

Audio system/video screen module/amplifier/radio/hands-free module - 2009 and earlier models

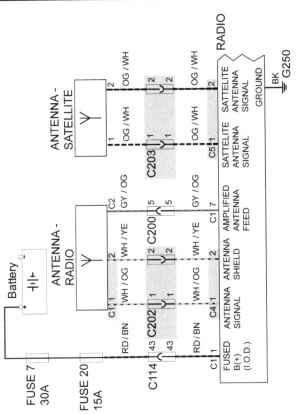

Audio system/antenna - 2009 and earlier models

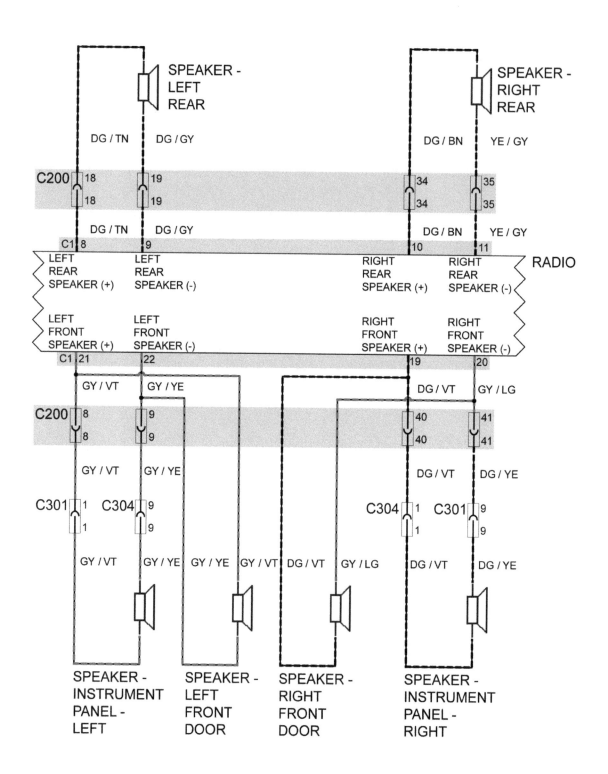

Audio system/speakers (base) - 2009 and earlier models

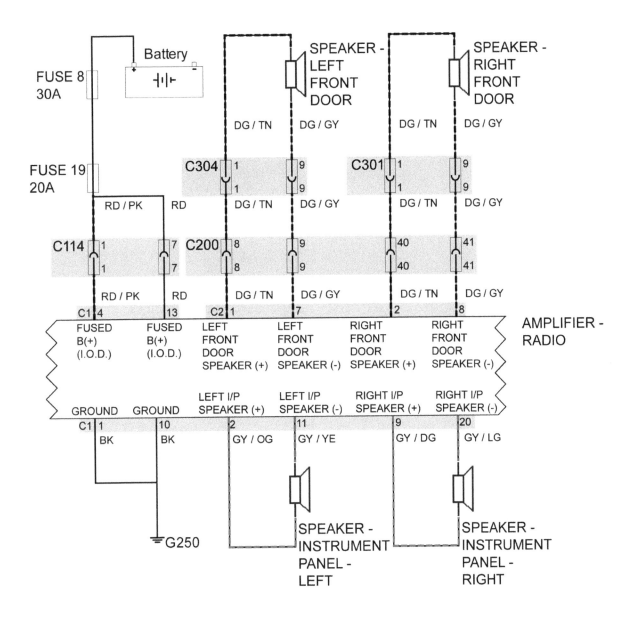

Audio system/speakers (premium) - 2009 and earlier models (1 of 2)

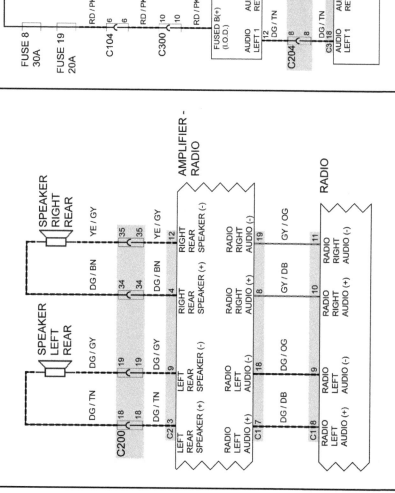

Audio system/radio/DVD/video screen module - 2009 and earlier models (1 of 2)

Audio system/speakers (premium) - 2009 and earlier models (2 of 2)

Audio system/hands-free, w/o navigation - 2009 and earlier models

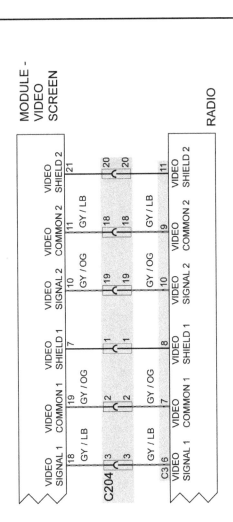

Audio system/radio/DVD/video screen module - 2009 and earlier models (2 of 2)

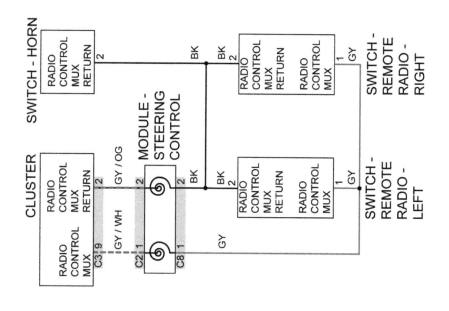

Audio system/steering wheel controls - 2009 and later models

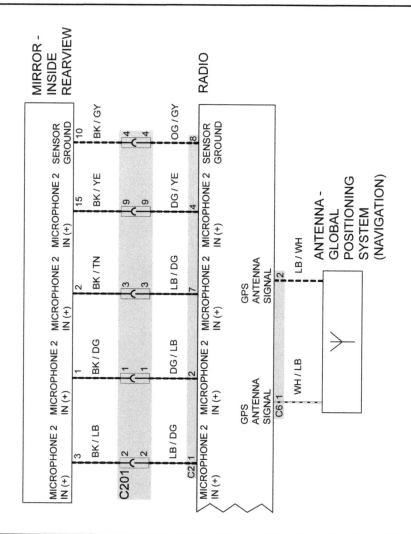

Audio system/hands-free, with navigation - 2009 and earlier models

Index

2.7L V6 engine, 2B-1
   Camshafts - removal, inspection and installation, 2B-7
   Crankshaft front oil seal - replacement, 2B-12
   Crankshaft pulley - removal and installation, 2B-11
   Cylinder head(s) - removal and installation, 2B-10
   Driveplate - removal and installation, 2B-14
   Engine mounts - check and replacement, 2B-15
   Exhaust manifold - removal and installation, 2B-10
   General information, 2B-3
   Intake manifolds - removal and installation, 2B-9
   Oil pan - removal and installation, 2B-12
   Oil pump - removal, inspection and installation, 2B-13
   Rear main oil seal - replacement, 2B-15
   Repair operations possible with the engine in the
      vehicle, 2B-3
   Rocker arms and hydraulic lash adjusters - removal,
      inspection and installation, 2B-4
   Timing chain and sprockets - removal, inspection and
      installation, 2B-5
   Top Dead Center (TDC) for number one piston -
      locating, 2B-3
   Valve cover - removal and installation, 2B-3
3.5L V6 engine, 2C-1
   Camshafts - removal, inspection and installation, 2C-6
   Crankshaft front oil seal - replacement, 2C-10
   Crankshaft pulley - removal and installation, 2C-10
   Cylinder heads - removal and installation, 2C-9
   Driveplate - removal and installation, 2C-12
   Engine mounts - check and replacement, 2C-12
   Exhaust manifold(s) - removal and installation, 2C-8
   General information, 2C-3
   Intake manifolds - removal and installation, 2C-7
   Oil pan - removal and installation, 2C-11
   Oil pump - removal, inspection and installation, 2C-11
   Rear main oil seal - replacement, 2C-12
   Repair operations possible with the engine in the
      vehicle, 2C-3
   Rocker arm assembly and hydraulic lash adjusters -
      removal, inspection and installation, 2C-4
   Timing belt - replacement, 2C-4
   Top Dead Center (TDC) for number one piston -
      locating, 2C-3
   Valve cover(s) - removal and installation, 2C-3
3.6L V6 engine, 2D-1
   Camshaft(s) – removal, inspection and installation, 2D-10
   Crankshaft front oil seal - replacement, 2D-6
   Crankshaft pulley - removal and installation, 2D-6
   Cylinder heads - removal and installation, 2D-12
   Driveplate - removal and installation, 2D-15

Engine mounts - check and replacement, 2D-15
General information, 2D-3
Intake manifolds - removal and installation, 2D-4
Oil cooler - removal and installation, 2D-13
Oil pans - removal and installation, 2D-13
Oil pump - removal, inspection and installation, 2D-13
Rear main oil seal - replacement, 2D-15
Repair operations possible with the engine in the
   vehicle, 2D-3
Timing chain cover, chain and sprockets - removal,
   inspection and installation, 2D-7
Top Dead Center (TDC) for number one piston -
   locating, 2D-3
Valve covers - removal and installation, 2D-3

A

About this manual, 0-5
**Accelerator Pedal Position (APP) sensor -
   replacement, 6-8**
Air conditioning
   and heating system - check and maintenance, 3-4
   compressor - removal and installation, 3-15
   condenser - removal and installation, 3-16
   receiver-drier - removal and installation, 3-16
Air filter
   check and replacement, 1-20
   housing - removal and installation, 4-8
**Air Injection system (2.4L PZEV engines) -
   component replacement, 6-20**
**Airbag system - general information and
   precautions, 12-16**
Alternator - removal and installation, 5-5
Antenna - removal and installation, 12-9
Anti-lock Brake System (ABS) - general information, 9-7
Automatic transaxle fluid and filter change, 1-29
Automatic transaxle, 7A-1
   Automatic transaxle - removal and installation, 7A-6
   Automatic transaxle overhaul - general information, 7A-7
   Brake Transmission Shift Interlock (BTSI) system -
      description, check, replacement and adjustment, 7A-4
   Diagnosis - general, 7A-2
   Driveaxle oil seals - replacement, 7A-2
   General information, 7A-2
   Shift cable - removal, installation and adjustment, 7A-3
   Shift lever - removal and installation, 7A-5
   Transaxle oil cooler - removal and installation, 7A-6
Automotive chemicals and lubricants, 0-20

B

Balljoints - replacement, 10-8
Battery
   cables - replacement, 5-4
   check, maintenance and charging, 1-17
   disconnection, 5-3
   removal and installation, 5-4
Blower motor resistor/power module and blower
   motor assembly - replacement, 3-12
Body, 11-1
   Body repair
      major damage, 11-3
      minor damage, 11-2
   Bumper covers - removal and installation, 11-7
   Center floor console - removal and installation, 11-19
   Cowl cover - removal and installation, 11-9
   Dashboard trim panels - removal and installation, 11-15
   Door - removal and installation, 11-10
   Door module, latch, lock and handle - removal and
      installation, 11-10
   Door trim panels - removal and installation, 11-9
   Door window glass - removal and installation, 11-12
   Fastener and trim removal, 11-5
   Front fender - removal and installation, 11-8
   General information, 11-1
   Hood - removal, installation and adjustment, 11-6
   Hood latch and cable - removal and installation, 11-7
   Instrument panel - removal and installation, 11-16
   Mirrors - removal and installation, 11-12
   Radiator grille - removal and installation, 11-7
   Repairing minor paint scratches, 11-2
   Seats - removal and installation, 11-18
   Steering column covers - removal and installation, 11-14
   Trunk lid, latch and support struts - removal and
      installation, 11-13
   Upholstery, carpets and vinyl trim - maintenance, 11-5
   Window glass regulators - removal and installation, 11-12
Booster battery (jump) starting, 0-18
Brake
   fluid change, 1-25
   system check, 1-20
Brake Transmission Shift Interlock (BTSI)
   system - description, check, replacement and
   adjustment, 7A-4
Brakes, 9-1
   Anti-lock Brake System (ABS) - general information, 9-7
   Brake
      disc - inspection, removal and installation, 9-12
      hoses and lines - inspection and replacement, 9-18
      light switch - check, replacement and adjustment, 9-21
      system - bleeding, 9-19
      vacuum pump - removal and installation, 9-22
   Disc brake
      caliper - removal and installation, 9-12
      pads - replacement, 9-8
   Drum brake shoes - replacement, 9-13
   General information, 9-2
   Master cylinder - removal and installation, 9-17
   Parking brake - adjustment, 9-21
   Parking brake shoes (models with rear disc brakes) -
      replacement, 9-20
   Power brake booster - check, removal and
      installation, 9-19
   Troubleshooting, 9-3
   Wheel cylinder - removal and installation, 9-16
Bulb replacement, 12-12
Bumper covers - removal and installation, 11-7
Buying parts, 0-11

C

Cabin air filter replacement, 1-23
Camshaft Position (CMP) sensor - replacement, 6-8
Camshaft(s) – removal, inspection and installation
   2.7L V6 engine, 2B-7
   3.5L V6 engine, 2C-6
   3.6L V6 engine, 2D-10
   Four-cylinder engine, 2A-10
Catalytic converter - replacement, 6-16
Center floor console - removal and installation, 11-19
Chassis electrical system, 12-1
   Airbag system - general information and
      precautions, 12-16
   Antenna - removal and installation, 12-9
   Bulb replacement, 12-12
   Circuit breakers - general information, 12-3
   Cruise control system – description and check, 12-14
   Daytime Running Lights (DRL) - general
      information, 12-16
   Electric side view mirrors - general information, 12-14
   Electrical
      connectors - general information, 12-4
      troubleshooting - general information, 12-1
   Fuse box/power distribution center (2015 and later
      models) - removal and installation, 12-17
   Fuses and fusible links - general information, 12-3
   General information, 12-1
   Headlight
      bulb - replacement, 12-10
      housing - replacement, 12-10
   Headlights - adjustment, 12-11
   Horn - replacement, 12-14
   Ignition switch - replacement, 12-6
   Instrument cluster - removal and installation, 12-6
   Multi-function switch - replacement, 12-5
   Power door lock and keyless entry system –
      description and check, 12-15
   Power seat - general information and procedures, 12-16
   Power window system - description and check, 12-15
   Radio and speakers - removal and installation, 12-8

Rear window defogger - check and repair, 12-9
Relays - general information, 12-4
Wiper motor - replacement, 12-7
Wiring diagrams - general information, 12-17
Circuit breakers - general information, 12-3
Control arm - removal, inspection, and installation, 10-7
Conversion factors, 0-21
Cooling system
    check, 1-19
    servicing (draining, flushing and refilling), 1-28
Cooling, heating and air conditioning systems, 3-1
    Air conditioning
        and heating system - check and maintenance, 3-4
        compressor - removal and installation, 3-15
        condenser - removal and installation, 3-16
        receiver-drier - removal and installation, 3-16
    Blower motor resistor/power module and blower motor assembly - replacement, 3-12
    Coolant reservoir/expansion tank - removal and installation, 3-9
    Engine cooling fans - check and replacement, 3-8
    Expansion valve (2015 and later models) - replacement, 3-17
    General information, 3-2
    Heater core - replacement, 3-13
    Heater/air conditioner control assembly - removal and installation, 3-14
    Orifice tube (2014 and earlier models) - removal and installation, 3-17
    Radiator - removal and installation, 3-9
    Thermostat - replacement, 3-6
    Troubleshooting, 3-2
    Water pump - replacement, 3-10
Cowl cover - removal and installation, 11-9
Crankshaft - removal and installation, 2E-12
Crankshaft front oil seal – replacement
    2.7L V6 engine, 2B-12
    3.5L V6 engine, 2C-10
    3.6L V6 engine, 2D-6
    Four-cylinder engine, 2A-7
Crankshaft Position (CKP) sensor - replacement, 6-8
Crankshaft pulley - removal and installation
    2.7L V6 engine, 2B-11
    3.5L V6 engine, 2C-10
    3.6L V6 engine, 2D-6
    Four-cylinder engine, 2A-7
Cruise control system – description and check, 12-14
Cylinder compression check, 2E-4
Cylinder heads - removal and installation
    2.7L V6 engine, 2B-10
    3.5L V6 engine, 2C-9
    3.6L V6 engine, 2D-12
    Four-cylinder engine, 2A-13

D

Dashboard trim panels - removal and installation, 11-15
Daytime Running Lights (DRL) - general information, 12-16
Diagnosis - general, 7A-2
Disc brake
    caliper - removal and installation, 9-12
    pads - replacement, 9-8
Door - removal and installation, 11-10
Door module, latch, lock and handle - removal and installation, 11-10
Door trim panels - removal and installation, 11-9
Door window glass - removal and installation, 11-12
Driveaxle oil seal (right side) - replacement, 7B-2
Driveaxle oil seals - replacement, 7A-2
Drivebelt check and replacement, 1-24
Driveline, 8-1
    Driveaxle - removal and installation, 8-2
    Driveaxle boot replacement, 8-3
    Driveaxles - general information and inspection, 8-2
    Driveshaft (AWD models) - removal and installation, 8-5
    Driveshaft center support bearing (AWD models) - replacement, 8-6
    Driveshaft universal and constant velocity joints (AWD models) - general information and check, 8-5
    Rear differential
        assembly (AWD models) - removal and installation, 8-6
        driveaxle oil seals (AWD models) - replacement, 8-6
        electronic clutch - removal and installation, 8-6
        pinion seal (AWD models) - replacement, 8-6
Driveplate - removal and installation
    2.7L V6 engine, 2B-14
    3.5L V6 engine, 2C-12
    3.6L V6 engine, 2D-15
    Four-cylinder engine, 2A-15
Drum brake shoes - replacement, 9-13

E

Electric side view mirrors - general information, 12-14
Electrical connectors - general information, 12-4
Electrical troubleshooting - general information, 12-1
Emissions and engine control systems, 6-1
    Accelerator Pedal Position (APP) sensor - replacement, 6-8
    Air Injection system (2.4L PZEV engines) - component replacement, 6-20
    Camshaft Position (CMP) sensor - replacement, 6-8
    Catalytic converter - replacement, 6-16
    Crankshaft Position (CKP) sensor - replacement, 6-8
    Engine Coolant Temperature (ECT) sensor - replacement, 6-11
    Evaporative Emissions Control (EVAP) system - component replacement, 6-18

Exhaust Gas Recirculation (EGR) system - component replacement, 6-18

General information, 6-3

Intake Air Temperature (IAT) sensor - replacement, 6-12

Knock sensor - replacement, 6-13

Manifold Absolute Pressure (MAP) sensor - replacement, 6-13

Manifold tuning valve solenoids (2.7L and 3.5L V6 engines) - removal and installation, 6-10

Obtaining and clearing Diagnostic Trouble Codes (DTCs), 6-5

Oil temperature sensor - removal and installation, 6-20

On Board Diagnosis (OBD) system, 6-5

Oxygen sensors - general information and replacement, 6-14

Positive Crankcase Ventilation (PCV) system, 6-19

Powertrain Control Module (PCM) - replacement, 6-16

Throttle Position (TP) sensor - replacement, 6-15

Transaxle speed sensors - replacement, 6-15

Transmission Range (TR) and transmission temperature sensors - replacement, 6-15

Variable valve timing solenoids (2.4L and 3.6L engines) - removal and installation, 6-10

Engine - removal and installation, 2E-7

Engine Coolant Temperature (ECT) sensor - replacement, 6-11

Engine cooling fans - check and replacement, 3-8

Engine electrical systems, 5-1

Alternator - removal and installation, 5-5

Battery - disconnection, 5-3

Battery - removal and installation, 5-4

Battery cables - replacement, 5-4

General information and precautions, 5-1

Ignition coils - removal and installation, 5-5

Starter motor - removal and installation, 5-7

Troubleshooting, 5-2

Engine, in-vehicle repair procedures

Four-cylinder engine, 2A-1

Camshaft(s) – removal, inspection and installation, 2A-10

Crankshaft front oil seal - replacement, 2A-7

Crankshaft pulley - removal and installation, 2A-7

Cylinder head - removal and installation, 2A-13

Driveplate - removal and installation, 2A-15

Engine mounts - check and replacement, 2A-16

Exhaust manifold - removal, inspection and installation, 2A-6

General information, 2A-3

Intake manifold - removal and installation, 2A-4

Oil pan - removal and installation, 2A-14

Oil pump/balance shaft module and chain - removal, inspection and installation, 2A-14

Rear main oil seal - replacement, 2A-16

Repair operations possible with the engine in the vehicle, 2A-3

Timing chain cover, chain and sprockets – removal, inspection and installation, 2A-7

Top Dead Center (TDC) for number one piston - locating, 2A-4

Valve clearance - check and adjustment, 2A-14

Valve cover - removal and installation, 2A-4

Variable Valve Actuation (VVAA) – removal and installation, 2A-17

2.7L V6 engine, 2B-1

Camshafts - removal, inspection and installation, 2B-7

Crankshaft front oil seal - replacement, 2B-12

Crankshaft pulley - removal and installation, 2B-11

Cylinder head(s) - removal and installation, 2B-10

Driveplate - removal and installation, 2B-14

Engine mounts - check and replacement, 2B-15

Exhaust manifold - removal and installation, 2B-10

General information, 2B-3

Intake manifolds - removal and installation, 2B-9

Oil pan - removal and installation, 2B-12

Oil pump - removal, inspection and installation, 2B-13

Rear main oil seal - replacement, 2B-15

Repair operations possible with the engine in the vehicle, 2B-3

Rocker arms and hydraulic lash adjusters - removal, inspection and installation, 2B-4

Timing chain and sprockets - removal, inspection and installation, 2B-5

Top Dead Center (TDC) for number one piston - locating, 2B-3

Valve cover - removal and installation, 2B-3

3.5L V6 engine, 2C-1

Camshafts - removal, inspection and installation, 2C-6

Crankshaft front oil seal - replacement, 2C-10

Crankshaft pulley - removal and installation, 2C-10

Cylinder heads - removal and installation, 2C-9

Driveplate - removal and installation, 2C-12

Engine mounts - check and replacement, 2C-12

Exhaust manifold(s) - removal and installation, 2C-8

General information, 2C-3

Intake manifolds - removal and installation, 2C-7

Oil pan - removal and installation, 2C-11

Oil pump - removal, inspection and installation, 2C-11

Rear main oil seal - replacement, 2C-12

Repair operations possible with the engine in the vehicle, 2C-3

Rocker arm assembly and hydraulic lash adjusters - removal, inspection and installation, 2C-4

Timing belt - replacement, 2C-4

Top Dead Center (TDC) for number one piston - locating, 2C-3

Valve cover(s) - removal and installation, 2C-3

3.6L V6 engine, 2D-1

Camshaft(s) – removal, inspection and installation, 2D-10

Crankshaft front oil seal - replacement, 2D-6

Crankshaft pulley - removal and installation, 2D-6
Cylinder heads - removal and installation, 2D-12
Driveplate - removal and installation, 2D-15
Engine mounts - check and replacement, 2D-15
General information, 2D-3
Intake manifolds - removal and installation, 2D-4
Oil cooler - removal and installation, 2D-13
Oil pans - removal and installation, 2D-13
Oil pump - removal, inspection and installation, 2D-13
Rear main oil seal - replacement, 2D-15
Repair operations possible with the engine in the vehicle, 2D-3
Timing chain cover, chain and sprockets - removal, inspection and installation, 2D-7
Top Dead Center (TDC) for number one piston - locating, 2D-3
Valve covers - removal and installation, 2D-3

Engine mounts - check and replacement
2.7L V6 engine, 2B-15
3.5L V6 engine, 2C-12
3.6L V6 engine, 2D-15
Four-cylinder engine, 2A-16
Engine oil and filter change, 1-15
Engine overhaul
disassembly sequence, 2E-8
reassembly sequence, 2E-16
Engine rebuilding alternatives, 2E-6
Engine removal - methods and precautions, 2E-6
Evaporative Emissions Control (EVAP) system - component replacement, 6-18
Exhaust Gas Recirculation (EGR) system - component replacement, 6-18
Exhaust manifold - removal and installation
2.7L V6 engine, 2B-10
3.5L V6 engine, 2C-8
Exhaust manifold - removal, inspection and installation, 2A-6
Exhaust system
check, 1-23
servicing - general information, 4-5
Expansion valve (2015 and later models) - replacement, 3-17

F

Fastener and trim removal, 11-5
Fluid level checks, 1-9
Four-cylinder engine, 2A-1
Camshaft(s) – removal, inspection and installation, 2A-10
Crankshaft front oil seal - replacement, 2A-7
Crankshaft pulley - removal and installation, 2A-7
Cylinder head - removal and installation, 2A-13
Driveplate - removal and installation, 2A-15
Engine mounts - check and replacement, 2A-16
Exhaust manifold - removal, inspection and installation, 2A-6

General information, 2A-3
Intake manifold - removal and installation, 2A-4
Oil pan - removal and installation, 2A-14
Oil pump/balance shaft module and chain - removal, inspection and installation, 2A-14
Rear main oil seal - replacement, 2A-16
Repair operations possible with the engine in the vehicle, 2A-3
Timing chain cover, chain and sprockets - removal, inspection and installation, 2A-7
Top Dead Center (TDC) for number one piston - locating, 2A-4
Valve clearance - check and adjustment, 2A-14
Valve cover - removal and installation, 2A-4
Variable Valve Actuation Assembly (VVAA) – removal and installation, 2A-17
Fraction/decimal/millimeter equivalents, 0-22
Front fender - removal and installation, 11-8
Front load beam (2015 and later models) - removal and installation, 10-18
Fuel and exhaust systems, 4-1
Air filter housing - removal and installation, 4-8
Exhaust system servicing - general information, 4-5
Fuel
level sending unit - replacement, 4-7
lines and fittings - general information and disconnection, 4-5
pressure - check, 4-4
pressure relief procedure, 4-3
pump/fuel pressure regulator/fuel level sending unit assembly - removal and installation, 4-6
rail and injectors - removal and installation, 4-9
tank - removal and installation, 4-5
General information, 4-2
Throttle body - removal and installation, 4-9
Troubleshooting, 4-2
Fuel system check, 1-23
Fuse box/power distribution center (2015 and later models) - removal and installation, 12-17
Fuses and fusible links - general information, 12-3

G

General engine overhaul procedures, 2E-1
Crankshaft - removal and installation, 2E-12
Cylinder compression check, 2E-4
Engine - removal and installation, 2E-7
Engine overhaul
disassembly sequence, 2E-8
reassembly sequence, 2E-16
Engine rebuilding alternatives, 2E-6
Engine removal - methods and precautions, 2E-6
General information - engine overhaul, 2E-3
Initial start-up and break-in after overhaul, 2E-17
Oil pressure check, 2E-4

Pistons and connecting rods - removal and installation, 2E-9

Vacuum gauge diagnostic checks, 2E-5

H

Headlight bulb - replacement, 12-10

Headlight housing - replacement, 12-10

Headlights - adjustment, 12-11

Heater core - replacement, 3-13

Heater/air conditioner control assembly - removal and installation, 3-14

Hood - removal, installation and adjustment, 11-6

Hood latch and cable - removal and installation, 11-7

Horn - replacement, 12-14

Hub and bearing assembly - removal and installation
    front, 10-8
    rear, 10-12

I

Ignition coils - removal and installation, 5-5

Ignition switch - replacement, 12-6

Initial start-up and break-in after overhaul, 2E-17

Input shaft oil seal - replacement, 7B-2

Instrument cluster - removal and installation, 12-6

Instrument panel - removal and installation, 11-16

Intake Air Temperature (IAT) sensor - replacement, 6-12

Intake manifold - removal and installation
    2.7L V6 engine, 2B-9
    3.5L V6 engine, 2C-7
    3.6L V6 engine, 2D-4
    Four-cylinder engine, 2A-4

J

Jacking and towing, 0-19

K

Knock sensor - replacement, 6-13

L

Lower control arm (rear) - removal and installation, 10-12

M

Maintenance schedule, 1-7

Maintenance techniques, tools and working facilities, 0-11

Manifold Absolute Pressure (MAP) sensor - replacement, 6-13

Manifold tuning valve solenoids (2.7L and 3.5L V6 engines) - removal and installation, 6-10

Master cylinder - removal and installation, 9-17

Mirrors - removal and installation, 11-12

Multi-function switch - replacement, 12-5

O

Obtaining and clearing Diagnostic Trouble Codes (DTCs), 6-5

Oil cooler - removal and installation, 2D-13

Oil pan - removal and installation
    2.7L V6 engine, 2B-12
    3.5L V6 engine, 2C-11
    3.6L V6 engine, 2D-13
    Four-cylinder engine, 2A-14

Oil pressure check, 2E-4

Oil pump - removal, inspection and installation
    2.7L V6 engine, 2B-13
    3.5L V6 engine, 2C-11
    3.6L V6 engine, 2D-13

Oil pump/balance shaft module and chain - removal, inspection and installation, 2A-14

Oil temperature sensor - removal and installation, 6-20

On Board Diagnosis (OBD) system, 6-5

Orifice tube (2014 and earlier models) - removal and installation, 3-17

Oxygen sensors - general information and replacement, 6-14

P

Parking brake - adjustment, 9-21

Parking brake shoes (models with rear disc brakes) - replacement, 9-20

Pinion seal - replacement, 7B-1

Pistons and connecting rods - removal and installation, 2E-9

Positive Crankcase Ventilation (PCV)
    system, 6-19
    valve check and replacement, 1-27

Power brake booster - check, removal and installation, 9-19

Power door lock and keyless entry system – description and check, 12-15

Power seat - general information and procedures, 12-16

Power steering pump (2014 and earlier models) - removal and installation, 10-16

Power steering system - bleeding (2014 and earlier models), 10-17

Power window system - description and check, 12-15

Powertrain Control Module (PCM) - replacement, 6-16

R

Radiator - removal and installation, 3-9
Radiator grille - removal and installation, 11-7
Radio and speakers - removal and installation, 12-8
Rear differential
  assembly (AWD models) - removal and installation, 8-6
  driveaxle oil seals (AWD models) - replacement, 8-6
  electronic clutch - removal and installation, 8-6
  pinion seal (AWD models) - replacement, 8-6
Rear main oil seal – replacement
  2.7L V6 engine, 2B-15
  3.5L V6 engine, 2C-12
  3.6L V6 engine, 2D-15
  Four-cylinder engine, 2A-16
Rear stabilizer bar - removal and installation, 10-11
Rear suspension knuckle - removal and installation, 10-12
Rear window defogger - check and repair, 12-9
Recall information, 0-7
Relays - general information, 12-4
Repair operations possible with the engine in the vehicle
  2.7L V6 engine, 2B-3
  3.5L V6 engine, 2C-3
  3.6L V6 engine, 2D-3
  Four-cylinder engine, 2A-3
Repairing minor paint scratches, 11-2
Rocker arm assembly and hydraulic lash adjusters - removal, inspection and installation, 2C-4
Rocker arms and hydraulic lash adjusters - removal, inspection and installation, 2B-4

S

Safety first!, 0-23
Seat belt check, 1-19
Seats - removal and installation, 11-18
Shift cable - removal, installation and adjustment, 7A-3
Shift lever - removal and installation, 7A-5
Shock absorber/coil spring assembly (rear)
  component replacement, 10-10
  removal and installation, 10-9
Spark plug check and replacement, 1-26
Stabilizer bar and bushings (front) - removal and installation, 10-5
Starter motor - removal and installation, 5-7
Steering column
  covers - removal and installation, 11-14
  removal and installation, 10-15
Steering
  gear - removal and installation, 10-15
  knuckle - removal and installation, 10-9
  wheel - removal and installation, 10-13
Steering, suspension and driveaxle boot check, 1-21

Strut assembly (front) - removal, inspection and installation, 10-5
Strut/coil spring - replacement, 10-6
Subframe - removal and installation, 10-17
Suspension and steering systems, 10-1
  Balljoints - replacement, 10-8
  Control arm - removal, inspection, and installation, 10-7
  Front load beam (2015 and later models) - removal and installation, 10-18
  General information, 10-3
  Hub and bearing assembly (front) - removal and installation, 10-8
  Hub and bearing assembly (rear) - removal and installation, 10-12
  Lower control arm (rear) - removal and installation, 10-12
  Power steering pump (2014 and earlier models) - removal and installation, 10-16
  Power steering system - bleeding (2014 and earlier models), 10-17
  Rear stabilizer bar - removal and installation, 10-11
  Rear suspension knuckle - removal and installation, 10-12
  Shock absorber/coil spring assembly (rear) - component replacement, 10-10
  Shock absorber/coil spring assembly (rear) - removal and installation, 10-9
  Stabilizer bar and bushings (front) - removal and installation, 10-5
  Steering column - removal and installation, 10-15
  Steering gear - removal and installation, 10-15
  Steering knuckle - removal and installation, 10-9
  Steering wheel - removal and installation, 10-13
  Strut assembly (front) - removal, inspection and installation, 10-5
  Strut/coil spring - replacement, 10-6
  Subframe - removal and installation, 10-17
  Tie-rod ends - removal and installation, 10-14
  Toe link - removal and installation, 10-11
  Trailing link - removal and installation, 10-10
  Upper control arm (rear) - removal and installation, 10-12
  Wheel alignment - general information, 10-18
  Wheels and tires - general information, 10-17

T

Thermostat - replacement, 3-6
Throttle body - removal and installation, 4-9
Throttle Position (TP) sensor - replacement, 6-15
Tie-rod ends - removal and installation, 10-14
Timing belt - replacement, 2C-4
Timing chain and sprockets - removal, inspection and installation, 2B-5
Timing chain cover, chain and sprockets - removal, inspection and installation
  3.6L V6 engine, 2D-7
  Four-cylinder engine, 2A-7

Tire and tire pressure checks, 1-14
Tire rotation, 1-19
Toe link - removal and installation, 10-11
Top Dead Center (TDC) for number one
 piston – locating
  2.7L V6 engine, 2B-3
  3.5L V6 engine, 2C-3
  3.6L V6 engine, 2D-3
  Four-cylinder engine, 2A-4
Trailing link - removal and installation, 10-10
Transaxle oil cooler - removal and installation, 7A-6
Transaxle speed sensors - replacement, 6-15
Transfer case, 7B-1
  Driveaxle oil seal (right side) - replacement, 7B-2
  General information, 7B-1
  Input shaft oil seal - replacement, 7B-2
  Pinion seal - replacement, 7B-1
  Transfer case - removal and installation, 7B-2
Transmission Range (TR) and transmission
 temperature sensors - replacement, 6-15
Trunk lid, latch and support struts - removal and
 installation, 11-13
Tune-up and routine maintenance, 1-1
  Air filter check and replacement, 1-20
  Automatic transaxle fluid and filter change, 1-29
  Battery check, maintenance and charging, 1-17
  Brake fluid change, 1-25
  Brake system check, 1-20
  Cabin air filter replacement, 1-23
  Cooling system check, 1-19
  Cooling system servicing (draining, flushing and
   refilling), 1-28
  Drivebelt check and replacement, 1-24
  Engine oil and filter change, 1-15
  Exhaust system check, 1-23
  Fluid level checks, 1-9
  Fuel system check, 1-23
  Introduction, 1-8
  Maintenance schedule, 1-7
  Positive Crankcase Ventilation (PCV) valve check and
   replacement, 1-27
  Seat belt check, 1-19
  Spark plug check and replacement, 1-26

  Steering, suspension and driveaxle boot check, 1-21
  Tire and tire pressure checks, 1-14
  Tire rotation, 1-19
  Tune-up general information, 1-8
  Underhood hose check and replacement, 1-19
  Windshield wiper blade inspection and replacement, 1-16

U

Underhood hose check and replacement, 1-19
Upholstery, carpets and vinyl trim - maintenance, 11-5
Upper control arm (rear) - removal and
 installation, 10-12

V

Vacuum gauge diagnostic checks, 2E-5
Valve clearance - check and adjustment, 2A-14
Valve cover (s) - removal and installation
  2.7L V6 engine, 2B-3
  3.5L V6 engine, 2C-3
  3.6L V6 engine, 2D-3
  Four-cylinder engine, 2A-4
Variable Valve Actuation Assembly (VVAA) – removal
 and installation, 2A-17
Variable valve timing solenoids (2.4L and
 3.6L engines) - removal and installation, 6-10
Vehicle identification numbers, 0-6

W

Water pump - replacement, 3-10
Wheel alignment - general information, 10-18
Wheel cylinder - removal and installation, 9-16
Wheels and tires - general information, 10-17
Window glass regulators - removal and
 installation, 11-12
Windshield wiper blade inspection and
 replacement, 1-16
Wiper motor - replacement, 12-7
Wiring diagrams - general information, 12-17

Haynes Automotive Manuals

NOTE: If you do not see a listing for your vehicle, please visit **haynes.com** for the latest product information and check out our **Online Manuals!**

ACURA
- **12020** Integra '86 thru '89 & **Legend** '86 thru '90
- **12021** Integra '90 thru '93 & **Legend** '91 thru '95
  Integra '94 thru '00 - see HONDA Civic (42025)
  MDX '01 thru '07 - see HONDA Pilot (42037)
- **12050** Acura TL all models '99 thru '08

AMC
- **14020** Mid-size models '70 thru '83
- **14025** (Renault) Alliance & Encore '83 thru '87

AUDI
- **15020** 4000 all models '80 thru '87
- **15025** 5000 all models '77 thru '83
- **15026** 5000 all models '84 thru '88
  Audi A4 '96 thru '01 - see VW Passat (96023)
- **15030** Audi A4 '02 thru '08

AUSTIN-HEALEY
Sprite - see MG Midget (66015)

BMW
- **18020** 3/5 Series '82 thru '92
- **18021** 3-Series incl. Z3 models '92 thru '98
- **18022** 3-Series incl. Z4 models '99 thru '05
- **18023** 3-Series '06 thru '14
- **18025** 320i all 4-cylinder models '75 thru '83
- **18050** 1500 thru 2002 except Turbo '59 thru '77

BUICK
- **19010** Buick Century '97 thru '05
  Century (front-wheel drive) - see GM (38005)
- **19020** Buick, Oldsmobile & Pontiac Full-size
  (Front-wheel drive) '85 thru '05
  Buick Electra, LeSabre and Park Avenue;
  Oldsmobile Delta 88 Royale, Ninety Eight
  and Regency; Pontiac Bonneville
- **19025** Buick, Oldsmobile & Pontiac Full-size
  (Rear wheel drive) '70 thru '90
  Buick Estate, Electra, LeSabre, Limited,
  Oldsmobile Custom Cruiser, Delta 88,
  Ninety-eight, Pontiac Bonneville,
  Catalina, Grandville, Parisienne
- **19027** Buick LaCrosse '05 thru '13
  Enclave - see GENERAL MOTORS (38001)
  Rainier - see CHEVROLET (24072)
  Regal - see GENERAL MOTORS (38010)
  Riviera - see GENERAL MOTORS (38030, 38031)
  Roadmaster - see CHEVROLET (24046)
  Skyhawk - see GENERAL MOTORS (38015)
  Skylark - see GENERAL MOTORS (38020, 38025)
  Somerset - see GENERAL MOTORS (38025)

CADILLAC
- **21015** CTS & CTS-V '03 thru '14
- **21030** Cadillac Rear Wheel Drive '70 thru '93
  Cimarron - see GENERAL MOTORS (38015)
  DeVille - see GENERAL MOTORS (38031 & 38032)
  Eldorado - see GENERAL MOTORS (38030)
  Fleetwood - see GENERAL MOTORS (38031)
  Seville - see GM (38030, 38031 & 38032)

CHEVROLET
- **10305** Chevrolet Engine Overhaul Manual
- **24010** Astro & GMC Safari Mini-vans '85 thru '05
- **24013** Aveo '04 thru '11
- **24015** Camaro V8 all models '70 thru '81
- **24016** Camaro all models '82 thru '92
- **24017** Camaro & Firebird '93 thru '02
  Cavalier - see GENERAL MOTORS (38016)
  Celebrity - see GENERAL MOTORS (38005)
- **24018** Camaro '10 thru '15
- **24020** Chevelle, Malibu & El Camino '69 thru '87
  Cobalt - see GENERAL MOTORS (38017)
- **24024** Chevette & Pontiac T1000 '76 thru '87
  Citation - see GENERAL MOTORS (38020)
- **24027** Colorado & GMC Canyon '04 thru '12
- **24032** Corsica & Beretta all models '87 thru '96
- **24040** Corvette all V8 models '68 thru '82
- **24041** Corvette all models '84 thru '96
- **24042** Corvette '97 thru '13
- **24044** Cruze '11 thru '19
- **24045** Full-size Sedans Caprice, Impala, Biscayne,
  Bel Air & Wagons '69 thru '90
- **24046** Impala SS & Caprice and Buick Roadmaster
  '91 thru '96
  Impala '00 thru '05 - see LUMINA (24048)
- **24047** Impala & Monte Carlo all models '06 thru '11
  Lumina '90 thru '94 - see GM (38010)
- **24048** Lumina & Monte Carlo '95 thru '05
  Lumina APV - see GM (38035)
- **24050** Luv Pick-up all 2WD & 4WD '72 thru '82
- **24051** Malibu '13 thru '19
- **24055** Monte Carlo all models '70 thru '88
  Monte Carlo '95 thru '01 - see LUMINA (24048)
- **24059** Nova all V8 models '69 thru '79

- **24060** Nova and Geo Prizm '85 thru '92
- **24064** Pick-ups '67 thru '87 - Chevrolet & GMC
- **24065** Pick-ups '88 thru '98 - Chevrolet & GMC
- **24066** Pick-ups '99 thru '06 - Chevrolet & GMC
- **24067** Chevrolet Silverado & GMC Sierra '07 thru '14
- **24068** Chevrolet Silverado & GMC Sierra '14 thru '19
- **24070** S-10 & S-15 Pick-ups '82 thru '93,
  Blazer & Jimmy '83 thru '94,
- **24071** S-10 & Sonoma Pick-ups '94 thru '04,
  including Blazer, Jimmy & Hombre
- **24072** Chevrolet TrailBlazer, GMC Envoy &
  Oldsmobile Bravada '02 thru '09
- **24075** Sprint '85 thru '88 & Geo Metro '89 thru '01
- **24080** Vans - Chevrolet & GMC '68 thru '96
- **24081** Chevrolet Express & GMC Savana
  Full-size Vans '96 thru '19

CHRYSLER
- **10310** Chrysler Engine Overhaul Manual
- **25015** Chrysler Cirrus, Dodge Stratus,
  Plymouth Breeze '95 thru '00
- **25020** Full-size Front-Wheel Drive '88 thru '93
  K-Cars - see DODGE Aries (30008)
  Laser - see DODGE Daytona (30030)
- **25025** Chrysler LHS, Concorde, New Yorker,
  Dodge Intrepid, Eagle Vision, '93 thru '97
- **25026** Chrysler LHS, Concorde, 300M,
  Dodge Intrepid, '98 thru '04
- **25027** Chrysler 300 '05 thru '18, Dodge Charger
  '06 thru '18, Magnum '05 thru '08 &
  Challenger '08 thru '18
- **25030** Chrysler & Plymouth Mid-size
  front wheel drive '82 thru '95
  Rear-wheel Drive - see Dodge (30050)
- **25035** PT Cruiser all models '01 thru '10
- **25040** Chrysler Sebring '95 thru '06, Dodge Stratus
  '01 thru '06 & Dodge Avenger '95 thru '00
- **25041** Chrysler Sebring '07 thru '10, 200 '11 thru '17
  Dodge Avenger '08 thru '14

DATSUN
- **28005** 200SX all models '80 thru '83
- **28012** 240Z, 260Z & 280Z Coupe '70 thru '78
- **28014** 280ZX Coupe & 2+2 '79 thru '83
  300ZX - see NISSAN (72010)
- **28018** 510 & PL521 Pick-up '68 thru '73
- **28020** 510 all models '78 thru '81
- **28022** 620 Series Pick-up all models '73 thru '79
  720 Series Pick-up - see NISSAN (72030)

DODGE
- **400 & 600** - see CHRYSLER (25030)
- **30008** Aries & Plymouth Reliant '81 thru '89
- **30010** Caravan & Plymouth Voyager '84 thru '95
- **30011** Caravan & Plymouth Voyager '96 thru '02
- **30012** Challenger & Plymouth Sapporro '78 thru '83
- **30013** Caravan, Chrysler Voyager &
  Town & Country '03 thru '07
- **30014** Grand Caravan &
  Chrysler Town & Country '08 thru '18
- **30016** Colt & Plymouth Champ '78 thru '87
- **30020** Dakota Pick-ups all models '87 thru '96
- **30021** Durango '98 & '99 & Dakota '97 thru '99
- **30022** Durango '00 thru '03 & Dakota '00 thru '04
- **30023** Durango '04 thru '09 & Dakota '05 thru '11
- **30025** Dart, Demon, Plymouth Barracuda,
  Duster & Valiant 6-cylinder models '67 thru '76
- **30030** Daytona & Chrysler Laser '84 thru '89
  Intrepid - see CHRYSLER (25025, 25026)
- **30034** Neon all models '95 thru '99
- **30035** Omni & Plymouth Horizon '78 thru '90
- **30036** Dodge & Plymouth Neon '00 thru '05
- **30040** Pick-ups full-size models '74 thru '93
- **30042** Pick-ups full-size models '94 thru '08
- **30043** Pick-ups full-size models '09 thru '18
- **30045** Ram 50/D50 Pick-ups & Raider and
  Plymouth Arrow Pick-ups '79 thru '93
- **30050** Dodge/Plymouth/Chrysler RWD '71 thru '89
- **30055** Shadow & Plymouth Sundance '87 thru '94
- **30060** Spirit & Plymouth Acclaim '89 thru '95
- **30065** Vans - Dodge & Plymouth '71 thru '03

EAGLE
Talon - see MITSUBISHI (68030, 68031)
Vision - see CHRYSLER (25025)

FIAT
- **34010** 124 Sport Coupe & Spider '68 thru '78
- **34025** X1/9 all models '74 thru '80

FORD
- **10320** Ford Engine Overhaul Manual
- **10355** Ford Automatic Transmission Overhaul
- **11500** Mustang '64-1/2 thru '70 Restoration Guide
- **36004** Aerostar Mini-vans all models '86 thru '97
- **36006** Contour & Mercury Mystique '95 thru '00
- **36008** Courier Pick-up all models '72 thru '82

- **36012** Crown Victoria &
  Mercury Grand Marquis '88 thru '11
- **36014** Edge '07 thru '19 & Lincoln MKX '07 thru '18
- **36016** Escort & Mercury Lynx all models '81 thru '90
- **36020** Escort & Mercury Tracer '91 thru '02
- **36022** Escape '01 thru '17, Mazda Tribute '01 thru '11,
  & Mercury Mariner '05 thru '11
- **36024** Explorer & Mazda Navajo '91 thru '01
- **36025** Explorer & Mercury Mountaineer '02 thru '10
- **36026** Explorer '11 thru '17
- **36028** Fairmont & Mercury Zephyr '78 thru '83
- **36030** Festiva & Aspire '88 thru '97
- **36032** Fiesta all models '77 thru '80
- **36034** Focus all models '00 thru '11
- **36035** Focus '12 thru '14
- **36045** Fusion '06 thru '14 & Mercury Milan '06 thru '11
- **36048** Mustang V8 all models '64-1/2 thru '73
- **36049** Mustang II 4-cylinder, V6 & V8 models '74 thru '78
- **36050** Mustang & Mercury Capri '79 thru '93
- **36051** Mustang all models '94 thru '04
- **36052** Mustang '05 thru '14
- **36054** Pick-ups & Bronco '73 thru '79
- **36058** Pick-ups & Bronco '80 thru '96
- **36059** F-150 '97 thru '03, Expedition '97 thru '17,
  F-250 '97 thru '99, F-150 Heritage '04
  & Lincoln Navigator '98 thru '17
- **36060** Super Duty Pick-ups & Excursion '99 thru '10
- **36061** F-150 full-size '04 thru '14
- **36062** Pinto & Mercury Bobcat '75 thru '80
- **36063** F-150 full-size '15 thru '17
- **36064** Super Duty Pick-ups '11 thru '16
- **36066** Probe all models '89 thru '92
  Probe '93 thru '97 - see MAZDA 626 (61042)
- **36070** Ranger & Bronco II all models '83 thru '92
- **36071** Ranger '93 thru '11 & Mazda Pick-ups '94 thru '09
- **36074** Taurus & Mercury Sable '86 thru '95
- **36075** Taurus & Mercury Sable '96 thru '07
- **36076** Taurus '08 thru '14, Five Hundred '05 thru '07,
  Mercury Montego '05 thru '07 & Sable '08 thru '09
- **36078** Tempo & Mercury Topaz '84 thru '94
- **36082** Thunderbird & Mercury Cougar '83 thru '88
- **36086** Thunderbird & Mercury Cougar '89 thru '97
- **36090** Vans all V8 Econoline models '69 thru '91
- **36094** Vans full size '92 thru '14
- **36097** Windstar '95 thru '03, Freestar & Mercury
  Monterey Mini-van '04 thru '07

GENERAL MOTORS
- **10360** GM Automatic Transmission Overhaul
- **38001** GMC Acadia '07 thru '16, Buick Enclave
  '08 thru '17, Saturn Outlook '07 thru '10
  & Chevrolet Traverse '09 thru '17
- **38005** Buick Century, Chevrolet Celebrity,
  Oldsmobile Cutlass Ciera & Pontiac 6000
  all models '82 thru '96
- **38010** Buick Regal '88 thru '04, Chevrolet Lumina
  '88 thru '04, Oldsmobile Cutlass Supreme
  '88 thru '97 & Pontiac Grand Prix '88 thru '07
- **38015** Buick Skyhawk, Cadillac Cimarron,
  Chevrolet Cavalier, Oldsmobile Firenza,
  Pontiac J-2000 & Sunbird '82 thru '94
- **38016** Chevrolet Cavalier & Pontiac Sunfire '95 thru '05
- **38017** Chevrolet Cobalt '05 thru '10, HHR '06 thru '11,
  Pontiac G5 '07 thru '09, Pursuit '05 thru '06
  & Saturn ION '03 thru '07
- **38020** Buick Skylark, Chevrolet Citation,
  Oldsmobile Omega, Pontiac Phoenix '80 thru '85
- **38025** Buick Skylark '86 thru '98, Somerset '85 thru '87,
  Oldsmobile Achieva '92 thru '98, Calais '85 thru '91,
  & Pontiac Grand Am all models '85 thru '98
- **38026** Chevrolet Malibu '97 thru '03, Classic '04 thru '05,
  Oldsmobile Alero '99 thru '03, Cutlass '97 thru '00,
  & Pontiac Grand Am '99 thru '03
- **38027** Chevrolet Malibu '04 thru '12, Pontiac G6
  '05 thru '10 & Saturn Aura '07 thru '10
- **38030** Cadillac Eldorado, Seville, Oldsmobile
  Toronado & Buick Riviera '71 thru '85
- **38031** Cadillac Eldorado, Seville, DeVille, Fleetwood,
  Oldsmobile Toronado & Buick Riviera '86 thru '93
- **38032** Cadillac DeVille '94 thru '05, Seville '92 thru '04
  & Cadillac DTS '06 thru '10
- **38035** Chevrolet Lumina APV, Oldsmobile Silhouette
  & Pontiac Trans Sport all models '90 thru '96
- **38036** Chevrolet Venture '97 thru '05, Oldsmobile
  Silhouette '97 thru '04, Pontiac Trans Sport
  '97 thru '98 & Montana '99 thru '05
- **38040** Chevrolet Equinox '05 thru '17, GMC Terrain
  '10 thru '17 & Pontiac Torrent '06 thru '09

GEO
Metro - see CHEVROLET Sprint (24075)
Prizm - '85 thru '92 see CHEVY (24060),
'93 thru '02 see TOYOTA Corolla (92036)
- **40030** Storm all models '90 thru '93
Tracker - see SUZUKI Samurai (90010)

(Continued on other side)

Haynes Automotive Manuals (continued)

NOTE: *If you do not see a listing for your vehicle, please visit* **haynes.com** *for the latest product information and check out our* **Online Manuals!**

GMC

- **Acadia** - see GENERAL MOTORS (38001)
- **Pick-ups** - see CHEVROLET (24027, 24068)
- **Vans** - see CHEVROLET (24081)

HONDA

- 42010 **Accord CVCC** all models '76 thru '83
- 42011 **Accord** all models '84 thru '89
- 42012 **Accord** all models '90 thru '93
- 42013 **Accord** all models '94 thru '97
- 42014 **Accord** all models '98 thru '02
- 42015 **Accord** '03 thru '12 & **Crosstour** '10 thru '14
- 42016 **Accord** '13 thru '17
- 42020 **Civic 1200** all models '73 thru '79
- 42021 **Civic 1300 & 1500 CVCC** '80 thru '83
- 42022 **Civic 1500 CVCC** all models '75 thru '79
- 42023 **Civic** all models '84 thru '91
- 42024 **Civic & del Sol** '92 thru '95
- 42025 **Civic** '96 thru '00, **CR-V** '97 thru '01 & **Acura Integra** '94 thru '00
- 42026 **Civic** '01 thru '11 & **CR-V** '02 thru '11
- 42027 **Civic** '12 thru '15 & **CR-V** '12 thru '16
- 42030 **Fit** '07 thru '13
- 42035 **Odyssey** all models '99 thru '10
- **Passport** - see ISUZU Rodeo (47017)
- 42037 **Honda Pilot** '03 thru '08, **Ridgeline** '06 thru '14 & **Acura MDX** '01 thru '07
- 42040 **Prelude CVCC** all models '79 thru '89

HYUNDAI

- 43010 **Elantra** all models '96 thru '19
- 43015 **Excel & Accent** all models '86 thru '13
- 43050 **Santa Fe** all models '01 thru '12
- 43055 **Sonata** all models '99 thru '14

INFINITI

- **G35** '03 thru '08 - see NISSAN 350Z (72011)

ISUZU

- **Hombre** - see CHEVROLET S-10 (24071)
- 47017 **Rodeo** '91 thru '02, **Amigo** '89 thru '94 & '98 thru '02 & **Honda Passport** '95 thru '02
- 47020 **Trooper** '84 thru '91 & **Pick-up** '81 thru '93

JAGUAR

- 49010 **XJ6** all 6-cylinder models '68 thru '86
- 49011 **XJ6** all models '88 thru '94
- 49015 **XJ12 & XJS** all 12-cylinder models '72 thru '85

JEEP

- 50010 **Cherokee, Comanche & Wagoneer Limited** all models '84 thru '01
- 50011 **Cherokee** '14 thru '19
- 50020 **CJ** all models '49 thru '86
- 50025 **Grand Cherokee** all models '93 thru '04
- 50026 **Grand Cherokee** '05 thru '19 & **Dodge Durango** '11 thru '19
- 50029 **Grand Wagoneer & Pick-up** '72 thru '91 Grand Wagoneer '84 thru '91, Cherokee & Wagoneer '72 thru '83, Pick-up '72 thru '88
- 50030 **Wrangler** all models '87 thru '17
- 50035 **Liberty** '02 thru '12 & **Dodge Nitro** '07 thru '11
- 50050 **Patriot & Compass** '07 thru '17

KIA

- 54050 **Optima** '01 thru '10
- 54060 **Sedona** '02 thru '14
- 54070 **Sephia** '94 thru '01, **Spectra** '00 thru '09, **Sportage** '05 thru '20
- 54077 **Sorento** '03 thru '13

LEXUS

- **ES 300/330** - see TOYOTA Camry (92007, 92008)
- **ES 350** - see TOYOTA Camry (92009)
- **RX 300/330/350** - see TOYOTA Highlander (92095)

LINCOLN

- **MKX** - see FORD (36014)
- **Navigator** - see FORD Pick-up (36059)
- 59010 **Rear-Wheel Drive Continental** '70 thru '87, **Mark Series** '70 thru '92 & **Town Car** '81 thru '10

MAZDA

- 61010 **GLC** (rear-wheel drive) '77 thru '83
- 61011 **GLC** (front-wheel drive) '81 thru '85
- 61012 **Mazda3** '04 thru '11
- 61015 **323 & Protegé** '90 thru '03
- 61016 **MX-5 Miata** '90 thru '14
- 61020 **MPV** all models '89 thru '98
- **Navajo** - see Ford Explorer (36024)
- 61030 **Pick-ups** '72 thru '93 **Pick-ups** '94 thru '09 - see Ford Ranger (36071)
- 61035 **RX-7** all models '79 thru '85
- 61036 **RX-7** all models '86 thru '91
- 61040 **626** (rear-wheel drive) all models '79 thru '82
- 61041 **626 & MX-6** (front-wheel drive) '83 thru '92
- 61042 **626** '93 thru '01 & **MX-6/Ford Probe** '93 thru '02
- 61043 **Mazda6** '03 thru '13

MERCEDES-BENZ

- 63012 **123 Series Diesel** '76 thru '85
- 63015 **190 Series** 4-cylinder gas models '84 thru '88
- 63020 **230/250/280** 6-cylinder SOHC models '68 thru '72
- 63025 **280 123 Series** gas models '77 thru '81
- 63030 **350 & 450** all models '71 thru '80
- 63040 **C-Class:** C230/C240/C280/C320/C350 '01 thru '07

MERCURY

- 64200 **Villager & Nissan Quest** '93 thru '01 *All other titles, see FORD Listing.*

MG

- 66010 **MGB** Roadster & GT Coupe '62 thru '80
- 66015 **MG Midget, Austin Healey Sprite** '58 thru '80

MINI

- 67020 **Mini** '02 thru '13

MITSUBISHI

- 68020 **Cordia, Tredia, Galant, Precis & Mirage** '83 thru '93
- 68030 **Eclipse, Eagle Talon & Plymouth Laser** '90 thru '94
- 68031 **Eclipse** '95 thru '05 & **Eagle Talon** '95 thru '98
- 68035 **Galant** '94 thru '12
- 68040 **Pick-up** '83 thru '96 & **Montero** '83 thru '93

NISSAN

- 72010 **300ZX** all models including Turbo '84 thru '89
- 72011 **350Z & Infiniti G35** all models '03 thru '08
- 72015 **Altima** all models '93 thru '06
- 72016 **Altima** '07 thru '12
- 72020 **Maxima** all models '85 thru '92
- 72021 **Maxima** all models '93 thru '08
- 72025 **Murano** '03 thru '14
- 72030 **Pick-ups** '80 thru '97 & **Pathfinder** '87 thru '95
- 72031 **Frontier** '98 thru '04, **Xterra** '00 thru '04, & **Pathfinder** '96 thru '04
- 72032 **Frontier & Xterra** '05 thru '14
- 72037 **Pathfinder** '05 thru '14
- 72040 **Pulsar** all models '83 thru '86
- 72042 **Roque** all models '08 thru '20
- 72050 **Sentra** all models '82 thru '94
- 72051 **Sentra & 200SX** all models '95 thru '06
- 72060 **Stanza** all models '82 thru '90
- 72070 **Titan pick-ups** '04 thru '10, **Armada** '05 thru '10 & **Pathfinder Armada** '04
- 72080 **Versa** all models '07 thru '19

OLDSMOBILE

- 73015 **Cutlass** V6 & V8 gas models '74 thru '88 *For other OLDSMOBILE titles, see BUICK, CHEVROLET or GENERAL MOTORS listings.*

PLYMOUTH

For PLYMOUTH titles, see DODGE listing.

PONTIAC

- 79008 **Fiero** all models '84 thru '88
- 79018 **Firebird** V8 models except Turbo '70 thru '81
- 79019 **Firebird** all models '82 thru '92
- 79025 **G6** all models '05 thru '09
- 79040 **Mid-size Rear-wheel Drive** '70 thru '87 **Vibe** '03 thru '10 - see TOYOTA Corolla (92037) *For other PONTIAC titles, see BUICK, CHEVROLET or GENERAL MOTORS listings.*

PORSCHE

- 80020 **911 Coupe & Targa** models '65 thru '89
- 80025 **914** all 4-cylinder models '69 thru '76
- 80030 **924** all models including Turbo '76 thru '82
- 80035 **944** all models including Turbo '83 thru '89

RENAULT

- **Alliance & Encore** - see AMC (14025)

SAAB

- 84010 **900** all models including Turbo '79 thru '88

SATURN

- 87010 **Saturn** all S-series models '91 thru '02 **Saturn Ion** '03 thru '07- see GM (38017) **Saturn Outlook** - see GM (38001)
- 87020 **Saturn L-series** all models '00 thru '04
- 87040 **Saturn VUE** '02 thru '09

SUBARU

- 89002 **1100, 1300, 1400 & 1600** '71 thru '79
- 89003 **1600 & 1800** 2WD & 4WD '80 thru '94
- 89080 **Impreza** '02 thru '11, **WRX** '02 thru '14, & **WRX STI** '04 thru '14
- 89100 **Legacy** all models '90 thru '99
- 89101 **Legacy & Forester** '00 thru '09
- 89102 **Legacy** '10 thru '16 & **Forester** '12 thru '16

SUZUKI

- 90010 **Samurai/Sidekick & Geo Tracker** '86 thru '01

TOYOTA

- 92005 **Camry** all models '83 thru '91
- 92006 **Camry** '92 thru '96 & **Avalon** '95 thru '96
- 92007 **Camry, Avalon, Solara, Lexus ES 300** '97 thru '01
- 92008 **Camry, Avalon, Lexus ES 300/330** '02 thru '06 & **Solara** '02 thru '08
- 92009 **Camry, Avalon & Lexus ES 350** '07 thru '17
- 92015 **Celica Rear-wheel Drive** '71 thru '85
- 92020 **Celica Front-wheel Drive** '86 thru '99
- 92025 **Celica Supra** all models '79 thru '92
- 92030 **Corolla** all models '75 thru '79
- 92032 **Corolla** all rear-wheel drive models '80 thru '87
- 92035 **Corolla** all front-wheel drive models '84 thru '92
- 92036 **Corolla & Geo/Chevrolet Prizm** '93 thru '02
- 92037 **Corolla** '03 thru '19, **Matrix** '03 thru '14, & **Pontiac Vibe** '03 thru '10
- 92040 **Corolla Tercel** all models '80 thru '82
- 92045 **Corona** all models '74 thru '82
- 92050 **Cressida** all models '78 thru '82
- 92055 **Land Cruiser FJ40, 43, 45, 55** '68 thru '82
- 92056 **Land Cruiser FJ60, 62, 80, FZJ80** '80 thru '96
- 92060 **Matrix** '03 thru '11 & **Pontiac Vibe** '03 thru '10
- 92065 **MR2** all models '85 thru '87
- 92070 **Pick-up** all models '69 thru '78
- 92075 **Pick-up** all models '79 thru '95
- 92076 **Tacoma** '95 thru '04, **4Runner** '96 thru '02 & **T100** '93 thru '08
- 92077 **Tacoma** all models '05 thru '18
- 92078 **Tundra** '00 thru '06 & **Sequoia** '01 thru '07
- 92079 **4Runner** all models '03 thru '09
- 92080 **Previa** all models '91 thru '95
- 92081 **Prius** all models '01 thru '12
- 92082 **RAV4** all models '96 thru '12
- 92085 **Tercel** all models '87 thru '94
- 92090 **Sienna** all models '98 thru '10
- 92095 **Highlander** '01 thru '19 & **Lexus RX330/330/350** '99 thru '19
- 92179 **Tundra** '07 thru '19 & **Sequoia** '08 thru '19

TRIUMPH

- 94007 **Spitfire** all models '62 thru '81
- 94010 **TR7** all models '75 thru '81

VW

- 96008 **Beetle & Karmann Ghia** '54 thru '79
- 96009 **New Beetle** '98 thru '10
- 96016 **Rabbit, Jetta, Scirocco & Pick-up** gas models '75 thru '92 & Convertible '80 thru '92
- 96017 **Golf, GTI & Jetta** '93 thru '98, **Cabrio** '95 thru '02
- 96018 **Golf, GTI, Jetta** '99 thru '05
- 96019 **Jetta, Rabbit, GLI, GTI & Golf** '05 thru '11
- 96020 **Rabbit, Jetta & Pick-up** diesel '77 thru '84
- 96021 **Jetta** '11 thru '18 & **Golf** '15 thru '19
- 96023 **Passat** '98 thru '05 & **Audi A4** '96 thru '01
- 96030 **Transporter 1600** all models '68 thru '79
- 96035 **Transporter 1700, 1800 & 2000** '72 thru '79
- 96040 **Type 3 1500 & 1600** all models '63 thru '73
- 96045 **Vanagon Air-Cooled** all models '80 thru '83

VOLVO

- 97010 **120, 130 Series & 1800 Sports** '61 thru '73
- 97015 **140 Series** all models '66 thru '74
- 97020 **240 Series** all models '76 thru '93
- 97040 **740 & 760 Series** all models '82 thru '88
- 97050 **850 Series** all models '93 thru '97

TECHBOOK MANUALS

- 10205 **Automotive Computer Codes**
- 10206 **OBD-II & Electronic Engine Management**
- 10210 **Automotive Emissions Control Manual**
- 10215 **Fuel Injection Manual** '78 thru '85
- 10225 **Holley Carburetor Manual**
- 10230 **Rochester Carburetor Manual**
- 10305 **Chevrolet Engine Overhaul Manual**
- 10320 **Ford Engine Overhaul Manual**
- 10330 **GM and Ford Diesel Engine Repair Manual**
- 10331 **Duramax Diesel Engines** '01 thru '19
- 10332 **Cummins Diesel Engine Performance Manual**
- 10333 **GM, Ford & Chrysler Engine Performance Manual**
- 10334 **GM Engine Performance Manual**
- 10340 **Small Engine Repair Manual,** 5 HP & Less
- 10341 **Small Engine Repair Manual,** 5.5 thru 20 HP
- 10345 **Suspension, Steering & Driveline Manual**
- 10355 **Ford Automatic Transmission Overhaul**
- 10360 **GM Automatic Transmission Overhaul**
- 10405 **Automotive Body Repair & Painting**
- 10410 **Automotive Brake Manual**
- 10411 **Automotive Anti-lock Brake (ABS) Systems**
- 10420 **Automotive Electrical Manual**
- 10425 **Automotive Heating & Air Conditioning**
- 10435 **Automotive Tools Manual**
- 10445 **Welding Manual**
- 10450 **ATV Basics**

Over a 100 Haynes motorcycle manuals also available

10/22

Haynes North America, Inc. • (805) 498-6703 • www.haynes.com